Changing the Face of the Earth

Changing the Face of the Earth

Culture, Environment, History

I.G. SIMMONS

Basil Blackwell

First published 1989

Basil Blackwell Ltd
108 Cowley Road, Oxford, OX4 1JF, UK

Basil Blackwell Inc.
432 Park Avenue South, Suite 1503
New York, NY 10016, USA

British Library Cataloguing in Publication Data

Simmons, I. G. (Ian Gordon)
Changing the face of the earth: culture,
environment, history
1. Environment. Influence, to 1987, of man
I. Title
304.2′8

ISBN 0-631-14049-2
ISBN 0-631-16351-4 Pbk

Library of Congress Cataloging-in-Publication Data

Simmons, I. G. (Ian Gordon). 1937–
Changing the face of the earth: culture, environment, history/I.G. Simmons.
p. cm.
Includes index.
ISBN 0-631-14049-2
ISBN 0-631-16351-4 (pbk.)
1. Human ecology – History. 2. Man – Influence on nature.
I. Title.
GF13.S56 1988 88-5031
304.2 – dc19 CIP

Typeset in 10½ on 12pt Sabon
by Joshua Associates
Printed in Great Britain at the University Press, Cambridge

Contents

Preface

This work is intended as a tribute to a book which was influential for me: W. L. Thomas's edited collection, *Man's Role in Changing the Face of the Earth*, published in 1956, when I was an undergraduate. I was awarded a College prize in 1957 and chose a copy of it; I have it still, with the marking to say that the hardback volume cost 94 old shillings. Since then, the study of human impact upon the biophysical systems of this planet has exploded with a force that is scarcely presaged even by the impressive bulk of the Thomas volume and so my own follow-up is bound to be inadequate in many places. Indeed I would not have dared to use the present title myself but since it was suggested by Professor David Harvey and enthusiastically taken up by John Davey at Basil Blackwell, I admit that it seems appropriate, though I have misgivings about it seeming pretentious.

In writing the book, I have tried to give some coherence to the mass of information by adopting an ecological-functional viewpoint and tying this wherever possible to the quantitative data that go along with that methodology, namely the study of energy flow through ecosystems. Thus this is an empirical study, with realist preconceptions and it maintains that perspective throughout; it is also a one-way study of human impact on the environment and not the other way round. I know that sticking to such a narrow epistemological foundation has its philosophical problems and that has made me want to write another book on environment as a human mental construction, to complement the 'facts' of this one. A second difficulty has been information control, for the material has come at all levels of detail, from recent research papers (mostly using different systems of measurement: I have mostly kept the originals for their internal comparability) to major historical syntheses at high levels of generalization. I have tried to make clear the broad trends while at the same time giving some particular example by way of actuality but from time to time, the one may have outrun the other, just as the organizing concepts sometimes get a bit submerged under my desire to put in an especially interesting example but one which doesn't quite fit the framework. I hope though that I have avoided the worst dangers of stichomancy. In fact, of course, any time is a bad time to write this type of book since there is always new and highly relevant information just coming out which ought to be treated in full but any account of which is going to appear dated by the time the production process has run its course. I decided that I would write the book, let it appear and hope that somebody else will improve upon it and so here it is.

There are many people who deserve a special word of thanks for their part in this work. It is historically fitting, as it were, to acknowledge the quality of the teaching at UCL in my student days and I remember gratefully that of Professor Sir Clifford Darby, Professor Eric Brown, and Professor Bill Mead, who all infused me with a sense of the necessity of historical depth to any story. At that time I was confirmed in my ways by Professor Jim Johnson (now of the University of Lancaster) and Professor Bruce Proudfoot (then of QUB, now of St Andrews): these two men were probably decisive in making me want to be an academic and I thank them very much for that. Another critical moment came during my PhD research when for a few months I sat at a microscope with Professor (then just Mr) Frank Oldfield in Leicester, and now in Liverpool. He taught me to recognize pollen grains but then and thereafter a great deal else besides; any breadth I might have acquired is due to attitudes instilled by him about the need not to be too specialized; again, I am grateful for that inspiration. The scholarly habits acquired at the University of California's Berkeley campus are doubtless apparent and I always think with awe of the example of Clarence Glacken, whose major book is always in the mind of anybody who embarks on history and on environment.

The actual text has been put together over about ten years and that period encompasses a spell at the University of Bristol, working alongside Professor Peter Haggett. To be encouraged by him to follow my own ways in a department rather inclined to the mathematical was only one of the privileges of that time and I am very grateful to Peter for many insights into the nature of intellectual endeavour and its communication to the profession and our students, along with exemplary lessons (not alas always followed) in the arts of patience and charity. In Durham, my three years as departmental Chairman (1984–87) were made less onerous by the immense support I had from colleagues in Geography. Even through a very difficult period in the history of this and other universities, I was able to spend some time writing, and this was only possible because of other people's restraint: thank you all very much. The final draft and all the other processes were accelerated by the award of a year's research leave by the University and this has made possible the appearance of the book one or two years ahead of what otherwise would have been the case.

At the more immediate, personal level, I am happy to thank Jean and Peter Eccleston for the loan of Star of Hope Cottage (what a lovely and indeed inspiring name) for a period of very productive work on the book; my good friend Rob Catty for the kind of help and encouragement that goes well beyond the terse notes in the medical record; and also Judith Catty for her cheerful scepticism about the ways of academics, which has often provided a much-needed sense of perspective. The first draft of the text was put on disk by Elizabeth Pearson, who can read my handwriting; many many subsequent editings were done by Catherine Reed, who can look cheerful even in the face of yet another batch of colour-coded alterations. Thanks to you both.

My wife, Carol, has of course been the first to hear of all the grumbles and the problems of the book, like everything else, without necessarily being the first also to hear the good news when something appeared finally to fit together or even be finished. But as always, the end-product is as much to her credit as to mine.

I. G. Simmons
Durham, June 1988

Acknowledgements

The author and publisher acknowledge with thanks those who have given permission for the reproduction of copyrighted material. Every effort has been made to trace copyright holders, and apology is made to any who may have been omitted. Permission has been granted for reproduction of figures by the following:

1.1 by C. E Merrill Inc.; 1.2 and 1.3 reprinted with the permission of Pergamon Books Ltd; 2.1 by Longman Group UK Ltd; 2.3 and 2.4 by *American Scientist*; 2.5 by J. D. Speth and copyright 1976 AAAS; 2.6 by CUP; 3.3 copyright © 1981 Columbia University Press, used by permission; 3.4 by Collins and by Roxby Archaeology Ltd; 3.5 by Leicester University Press (after an original by J. B. Wheat in R. S. MacNeish (ed.) *Early Man in America*, San Francisco: Freeman, 1972, 80–88); 3.9 by *Scientific American* and W. H. Freeman and Co; 3.11 by C. A. Simenstad and copyright 1978 by the AAAS; 3.12 by P. S. Martin and copyright 1973 by the AAAS; 3.13 by the University of Arizona Press, copyright 1984; 4.2 by Oxford University Press; 4.5 by Elsevier Science Publishers B. V.; 4.7 by courtesy of the editor of *Pacific Viewpoint*; 4.9 by R. T. Matheney and copyright 1976 by the AAAS; 4.11 by Academic Press; 4.12 by McGraw Hill Book Company Australia Pty Ltd; 4.13, 4.14 and 4.15 by G. E. B. Morren and Academic Press; 4.16 by CUP; 4.17 by Elsevier Science Publishers B. V.; 4.18 by M. A. Little and Academic Press; 4.20 by CUP; 4.21 by Gebruder Borntraeger VB, Stuttgart; 4.22 by M. B. Coughenour and copyright 1985 AAAS; 4.23 by J. M. Dent/ Weidenfeld and Nicolson Ltd; 4.26 by Blackwell Scientific Publications Ltd; 4.27 by the Foundation for Environmental Conservation; 4.28 by R. Shepherd, The trustees of the British Museum, and Academic Press; 4.29 by Longman Group UK Ltd; 4.30 by CUP; 5.1 by Edward Arnold; 5.2 by courtesy of the Ironbridge Gorge Museum; 5.5 reproduced by permission of Penguin Books Ltd; 5.6 W. H. Freeman and Co., reprinted with permission; 5.8 by courtesy of the International Institute for Applied Systems Analysis, Laxenburg; 5.10 by Edward Arnold; 5.11 reprinted by permission of © John Wiley and Sons Ltd; 5.12 and 5.13 by Edward Arnold; 5.15 by W. M. Lewis and Springer-Verlag, Heidelberg; 5.16 by courtesy of K. Newcombe; 5.17 © 1984 by the United Nations University; 5.21 by CUP; 5.22 and 5.23 by OECD, Paris; 5.24 by Elsevier Science Publishers B. V.; 5.25 reprinted by permission of © John Wiley and Sons Ltd; 5.28 by Arnold Bergstraesser

Institut, Freiburg; 5.29 Unwin Hyman and the *Journal and Proceedings of the Western Australia Historical Society* 2, 1939; 5.31 by Elsevier Science Publishers B. V.; 5.33 by UNESCO, Paris; 5.34 by the Swedish National Research Council; 5.35 by OUP; 5.38 by *The Ecologist*, Worthyvale Manor, Camelford, Cornwall, UK; 5.39 and 5.40 by Elsevier Science Publishers B. V.; 5.42 courtesy of D. A. Peel and reprinted by permission from *Nature*, copyright © Macmillan Magazines Ltd; 5.43 by Edward Arnold; 5.44 by Elsevier Science Publishers B. V.; 6.2 by *New Scientist*; 6.4 by Pergamon Journals Ltd; 6.5 by Edward Arnold; 6.6 reprinted by permission of © John Wiley and Sons Ltd.

Permission has been granted for reproduction of epigraphs by the following: to chapter 2 (from Nadine Gordimer, *The Conservationist*, 1974) by Jonathan Cape Ltd; to chapter 3 (from Denys Thompson (ed.) *Distant Voices: Poetry of the Preliterate*, Heinemann, 1978) by Denys Thompson; to chapter 5 by Aris and Phillips Ltd; and to chapter 6 by Methuen and Co.

Permission has been granted for reproduction of quotations from T. S. Eliot and Ted Hughes by Faber and Faber Ltd; excerpt from 'The Dry Salvages' in *Four Quartets*, copyright 1943 by T. S. Eliot, renewed 1971 by Esme Valerie Eliot, reprinted by permission of Harcourt Brace Jovanovich. Extracts from the Authorized (King James) Version of the Bible, the rights of which are vested in the Crown in perpetuity within the United Kingdom, are reproduced by permission of Eyre and Spottiswoode Publishers, Her Majesty's Printers, London.

Units and References

Readers will note that although the metric system is usually used in this book, some imperial equivalents have occasionally been given for ease of comprehension. In the case of units of energy, both SI units (based on the Joule) and the older calorie system have been used, depending on the practice of the original author; internal comparative consistency was thus retained. To convert Joules to calories, multiply by 0.239; for the reverse, multiply by 4.184. Multiples used commonly are k (10^3); M (10^6), G (10^9), T (10^{12}) and P (10^{15}). Others are explained at the point of use. A rough average figure for human metabolism is an energy consumption of 3000 kcal/person/day.

In an attempt to make the book readable, no scholarly apparatus has been inflicted on the text. A *Further Reading* section at the end of the volume contains references to all the material that has been used, with some indication of especially important sources.

General Introduction

The stellar universe is not so difficult of comprehension as the real actions of other people.

Marcel Proust, *Remembrance of Things Past: The Captive* (1923)

Imagine that in a space-ship we can rove among the stars and find points where our remote sensing technology can pick up the light reflected from earth about 2 million years ago and then zoom in, capturing images of the land surface every century or so until the very recent past. These images could be made into maps of the cover of the land surfaces and the condition of shallow waters. Over the Pleistocene period (1.8–0.01 million years ago) certain changes would be manifest, notable among them the great waxing and waning of ice sheets, with consequent adjustments of belts of natural vegetation and the proportions of land and sea. By about 0.01 million years ago (10 000 BC) the ice had largely withdrawn to the polar regions, with vegetation belts adjusting to the new interglacial period. But soon after that date, into the Holocene or 'postglacial' period, our equipment (if suitably sensitive) would begin to pick up traces of change which happened because of the activities of one species of animal: *Homo sapiens*. These early traces (and we might possibly have overlooked some even earlier ones) grew, as did the numbers of humans, until the most recent frames would show a world where human manipulation of the earth's surface is very strong indeed. For example, in *c.* 8000 BC there is a little land devoted to food production, rather than food collection, by humans; now there are 1472 million ha of cropland. In AD 1800 the area under urban use held 2.5 per cent of humans and now it houses 42 per cent. This book is about the history of such changes, and its basis is the gathering of empirical facts about the changes of the sort that our space-derived images would provide, were they not imaginary. But a unique feature of *Homo sapiens* is that observed facts also have meaning so we must provide a framework in which this information can escape from its status as mere isolated words or numbers and become a pattern. This pattern may be a clue to the kind of underlying regularities in nature and human activity that we call theory; equally it may lead to a set of tools for discussing the future, the more so since we may to a large extent choose the type of future we want.

Thus there are three layers: we start out with a discussion of ideas about the relation of man and nature, and of models of nature itself, and end with asking whether the main body of material has any lessons for the future. In between, chapters 2 to 6 tell stories of the changes humans have wrought in their surroundings, deriving from the first chapter a set of organizing principles to give the stories some structure. History, of course, relies very much on dating but this book is not the place to discuss the problems thereof. Before the age of documents all dates must be assumed to have a margin of error; more recent ones are likely to be valid, though even here pre-scientific evidence can be wrong. But little of the story depends on an exact year: where it does then the time is known very well, as with Fermi's controlled chain-reaction in Chicago in 1942.

Ideas about the relationship of humanity and nature

In view of the self-conscious nature of humankind, it is scarcely surprising that the formulation of abstract ideas about the relation of man and his environment has emerged from the everyday reality of getting a living. So far as the archaeological record permits, we can see evidence of such abstractions from the very early days of human cultural development. Numerous instances of practices of a non-material kind suggest that inferences were being drawn about the relationship of the human population with, usually, a major means of subsistence such as wild animals; the Upper Palaeolithic cave paintings of southern France and northern Spain include pictures which must have represented human conceptions of their relationships with the non-human world.

We may surmise that there have been, through the millennia, various reasons for such abstractions. For a first example, there is the long-standing urge to explore and account for the origins of our universe and of our species. Cosmogenesis itself and the origin of mankind were in early times enshrined in myths, which have to some extent now been replaced with the apparently more objective (although equally fantastic when we consider the models of phenomena like the 'big bang' and the 'black hole') findings of science. Then there is the desire to understand people's relationships with nature and with some kind of Absolute. This is exemplified in the way in which some hunters found a need to propitiate either the spirit of the actual animal they killed or a tutelary deity of all animal spirits in order to avoid 'blame' for their act and to ensure the future supply of food. The right type of relationship with natural phenomena (and their Superiors if they have them), will also bring prosperity and fertility to the human communities. The historian of religion Mircea Eliade describes the way in which nature often attracted a religious focus, and even in the Christianized Europe of the sixth century, St. Martin of Braga complains of the habits of peasants who burn candles before rocks and plants and honour streams by placing bread upon their waters. Literate theology developed a Christian eschatology permeated by an environmental attitude which saw the earth as over-populated and full. Other Christians, by contrast, had a sense of the recreational powers and potentials of nature: Petrarch may have been the world's first tourist when he climbed Mont Ventoux in Provence in 1336 so as to contemplate the wonders of creation and the human place within its order. Not that pagan attitudes were entirely lost: besides the votive offerings denounced by St. Martin, there was a healthy fear of the forests, the moor, the fen and the sea: indeed all those places we now call wilderness. There was for those people no assurance that nature would not reclaim their fields and villages as the author of *Beowulf* lamented:

> . . .their time ended
> their houses, their halls, empty and still.

The long literary history of cultural attitudes to nature in the West from classical antiquity to the eighteenth century has been reviewed in Clarence Glacken's magisterial opus of 1967, *Traces on the Rhodian Shore*. From a great mass of detail, he extracts and presents for our consideration the relatively simple notion that throughout this period three themes have been persistent, although their exact form has had different modes of expression. The first of these is the idea of a designed earth: one especially fitted for the human species. To a great extent, this is part of the wider concept of teleology, i.e. the concept of an overall creation with a particular purpose which was usually divine, but it represents an attempt on the part of the western thinkers to create an holistic framework which was taken much further when the models of evolution and of ecology became subjects for great debates.

The second great theme is that of environmental influence on culture. This derives initially from the contrast between nature and custom in different places and came to be used in interpreting the great array of human cultural and biological differences. It is not totally incompatible with the idea of a designed earth and attracted thinkers such as Thomas Malthus (1760–1834, plate 1.1), who emphasized not only the influence of different environments but also the limitations which the earth imposed on social development. Malthus helped to keep alive an idea of great antiquity that was born again in the nineteenth century as determinism. The father of this generation of offspring seems to have been Carl Ritter (1779–1859) whose theme was that the physical environment was capable of determining the course of human development. His ideas were strengthened by the publication of Charles Darwin's *Origin of Species* in 1859, with its emphasis on the close relationships of organisms and their habitats, and the notion of the pressures of natural selection (plate 1.2). Thus arose a 'scientific' type of environmental determinism which accounted for such features as migrations and the national characteristics of particular peoples. The names of Freidrich Ratzel (1844–1904) and Ellen Churchill Semple (1863–1932) are associated with the most outspoken expression of these ideas. The somewhat broad-brush approach of these writers was modified by geographers like Ellesworth Huntington and

Plate 1.1 Thomas Malthus (1766–1834). A founder of the study of the relations between population and resources; a strong influence on the ideas of Charles Darwin. This and other photographs of individuals remind us that ideas about the environment originally came from people who in many ways were like you and me. (The Mansell Collection)

Griffith Taylor. Huntington (1876–1947) tried to seek out objective evidence of the effect of the physical environment, and in particular climate which he regarded as an important influence on human behaviour. He suggested that the 'best' climates for work were those in which there was variety and in which the temperatures fell within a certain range, and wrote of the correlation between a stimulating climate and high civilization based on industry in the UK and New England. Taylor (1880–1963) was even more careful to gather accurate data about the environment and to relate these to his idea of human habitability, especially in Australia. He tended to play down socio-economic factors ('cultural accretions have little relevance . . . in a discussion of the importance of environmental control'), but instead suggested that although the physical environment led inexorably in a particular direction, societies could control the rate at which they progressed through the various stages of development. This was likened to a traffic control system which determined the rate but not the

Plate 1.2 Charles Darwin (1809–82). Credited with the first publication of the theory of organic evolution, still a principal pillar of biological thought. (The Mansell Collection)

direction of progress, and so it became known as 'stop and go determinism'.

The nineteenth century had also seen the rise of an alternative to determinism, usually labelled 'possibilism', associated especially with the French school of geography founded by Paul Vidal de la Blache (1845–1918). These scholars saw in the physical environment a series of possibilities for human development, but argued that the actual ways in which development took place were related to the culture of the people concerned, except perhaps in regions of extremes like deserts and the tundra. The historian Lucien Febvre (1878–1956) set out to demolish the determinist argument by asserting the initiative and mobility of man as against the passivity of the environment, and regarded other humans as part of the environment of any group because they contributed to the formation of the next group's cultural surroundings, or *milieu*. Among those influenced by this type of thinking was H. J. Fleure (1877–1969) who tried to formulate world regions based on human characteristics rather than the traditional climatic–biotic regions. So he brought forth a scheme which included 'regions of effort', 'regions of hunger' and 'industrialized regions', to name a few. But because he recognized in his scheme regions of 'lasting difficulty' and 'debilitation' he too saw the relevance of the influence of the physical environment. Possibilism has also been influential in the rise of the school of cultural geography associated with the name of Carl Ortwin Sauer and the University of California at Berkeley, and with the development of the idea of human ecology. The founder of this latter notion was H. H. Barrows (1877–1960) of the University of Chicago, who stressed the importance of the type of relationship between human and physical factors, and especially wanted to study the physical elements only when these were relevant to human activity. Some historians, too, have moved in a more ecological direction. Most notable are the *Annales* school under the leadership of F. Braudel, who delineated the eighteenth century as a watershed of biological regimes. Even at the end of that period, he postulated, vast areas of the earth were still a Garden of Eden for animal life, and the *Jungle Book* could have been written about almost any part of the globe. He too made points against technological determinism.

In summing up these two themes we ought to ask whether new ecological knowledge and thinking unknown to the writers mentioned above affects our views on determinism and its alternatives? A new evaluation might run something like this: the determinists and the possibilists both have useful points to make but they are valid at different scales. The environmentalist movement started in the 1960s has shown quite distinctly that there is an overall limit to certain kinds of human economic activity in terms of the biophysical persistence and resilience of the planet's systems. So there is a physical or ecological envelope, but within this, human technology and knowledge allow a variety of adjustments to the resources of the planet. At the very largest scale we can be determinist, whereas at more local scales we can see the virtue of possibilism. Nevertheless the technological achievements of man are such that we can now perhaps contemplate the replacement of some of the biophysical systems of the planet with man-made ones, providing that enough energy is available, thus escaping for ever the possibility of determinism's being correct; on the other hand, knowledge of the biospheric envelope might make us even more intent upon living within it. This is the new determinism of the environmentalists.

Alongside the debate we have just reviewed, a third ancient strand of thought identified by Glacken has been worked out in considerable detail. This is the set of views that sees man empirically as the modifier of nature but where nature in turn affects the perceptions of human societies. The evolving western world-view[1] has entrained the notion

[1] This term encapsulates the idea of a general conception of the world, often implicit rather than explicit, which contains a system of values, such as the western ideas that the world consists of materials for human use and that problems will be accessible to scientific and technological solutions. It is a translation of the German word *Weltanschauung*; a better but longer version might be 'world-outlook'.

that nature may be seen as a set of resources to be used to satisfy human demands, and so in the course of centuries there have been many changes in the genetics of species and manifold alterations of ecological systems at all scales. Glacken has traced the descent of the basic idea from its state where God was an artisan who left the creation unfinished and man became the collaborative finisher (by draining marshes, reclaiming wild lands and generally making the rough places plain), through its optimistic development in the seventeenth and eighteenth century with men like J. Ray (1627–1705) and G. L. de C. Buffon (1707–88) forecasting a tide which would carry men to ever greater heights of perfectibility, to a nineteenth century pessimism when the deleterious effects of some of the efforts of the finishers became all too obvious. In more recent times, we associate this approach to man–nature relationships with the names of Mary Somerville (1780–1872) and more especially with George Perkins Marsh (1801–82). Marsh was a US diplomat who lived abroad for much of his life and his book *Man and Nature, or Physical Geography as Modified by Human Action* (1864) highlighted what he saw as the upsetting of a balance within nature by unwise human action. Thus he was not only a chronicler of man-induced environmental change but a judge of it as well. His influence has persisted in many fields of study and in 1954 a commemorative symposium was held, published in 1956 as *Man's Role in Changing the Face of the Earth* (edited by W. L. Thomas). In many ways, Marsh can be seen as the immediate antecedent of today's environmentalists, and the type of work found in the 1956 volume has been expanded many-fold to give us more details of human impact upon the environment for both the past, where we may cite the work of the Danish palaeoecologist Johannes Iversen as founding a new paradigm for environmental prehistory, and the National Environmental Protection Act (NEPA) of 1969 with its requirement of Environmental Impact Statements seems a logical successor to Marsh's concerns. It is squarely in the tradition of the human role in changing the face and the functioning of nature, with an emphasis on empirical historical evidence and allied to a concern for the future, that this present book is put forward. There is unfortunately no space to discuss the fascinating question of whether 'the environment' only exists as a mental construct. For present purposes we shall assume it is a material reality.

So far, only western philosophies and world-views have been mentioned. At an abstract level, non-western views may be very different from those of the West, especially those which emphasize quietism and non-interference, or which reject the western dualism of man and nature in favour of equality of standing and value and where change is to be expected of both. But the results in operational terms do not seem to be very different from those in the West. Deforestation under western materialism can be matched with deforestation under Buddhism and Taoism; conservation of forest lands for hunting or as sacred groves in Europe and the Mediterranean can be set alongside the sacred trees of India (*Ficus religiosa* was worshipped as early as the third or fourth millennium BC), and the forest sanctuaries of Japan and the Philippines which may be pre-agricultural in origin. It is therefore scarcely possible to equate non-western philosophies with different environmental outcomes at those levels of generalization. This is not to say that particular societies did not achieve very delicately balanced relationships with their surroundings.

As far as the present is concerned, we should add that much of the world is under the sway of the western world-view: it has been adopted directly and consciously by states which want to 'modernize', accepted as an inevitable part of aid policies by others, and taken in unknowingly by yet more as part of the price of allowing the operations of a transnational company. Alternatives to the western outlook are found mostly in the western countries themselves where a particular level of affluence has allowed the rise of 'alternative' lifestyles based on a low-impact relationship with nature. But no great harm to truth is done if we regard the western world-view to be today's most strongly pervasive set of operational processes. As J. M. Roberts puts it, 'the story of western civilisation is now the story of mankind.'

Organizing concepts

Towards a framework

The sequence of intellectual and historical ideas just related does not go far enough for present purposes since it fails to provide a suitable vehicle for bringing into a coherent perspective the impact of the human species upon the rest of the world. Even the three traditions reconstructed by Glacken leave a lot of scope to delineate a more exact methodology: if we consider the idea of man creating a second world within the world of nature as being germinal for this book, we are still left with at least two options as to how to use this idea. The first is a nature-based framework in which we look primarily at natural systems (on a regional basis, for example) and then consider *Homo sapiens* as a different kind of animal with the capacity to alter those systems consciously. Human use becomes one more factor in the analysis of ecological dynamics. The obvious alternative is a humanist stance: a cultural and behavioural viewpoint where *Homo* is not related to other animal levels at all, but is rather the measure of all things and indeed may sometimes be regarded as completely different from the rest of nature. Although the existence of the rest of the planet is acknowledged, it is from the standpoint of the old statement for discussion in examinations: 'natural resources are cultural appraisals'.

Common to both these strands of thinking is an acceptance of the idea that the human species is different in kind from the rest of the animal world, let alone plant life and the inorganic materials of the planet. This notion is called dualism and is a basic feature of the western intellectual tradition, with Descartes having given it a significant boost in the seventeenth century. In this book we shall accept dualism as a convenient vehicle. For the bulk of the descriptions and discussions, it is assumed that the hominid characteristics of self-consciousness, leading to cultural diversity, sense of social purpose and concern for the future, mark us off from our nearest animal relations.

Acceptance of dualism does not get us much further in the search for a suitable framework for a history of human impact. Starting with nature-based methodologies, the most suitable appears to be the dynamic models developed by the science of ecology (which is the study of the interactions of organisms among themselves and with their inorganic matrix), especially since we are concerned with change through time. The idea of the exchange of energy and matter between organic and inorganic elements of a given portion of the earth's surface is sufficiently emphatic about the changing relationships between components for it to be called a system, in this case an ecosystem, to be discussed in more detail on pp. 10–19. Where human beings are active agents in ecosystems, then the field of study is sometimes called human ecology and the resulting systems have been given a variety of labels such as anthroposystems, socionatural systems and, in a more specialized case, agroecosystems. Examined more closely, the idea of a humanized ecology does present some difficulties in use and somehow, for the purposes of this book, we have to find a way of subsuming a great deal of historical and near-recent empirical data into a framework which will contain and order it all. We need also to remember that the type and quality of the information will change through time, and to take note of both scientific and humanistic approaches so as, for example, to meet the needs of the emerging discipline of environmental history. Since selectivity is necessary, not only of examples, but of approach, I propose to adopt the following guidelines:

1. The western idea of the dualism of man and environment will be employed.
2. The data will be fitted into an ecosystem-based approach which will have a biological basis, but which will wherever possible include human effects in their unique historical contexts, rather than simply as herbivores or carnivores. The cultural factors underlying particular choices involved in environmental manipulation will be mentioned where possible, but the emphasis is on empirical narrative rather than highly abstract explanation.
3. In order to provide some continuity from one historical era to the next, the language of ecosystem studies and in particular the study of the flows of energy (quantified where possible) will be used, bearing in mind the limitations of this approach as an explanatory (as distinct from descriptive) tool.

Thus it becomes necessary to acquire the vocabularies belonging to the study of the ecosystem, and to energy analysis and the role of energy in society: it is to these areas we now turn.

The ecosystem as a conceptual tool

To help our discussions of the impact of human societies upon nature, we need to be able to characterize an area of natural landscape and its processes, and then within the same conceptual frame to detail the impact upon it of human activities and how our societies have continued to live within the new system thus created. Here it is fortunate that the idea of the *ecosystem* as a functional unit has been strongly developed in recent decades. The ideas to be used derive largely from the Lindeman model of ecosystems, articulated in 1942, as developed by the work of Eugene and Howard T. Odum. Eugene Odum gives a useful formal definition of an ecosystem:

> Any unit that includes all of the organisms in a given area interacting with the physical environment so that a flow of energy leads to . . . exchange of materials between living and non-livng parts within the ecosystem . . . is an ecosystem.

We may note firstly that 'all of the organisms' are included, so that we may discuss the role of mankind; secondly that the flows of energy and of materials can often be measured so that we can quantify some of the ecosystem characteristics (table 1.1).

One of the potentials of the concept is that it is scale-free and it can be visualized at any spatial scale: we can think of the ecosystem of the pool of water in the crook of a tree branch (or even smaller systems), we can delineate major world biomes like the tundra or the savannas as ecosystems, or we can even consider the entire globe as a single ecosystem. At all levels below that of the whole planet, though, boundary problems often occur which can usually be resolved only arbitrarily. We can identify the tundra, for instance, and measure many of its ecological characteristics, but we must take account of the fact that many of the animals are migratory and spend only the summer period there. Any evaluation of a pond must similarly recognize that a duck may be only a sporadic visitor which spends some of its time in another ecosystem. A field and a wood may appear to be separate systems but be quite strongly linked by surface runoff or the movements of animals. Where *Homo sapiens* is involved, virtually no ecosystem now has closed boundaries.

The flow of radiant energy from the sun forms the foundation of most life in its familiar forms, so the capture of energy by the process of photosynthesis and the subsequent fate of the chemical energy thus 'fixed' is of considerable importance in ecosystem studies. The detailed biochemistry and energetics of the process can be studied elsewhere: here we note that the immediate outcome is the growth, survival and reproduction of green plants. These form the starting point of nearly every ecosystem, natural or man-made, past or present, and the ways in which the consequences ramify into human societies are explored in figure 1.1. What seems remarkable is the small proportion of the sun's energy incident upon the earth which becomes plant organic matter: about 520×10^{22}J of energy hit the top of the earth's atmosphere every

Table 1.1 Characteristics of the ecosystem concept

Characteristic	*Quantifiable Yes*	*No*	*Comments*
1 Energy flow	x		Usually expressed in kilocalories (kcal) or joules (J) unit area/unit time. Cannot be cyclic
2 Nutrient flow	x		Usually as grams/area/time. Often 'cyclic' in natural systems
3 Productivity	x		Usually as grams dry matter/area/time, i.e. a rate. Is the outcome of growth resulting from interaction of 1 and 2
4 Population dynamics	x		Is expressed as individuals of a taxon (e.g. a species). Includes concepts of abundance
5 Succession	?		Progress of the system through time from its inception. Quantifiable by reference to other features, but not as such
6 Diversity	x		Variety and abundance of species. Quantified by a number of possible indices
7 Stability		x	Used in the sense of resistance to perturbation
8 Degree of modification		x	Historical data often needed. Broad volume or index-number description often possible, but not rigorous measurement

Characteristics 1–7 can all be measured for man-made and man-modified systems as well as natural ones.

year, of which 100×10^{22}J reach the surface of the earth and are of a suitable wavelength for photosynthesis. Some 40 per cent of this is reflected back into space, leaving 60×10^{22}J as the pool for photosynthesis. The plants use up a lot of this energy in their own biochemical processes, so approximately 170×10^{19}J appear each year as plant matter, much of it in the form of oceanic phytoplankton. The ratio of organic matter production to incident photosynthetically active radiation is about 0.2 per cent. Viewed anthropocentrically, photosynthesis thus appears rather inefficient but by way of perspective we might think that it is remarkable that it exists at all, and that we are unable to replicate it in the laboratory, let alone improve upon it; presumably evolution has produced plants that maximize for some variable other than energy efficiency. The energy fixed by plants follows one of two paths. If the plant dies, then the energy-rich organic matter is fed upon and broken down by the soil flora and fauna (or their aquatic equivalents), releasing heat; if it is eaten by a plant-eating (herbivorous) animal, then some of the energy ends up as animal tissues, some is again lost as heat, for example in the movement of the creature to get more food. The body of the organism is eventually given over to the decomposers or it may become the food of a carnivore. So

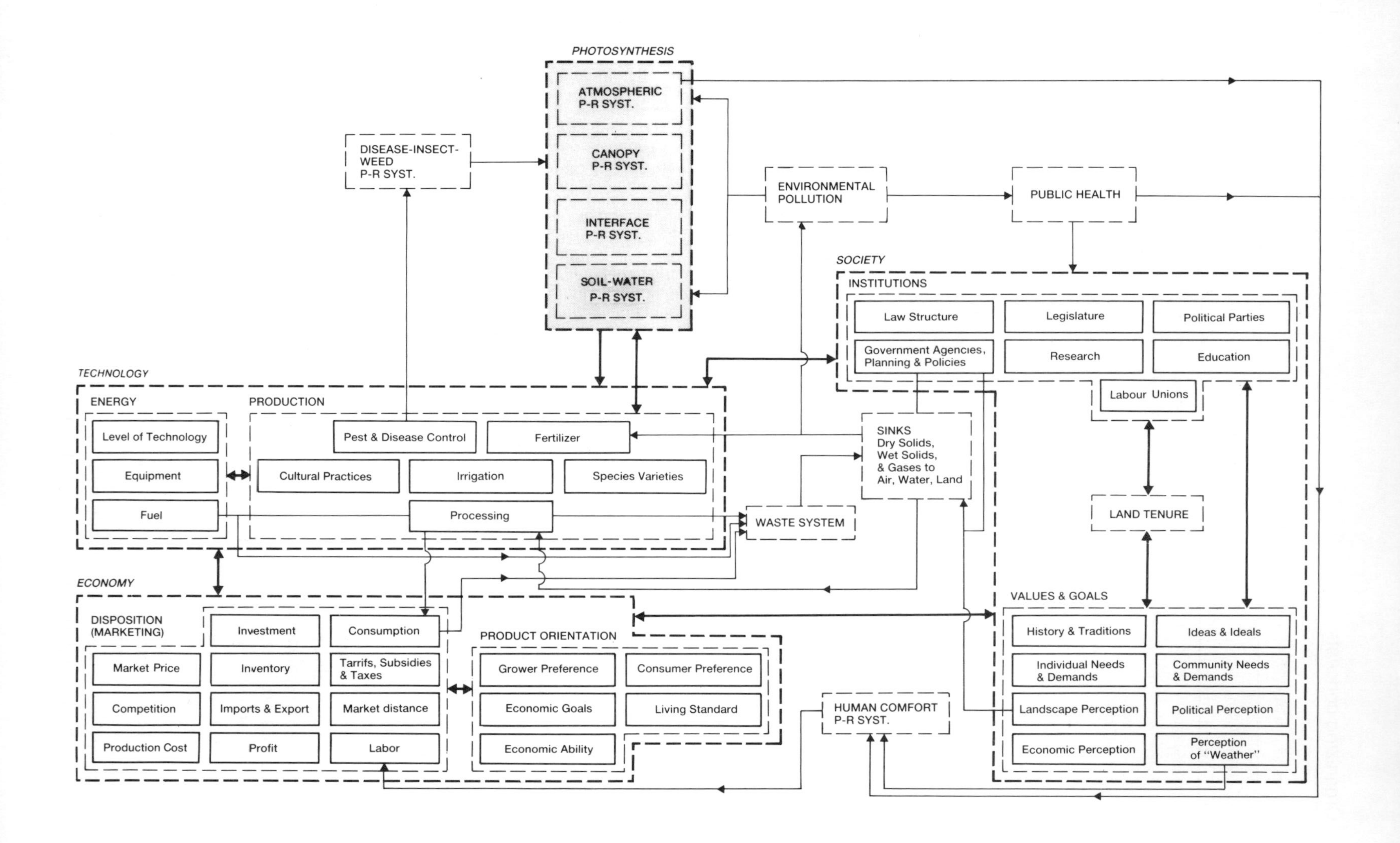
PHOTOSYNTHESIS
ATMOSPHERIC P-R SYST.
CANOPY P-R SYST.
INTERFACE P-R SYST.
SOIL-WATER P-R SYST.
DISEASE-INSECT-WEED P-R SYST.
ENVIRONMENTAL POLLUTION
PUBLIC HEALTH
SOCIETY
INSTITUTIONS
Law Structure
Legislature
Political Parties
Government Agencies, Planning & Policies
Research
Education
Labour Unions
LAND TENURE
VALUES & GOALS
History & Traditions
Ideas & Ideals
Individual Needs & Demands
Community Needs & Demands
Landscape Perception
Political Perception
Economic Perception
Perception of "Weather"
TECHNOLOGY
ENERGY
Level of Technology
Equipment
Fuel
PRODUCTION
Pest & Disease Control
Fertilizer
Cultural Practices
Irrigation
Species Varieties
Processing
SINKS Dry Solids, Wet Solids, & Gases to Air, Water, Land
WASTE SYSTEM
ECONOMY
DISPOSITION (MARKETING)
Investment
Consumption
Market Price
Inventory
Tarrifs, Subsidies & Taxes
Competition
Imports & Export
Market distance
Production Cost
Profit
Labor
PRODUCT ORIENTATION
Grower Preference
Consumer Preference
Economic Goals
Living Standard
Economic Ability
HUMAN COMFORT P-R SYST.

we can envisage the energy forming a chain with a number of branches but eventually all ending up as heat. It is still energy, but it is 'bound' energy in a dispersed form and can do no work: it is radiated out to space and thereafter plays no direct role in sustaining life. The lesson learned here is a fundamental one: all forms of energy on the earth end up as heat which can do no work. This applies to human-directed systems no less than to natural ones.

Living tissue also comprises a number of chemical elements which combine with carbon to constitute organic matter. Of these, only oxygen and hydrogen are normally freely available in large quantities, and the others circulate between reservoirs ('pools') on varying scales. The pools usually include a non-living stage so the circulation of these elements is called a biogeochemical cycle. Study of these flows usually reveals two cycles of rotation. The first has a very slow turnover and involves a very large pool of the element stored in the atmosphere or in the earth's crust. By contrast, there is a smaller pool which undergoes rapid exchange between an organism and the immediate components of its environment, both living and non-living. The quantities and fluxes involved in many biogeochemical cycles are known at the world scale, and much research has elucidated the fine details for individual ecosystems. But we will note that the flow of such mineral nutrients (the overall term for these elements) is genuinely cyclic, unlike energy, even though the time-scale is geological rather than anthropocentric in some parts of a biogeochemical cycle. In the rapid exchange part of the cycle we should note the importance of the decomposers in any ecosystem, for they break down complex organic matter into the simpler inorganic forms from which uptake of the mineral nutrients by plants is possible once more.

In spite of the immensity of the quantities of matter in the global cycles, mankind has been able to contribute to them in significant ways. Conversion of land from forest to grassland or cropland frees carbon, as does the combustion of fossil fuels. Together, these have contributed to a rise of 30 per cent in the concentration of atmospheric CO_2 in the last 125 years. Another source of carbon in the atmosphere is methane, where the production from such sources as cattle and wet crops like *padi* rice is now twice that from the world's swamps and natural wetlands. The emissions of the oxides of nitrogen from human activities now equal those from all natural sources, as do those of sulphur compounds. For metals, the ratio of human to natural mobilizations is 333 for lead, 38 for zinc and 20 for cadmium. To be added to these concentrations and diversions of natural materials are the synthetic substances like pesticides and CFCs (chlorofluorocarbons) which are caught up in the global cycles of other elements.

The flow of energy and the cycling of nutrients come together in the production of organic matter, and this material represents an integration of many factors like climate, nutrient availability (especially in the sea), the age of the individual organism and of the ecosystem. The actual amount of organic matter per unit area per

Figure 1.1 The crucial physiological process of photosynthesis, so central to life on Earth, exists in a web of many other processes of human origin and direction. ('P-R system' in some boxes refers to Photosynthesis minus Respiration, the product of which is Net Productivity: see text). *Source*: D. Greenland, *Guidelines for Modern Resource Management*. Columbus OH: C. E. Merrill, 1983.

unit time is called biological productivity and that of plants the net primary productivity (NPP). Research under the aegis of the International Biological Programme of the 1960s and 1970s has yielded a reasonably accurate picture of the world distribution of NPP (table 1.2). The data for NPP exhibit few surprises, except perhaps:

1 that the open oceans are so unproductive: a virtual desert;
2 that coral reefs and estuaries are very productive, almost as high as the highest terrestrial systems;
3 that agriculture is not particularly productive compared with many natural ecosystems. But these values are measures of all plant life, irrespective of its value to human societies; thus the lower productivity of a useful crop is clearly more attractive in the short term than the higher productivity of a plant that cannot be eaten, sold for fibre, or used as an animal foodstuff.

The rate of production of animal tissue per unit area is termed secondary productivity, and although much research has been done, broad-scale pictures are much slower to emerge. Given that animals are usually selective consumers, that they use energy in searching for food, and that warm-blooded animals require a lot of energy to maintain a constant temperature, it is not surprising that 10 per cent at most of the NPP of a terrestrial ecosystem will appear as herbivore tissue. It is scarcely remarkable therefore that carnivores are often relatively rare animals.

The structure and function of an ecosystem are ordered by a set of control systems which are generally referred to as 'feedback' and which in the natural state involve the interplay of a variety of biological, chemical and physical stimuli. The feedback loops may be negative, analogous to a thermostat, or positive, analogous to the spread of an epidemic, and more than one loop may operate at any one time. These regulatory flows do not usually produce complete stasis, but a system

Table 1.2 Net primary productivity of major biome units, 1970s

Ecosystem type	*Surface area* ($\times 10^6$ km^2)	*NPP* ($\times 10^{15}$g or 10^9t)	*Mean NPP* g/m^2/yr
Forest	31	48.7	1570
Woodland, grassland and savanna	37	52.1	1408
Deserts	30	3.1	103
Arctic–alpine	25	2.1	84
Cultivated land	16	15.0	937
Humanized area	2	0.4	200
Other terrestrial (sclerophyll, wetlands)	6	10.7	1783
Lakes and streams	2	0.8	400
Marine	361	91.6	254
Subtotal terrestrial	147	132.1	899
Subtotal aquatic	363	92.4	255
World Total	510	224.5	440

Source: adapted from P. M. Vitousek et al., 'Human appropriation of the products of photosynthesis', *BioScience* 36, 1986, 368–73.

which oscillates within a wide tolerance zone. There is perhaps an image of ecological systems always behaving as if to return to some former equilibrium state, like the 'ecological balance' sometimes found in eco-political literature. In fact the behaviour of ecosystems is more likely to be open-ended: as they change through time their exact course cannot be foretold. As an ecosystem gets older it develops a complex web of links and loops and if human societies disturb the system it is not at all easy to restore that web, should that ever be desired. In fact, humans are rather more accustomed to systems which have two states, on and off, and which are controlled rather like machines by a single regulator. Thus the science of managing ecosystems (if by science we mean the attainment of temporal predictability) is not yet highly developed. Systems which are subject to strong direction from humans are normally managed as open-ended entities, where the goal is usually to move on to a new state rather than to return to a former one, and where the feedback loops tend to be positive (i.e. the deviations are positive and self-amplifying) and often cognitive rather than physio-chemical in nature.

Ecosystems are not static in time. In the northern hemisphere, for example, many of them started anew after the Pleistocene glaciations; new ecosystems begin to assemble on land emerging from the sea (as happened recently on volcanoes such as Krakatoa and Surtsey), on terrain recently deglaciated as in Arctic Canada, and on the cooled lavas of continental volcanoes, to say nothing of land made bare by human activities and then abandoned, like agricultural fields or tip heaps. Ecosystems normally progress from relatively simple systems with a low species diversity and short simple food chains, to a condition of maturity with a higher diversity, complex food webs, and the capacity to renew themselves if subjected to external perturbations. This latter condition is sometimes called the climax and the preceding phases seral stages, the whole process being called succession. It is now recognized that the idea of climax as a seemingly static endpoint to ecosystem succession is rather simplistic and the term is less often used than 20 years ago; we might perhaps better talk of mature ecosystems, recognizing that these are in equilibrium with prevailing environmental conditions provided we have the evidence for their history which suggests that they are not undergoing major transformations.

Eugene Odum has led us to the idea that the world consists of a mosaic of ecosystems at different successional stages: some mature, some at an early phase, some a mixture of the two, and others being inert systems devoid of life, like volcanoes, glacial outwash and perpetual ice and snow. At any one time, these are in some kind of balance, though exactly what the nature of that balance might be is not entirely clear. But in a natural world successional stages are becoming mature, fresh inert areas are always being created, and retrogression occurs in the mature systems with events like windthrow and fire in forests.

Diversity is another measurable characteristic of ecosystems. It is, however, a rather more complex measurement than at first appears, since it has to be related to more features of the ecosystem than the simple count of the number of species per unit volume of space which is usually termed richness. It can also be argued that not only should it measure the presence of species but the evenness of the distribution of the relative abundances of the individuals too, since a common species exerts much more influence on the nature of the ecosystem than one which is extremely rare. So a number of indices of diversity have been built up, although they are never a measure of the whole ecosystem but only of certain components' functions.

Stability too is capable of being measured in different ways (table 1.3), but the meaning taken here is that of resistance to and recovery from external perturbations (i.e. inertia or resilience). At one time, diversity was linked to stability in an almost dogmatic way: the higher the diversity in an ecosystem, the greater its stability. So it was held that a greater complexity of food webs must necessarily mean continued stability should one species disappear. But more evidence has brought examples of other relationships: that of the key species, for example, the disappearance of which allows the breakdown of an otherwise highly complex and stable system. So diversity and stability are not so closely linked in all systems, and there are a number of other parameters which can be important in a given ecosystem: connectance (a measure of interspecific interaction), evenness (the distribution of abundance), and variability (the variance of population densities over time), are examples. Stability is hard to measure in any objective way but R. Margalef has likened the resilience concept to the application of energy to a ball-bearing in a saucer: flicked a certain amount, the ball-bearing will return to the centre of the saucer; flicked harder, it will surmount the rim and come to a new equilibrium position somewhere else. So perhaps measurements of impact intensity might eventually tell us about this phenomenon objectively: hit a rainforest with one bulldozer and it recovers; maul it with ten of them and 100 beef cattle and it turns to grassland. So far, however, we do not have rigorously made measurements of that type of stability. But the very concept of stability has the inherent defect that it assumes that some kind of equilibrium state is the norm. More sophisticated ecological studies now show that heterogeneity through space and time may be characteristic of many ecosystems, and that the systems are likely to be non-linear in their behaviour. Some show a great deal of instability in the short-term, with cyclic fluctuations of high inequitude in their populations, but nevertheless the system persists for hundreds of years. The ecologist C. A. Holling has pointed out that if there is an analogy here for human ecology in a world whose behaviour contains a lot of unknown quantities, then the best strategy is to keep as close to a stability boundary as possible, so that it is possible to move easily to a new position when new information becomes available or when other conditions dictate a shift. Thus the ingredients for persistence are learning to live with disturbance, variability and uncertainties.

Table 1.3 Some meanings of 'stability' in ecology

Type	*Graphical representation*	*Properties*
Constancy	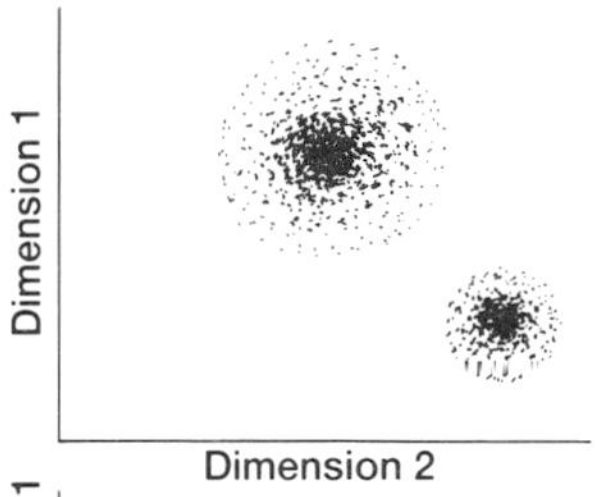 Dimension 1 / Dimension 2	A lack of change in some parameters of a system, e.g. number of species, taxonomic composition or features of the physical environment
Inertia or resilience	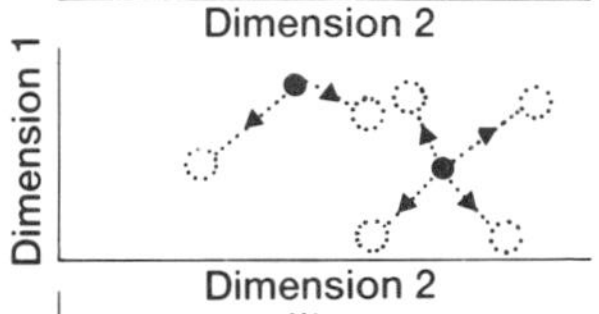 Dimension 1 / Dimension 2	Ability of a system to resist external perturbations. Is sometimes called resilience. Can be measured for a number of parameters additional to those above
Elasticity	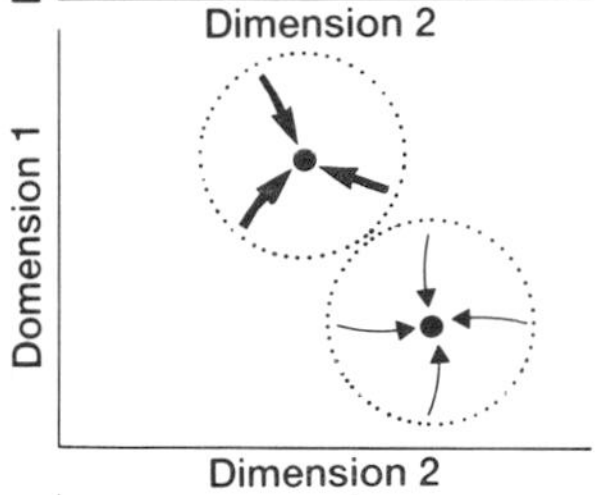 Domension 1 / Dimension 2	The speed at which the system returns to its former state following a perturbation
Amplitude	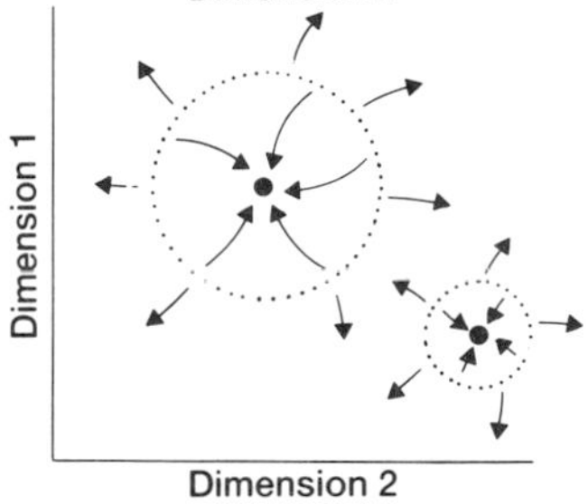 Dimension 1 / Dimension 2	The area over which a system is stable. It has a high amplitude if it can be considerably displaced from its initial state and still return to it
Cyclic stability	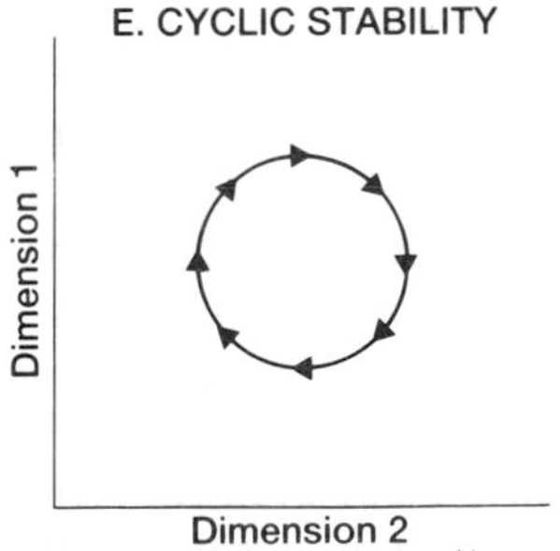 E. CYCLIC STABILITY Dimension 1 / Dimension 2	The property of a system to cycle or oscillate round a central point or zone, as in oscillating predator–prey systems
Trajectory stability	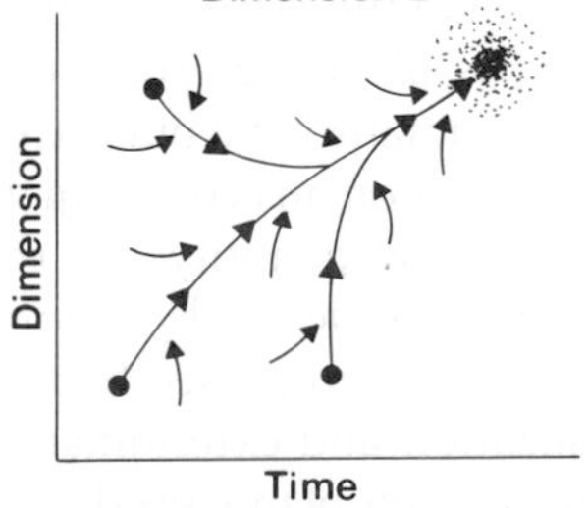 Dimension / Time	The property of a system to move towards some final end point or zone despite differences in starting points. A single 'climax' may be reached despite different initial beginnings

This is not a classification, for the terms are not comparable, merely a set of descriptive terms.

Source: G. H. Orians, Diversity, stability, maturity in natural ecosystems. In W. H. Van Dobben and R. H. Lowe-McConnel (eds), *Unifying Concepts in Ecology*. The Hague: Junk, 1975, 139–50.

Lastly, it is vital to know the degree of man-made modification that an ecosystem possesses. There are two main difficulties here: firstly we lack knowledge about the ecological history of many of the world's present ecological systems; even where they seem quite natural, we often have the suspicion that they are in a successional state and responding to some past influence of which we have no evidence. Secondly, there is as yet no simple and objective way of characterizing the degree of change which an ecosystem has undergone at human hands, partly at least because the response is so varied. If an ecosystem always became less diverse in species as the result of human impact upon it then we might have the basis for measurement, but the opposite is known to happen. Being unable to put a number to it, however, does not vitiate the concept and we shall frequently use the idea of the degree of manipulation of an ecosystem away from its pristine state.

The manipulative effects of man are of two basic types. There are first of all transient perturbations, of which examples are a fire, an application of a poison, an episode of deforestation or shifting cultivation, or the harvest of a fish population. This produces many kinds of effect in detail but overall it tends to put back the progress of successional stage to an earlier phase. It can, of course, go further and reduce the ecosystem to an initial state and become a permanent effect, as when forest clearance is succeeded by soil erosion and virtually nothing grows on the resulting badlands.

Secondly, there are chronic changes, which are long-term shifts to permanent new conditions. Here are the complex patterns in space and time of human influences which form much of the theme of this book. Examples are the long term effects of biocides, of changes in the composition of the atmosphere, irrigation, the construction of polders and of the growth of cities. Normally we think of these changes as reducing stability and diversity but while numerous examples of these can be found, there are others to the contrary. Where, for instance, humans have created new structures in an otherwise uniform environment, the biological diversity may well be increased: in the suburbs of a desert town the water supply and the garden plants will provide for a higher diversity and biomass of birds than previously existed. But in the long term, the effect of mankind globally has been to reduce NPP, even though the lower levels are culturally more acceptable than the pristine status.

If we put these two types of change together and think of the processes of change which humans have put into action, we can see that there is no man-made effect on an ecosystem which is exclusively ours except by combination and quantity. All these effects can, in principle, be caused by other species or by abiotic agents. Shifts to earlier stages of succession are often caused by natural disasters, fire is set by lightning, colonization and extermination occur naturally, monocultures are produced when plants produce toxins that inhibit the growth of other plants (allelopathy) and by competitive shading, and eutrophication which is after all a natural process of the enrichment of water by nutrient inflow, resulting eventually in silting up and the formation of dry land. So what is human is the variety and combination of effects in mass, space and time; even plants produce natural toxins though not on the scale of which our industries are capable.

One other aspect of the ecosystem concept needs a brief exposition. Looked at from the perspective of the structure of science, it is clear that it deals with a very complex level of organization, involving as it does varying combinations of large molecules, water, mineral nutrients, energy, plants, animals, fungi, bacteria, viruses, climatic factors and very probably human societies as well. It is a feature of levels of organization that knowledge of the properties of one level does not enable an observer to predict the behaviour of the next. Thus, information about the physics and chemistry of hydrogen and oxygen does not allow any predictions about the freezing properties of water, for instance. It is equally the case that even comprehensive knowledge of the biochemistry of cells does not explain the functioning of organs, and certainly not all psychologists would allow that the individuality of human personality will ever be explicable in terms of chemistry and physics. The presence of characteristics which suggest that the whole is necessarily more than the sum of its parts is referred to as an emergent property, and in the case of ecosystems is yet another reason for their unpredictable behaviour. Any study of a system by scientific means must necessarily be of the measurable parameters exhibited by the components, but is unlikely to encompass the emergent properties which are nevertheless real. The nature of these properties and their treatment by epistemology is subject to a great deal of debate which can end up in metaphysics; here suffice it to say that when everything that can be measured has been measured there often seems to be something left over and use of the ecosystem concept cannot always ignore this residual quality.

Our conclusion at this point is that the ecosystem idea subsumes several lineaments of thinking, some of which are quantifiable and have been measured, others of which are potentially quantifiable, and others still which do not appear to be amenable to numeration. But the overriding concept of an interactive system remains: a system in which from time to time and place to place different components of the system will dominate it. Thus, near the South Pole the abiotic parts of the system are clearly dominant; in a pristine rainforest the living non-human elements set the scene for all else; in the Ginza district of Tokyo the hand of man has removed all except the grosser elements of the lithosphere and atmosphere and replaced them with his own creations.

Energy and society

The discussion of ecosystems showed how important in their structure and function is the flow of energy through them. Since human societies are functional components of ecosystems (which now means virtually everywhere on this planet, one way or another) they tend to divert some of the flows and transformations, organizing them in a purposeful way, for example by growing fields of cereals which maximize food energy output but minimize labour input, or extracting oil from the earth's crust and using it to power machinery which digs holes to extract coal which then powers machinery to make oil drilling equipment, and similar evidence of material progress. The physical and biological systems of the planet are evaluated by culture and turned into resources, which are then transformed into energy, goods and services. Access to extra-somatic energy (i.e. energy outside that of the human body itself) is important for all the stages of such a process. None of the components of that system is quite so simple, and J. W. Bennett's diagram (figure 1.2) of the interaction adds in some of the complexities and emphasizes the linkages which provide feedback or intra-system communication. The outcome of human transactions with nature has been incremental access to usable energy through time. Each stage of technological evolution has provided a greater net energy surplus (i.e. spare energy after basic metabolic needs have been met) than that of previous phases. Hence, more material success and more environmental change.

The nature of energy transformation is always subject to the laws of thermodynamics and no amount of human ingenuity has led to an escape from these laws. Energy exists in two states: free energy,

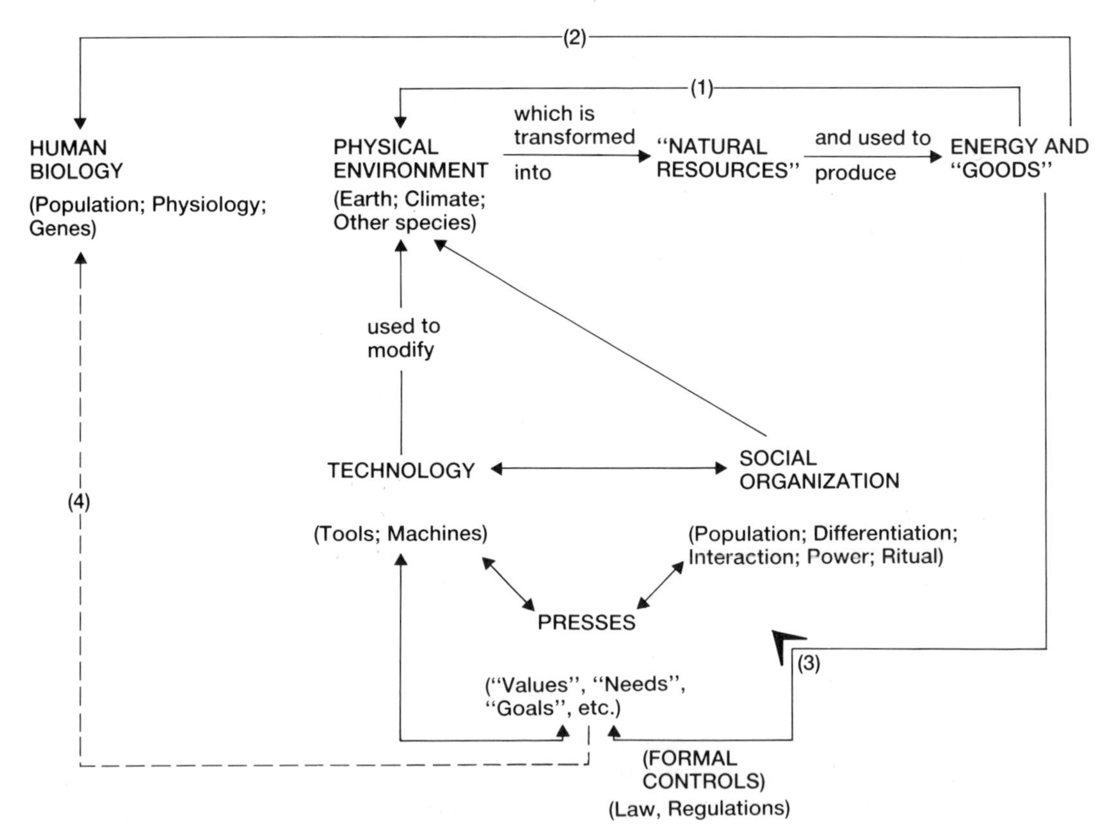

Figure 1.2 The ecology of the human use of the earth is subject to many processes of both the natural and social kinds. These are so interwoven that either environmental or cultural determinism seem to be considerable over-simplifications. *Source*: J. W. Bennett, *The Ecological Transition: Cultural Anthropology and Human Adaptation*. Oxford: Pergamon Press, 1976.

which can be used by man (and other living organisms) to do work, and bound energy, to which we have no access. A piece of coal represents free energy and can be burned to make steam which does useful work; the energy of the coal is transformed to heat and cannot be used again to do work. The difference between free (or 'available') and bound (or 'unavailable') energy is also a measure of order: free energy implies a form of structure or order, bound energy a state of chaos. (The economist N. Georgescu Roegen likens the former to a shop in which the meat is all on one counter, the vegetables on another and so on, and the latter to the same shop after being hit by a tornado.) The quantity of bound energy in a system is called its entropy and it appears to be a condition of the cosmos that entropy (i.e. the quantity of bound energy or, alternatively, its disorder) is bound to increase. Living organisms appear to evade this constraint in the short term but do so only by feeding upon low entropy (i.e. free energy) from another part of the system. All human activities, it can readily be seen, in the long run increase the quantity of entropy in the Universe, and can only be sustained while there is low entropy upon which to feed: notably fossil fuels, the atomic nucleus, and sunlight, though each of these has concomitant problems of usage.

At a less cosmic scale, the study of energy flow through human-controlled systems is clearly capable of yielding results of both academic interest and practical value. Complex interactions involving the materials of both ecology and of economics can be modelled, and the computer has allowed the simulation of such systems under a variety of conditions, leading to predictions of future states. Needless to say, the condition of the energy in the system is as important as its amount, since all systems are 100 per cent efficient at turning free energy into heat. But the quantification of the data for a good energy analysis is difficult: the flexibility and rapid changes within human ecosystems, and the sudden and perhaps evanescent growth of a new pathway make modelling difficult. The work of H. T. Odum and his school in modelling the Florida coast goes the furthest of any of this kind of work, but some workers have found difficulty in, for example, a two-way translation between ecological energy and money via the commercial price of energy from such as diesel oil or petrol (gasoline). Some energetics analyses of social phenomena are perhaps good analogies but are rather too deterministic to be anything else. Nevertheless, the work of J. Zuchetto and A.-M. Jansson in formulating an energy-based systems ecology study of the island of Gotland in the Baltic Sea is indicative of the usefulness of the approach both in terms of empirical description and the modelling of futures. Studies with slightly different aims but a similar conceptual framework have been carried out for Hong Kong (see chapter 5), Hawaii, and Obergurgl in Austria.

But given the fact that no organism, no ecological system and no economic system can exist without an input of free energy, and that the system functions by transforming energy, it is not surprising that some scholars have seen energy flow patterns as a fundamental determinant in society. The anthropologist L. A. White saw culture as an extra-somatic mechanism for capturing energy and putting it to work: for him culture is essentially energy per capita per unit time. He originated the idea of dividing human history into cultural periods according to human use of energy (see below), and these ideas were much amplified by F. Cottrell who was keen to exploit the idea of energy surpluses which might provide positive feedbacks which in turn generated further surpluses. Such relationships would, he thought, affect phenomena such as the design and distribution of human settlements; later, Sjoberg amplified such ideas to a macro-level in postulating that the shift from biological to geological sources of energy was the dynamic behind the breakdown of feudalism, and the historian R. Adams goes further and reckons that social and political power derives from energy form and flows. As more energy becomes available to a society, control over it becomes disproportionately concentrated in a few hands. Such a set of views led the 'alternative' thinker Ivan Illich to the proposition that democracy was only possible when nobody travelled any faster than the speed of a bicycle.

Not all scholars would go so far as White and Adams: they reject the notion that energy flow can be seen as a theoretical construct that explains human behaviour, arguing in particular that its study does not explain how systems transform themselves. But the political economy of energy, for example, includes suggestions that value in the economists' sense should derive from the supply of useful energy, rather than other less physical bases such as labour and capital. Marx and Engels, interestingly, said very little about energy and its availability in spite of having read the pioneer political work of Sergei Podolinsky, for they were perhaps unwilling to consider a future in which abundance might be limited by merely physical considerations. Not until L. A. White (see above) did Marxists revive the question of the linkages between energy transformation and the nature of society. Now, writers of all shades of opinion are generally conscious of it, and of the fundamental changes that the future might bring. The effective use of solar energy as an alternative to fossil fuels, for example, might require technological and demographic decentralization, since it would be very difficult (given the foreseeable technology) to collect solar radiation at the level required to support a typical twentieth-century urban-industrial area. On the other hand, nuclear fusion based on hydrogen (still at least 20 years away) might require even greater technological and population concentration than we now have, and this might in turn lead to socially and institutionally intractable difficulties. Given the centrality of energy in society it is scarcely remarkable that myths arise: each new technological development has been interpreted as the provider of unlimited supplies at a low price and indeed a harbinger of miracles of various sorts.

In such a context, we need always to remember that energy analysis can only measure, account for and perhaps predict in the limited field in which its units are relevant. But there is no need to apologize overmuch for a framework which links together among other things solar input, photosynthesis, industrialization, the CO_2 problem, and nuclear weapons; it is a convenient starting point that highlights the importance of, and the links between, human societies and their surroundings. At the very least, it looks as if access to very little extra-somatic energy is disabling to people, as is (probably) access to too much. There may be a golden mean for social and economic development.

Some of the interest and concern over energy in society comes, as it were, full circle when the relationship of non-human energy flows to those controlled via technology is concerned. Generalizing rather broadly, we can see that the energy embedded in the latter has been used to diminish the energy flowing through the former, since the transformations of fossil and nuclear energy place stresses upon natural and semi-natural ecosystems which in turn lower their biological productivity. This situation is the outcome of millennia of social change in which the physical environment has been progressively absorbed into the cultural world of perception and cognition, with an ensuing use of technology to satisfy the ever larger human populations with their increasing demands for comfort, wealth and nutrition. The social networks have become much larger and the communications webs more complex, and local self-sufficiency has tended to break down, so that nearly all the globe is material for the satisfaction of wants. Sir Fredrick Soddy, a Nobel Laureate in chemistry, summed it up in 1933 when he said, 'If we have available energy, we may maintain life and produce every material requisite necessary.'

Given all these ideas about energy flow, we can use them to construct a layout for the book. The chapters which form its core are each characterized by a stage of socio-economic evolution (figure 1.3) whose defining criteria are the energy sources to which society had access (table 1.4), for this is a powerful measure of their relationships with their environment, since largely they set the bounds to which the environment can be altered.

Climate and history

One of the difficulties of using a criterion like a society's access to energy sources as an organizing concept is that it is likely to minimize

Table 1.4 Energy sources available at different economic stages, each referring to a chapter (2–6) of this book

	Primitive man (2)	*Advanced hunters* (3)	*Agriculturalists* (4)	*Industrialism* (5)	*Nuclear age* (6)
Fire	X	X	X	X	X
Domestic animals			X	X	X
Wind			X	X	X
Water			X	X	X
Fossil fuels				X	X
Nuclear					X

N.B. Fossil fuels had a low level of use in some largely agricultural societies.

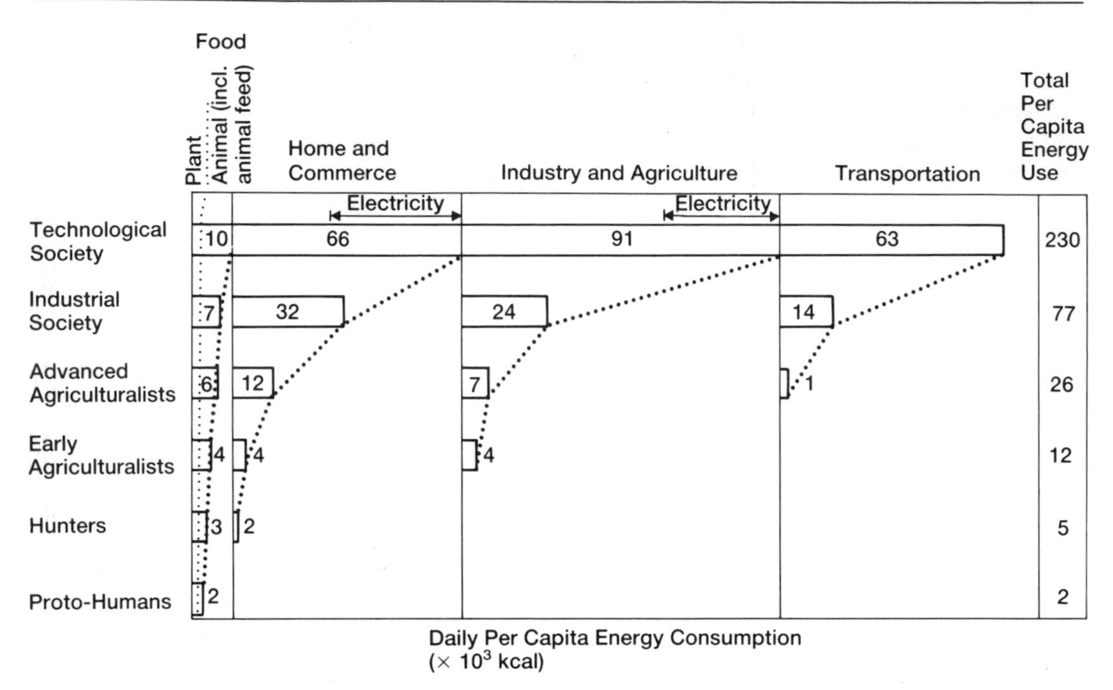

Figure 1.3 The quantities of energy used by human groups at various historical stages, and the purposes for which it was used. In this book, ch. 2 deals with the lowest group on the diagram; ch. 3 with the 'hunters', ch. 4 with both types of agriculturalist, and ch. 5 with the remaining groups. Ch. 6 deals mostly with alternatives to the 'technological society' but also considers an integral part of it in the shape of nuclear energy.
Source: J. W. Bennett, *Ecological Transition*.

the role of environmental factors. There are the obvious features of a particular place like landforms, rocks and the soil–vegetation complex, but a dynamic environmental component like climate is likely to be ignored. A short discussion of the relations between climatic change and environmental impact may therefore be useful.

At the very large scales of space and time, climate and its shifts need very little emphasis. Obviously, climatic distribution produced the major post-Pleistocene biomes which were host to different types of hunter–gatherer economy and which have strongly affected the crop combinations of agriculture past and present. But what of relatively minor shifts of climate? What influence can these exert on the degree of environmental impact that a society makes? Given that such effects are likely to interact with many other factors of, for example, an economic, social, technological, medical and demographic nature, can we even measure the effects of fluctuations such as the relatively warm period of the Middle Ages (AD 1000–1400) in Europe, or the succeeding Little Ice Age (AD 1400–1800) upon economies and hence upon impact? Can we go so far as to assert, as have some historians, that the coincident collapse of the centralized empires of Rome, Eurasia and China in the fourth century AD had a climatic cause, or that climatic shifts were immediately to blame for the demographic, economic, political and medical disasters of the fourteenth century in Europe, or for the 30 years of economic depression that followed the Napoleonic Wars? Or even that a slight raising of annual temperatures in England between 1540 and 1870 was reflected in the population growth rate and hence, presumably, in a demand for more food and a resulting increase in environmental manipulation somewhere?

Detailed analysis of these and similar questions seems to produce the reply that rarely if ever can such questions be answered unequivocally. The reality of the primary impact of climate upon all kinds of human lifeways cannot be disputed, but other factors make it difficult to construct models which are capable of being tested, let alone provided with consistent and reliable data for the past. The only places where a good case can be made are in marginal areas for human settlement: the abandonment of the Norse settlements in Greenland in AD 1500 comes rather neatly after the first 100 years of the Little Ice Age, for example. Likewise between AD 1200–1700 parts of the uplands of southern Scotland were allowed to relapse back to unenclosed pasture after being in cereal agriculture. It is argued that such events must have been fuelled by a long-term dynamic like climatic change rendering impossible the cultivation of oats, but it has also to be admitted that socio-economic causes like a change in farming objectives might be equally significant. The examples of the various droughts on the Great Plains of the USA might seem outstanding, especially those of the 1890s and 1930s. But the land use changes which followed them were not permanent, partly because the rainfall quantity seems to be cyclical and partly because, in the 1930s particularly, the disaster could be borne by the national economy as a whole (via credit extensions, federal funds and the like), thus tending to stabilize the land use pattern, just as families in Sahelian Africa today are integrated into the world economy via food relief programmes with subsequent effects upon the regional farming systems. Because the High Plains are now farmed by means of pumped fossil water and mechanically applied fossil fuel energy, a drought that coincided with a jump in energy costs might be the severest stress the system has yet faced.

In general, therefore, we can conclude that up until the present, the economic effect of climatic change (beyond the global scale differences) has been slight, perhaps even negligible and certainly very difficult to detect from the available evidence. This does not mean that future shifts (see e.g. pp. 332–3) may not be important, still less that unpredictable variability may not be more important than more gradual secular modulations.

Disasters in history

A useful addendum to the discussion of climate is that of the impact of disasters such as earthquakes, floods, famines and plague upon environmental manipulation. At their worst, they produce something like an early stage of ecological succession, for all natural and man-made phenomena are wiped out and there is virtually bare ground. Most of them have, at least locally, reduced human populations and hence the pressure to alter the local environment; thus after a plague, areas of agricultural land may revert to grassland or scrub if not cultivated. But in the long term, few disasters seem to have checked the processes of change for very long. The most significant events seem to have been earthquakes, floods and disease: the first caused losses of 0.02–0.09 per cent of the continental population in individual instances in the period 1400–1799, with an impact ranking of China, India, Europe and the Near East. Some very large disasters included an earthquake and landslide in China in 1556 with a death toll of 830,000, and an inland flood in China in 1642 with over 300,000 dead. In 1737 Calcutta was visited by a hurricane and a flood wave which killed 300,000 people. Famines seem to have been endemic in China throughout recorded history until after the 1949 revolution, and in each of them about a quarter of the population are recorded as eating bark and grass, so that environmental impact is thus intensified at such periods since debarked trees will die. One scourge which seems to have had a greater effect in Europe than Asia is plague, which was often the cause of the desertion of whole tracts of countryside as well as cities. E. L. Jones (whose work forms the basis for this paragraph) thinks that such diseases spread from Asia where they were endemic and debilitating to a non-immune Europe where they were epidemic and often fatal. But even when 23 per cent of a European population died as in a visitation of plague in Russia and Lithuania in 1703, the retrogressive effects were soon erased: we would have a hard job tracing them in the landscape of today.

Types of human impact

While it is convenient to think of humans as components of ecosystems, there can be no doubt that they differ from other organisms in their power to manipulate many of the other components of the system. For a start, human activity has created new genotypes in both plants and animals by the process of substituting human for natural selection, and we have created new ecosystems by a variety of processes. The biological success of the human species now means that *Homo sapiens* has the highest biomass of any animal ($c.100 \times 10^6$ t dry weight or 6×10^{14}kcal); and a population growth rate of nearly 2 per cent per annum is fast, although longevity makes for slow turnover of the biomass. Human control of energy sources means that some 8–10 per cent of the energy at the herbivore and carnivore trophic levels is now from stored ('fossil') energy rather than recent solar radiation and the fluxes of energy which result from industrial processes have in recent years been equivalent to about 10 per cent of the natural flows of photosynthesis and respiration. This may seem a low proportion but we shall see that the capability of this 10 per cent to affect the rest of the ecosphere can be very high: we now manipulate or appropriate some 40 per cent of the continental NPP (see p. 340). One of the consequences is that many of the earth's ecosystems now depend for their continuing function upon the integrity of human societies.

The types of environmental impact which result from all this access to energy sources will emerge in greater detail in the course of this work, but it may be helpful to think of a few generalizations in order to give some sort of vocabulary for what is to follow. One of the general processes which has affected nature since the Neolithic Revolutions has been that of *domestication*: the reshaping of a plant or animal species in the mould of what a particular human culture wants from the organism, rather than that which nature has by evolution provided. Culture needs to be stressed because different societies have wanted various things of the same wild species: consider for instance the way in which a few species of wolf have been bred into the multiplicity of domestic dogs from wolf-hound (a curious irony) to various lap-dogs. By definition domestication involves permanent changes in the genetic material of the species, rather than merely a phenotypic and temporary alteration, and is also a continuing process. While we should not underrate the apparent surges of domestication in prehistoric times which gave us the ancestors of so many of today's important food plants and animals, further species have been brought into the human nexus at intervals ever since and there has been a policy of continuous improvement (with a few recalls) to those species which were early domesticates. This activity has accelerated greatly since the discoveries of science enabled plant and animal breeding to have a firm theoretical base rather than rely entirely on empirical experience. One of the side-effects of domestication has been the labelling of some non-domesticates as weeds and pests, and therefore agricultural ecosystems have been simplified in order to provide the best conditions for the domesticated species: weeds have to be hoed out, other herbivores scared away or killed, and carnivores extirpated if at all possible. Such actions are just part of the general trend towards *simplification* of ecosystems which has marked human interaction with many of them. This has often been accidental: a small oceanic island which receives a few humans, some goats and cattle, and a number of rats from the brief visit of a sailing ship in the seventeenth century is scarcely the object of a planned campaign of land use changes. Nevertheless, many plant species will become extinct because they are grazed out by the domesticated herbivores, and several species of animal will be lost because of the success of the rats at eating their young, or their eggs, or even the adults of, for instance, flightless birds. The same process is in fact at work where modern industrial contamination in the air causes the loss of moss and lichen species around cities and can proceed as far as *obliteration* in some cases: downwind from a source of concentrated toxic fall-out, there may be practically no life. Although systematic excavations are rarely undertaken, there is unlikely to be much life underneath the concrete pad surrounding an airport terminal. So although biologically inert places are rare on the surface of the earth, man has without doubt increased their area.

By contrast, some human activities have tended towards a *diversification* of the variety of species in a particular place. We have noted above the effects of domestication in producing strains which are suited for a special place or activity, and to this we add the deliberate importation of species into places where they are not native, in order to be the foundation of an economy: think of cattle in North America. To this must be added the accidental transportation of many species to foreign shores where they have flourished, though not all do so. The example of the European weeds which successfully flourished in the wake of political colonization to the point of being called 'the white man's footprint' (species of *Plantago*) is one such; a more complicated example would be that of a species deliberately introduced but meant to be kept in a confined environment: a pet for instance. This may then escape and establish breeding colonies in the wild and become part of the naturalized fauna. In Florida, several pet species (gerbils are one such) have become noticeable agricultural pests this way; in southern Britain, a species of parakeet is now seen as a threat to commercial fruit trees. The antiquity of all introductions cannot be established but the evidence points in the direction of showing that as long as people have moved around they have taken plants and animals with them into new places.

From time immemorial, too, humans must have noticed the changes which they and their economies had wrought, and from time to time they have wished to protect some species or some ecosystems against change for one reason or another. So we have a long history of attempts at *conservation*, which may have started with hunter–gatherers selectively killing members of a species so as to avoid the taking of gravid females and young individuals. At all events many cultures early in their history seem to have established game parks in which animals were preserved for hunting by the ruling classes, and the last 100 years have seen the rise of strong movements advocating the protection of species of plants and animals, of ecosystems, of scenery, and even of the whole continent in the case of Antarctica.

Terminology of impact studies

For the purposes of this book, we might agree that human-directed impact upon nature has been of two types (not always easily distinguishable in the past): purposive and accidental. Today we might call the purposive category 'environmental management', and if we were starting to define terms *ab initio* then we could use that term for deliberate actions, and 'environmental impact' for accidental changes. But in fact 'environmental impact' has come to have a technical usage (under the US National Environmental Protection Act and its equivalents elsewhere in the world) which subsumes both purposeful and accidental categories. So as not to try and impose a whole new set of definitions in the face of those already in use, in this work the term 'impact' will be used comprehensively to include all aspects of change concomitant with human societies, but also specifically (the context should make it clear) to unplanned alterations. If the term environmental management is used, then deliberate behaviour is assumed. I have tried hard to avoid using the word 'impact' as a verb.

CHAPTER TWO

Primitive Man and his Surroundings

Come to think of it all the earth is a graveyard, you never know when you're walking over heads – particularly this continent, cradle of man, prehistoric bones and the bits of shaped stone . . . that were weapons and utensils

Nadine Gordimer, *The Conservationist*, 1974

Mankind evolving and spreading

The scope of this chapter

To begin at the beginning is not all that easy since it is not yet certain where and when our species evolved. True, the evidence is accumulating rapidly and reasonable hypotheses can be made, but it is equally apparent that there are enormous gaps in the fossil record which make the delineation of mankind's ancestry somewhat uncertain. Since, too, an evolutionary descent implies a continuous process, the point of joining this process (for our purposes) must be to some extent arbitrary. There seems little point in discussing extinct Tertiary and early Pleistocene primates even if they were the animals that led eventually to the genus *Homo*. It seems best therefore to begin a discussion of primitive man at about the time of evolution of our genus (*c.* 2 Myr),[1] and to end when that major cultural and ecological change, the ability to produce and control fire (*c.* 0.5 Myr), was firmly established.

The evolution of the genus Homo

The evolutionary tree of man during the late Tertiary and the Pleistocene years is still a matter for research and discussion, and its chronology although much improved by K-Ar and fission track dating is still uncertain in parts. The lineage and its chronology forms the core of figure 2.1, where it can be seen that an ancestral primate apparently gave rise to two lines of hominids during the Tertiary. The main line is called *Homo habilis* in its early stages, and becomes of most interest to us at 1.8–1.5 Myr when palaeo-anthropologists recognize from bone remains a later species, *H. erectus*. In due

[1] Abbreviations used in dating: Myr – million years ago; Kyr – thousand years ago; BP – before present. 'The present' in radiocarbon dating is AD 1950. Normal usage of BC/AD is made: any date without a suffix is AD.

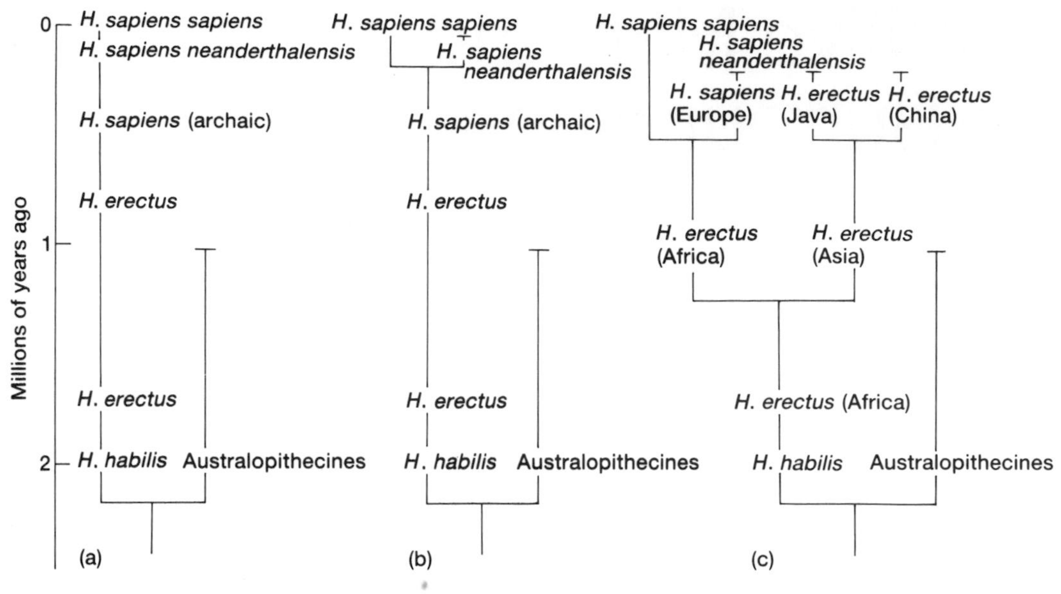

Figure 2.1 Three alternative pathways of human evolution, according to thinking in the 1980s. Further discoveries of skeletal remains and improved analytical methods will no doubt change the picture in the near future. The story of this book is mostly concerned with the period of *Homo erectus* and after, i.e. the last 1.75 million years.
Source: R. A. Foley, *Another Unique Species: Patterns in Human Evolutionary Ecology*. Chichester: Wiley, 1987.

course, perhaps around 120 Kyr, an early modern man (*Homo sapiens*) appears. Full anatomical modernity (*Homo sapiens sapiens*) appears about 90 Kyr in Africa and some 45 Kyr in Europe and Western Asia. This however carries us beyond the boundaries of the present chapter, so that it is the environmental relations of *Homo habilis* and especially *Homo erectus* which concern us now.

The setting of early Homo

It seems undisputed that Africa was the place of emergence of the earliest *Homo* stock: an upright bipedal hominid with a brain capacity of 650–750 cc (cf. the contemporary brain of man at 1400 cc), who shared the landscape with his cousins of Australopithecine type. These Australopithecines and early men lived in a fairly wide range of environments within the belt of tropical and near-tropical country bounded by Ethiopia to the north and the Republic of South Africa in the south (figure 2.2). In East Africa most of the sites are in tectonically isolated basins within rift valleys; in South Africa they are in upland plateaux and plains landscapes. In these two areas, lakeshores, delta marshes, seasonal stream channels, river banks, sinkholes in karst country and large fissures in limestone encompass the range of fossil localities. On a large scale, they share all the characteristics of a savanna environment, with open as well as wooded vegetation and alternating wet and dry seasons. At more local scales it appears that most of the sites were at the interface between open and closed vegetation, whether along a lakeshore or a stream or at a sinkhole; further, the sites were located amongst complex mosaics of environmental types, thus enhancing the variety of resources which were available (figure 2.3).

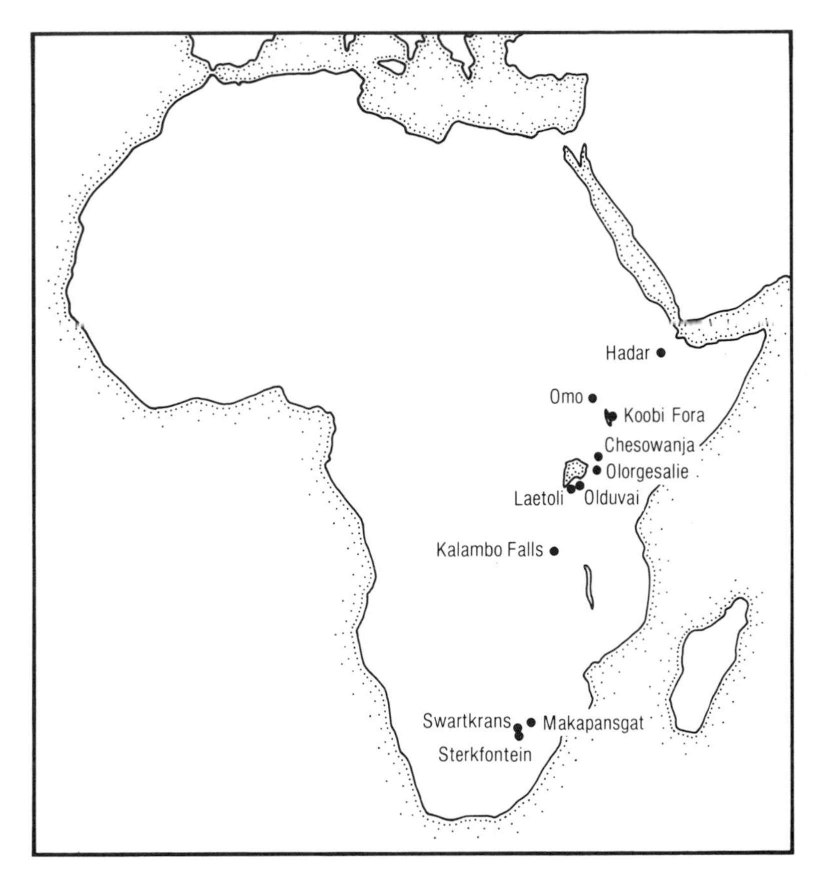

Figure 2.2 An outline map of Africa showing the important sites for the study of the human lineage during Plio-Pleistocene times.

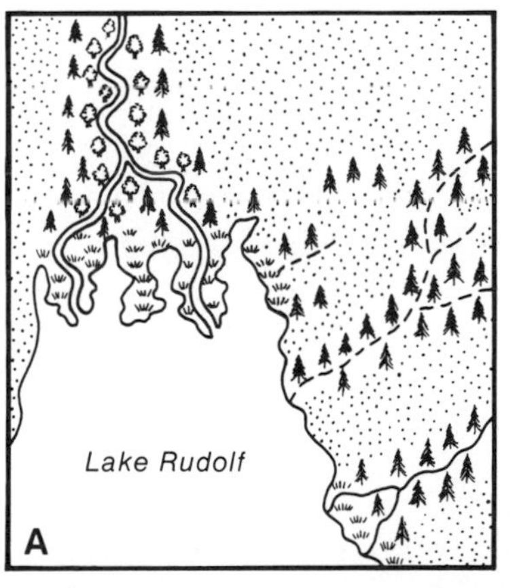

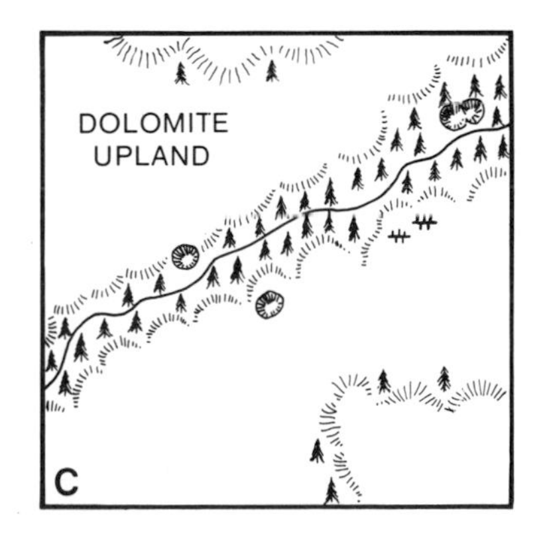

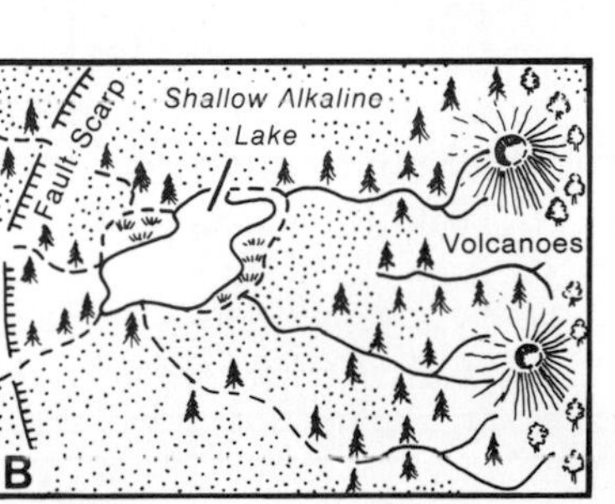

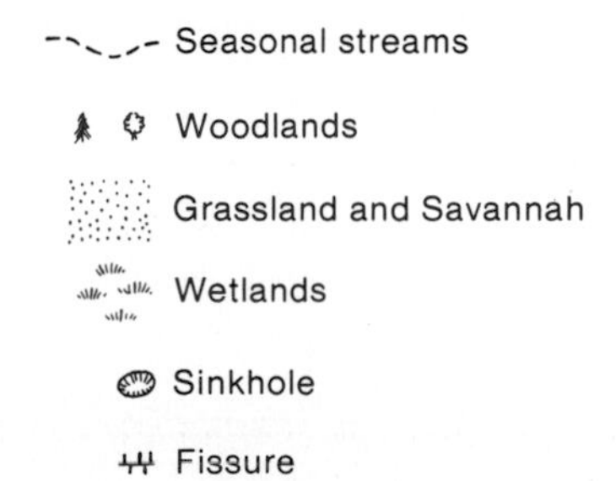

Figure 2.3 The environments occupied by early hominids in Africa. (A) Pliocene fossil localities near Lake Rudolf, with variety of stream channel and lakeshore sediment; (B) Olduvai Gorge with seasonally dry mudflats around the lake and overbank deposits near intermittent streams; (C) Australopithecine sites in the Transvaal where bones are preserved in sink-holes and fissures in the dolomite limestone. *Source*: K. Butzer, 'Environment, culture and human evolution', *American Scientist* 65, 1977, 572–84.

The spread of early Homo

Homo erectus seems to have superseded *H. habilis* in East Africa, apparently by 1.6 Myr, but did not remain confined to Africa, for *H. erectus* fossils of the period immediately thereafter turn up in several other parts of the world, notably in Mediterranean Europe, China and Indonesia, and it is clear that early man was quite widely spread with perhaps a temporal concentration of remains from around the 0.5 Myr period. This was one of the times during the Pleistocene which saw a great glaciation of the northern hemisphere, which may have caused a southward displacement of climatic belts. But man was clearly not driven out of Africa by such episodes and neither, we may suppose, were Pleistocene Europe and Asia a particular attractions. Some other reason for the immigrations must be sought: one author has suggested that the dispersal of *Homo erectus* is associated with the radiation from Africa of a fauna in which grazing mammals were predominant. This dispersal had its origin in the disruptions of the Pleistocene and was made possible by the environmental variety created then; the same diversity also provided a good set of opportunities for *Homo erectus*. After 0.5 Myr (and perhaps earlier), mankind had control of fire (plate 2.1) and this at the very least made more palatable the winters of Europe and China. Neither *H. habilis* nor *H. erectus* appear to have colonized the rainforests of Africa, which came with the Sangoan culture (45 000 BP) possessed by a subspecies of *H. sapiens*.

Plate 2.1 Thousands of years of human-induced fire can change the form and species composition of many ecosystems; this process represents too the first revolution in humanity's access to natural sources of energy that can be used to manipulate the environment. This example is from Australia, where the activities of aboriginal people in this regard are well documented. (Australian Overseas Information Service, London)

Plate 2.2 A cradle area of the genus *Homo*: the Rift Valley of Kenya. In an area such as this, the control of fire may well have been learned for the first time. (Photo: C. Weaver/Ardea, London)

Technology and environmental lineages of early man

The history of human technology appears to start around 2 Myr at Olduvai in East Africa with a kit of stone tools which are described as choppers; the material flaked off may have been used as an implement as well. This industry (called the Olduwan) lasted about 1.5 million years and was replaced at Olduvai by a more diverse tool kit associated with remains of *Homo erectus* with whom it seems to have diffused outwards (plate 2.2). It is called the Acheulian and is diagnosed by the presence of hand-axes which are often flaked all the way round, together with choppers, cleavers, chisels, scrapers, awls, anvils and hammers, all made from stone. There is the speculative possibility that bone was used widely as raw material for tools, and since in South and East Asia the earlier Olduwan type persists longer than in Africa, there is also discussion as to whether bamboo was an important raw material. It would have left no remains except in unusually preservative environments.

It is accepted that the main use of this technology was one or all of hunting wild animals, butchering them when caught, and scavenging on carcasses. Modern demonstrations have shown, for instance, that a medium-sized antelope can be skinned and dismembered quite quickly using sharp-edged flakes held between fingers and thumb. The tools tell us nothing about the role of plant foods in the diet. An evolutionary reduction in jaw size and certain types of tooth wear suggest that *Homo* was a meat-eater if even he/she was not totally reliant upon a diet of flesh.

Analysis of the deposits in which hominid fossils have been found, together with remains such as bones, allows us a measure of reconstruction of the ecosystems in which the hominids lived. In Africa, for example, *H. erectus* and *H. habilis* showed an apparent preference for sites near water and also places of considerable micro-environmental variety: lakeside flood plains, the beds of seasonal streams, deltas, river flood plains and meanders were all popular. Such places would presumably have given access to freshwater (we assume that no receptacle for carrying it had been discovered), and would have been near strips of trees as well (akin to contemporary 'gallery forest') so that such habitation sites would have had easy access to water, shade, plants and the game which would also have needed to come to water. Some writers have claimed that they can discern temporal trends in habitation sites, e.g. towards stream channel sites and cooler, more forested uplands after 1.5 Myr as if *H. erectus* was a more versatile colonizer than any predecessor. Overall we get a picture of small groups of people (the most popular number suggested is 25), with a simple material culture, living at the edge of forest and eating a diet which had meat in it. Animal bones are often found in association with the stone tools and, occasionally, with fragments of *Homo* skeletons. These accumulations appear usually to have resulted from active hominid effort: they seem to be neither chance assemblages, nor the work of scavenging animals of other species. In the lower beds at Olduvai distinct butchery sites have been postulated from the clusters of stone implements; the bones themselves range from mice to elephants but with a predominance of middle-sized antelopes and pigs. It seems a likely hypothesis that animals up to 30 kg in weight were hunted but that bones of animals over 30 kg result from scavenging activities. At the Makapansgat site in South Africa, the collection of animal bones suggests that the hominids were scavenging upon bovids killed by lion and cheetah since those killed by hyaenas or leopards would have been much more difficult to approach, given the close attention to a kill characteristic of a pack of hyaenas and the leopards' habit of carrying their prey up into the trees. But at Bed II at Olduvai we have evidence of the slaughter of a large number of the now extinct giant African buffalo, *Pelorovis*. The numbers of bones are such that one writer speaks of the 'successful decimation of animal herds'.

With the Acheulian tool-kit and the radiation of *Homo erectus* more sites are found, although most of them have not so far added greatly to the quality of new knowledge about environmental relationships. Most of the sites are associated with areas of open landscape in the shape of beaches or grasslands on river terraces, and the occupation of caves is noted as well. The full range of fauna seems to have been exploited including at Torralba (Spain) evidence of the killing and on-site butchery of at least 30 elephants as well as horses, oxen and rhinos. The nearby Ambrona site of the same age (400 Kyr) yielded some 40 elephants, but at both places the stratigraphy does not allow us to infer unequivocally that these were single episodes; that butchery took place and that some of the choicest cuts (and the skulls) were carried off, seems highly likely. Small pieces of worked wood (mostly fragments 2–27 cm long) have, by comparison with spears found at Clacton, England, and Lehringen, Germany, been interpreted as fragments of wooden spears, which with fire-hardened tips would have been adequate to pierce even the elephants' hides. Along with the Clacton spear, these are the oldest known humanly worked pieces of wood. An elephant tusk which had been whittled to a point was found as well.

The regional context of the site (figure 2.4) allows one interpretation that these were seasonal sites occupied during the spring and autumn migrations of the animal herds. The people presumably lived somewhere else: the presence of fire but the lack of hearths suggests either that fire was used in the hunt or that fire was brought to the butchery site for cooking purposes. We assume that in summer and winter the focus of the hominids' activity was elsewhere, as figure 2.4 depicts. So a small group of *Homo erectus* carried out a twice-yearly cull of migrating herds as they passed through a narrow valley; given the likely abundance of the animals we can agree with Butzer when he says of this site that 'man . . . for all practical purposes was unable to disturb the biological equilibrium.' However an overall context of the uncertainty of our data is emphasized by later interpretations of the material at Torralba, with L. R. Binford arguing that, 'it cannot be overstressed that the involvement by hominids with the carcass remnants at the site of Torralba is a very minor aspect of the overall behaviorally relevant materials found there.'

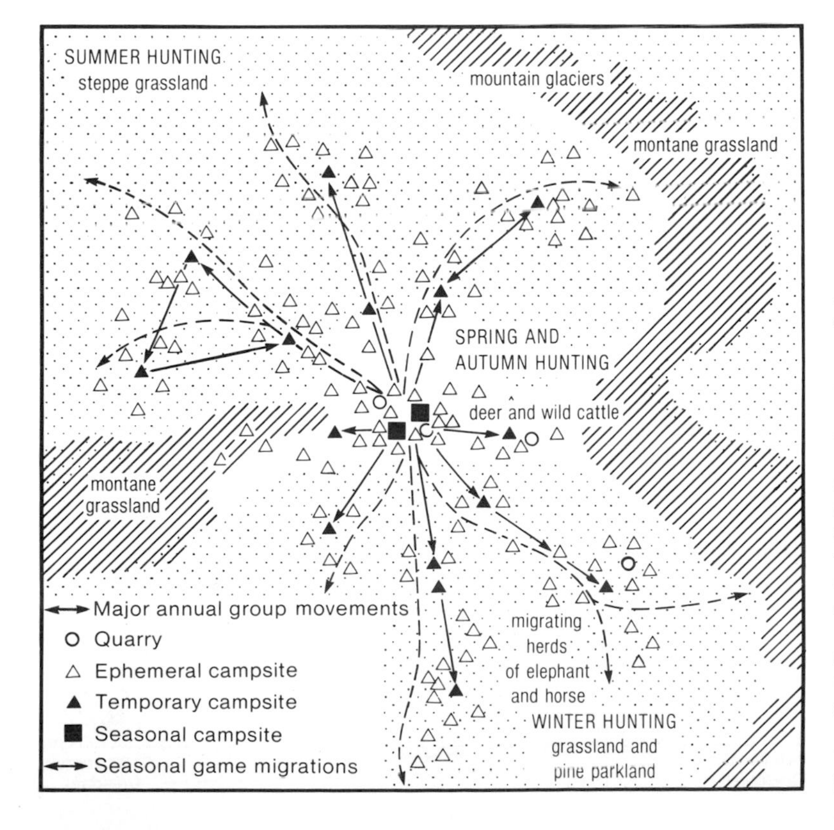

Figure 2.4 A reconstruction of the yearly hunting patterns at Torralba and Ambrona in mid-Pleistocene times. In spring and autumn, the hunters concentrated around the mountain passes through which the animals had to migrate; in summer and winter, smaller bands of hunters followed both the availability of water and smaller herds of beasts. Later interpretations of the sites suggest a more natural origin for the archaeological finds. *Source*: K. W. Butzer, 'Environment, culture and human evolution'.

At Olorgesailie in Kenya, the prey was predominantly medium-sized bovids and equids, but bone remains in one bed (*c.* 480 Kyr) suggest the slaughter of a troop (40+ adults and 15+ juveniles) of giant baboons of the now extinct species *Theropithecus oswaldi* (possibly by surrounding them at night), although again it is conceivable that the bones may have accumulated over more than one season: the stratigraphy is not definitive.

Given the most superficial knowledge of near-recent hunters and gatherers we would expect that these early men moved their camps from time to time, following their resources as they became available seasonally. Comparison of the Olduvai fauna with the catches of present-day Bushmen (figure 2.5) suggests that the Olduvai sites were occupied either early or late in the dry season: we have yet to find evidence of the localities occupied and animals gathered at other times of the year. Good evidence is lacking, too, on the nature and importance of plant material in the hominids' diet. Analogy with near-recent groups in the tropics suggests that the collecting of edible vegetable foodstuffs must have been a crucial factor in survival; convergence of the post-canine dental anatomy of bears, pigs, Australopithecus and early *Homo* has allowed the postulation that below-ground plant resources (especially rhizomes) were not ignored. The digging stick could have provided the functional equivalent of the bear's claws or the warthog's snout.

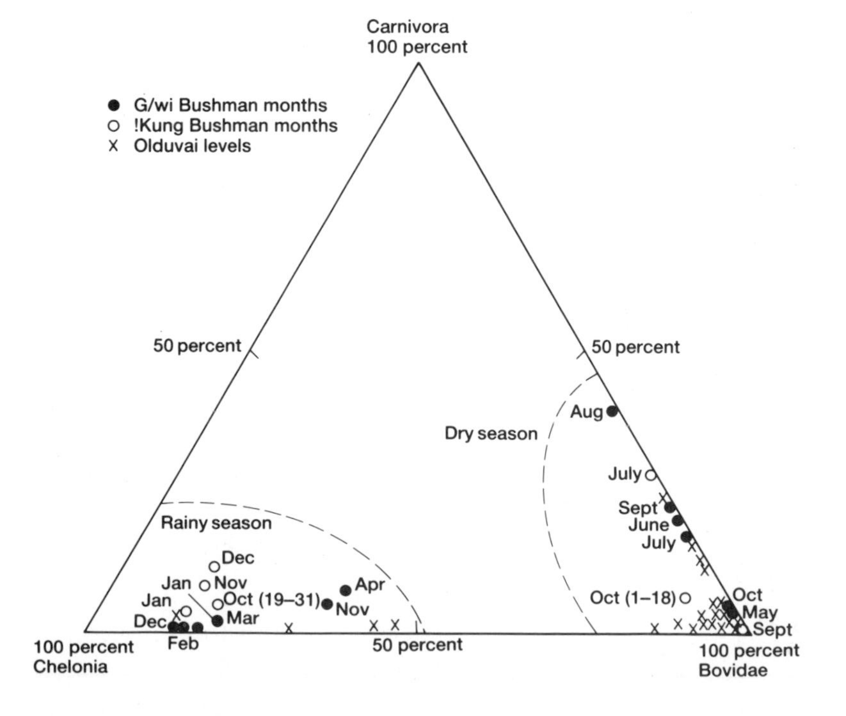

Figure 2.5 A way of estimating the type of game caught in the African Pleistocene by comparing the remains at Olduvai with the catch of G/wi and !Kung Bushmen at different seasons. A difference in catch between the dry and the rainy seasons is apparent. This means that effort could be concentrated on a limited number of species at one time, with the possibility of affecting its capacity to reproduce itself.
Source: J. D. Speth and D. D. Davis, 'Seasonal variability in early hominid predation', *Science* 197, 1977, 441–5.

Did these early men affect their environment?

Perhaps the only sensible answer is that we do not know. The total hominid population of the world must have been very low: one estimate based on the density of hunters in more recent times is 70 000–1 million for the Australopithecines and 1.7 million for *Homo erectus*. However, the study of faunal trends has led P. S. Martin, whose ideas we shall encounter again in the next chapter, to suggest that in tropical Africa during the later Acheulian period some 26 genera of animals became extinct out of a fauna of some 66 genera of mammals whose adult weight was over 50kg ('megafauna'). Martin asserts, for Africa as elsewhere, that man was responsible for those extinctions, being a new type of predator against whom the animals (whose populations may have been declining anyway because of environmental change) had no defensive behavioural adaptations. This idea of 'Pleistocene overkill' was disputed by L. S. B. Leakey, who argued that the definition of the genera is not sufficiently precise in extinct African mammals for the numbers to be usable in calculations of extinction rate. He also thought that a period of drought was a more likely cause of the animals' demise, and finally that if such a decimation (actually Martin suggests that 39 per cent of the fauna became extinct, so a 'quadrimation' might be more accurate) took place in the Acheulian, why did later hunters not greatly thin out the modern fauna which, as we know from early travellers' reports, abounded in Africa before the advent of the European explorers, missionaries and colonists?

There is, too, the evidence from deposits of the Hoxne (Holstein) interglacial (220–200 Kyr) in England (at Hoxne and Marks Tey) and in Germany (at Münster-Brelöh), which although ambiguous may be significant in this context. During a temperate forest stage at each of these sites, the pollen-analytical evidence reveals an episode of forest recession. At Marks Tey (Essex), for instance, woodland containing hazel, yew and some elm disappears rather abruptly, although the wetter parts of the forest, containing alder, were unaffected. The cleared areas were replaced by grassland and to a lesser extent by birch and pine. The phase probably lasted about 350 years, before revegetation by oak and hazel took place. At Marks Tey, microscopic fragments of charcoal were found at the appropriate horizons and at Hoxne larger pieces of burnt wood were discovered, as were stratigraphically correct Acheulian tools. No evidence of hominid bones has turned up at any of the sites, and the investigators conclude that there is no direct evidence to connect the forest recession with human activities although they concede that a climatic cause is unlikely. If natural forest fires were responsible, then it seems odd that they should have happened about the same time at widely separated places like East Anglia and Germany. Why, also, did the forest take so long to regenerate, for we know that birch and hazel are normally rapid invaders of grassland? It is possible that natural causes are a sufficient explanation, for once the forest was open, the browsing activities of elephants or hippos might have kept it that way. But if we allow ourselves a little freer rein of speculation than the original authors, then it is not too difficult to see the effects of a fire-possessing human culture (presumably *Homo erectus*) in these episodes, particularly given the charcoal and the tools in the lake deposits at Hoxne. There is no need necessarily to postulate deliberate forest clearance, although 350 years begins to look like something less than an accident.

If for the moment we side with the late Louis Leakey on the overkill question and with the botanists on the Hoxnian, then the rest of the evidence points to the conclusion reached by K. Butzer for the early species of *Homo* that 'Man–land relationships during the early Pleistocene were, then, one-sided, with the hominid populations but a minor ecological factor in their environment.' For *Homo erectus* he avers that they 'lacked the technology and numbers to modify the environment in any significant way.' Even where they could control fire, 'it is improbable that any appreciable, large-scale influence was executed on the natural vegetation during the course of the early and middle Pleistocene.'

These statements represent one interpretation of our knowledge at present. They may eventually require revision: one authority has noted that in Zaire in recent times, 75 per cent of the available meat to hunters is in the form of big plant-eaters like elephant, hippo and buffalo. If both they and the predators are protected in Parks and Reserves, then the herbivores quickly exceed the carrying capacity of their environment. Could it be that *Homo* replaced the sabre-toothed tiger as the significant predator of large animals? And any conclusions about early man's impact on his environment certainly need consideration against the background of the history of human ability to generate and control fire.

The history and ecology of fire

Evidence for fire associated with primitive man

A crucial question is, 'what constitutes acceptable evidence of early man's possession of fire?' We must be careful of negative evidence, especially in the tropics, for wood burned at a campfire might leave virtually nothing but fine ash, and the chances of its being preserved from 100 Kyr are not high. Even in the Upper Palaeolithic when fire is well documented, much of the evidence comes from caves rather than open sites, which points to preservation as a differentiating factor. So negative evidence is inconclusive in this instance. For early periods in the history of mankind there is good evidence of cone-shaped masses of burnt clay, with a shallow concrete base. Such features might have had natural causes but could equally have resulted from the intentional ignition of tree stumps in order to preserve fire. Sites like this are found at Chesowanja (Kenya) associated with *Homo erectus* at

1.5 Myr, and in the Middle Pleistocene of the Middle Awash valley in Ethiopia. Many more sites, more clearly interpretable as resulting from hominid activities, turn up *c.* 0.5 Myr. Fire is definitely associated with *Homo erectus* in their spread from the heartland into Europe and Asia. Thus definite evidence of human control of fire turns up at places like Torralba-Ambrona (400 000 BP), Zhoukoudian in China (400 000 BP) and Terra Amata in Mediterranean France (300 000 BP), Vertesszollos in Hungary (100 000 BP) and Westbury-sub-Mendip in England (500–400 000 BP). At Escale, near Nice, fire is variously reported at 600 000 BP and close to 1 Myr. The pattern has a neat simplicity: *Homo erectus* radiating out from Africa at *c.* 1 Myr and finding fire an indispensable aid to colonizing the chillier regions of the earth, especially after 600 Kyr in Eurasia, when there was a major glacial advance. If fire was fully controllable by *c.* 0.5 Myr then presumably the origins of its use must be sought much earlier, so we might expect good evidence from Africa older than 1 Myr.

Where good evidence of fire is found it tends to be at settlement sites or in association with animal bones. In the original interpretation of Torralba, the elephant bones were found in a deposit laid down in marshy conditions which were seasonally dry enough that charcoal from 10 successive marsh fires could be detected. There were no hearths showing multiple use, so presumably the fire was either used as part of the killing process or was brought to the animals to cook them: an early illustration of Zipf's principle of least effort. At Terra Amata, careful excavation revealed 21 separate living floors each with a hearth in a shallow pit, some about 0.3 m in diameter and some much larger. This site was occupied in late spring and early summer as part of a seasonal round which brought them to the Côte d'Azure for birds, turtles, some sea-food, and mostly young individuals of rhino, stag, oxen, ibex, boar and elephant; whether these *Homo erectus* groups had a discernibly French taste for undercooked tender meat or simply lacked the desire to take on large beasts we shall never know. An extension of the use of fire to technology is seen in its use as a hardener of bone and especially wooden implements of which a number, mostly spears, have turned up from this period in Europe and which seem indeed to have been the main weapon used to kill animals as large as elephants. A fire-hardened antler tip also appears at Zhoukoudian as well as charred bones from cooking, and at that site we get a tiny glimpse of the role of plants in the diet with the finds of the berries of *Celtis* and various tubers; wood from hazel and oak are also present in the cave deposits at cultural levels.

The ecology of fire

We lack evidence, during the period before the emergence of *Homo sapiens*, for the use of fire outside the settlement site. Once again, negative evidence is not to be ignored and at this point it seems appropriate to consider the ecology of fire in a wider context in view of the claims made for it as the first great revolutionary agent in mankind–environment relations.

Many writers have pointed out that making the acquaintance of fire would not have been difficult for hominids in Africa. The combination of living near the Rift Valley with its volcanic activity (attested by the many tuffs which have in some places preserved the archaeological sites), and in a climate in which lightning during the dry season can set the vegetation alight, must have repeatedly drawn the attention of early people to this phenomenon. We may perhaps eventually know whether they first perceived it in a settlement-related sense as a method of cooking food and providing safety at night, or in an environment-related sense as something before which animals fled so that hunting of large animals was somewhat easier and the picking up of small game greatly facilitated. But it may be unrealistic to expect that such a separation ever existed, let alone hope that we might detect it at this distance in time. It is an important stage, however, when a society can produce fire as well as collect it, but possible early methods like wood-on-wood friction are unlikely to leave much archaeological trace.

The natural and semi-natural habitats of recent times in which fire is a regular occurrence display definite ecological characteristics. Ground temperatures in grassland fires reach about 100°C and in forest fires about 150°C. The effects of the heat upon individual plants are variable according to their anatomy and stage of growth but normally the above-ground parts of all except mature trees are killed. Thus the vegetation composition shifts towards fire-adapted species, such as those that can effectively regenerate from underground parts (like the heather of Europe, *Calluna vulgaris*), or which have evolved protective mechanisms like a thick bark as in the giant redwoods of Californa. Some plants (called pyrophytes) are adapted to the point where they require fire to regenerate. For example, heat is needed to volatilize inhibitory chemicals from a seed-coat or to open the scales of the cones of a coniferous tree like the jack pine (*Pinus banksiana*) of North America whose cone only opens if subjected to temperatures above 120°C but the seeds of which when shed remain viable even at 1000°C. Animals have rather shorter-oscillation responses: they flee from fire and either escape or are roasted, with an intermediate stage for some insects whose flying ability is impaired, which allows easier predation. Post-burn vegetation is different from its pre-burn composition after repeated firing: woody species that do not sprout from their roots are either eliminated or reduced to a thin stand; herbs are usually favoured over grasses in both annual and perennial grasslands, and in dry grasslands and steppe annuals may gain at the expense of perennials. However, because the lower leaf layers of a grassland stand may respire more than they photosynthesize, fire will lead to a recovery stage (provided basal shoots survive) in which the NPP of the post-burn vegetation is higher than that of the original plant cover. Due to the mineralization of organic matter by the fire, post-burn vegetation is relatively rich in protein and minerals so may well selectively attract grazing and browsing herbivores: the quality and quantity of the forage available to herbivorous ungulates in deciduous woodlands improves in the order of 300–700 per cent during the few years following a burn.

These various features of fire might well confer adaptive advantages upon hominids and early men; figure 2.6 summarizes some of the positive values it could have had. Emphasis might perhaps be properly placed on the enhancement of the opportunities for catching insects, rodents and larger mammals and even gathering honey from bees' nest in fallen trees. In addition the 'cooking' of many plants reduces the toxins within them and makes them available as human food, as well as reducing the quantities of bacteria and fungi. Protection of a home base, and of meat from other scavengers, would also have been enhanced by the possession of fire. Indeed, J. D. Clark and J. W. K. Harris suggest that fire as much as any other factor (such as meat-eating, food-sharing or new forms of sexual behaviour) may have been responsible for the formation of the characteristic family units of human society. In that fire not only achieves a visible qualitative transformation of chemical material into actual tangible heat, but also can be the small start of something very large (one fire-brand can be the beginning of the ignition of a whole forest stand or a great swath of savanna), then it perhaps points forward to later endeavour.

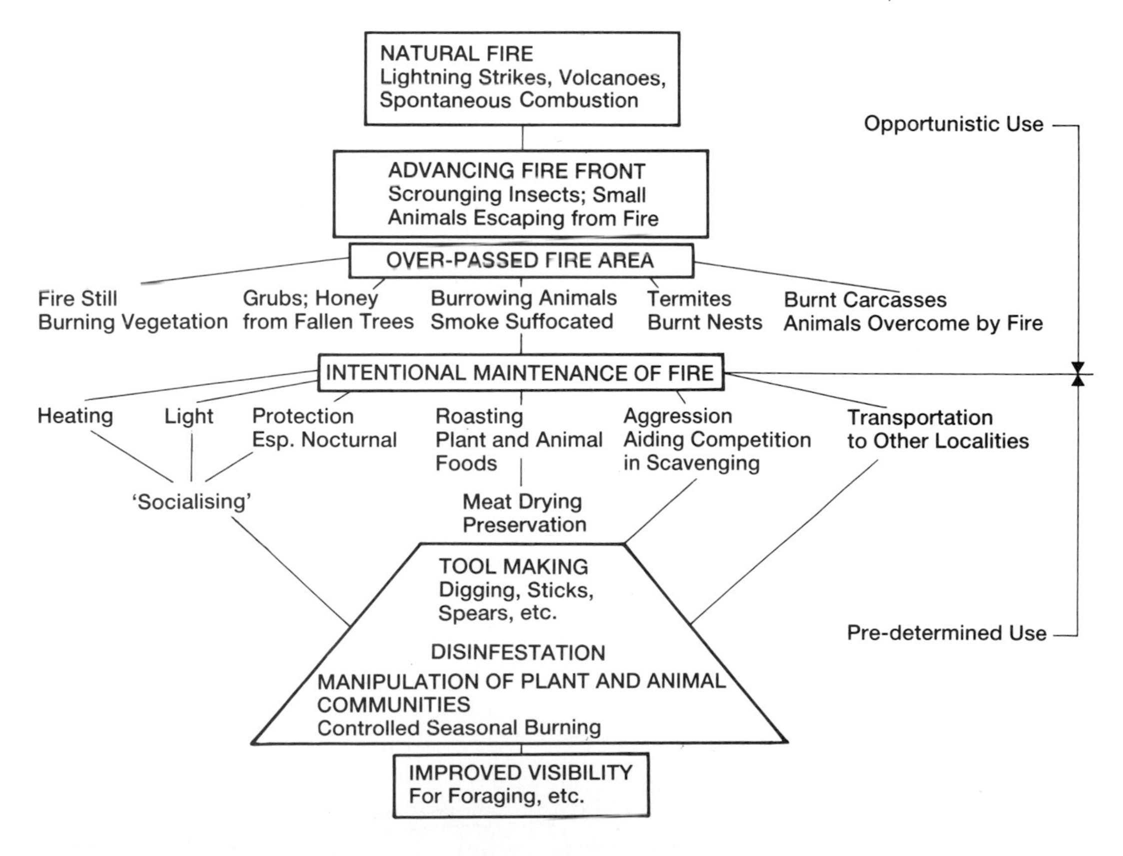

Figure 2.6 A chart of the usefulness of fire to early hominids, on the assumption that fire was intentionally maintained and that thereafter there was both planned and opportunistic uses.
Source: J. D. Clark and J. W. K. Harris, 'Fire and its roles in early hominid lifeways', *The African Archaeological Review* 3, 1985, 3–27.

Resources and environment of primitive man

The evidence for firm rather than provisional statements about these topics, let alone the equally interesting dimension of hominid population dynamics, does not exist so many of the criteria mentioned in the Introduction as characteristic of human impacts upon their environments do not exist. Most ecosystems remained in a natural condition unless they were fired more frequently by hominids or early humans, or unless the population of a key animal species was affected by large-scale slaughter. Domestication, then, did not exist and neither did the need for conservation; diversification is a possible consequence of fire if it creates a mosaic of habitats to replace a more uniform ecosystem; obliteration is not possible except for a few square metres round a hearth or similar focus, and simplification (by the local or possibly larger-scale extirpation of an animal population) likely to be temporary. This phase is then one when humankind and its forebears sat lightly upon the land, but in which we can see the first faint marks of a presence which was eventually to become ubiquitous in its ability to transform the natural scene.

Advanced Hunters

Hunter, lift up your heart, glide, run, leap and track,
the meat is in front of you, the huge piece of meat
the meat that walks like a hill
the meat that rejoices the heart
the meat that you will roast at your hearth

Chorus

Yoyo, Elephant hunters, take up your bow!
(Okoa Pygmy song, central Africa)

Evolution and culture

The scope of this chapter

Any reckoning of our past will show that humans have spent most of their history as hunters and gatherers, and this chapter aims to explore the environmental relationships and impact of that way of life in its full flowering. Thus it has to be chronologically very catholic, for material of relevance will come from such diverse sources as Palaeolithic archaeology and near-recent ethnography. If mankind was once (perhaps at about 10 000 BC) 100 per cent hunters and gatherers, then it is now more like 0.001 per cent, and the replacement of hunting by other modes of life has occurred at practically every period from the first agricultural communities (see chapter 4), to the present, in for example the northern tundras.

The terms 'hunter' and 'hunter gatherer' are used interchangeably in this book and should be read to subsume all the possibilities within 'hunter–gatherer–fisher'. The key point is that these people were and are food collectors and not food producers. As an energy source they rely on the sun in the form of recently grown organic material; to manipulate the energy and matter flows to themselves they use only fire and the muscular energy of their own bodies, together with that of a single tamed animal, the dog, which is found around 10 000 BC onwards in hunting societies. Yet because solar energy is so diffuse a source even when concentrated into the chemical energy of plants, the density of hunters has never been large, and an average for the world of 26 km^2 of land per head is often quoted, with the figure falling as low as 250 km^2 per head for the dry interior of Australia.

From the Pleistocene to the Middle Holocene

In this phase, anatomically modern man, *Homo sapiens sapiens*, became the sole representative of the genus and further evolution has been cultural rather than physical. This subspecies replaced earlier *Homo* groups such as the Neanderthals (which flourished first in Africa at about 90 Kyr) in much of the Old World, and in the case of Australia and the New World was the very first species of hominid to arrive. Dates vary, but for much of the world the process of colonization by the new men had taken place by 40 Kyr. The Americas are conventionally thought to have lagged behind and received their population from northern Asia at about 12 Kyr.

This species had good binocular vision and a delicate grip, an ability to live in various temperature and UV regimes and an immunological adaptiveness which allowed populations to survive parasites and diseases like malaria, which dispersed with *Homo sapiens* out of Africa. The social organization which allowed this remarkable spread and consolidation in most of the environments of the world during the last phases of the Pleistocene (with its attendant climatic upheavals) and the subsequent adjustments of sea-level and climate in the Holocene (from 10 Kyr/10 000 BP onwards[1]) was simple. Bands were probably 25 people strong and belonged to a dialect group some 500 strong. The breeding population of about 175 exercised, many interpreters think, strong population control, killing as many as 15–50 per cent of the children born alive.

This did not prevent a cultural explosion in Europe from 35 Kyr onwards, when the ability to make fire by percussion (see fig 3.1), the spear-thrower, the bow, and the skill to produce 300–420 cm of cutting edge from 2 kg of stone (compare the Olduwans at 50 cm and the Neanderthals at 100 cm) all evolved. Like their ancestors, their control over fire was very good and as preliminary to more detailed discussion we should note that nearly all the vegetation types established in the Holocene will burn at some season or other; even the *terra firme* forests of Amazonia have been subject to fire at random intervals throughout that period (plate 3.1).

The rise and fall of the hunters

At about 10 000 BP the entire human population of the earth was hunter–gatherers and lived basically as their ancestors had done. So for 25 000 years from their occupation of the entire inhabitable area of the earth at 35 000 BP, the hunters had undisputed tenure of it. But no sooner had the post-glacial adjustments begun to take place than the way of life began to retreat in the face of agriculture. The only way of life once practised by 100 per cent of the human species did not persist. Whereas at 10 000 BP/8000 BC the ecumene was occupied virtually only by hunters, by 500 BC there had been supersession in a large number of areas although in some places hunting was carried on as an accession to an agricultural mode of existence. By AD 1 this pattern had been further intensified, and by AD 100 only North

Figure 3.1 A chart of the extraordinary efflorescence of technological invention during the Upper Palaeolithic, i.e. after the Mousterian. The first occurrence is of course the time when the first evidence of the technique is found, and some depend upon rather fine points of interpretation, like the number of stone tool types.
Compiled from a number of sources.

[1] For the period after 10 000 years ago, the notation 10 Kyr etc is replaced by 10 000 BP (before present). This notation is used down to 0 BC/AD (1950 BP) after which AD years are given without any tag.

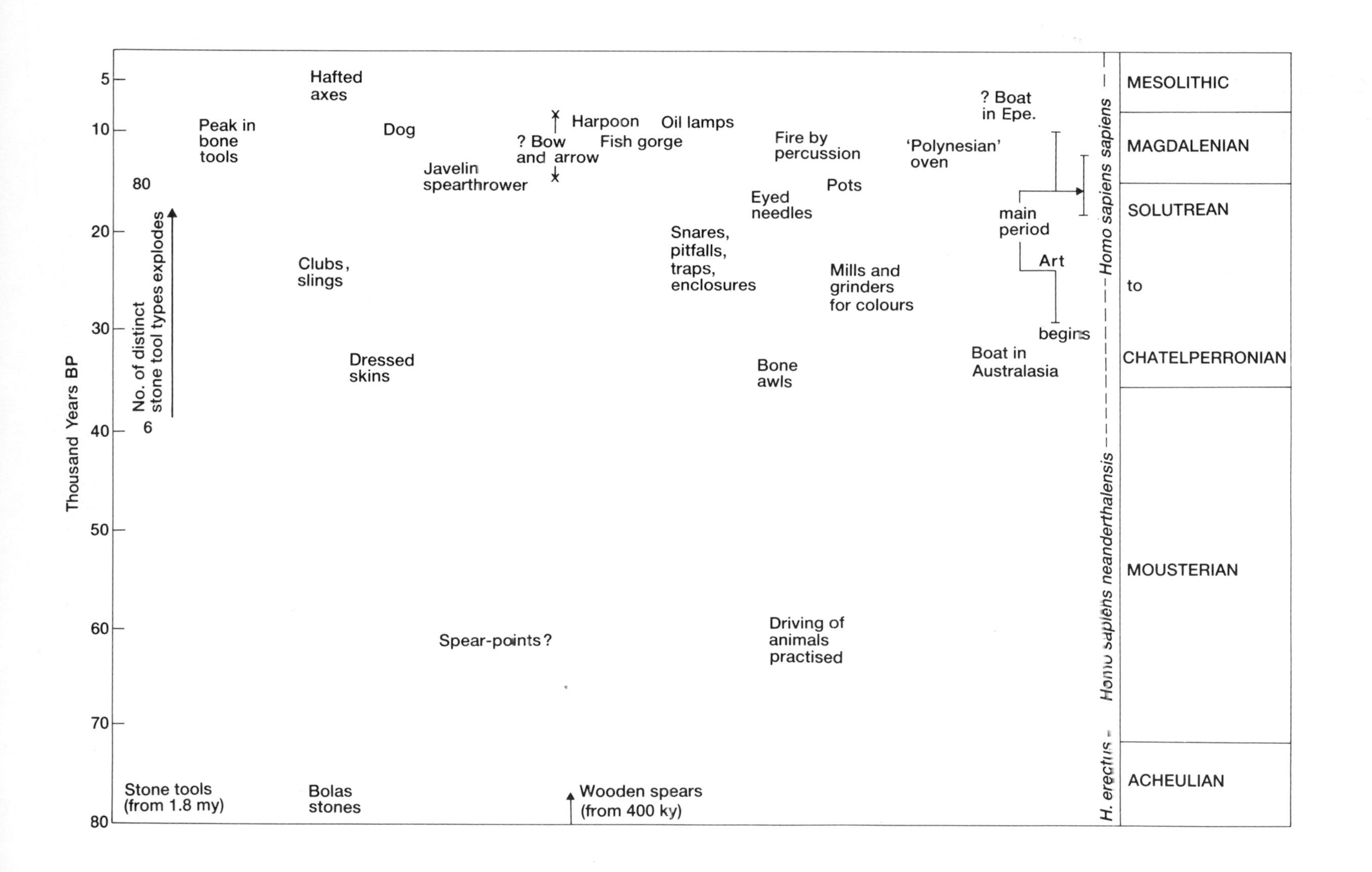
Thousand Years BP
5
10
20
30
40
50
60
70
80
80
No. of distinct stone tool types explodes
6
Peak in bone tools
Stone tools (from 1.8 my)
Hafted axes
Clubs, slings
Dressed skins
Bolas stones
Dog
Javelin spearthrower
? Bow and arrow
Spear-points ?
Wooden spears (from 400 ky)
Harpoon
Fish gorge
Oil lamps
Snares, pitfalls, traps, enclosures
Eyed needles
Fire by percussion
Pots
Bone awls
Driving of animals practised
Mills and grinders for colours
'Polynesian' oven
? Boat in Epe.
main period
Art
begins
Boat in Australasia
Homo sapiens sapiens
Homo sapiens neanderthalensis
H. erectus
MESOLITHIC
MAGDALENIAN
SOLUTREAN
to
CHATELPERRONIAN
MOUSTERIAN
ACHEULIAN

Plate 3.1 An area of Mesolithic activity in upland England where the treelessness and the peat growth are associated with the burning activities of the prehistoric hunter-gatherers. (Photo: Author)

America (outside the south-east), northern Siberia, and Australasia were left as areas where hunting predominated. At AD 1500 perhaps only 15 per cent of the earth's surface and 1 per cent of its population were hunters (albeit with some extensive strongholds, see figure 3.2) and in recent times the hunters occupy only sporadic interstices in the continental surfaces and comprise a bare 0.001 per cent of its population: the hunters' dominion over palm and pine is gone.

The reasons for this decline lie at the first sight in the superiority of agriculture as a source of nutrition and wealth, and its advantages will be discussed in due course. These must have been considerable, for hunting and gathering appeared to carry in it few apparent seeds of its own destruction. Malnutrition was, apparently, rare and starvation infrequent; chronic diseases were not frequently found and in the recent past at any rate predation was an insignificant factor in human mortality, though it may have been more so in the more distant past. Infectious diseases were variable in their incidence, perhaps depending upon the complexity and diversity of the local ecosystems. So in these terms, it seems likely that hunters saw advantage in the new techniques of subsistence and adopted them willingly or by conquest or by marriage because they represented a clear advance, rather than because their own way of life was seriously lacking acceptability. But others, we know, succumbed in the face of invasive cultures with superior force, new infectious diseases and a colonizing mission, as in much of North America and Australasia. Yet nowhere does it appear certain that environmental breakdown forced hunters into agriculture, though population growth might have brought hunters to the

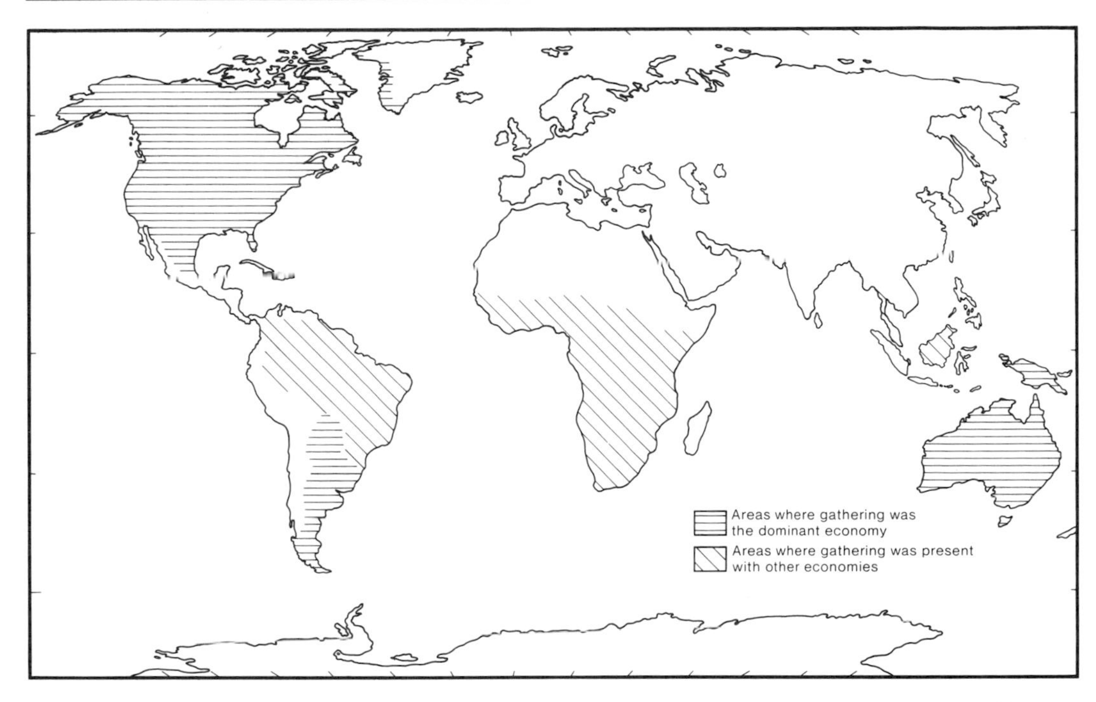

Figure 3.2 In 1492 AD, large portions of the world were still dominated by a food-collecting rather than a food-producing economy or at the least existed alongside agriculture. The advent of European colonization was the main reason for change in the Americas and Australia. *Source*: C. F. Bennett, *Man and Earth's Ecosystems*. Chichester: Wiley, 1975.

point where more intensive food 'production' was necessary. A simple adoption of this transition is unlikely to be the correct explanation of the demise of hunting, which like most other changes may have had an environmental component in some places but not in others.

Environment, past and present

Hunters then and now

The locations and names of the main present-day hunting groups are given in figure 3.3. So far as is known these are the only remaining people who are solely hunters and gatherers and who carry out neither cultivation nor pastoralism. However, many have contact with other economies: some Bushmen spend periods as hired labour on European-culture farms, for example, and the Mistassini Cree Indians of the Athabaskan group trap fur-bearing animals which they then exchange for cash and store goods. Even of these 'few isolated pockets' as G. P. Murdock puts it, the distribution is world-wide and comprises a variety of ecosystems: Arctic coast and tundra, Kalahari and other deserts, tropical and temperate forests are all represented. Though largely superseded, the hunting way of life still proves its adaptability to a diversity of ecosystems.

Although later we shall use data collected from these groups in discussions of the environmental relationships of hunters, they are probably not typical of the situations which existed when all or most

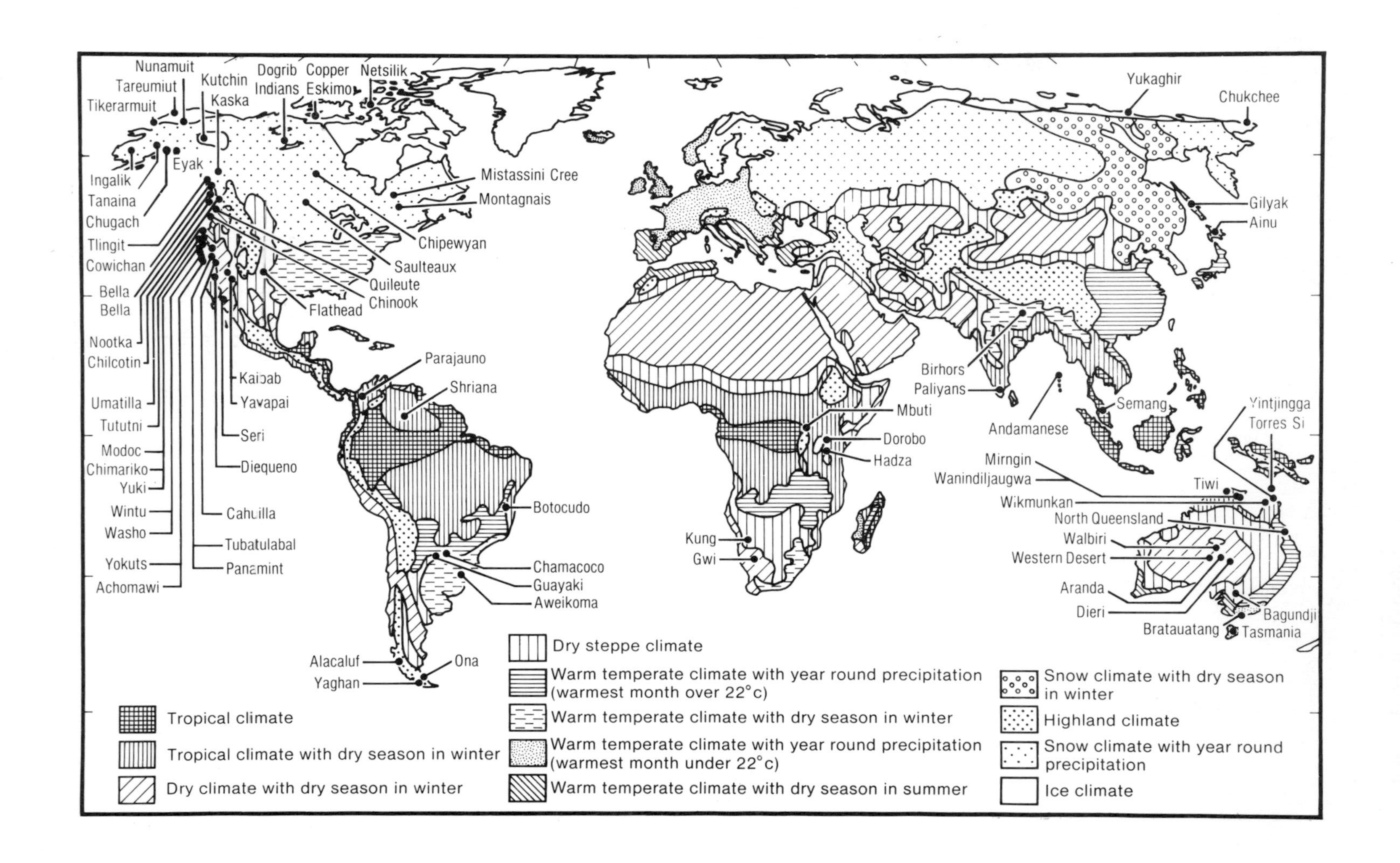

Nunamuit
Tareumiut
Tikerarmuit
Kutchin
Kaska
Dogrib Indians
Copper Eskimo
Netsilik
Eyak
Ingalik
Tanaina
Chugach
Tlingit
Cowichan
Bella Bella
Nootka
Chilcotin
Umatilla
Tututni
Modoc
Chimariko
Yuki
Wintu
Washo
Yokuts
Achomawi
Mistassini Cree
Montagnais
Chipewyan
Saulteaux
Quileute
Chinook
Flathead
Kaibab
Yavapai
Seri
Diequeno
Cahuilla
Tubatulabal
Panamint
Parajauno
Shriana
Botocudo
Chamacoco
Guayaki
Aweikoma
Alacaluf
Yaghan
Ona
Yukaghir
Chukchee
Gilyak
Ainu
Birhors
Paliyans
Andamanese
Semang
Mbuti
Dorobo
Hadza
Kung
Gwi
Yintjingga
Torres Si
Mirngin
Wanindiljaugwa
Tiwi
Wikmunkan
North Queensland
Walbiri
Western Desert
Aranda
Dieri
Bagundji
Bratauatang
Tasmania
Tropical climate
Tropical climate with dry season in winter
Dry climate with dry season in winter
Dry steppe climate
Warm temperate climate with year round precipitation (warmest month over 22°c)
Warm temperate climate with dry season in winter
Warm temperate climate with year round precipitation (warmest month under 22°c)
Warm temperate climate with dry season in summer
Snow climate with dry season in winter
Highland climate
Snow climate with year round precipitation
Ice climate

Plate 3.2 Near-recent hunter-gatherers in Arnhem Land, in the north of Australia. Though good clues to the environmental relations of this way of life, we have to remember that their ways are unlikely to have remained totally unchanged in the face of post 19th-century influences. (Photo: © Jean-Paul Ferrero/Ardea, London)

of the human population were hunters. If they have survived at all, we might argue, then they cannot be typical, for the typical have all vanished. They must possess some unusual cultural feature, or live in a very inaccessible place, to have survived in any form (plate 3.2). Again, most of them have had their way of life affected by recent contact with other economies, so that even if they still hunt and gather for a living, their culture and its attendant environmental relations are not 'pure'. The availability of store food and welfare cheques in Canada, for example, must lessen the pressure to hunt, and many Australasian aborigines find temporary work from time to time in ranching or extractive industry. So while ethnographic information from present and near-recent hunting groups is invaluable, its extrapolation into history and prehistory must be done with considerable care.

One footnote to the story of decline: in a few places the value of traditional foods and hunting practices is being reassessed. In Alaska, for example, the very high cost of imported food (two or three times the Seattle price) to native groups has made them reassess the virtues of seal, salmon, tomcod and reindeer. The government of Quebec supports a native hunter programme in Schefferville, helping Naskapi Indians keep alive the techniques of taking caribou but providing them with air transport to suitable locations. Australian aborigines wanting to live a more traditional life establish 'out-stations' with traditional patterns of society but because they have to be relatively

Figure 3.3 The world distribution of recent hunting and gathering groups. Their names are superimposed on a climatic classification to remind us of the ubiquity of that way of life in the past. *Source*: B. Hayden, 'Subsistence and ecological adaptations of modern hunter-gatherers', in R. S. O. Harding and G. Teleki (eds) *Omnivorous Primates*. New York: Columbia University Press, 1981, 344–421.

sedentary they tend to over-use native biota, especially in very dry years. Sympathetic whites are now working with aborigines to establish native plant horticulture of a type which is compatible with the re-established Aboriginal lifestyle.

Economies developed: the range of plant and animal resources

In this section we consider first some of the extinct economies of the Pleistocene and the early Holocene. In conditions everywhere different from those of today, a number of distinctive and apparently successful cultures developed, often in the sort of rather bleak environments which most of us would not relish as our habitat. Secondly, we consider the outstanding features of the economy of some of the near-recent groups of hunter–gatherers whose way of life has been well documented, and end with a more detailed rehearsal of the economy of a much-studied extant group, the Bushmen of the Kalahari desert and its fringes.

The evidence for the life and times of Neanderthal men of Mousterian culture during the last major phase of Pleistocene glaciation is spatially widespread (figure 3.4), but some regions present a more detailed picture than others and moreover show the transition from the Middle to Upper Palaeolithic. South-west France, for example, was the habitat for both Mousterian and a variety of Upper Palaeolithic cultures but there seem to have been differences in the use of it. The Mousterians seem to have used a wide variety of animal species. Their remains in caves and rock shelters rarely show a dominance of one species of animal prey: reindeer, horse, bovids and red deer all occur frequently and must have provided the bulk of the meat eaten; though smaller mammalian species do turn up at Mousterian sites, like ibex, chamois, roe and fallow deer, wild pig and fox. Some may have been caught for fur rather than for flesh. When we come to the Upper Palaeolithic in the same region, there seems to be a focus on only one species of large mammal herbivore. Thus in most sites 90 to 99 per cent of the bones are of one species, and in Magdalenian times (15 000–9 500 BP) the chosen animal at *c.* 95 per cent of sites was the reindeer, with red deer coming an extremely poor second, although in the Magdalenian of Cantabria (northern Spain) this is reversed and red deer becomes the dominant exploited species. But at the same time there is the first evidence that other animal groups are significant elements in the diet: in the Upper Palaeolithic enough bird bones appear to allow the inference that they were eaten in some quantity. The other resource group now appearing is fish. The remains are dominated by salmon, but six other species including eels occur as well and a number of settlement sites are found conveniently close to rivers. It has been hypothesized that in some of the foothill valleys of the Pyrenees salmon became an important element in the diet because the relatively small bottle-shaped valleys could not support a herd of reindeer large enough to feed the Palaeolithic groups whose territory this was. Plant materials

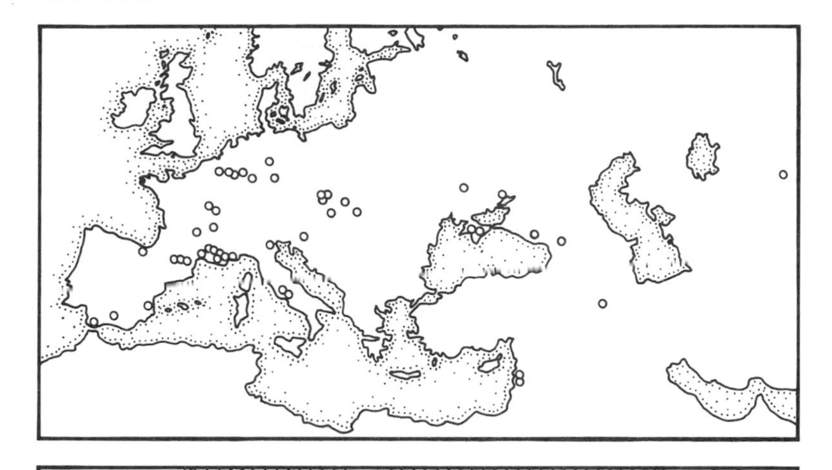

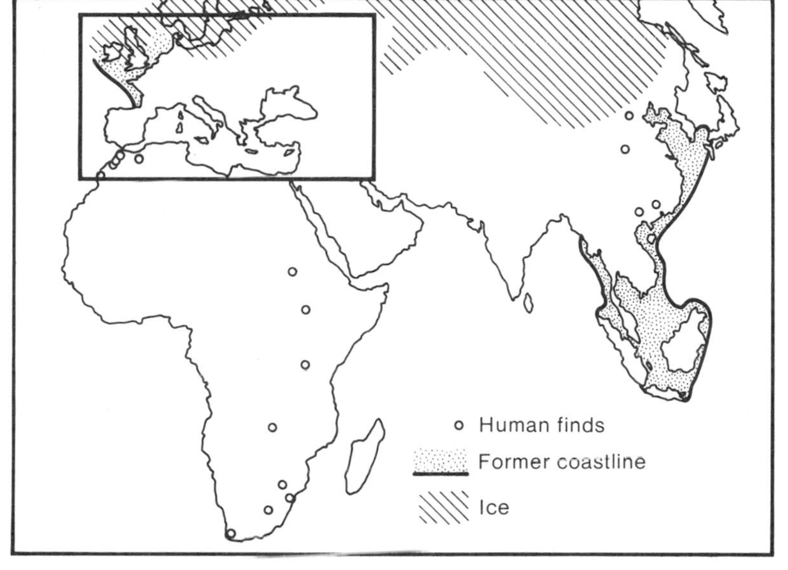

Figure 3.4 Human remains of the Neanderthal period for Eurasia, with (top) an enlargement of the European region. The African distribution suggests a continuity with earlier forms of *Homo* and even older hominids. *Source*: J. A. J. Gowlett, *Ascent to Civilisation*. London: Collins, 1984, 100–1.

are also found at some Palaeolithic sites in Europe though never in quantities which allow the interpretation of environmental impact, wood possibly excepted.

In South Africa, certain economic changes have been postulated by scholars which relate to the transition from the Middle Stone Age (MSA) to Late Stone Age (LSA) periods (*c.* 40–30 Kyr), probably the time of the replacement of archaic by modern forms of *Homo sapiens*. It appears that MSA people hunted more dangerous species such as wild pig, that they killed gregarious species such as eland by driving herds over cliffs or into traps, and that they took a greater proportion of the animal biomass than their LSA predecessors. But neither group seems to have been successful at killing prime-age adults of dangerous species like buffalo. In Eurasia there is evidence of successful adaptation to more extreme environments during Upper Palaeolithic times. Excavations of huts containing hearths in the Dnestr valley of the Ukraine show a culture in which mammoth bones were a central structural element (evidently used structurally in tents made of skin) as well as, possibly, a substitute for wood as fuel. These were winter sites and used for the hunting of cold-climate species, not only the mammoth, but musk-ox, wild horse, reindeer, woolly rhinoceros, bison, elk, fox and wolf. An apparently domesticated individual of wolf was also found. Although these sites bear impressive testimony to the ability of human groups to exist in severe conditions at places like Molodova, Mezin, Pushkan and Kostlenki, it is postulated that the mammoth bones which were so important were scavenged as well as hunted, and so accumulated from generations of animals rather than being the result of an intensive burst of hunting. Reindeer and horse were probably the most important foods. Upper Palaeolithic sites go as far north as 65° in the USSR; in the northern Urals a few flints of that period were associated with the bones of mammoth, reindeer and polar bear.

Archaeological evidence coupled with the reports of early European travellers has shown us some details of another kind of hunting economy, in the shape of the killing of large mammals by Indians and palaeo-Indians on the grasslands of the High Plains of North America during the Holocene period. The exact date of man's entry to North America is still disputed but at any rate by 12 000 BP a stone tool culture was established in North America and was diffusing southwards. On the Plains the earliest inhabitants were eclipsed in about 10 000 BP by sets of hunters who adapted to an altered climate and to a different fauna. These groups became the Plains Indians encountered by the European explorers, traders and settlers.

In the early days of Plains hunters, the quarry was chiefly bison (buffalo), but the bones of the North American Plains camel *Camelops* are found at some sites as well. At the Casper[2] sites in Wyoming, dating from 8 000 BC, a buffalo kill of at least 74 animals but perhaps 50 per cent more, also contained the remains of *Camelops* in much smaller numbers. In this case, the animals were enmired in a shallow pond in an interdunal slack surrounded by crescentic sand dunes. Piles of stones leading to the pond suggest markers along a planned drive route.

In later years, *Camelops* has disappeared, and the evidence is of hunting of buffalo, first the now extinct *Bison occidentalis* and then the modern species *Bison bison* (figure 3.5). The former species was the quarry at the Olsen–Chubbuck site in south-east Colorado, dated at 8200±500 BC. Here the hunters stampeded a herd of buffalo into a narrow canyon or *arroyo*, some 2–3 m wide and 2 m deep. The age of the calves suggests this happened in late June. The bone remains extended some 57 m along the arroyo and about 190 animals were killed (a typical herd would number 200–300 at that time of year), some of them being wedged so fast in the bottom of the canyon that they could not be extracted for butchering and thus must have rotted *in situ*. The bone analysis shows that 57 per cent of the animals were

[2] These sites are named after their landowners, not the present place names.

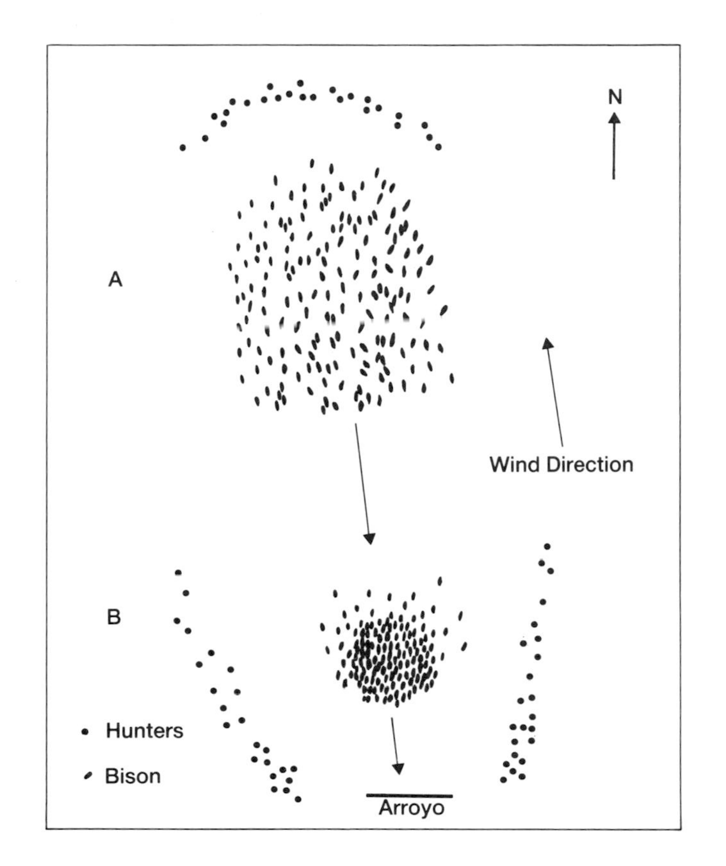

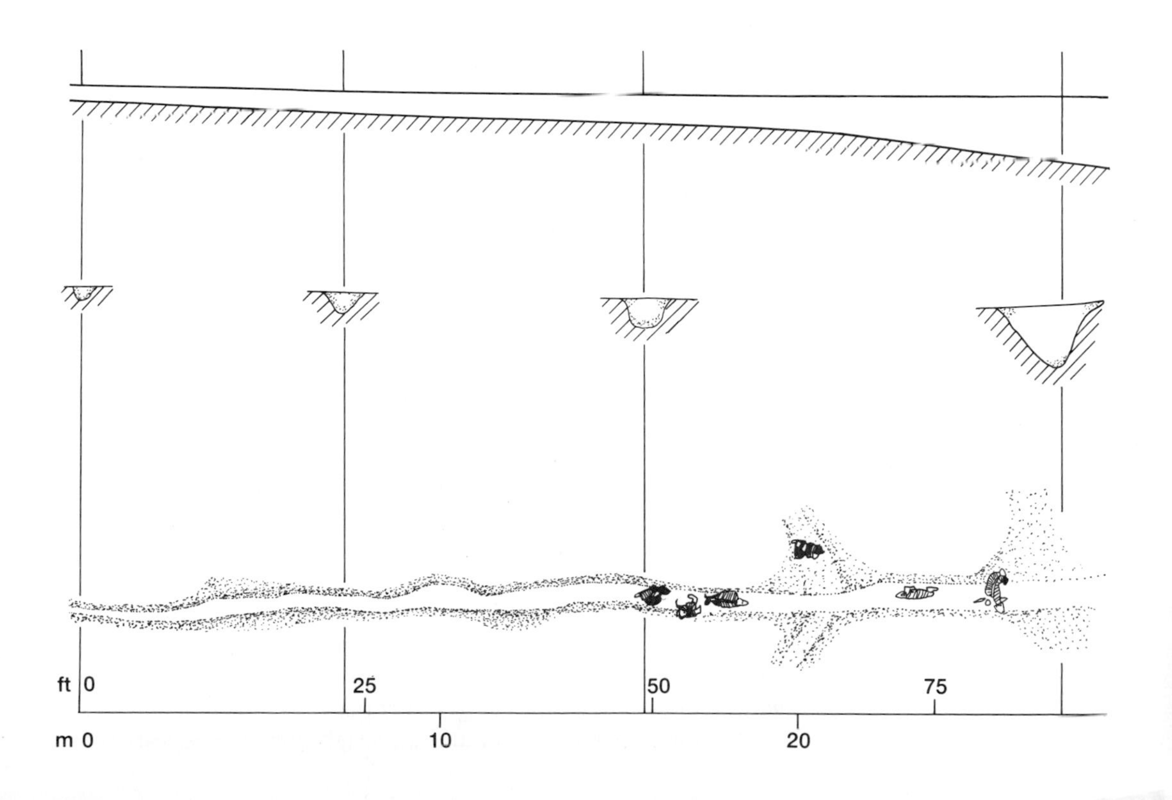

Figure 3.5 Process and effect in the Bison hunters of North America. The top shows a plan view of the way in which the hunters drove a group of bison towards a narrow canyon (*arroyo*); the lower shows a sectional view of the *arroyo* and the position of some of the skeletons found upon excavation. *Source*: J. Taylor, 'The earliest hunters, gatherers and farmers of North America', in J. V. S. Megaw (ed.) *Hunters, Gatherers and First Farmers beyond Europe*. Leicester: The University Press, 1977, 199–224.

mature, 37 per cent immature and 6 per cent juvenile, but no foetal remains were found, either because no gravid animals were killed or because the foetuses were taken away as a particular delicacy. The position of projectile points in the carcasses makes it clear that hunters were placed on the flanks of the herd and that the animals were run at right angles to the arroyo and, failing to leap it, the early beasts fell in and would have been killed by the weight of the following animals.

From about 4500 BC comes the Hawker site in Wyoming when *B. occidentalis* were trapped in a canyon, but this time the arroyo was used as a channel up which to drive the animals until a knick-point formed an end buffalo wall. For 83 m before this wall, the sides of the arroyo were too steep for bison to climb. This was an autumn kill but only of the males which form separate herds except at mating time. Each time this trap was used (and there is evidence of several episodes of usage), 15–30 buffalo were killed, including some young animals. A variation on this type of trap was seen on the upper Green River in Wyoming in the first millennium AD where the width of the upper end of the arroyo was such that it could be made into a corral by means of a fence hung with hides, the area of the trap being an oval about 10×15 m, in which 10–20 animals were killed with arrows, again in autumn. The pound was successful because buffalo will not jump anything they cannot see through and so a hide fence will contain them. Such a method is echoed at the Ruby site in Wyoming (AD 280) where a 10×15 m pound had a double line of stakes, suggesting that the spaces were filled with brushwood and sagebrush, perhaps with skins laid over. Here a drive line marked by post-holes can be traced, and at 13 m from the enclosure it has a sharp bend which obscured the view of the pound until the last moment. Again, 20–25 animals would be the most likely catch.

At one time during the years AD 1670–1740 another group of buffalo hunters used a spectacular site in central Wyoming, where a bluff 14 m high at Glenrock was utilized. The vertical part of the bluff is restricted, and the animals had to be carefully controlled: a metre or two either side and they would probably escape down steep but not lethal slopes. In this case, the drive lane suggests the control of the herd's movements for 1.5–4.5 km before it was finally stampeded over the cliff. The bones suggest that this and similar bluffs might be used in rotation: as one author points out, about 100 buffalo stomachs left to rot might not be approachable again for a year or two, even to people less easily made queasy than ourselves.

To summarize, the evidence suggests that kill operators were bringing in buffalo to kill sites from several kilometres around each site, usually but not always in the autumn, i.e. before the lean season for human food. Fire was often used in this as in other types of buffalo hunting. So men were reaching out and removing several hundred cows and calves from the breeding population and perhaps doing this in adjacent areas in the same season. Although doubtless plants were eaten, the buffalo was a staple, and although nobody suggests that the hunters were responsible for anything like the reduction in buffalo numbers that came later, we may surmise that the population dynamics of herds might often have been affected in a long-term sense. Where *Camelops* was concerned, there is the possibility that such hunting was one of the factors that led to its demise.

A rather similarly specialized economy was practised before 4200 BC on the *puña* of the Andes. Here, formerly generalized hunting became focused on specialized hunting of the native camelids of the *Llama* genus. This specialized hunting was based on an intimate knowledge of the social and territorial behaviour of the llama. One adult male and 4–7 females occupy a defended territory which is stable in space and time. So if only a certain number of animals were killed and the territories not destabilized, then there would be a constant supply of meat. This probably meant hunting by the ambush of individual animals rather than the mass killing which would have redistributed their territories.

From the recent past, we have good ethnographic evidence for the adaptation of hunters, gatherers and fishers to all kinds of natural environments: as in the far past the flexibility with which these people made use of different resources is remarkable, as is the fact that virtually no set of habitats apart from the entirely waterless and permanently snow-covered were unoccupied. At one end of the environmental spectrum, the equatorial rainforests have been settled by groups who form local bands of about 20 people and have a strongly vegetable diet, with items like wild yams, nuts and fruits being important in the diet. Hunting is practised since meat is much prized: the Senang of Malaysia used the bow to hunt pig, monkey, birds and bats, and also speared fish, and entangled birds in sticky traps. The Senang bands each had a territory of about 50 km^2. In central Africa, the Pygmies use nets for deer and pig, construct deadfalls and spiked pits for big game, and also spear elephant. Much of their plant food comes from exchange with agricultural groups. In Amazonia, groups like the Yanomama who practise agriculture nevertheless seem to depend upon hunting and fishing for their protein supply, and their relations with these foodstuffs have been held to be at the root of behavioural traits such as warfare, taboos and sexual politics. In the drier zones surrounding the tropical forests, such as tropical savanna and scrub, and low-latitude deserts, water is the key to human survival and the economies are often based on following herds of ungulates as they migrate to perennial water sources. A much-studied group (see p. 59) is the !Kung bushmen of the Kalahari, whose staple food is the mongongo nut although they also take a wide variety of other plant and animal foods. Other groups of this zone use a variety of hunting techniques, including brushwood fences set across country with pitfall traps at the gaps, fire-drives, and the use of poisons (derived for example from *Euphorbia* species) on arrows and in the animals' drinking water. A variant of this group were the Paiute of North America, who lasted until the nineteenth century. They had to survive long cold winters and so had permanent winter settlements with hearths and storage pits; their main food at this season was pine nuts. They also collected ripe grass heads, bulbs and roots at appropriate times, and were known to divert snow-melt water channels onto areas yielding those products, using small ditches and dams. The winter was also a hunting season: animals such as wild sheep, goats, and rabbits acted as food stored 'on the hoof'.

The highest density of hunter–gatherer populations seems to have been achieved in areas of sclerophyll vegetation, like those of the west coast of North America. Foods such as acorns and buck-eye nuts were plentiful and the techniques for extraction of the tannin well-developed, and a variety of other plant and animal foods could be utilized. The seasonal aridity of the areas made it easy to burn the vegetation in late summer, and this practice favoured the oak trees and provided ample nutritious forage for deer. It is not surprising therefore that bands were relatively large and territories relatively small, thus achieving high population concentrations. In northern forests, the environment supported rather low densities of people, partly because of the difficult winter period against which fish and game had to be stored, usually by drying. If caught in winter, meat and fish were of course easier to store, in natural deep-freezers made of stones to keep out scavengers such as the wolf. Groups like the Cree and Montagnais Indians lived mostly on the meat of animals like moose and beaver, along with porcupine, hare and eels. Drying and smoking were important ways of bridging critical seasons of the year and a food intake of 4000 kcal per head per day has been estimated for the Montagnais in the mid-seventeenth century. Further north, groups largely dependent on caribou, like the Chipewayans west of Hudson's Bay, relied on subsidiary foods in years when the caribou migration routes passed them by at a distance. On the coasts, the living was easier, principally because of the abundance of salmon, so that the Indians of Oregon, Washington, British Columbia and southern Alaska developed a very rich sedentary society, but supplemented the fish with seals, whales, sea lions, shellfish and seaweeds, with inland hunts for bear and deer to supply skins.

At the other end of the spectrum from the equatorial forests is the Arctic tundra. Inhospitable though it is, Inuit populations of North America, and their Eurasian equivalents, managed to adapt themselves to its constraints (figure 3.6). The Inuit came to be largely dependent on coastal animals (walrus, seal, fish, whales), but inland groups took caribou in large numbers during the animals' annual migrations and also fish from lakes and rivers. In summer, both groups might eat berries and other plant foods, and in spring the eggs and chicks of nesting birds. But no great density of population could ever be thus supported in this environment; before European impact on any scale, there were probably never more than 10 000 Inuit between Alaska and Baffin Island.

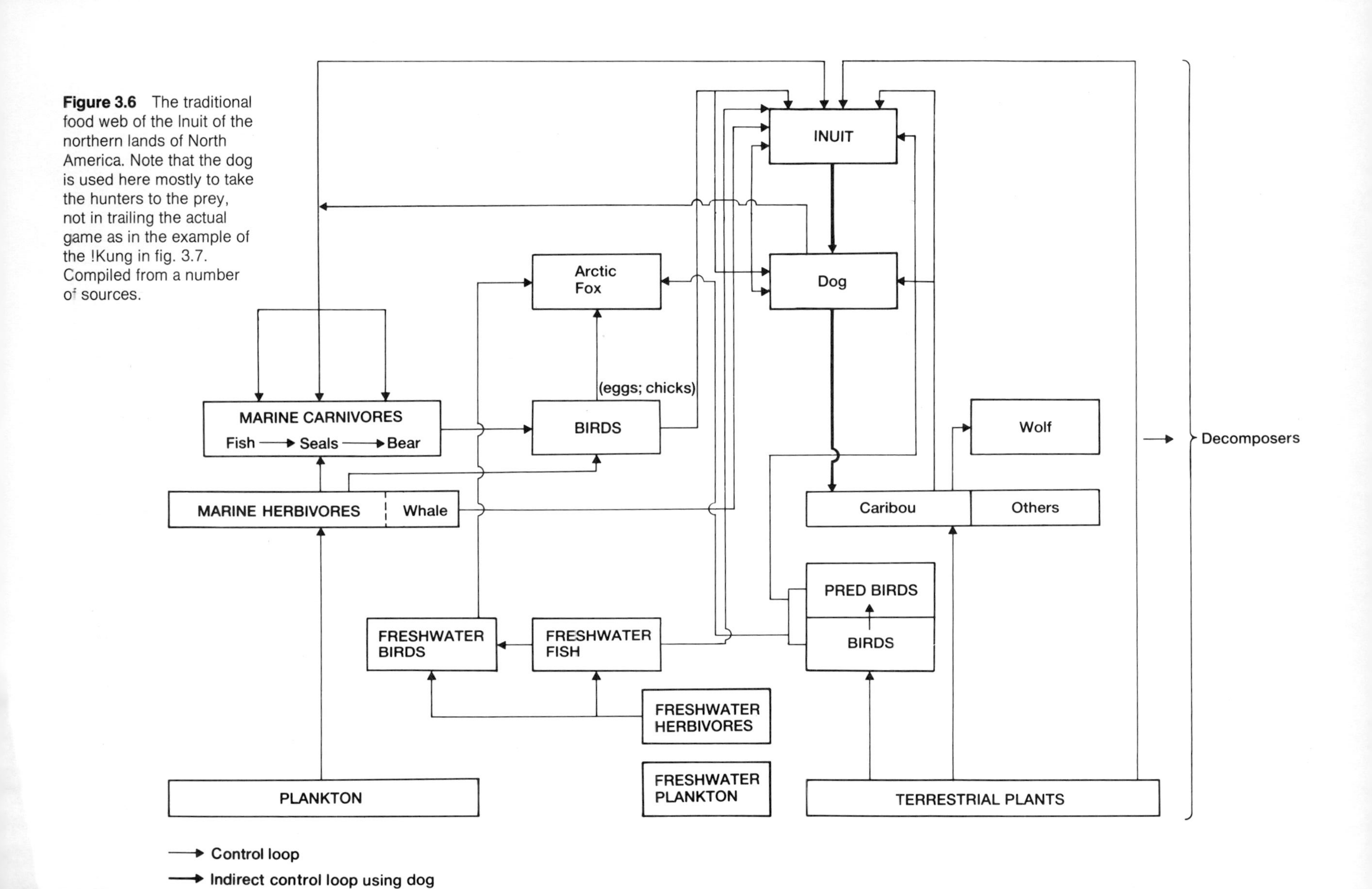

Figure 3.6 The traditional food web of the Inuit of the northern lands of North America. Note that the dog is used here mostly to take the hunters to the prey, not in trailing the actual game as in the example of the !Kung in fig. 3.7. Compiled from a number of sources.

In near-recent times, hunting and gathering economies have been present in forests as well as the more open and diverse environments so far described. Tropical forests in particular seem a relatively unpromising habitat for such a way of life, for most of the biomass is in the trees. But by being essentially vegetarians and then adding such meat as can be procured, forest peoples came to have a stable and persistent way of life. One such group are the Pygmies of the Zaire basin of central Africa, and among them the BaMbuti group of the Ituri forest are especially widely known from the work of Colin Turnbull. These people have the dog as a domesticate and utilize a wide variety of wild plant foods such as roots, fruits and mushrooms; 'small' animals and their products such as termites, grubs and snails are also collected, as is honey, which is an important seasonal food. Larger animals hunted (by the men) include antelope, monkey and chimps, chevrotain, wild boar and elephant. Spears are used for elephant, bow and arrow for the primates, and net hunting for most other mammals; occasionally animals are snared but pits are rarely used. Fish, though abundant, are rarely taken. There seems no doubt that net hunting (using a natural twine from a vine) is the most efficient method and that it procures the largest quantities of meat with the greatest certainty. Compared with the bow and arrow, however, it quickly scares animals away from the vicinity of a camp and so net hunters must travel up to several days march away from their families. A successful net driver may, for instance, deplete an area of deer for two months, so this hunt may be undertaken only at the start of the honey season when an alternative desirable food is becoming available. In general, though, there is an abundance of meat to the point where there are no cultural restrictions upon the killing of animals other than some clan totem restrictions of a rather vague kind. But it is considered a crime to kill or hunt animals unless they are needed for subsistence; cultural sanction to some hunting, according to Turnbull, is given by the belief that there is some formal opposition between the human and animal worlds and so people feel the need to assert the superiority of the former.

What generalizations can be drawn about recent and near-recent hunter–gatherers? We may note that people were never completely specialized in terms of their food sources: subsidiary foods were rarely neglected entirely so that there was usually a buffer if the preferred food failed. Clearly, though, the success of a group depended greatly on its success in gathering its main food item or items. In terms of foods utilized we notice that the pattern of subsistence has some correlation with latitude. R. B. Lee has pointed out that in the highest latitudes (above 60°), hunting was the dominant mode of subsistence (in 6 out of 8 societies which he surveyed); in the cool to cold latitudes 40–59°, fishing was dominant (14 out of 22 societies); and in the warm heat of 0–39°, the gathering of plant food was paramount (25 out of 28 societies). But hunting, though not dominant at all levels, did make a stable contribution to most diets: all hunting and gathering societies derived at least 20 per cent of their diet from the hunting of mammals. The subsistence of these groups can perhaps be summed up as representing what was for a long time a persistent and stable way of life based on eating as much vegetable food as was needed and as much meat as was possible. Even where relationships with animals dominate the account, the key position of plant foods must not be forgotten.

A more detailed example of the environmental relations of a modern hunter–gatherer group is provided by the !Kung Bushmen of Botswana and Namibia, studied intensively by R. B. Lee in the 1960s. The unit upon which he reported lived in an area of bushveld savanna and fixed dunes with an average rainfall of 250 mm/yr (10″) which is, however small, very valuable from year to year. The flora has a diversity of about 220 species and the main ecological constraint from the human point of view is the extreme scarcity of surface water. Water holes, tree hollows and fleshy roots form anchor points in the daily existence of the !Kung, with the filling of water containers a daily event. The technology of the Bushmen is simple: a variety of hunting tools including the bow and poison arrow, a carrier, the digging stick and the ability to use fire comprise much of the repertoire. The dog is sometimes a possession and sometimes not, but it is used as an aid to hunting if it is available. In these circumstances, the small bands of Bushmen live a nutritionally secure life, gained without too great an effort (figure 3.7). The key to this is the gathering of plant material: some 105 species provide edible material. Of these, 29 give fruits, nuts and berries, 41 roots and bulbs, 18 edible gum, and others give beans, melons and leafy material. Some 14 of these species yield about 75 per cent of the calories in the vegetable diet and one of them

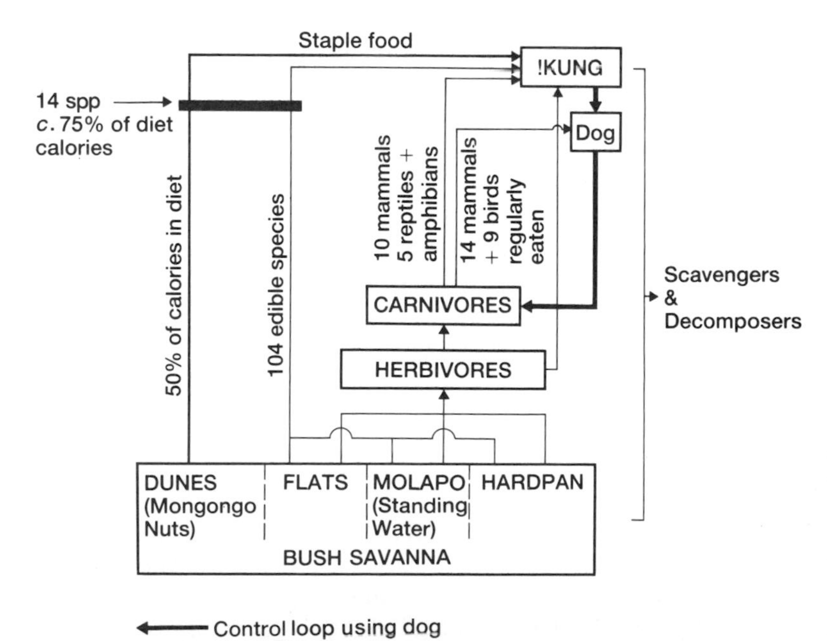

Figure 3.7 The traditional food web of the !Kung bushmen of the Kalahari desert. This type of diagram is basically qualitative and thus does not easily show the status of the mongongo nut as a staple.
Source: Based on numerical and verbal data in R. B. Lee, *The !Kung San. Men, Women and Work in a Foraging Society*. CUP, 1979.

nearly half. This is the mongongo nut, *Ricinodendron rautauenii*, (transliterated picturesquely from the !Kung as II"Xa) which forms both a staple and a stable food. It grows in linear groves along the crest of fixed dunes and in irregular groves on sandy soils of major rock outcrops. The kernel of the nut is about the size of a small hazel nut and is highly nutritious, so that 200 nuts/day give 2500 kcal and 77 g of protein. The nuts fall in April, but having a hard outer shell do not then rot and so remain on the ground until eaten by people or wild animals. In the territory of the group studied, a 62 per cent utilization rate of the nut production would give everybody 1000 kcal/day on average, so its status as a staple can hardly be denied. The mongongo does not grow to maturity except in groves so that while occasional seedlings will spring up from viable nuts that have been transported by the !Kung, no new groves result from either deliberate or accidental actions of the Bushmen, who say 'Why should we plant when there are so many mongongos in the world?' Meat is an important element in the diet and forms practically the commonest topic of discussion. The proportion of meat in the diet rarely falls below 20 per cent by weight and accounts for almost 40 per cent of calorie consumption, averaging some 2 kg per head per day. Some 55 species are deemed edible and are hunted by snaring, by mobile hunting and by the use of fire to asphyxiate warthogs, anteaters and the ant bear.

There is nothing in the Bushmen's way of life to indicate an unstable relationship with the environment. It is worth noting, though, the general principles which emerge as conditions for their success. They must be mobile and be able to cover a wide area to find sufficient food: in the case of the !Kung each able-bodied person has to cover some 2000–3500 km/yr on foot in the quest for food and water. The environment sets rather definite upper limits on group size and so because of regional and annual fluctuations in resources, group structures must be adaptable. Among other things, there develops a set of elaborate rules for sharing among groups. Lastly, the need for movement sets limits on material possessions. We ought also to note that Bushmen with access to mongongo nuts are much better off than the other Bushmen who do not have that resource, and that their population density may be artificially low because a number of them have sought work in Namibia and Botswana.

Energy flows in hunting societies

General considerations

Since access to energy sources is taken as one of the organizing concepts of this book, it is essential to look at the energy flows of the ecosystems inhabited by hunter–gatherers, and the ways in which those societies have made use of it. Hunters utilize the recently fixed energy of the sun: such organic matter as they gather represents solar radiation which has not been stored very long. Indeed, most of it is 'this year's' energy in the form of parts of plants or animals, though some 'older' energy in the form of trees, the roots of perennial plants, or the bones of long-lived animals may also be used. Other manifestations of the sun's immediate energy input to the earth are not directly utilized, though the relation of tides to marine food resources must have been known by every beachcomber through history. With the relatively simple material cultures, a basic equation of life is the balance between energy expenditure in gathering food and the nutritive value (especially its calorific content but also its protein yield) of the edible materials collected. For a few groups, a similarly fundamental role is played by water.

The energy costs of the food system of hunter–gatherers can be grouped under a number of headings and are relatively simple to tabulate, unlike those of industrial societies. The first need is the energy required to locate a food supply, be it plant or animal. Expenditure of energy is relatively high here, and although practice and expertise reduce this expenditure, it may be raised again if over-use of the resources forces the group to travel long distances from their camp. Little quantitative work has been done on the probabilities of success in finding food on a given collecting trip, although Lee reports that the success rate of !Kung hunters was 23 per cent, i.e. one kill for every 4 man-days of hunting, and made a net contribution to the energy balance of the total group (5100 kcal + 2539 g protein per hunter per day), though not as effective as the gatherers of mongongo nuts with their 100 per cent success rate. Coastal food-collectors in northern Australia had various recorded probabilities of success per trip: the probability for wallaby was 0.25, but for fish with hook and line 0.65; for wild yams and for shellfish 0.9, and 1.0 for the kernels of cycads. Most such data are collected over too short a period to reveal some of the variations in the probabilities of finding foods, particularly the variations between good years and bad. Figure 3.8 shows a simple example: that of the distribution of a major food of the G/wi bushmen, the *tsama* melon, in relation to a campsite. In a poor year, the search must be extended widely, necessitating the expenditure of more energy for a poorer yield (0.244 kg/ha/day) than in a prolific year where less energy nets a high yield (1.29 kg/ha/day).

Actual gathering and fetching of the food sources will usually involve a high proportion of the energy expended by the collector group since this is the phase of travel or chase, climbing and killing. In the case of large animals, outright death is less common, so that a large mammal has to be tracked to benefit from its greater yield. In this context, the various types of trap can be seen as energy-efficient provided that they do not take too much labour to construct; however, large pitfall traps can presumably be used more than once, as can the deer fences put up over considerable distances by the Ainu of Hokkaido. The very early domestication and subsequent widespread use of the dog can be seen as a hunting aid designed to bring in

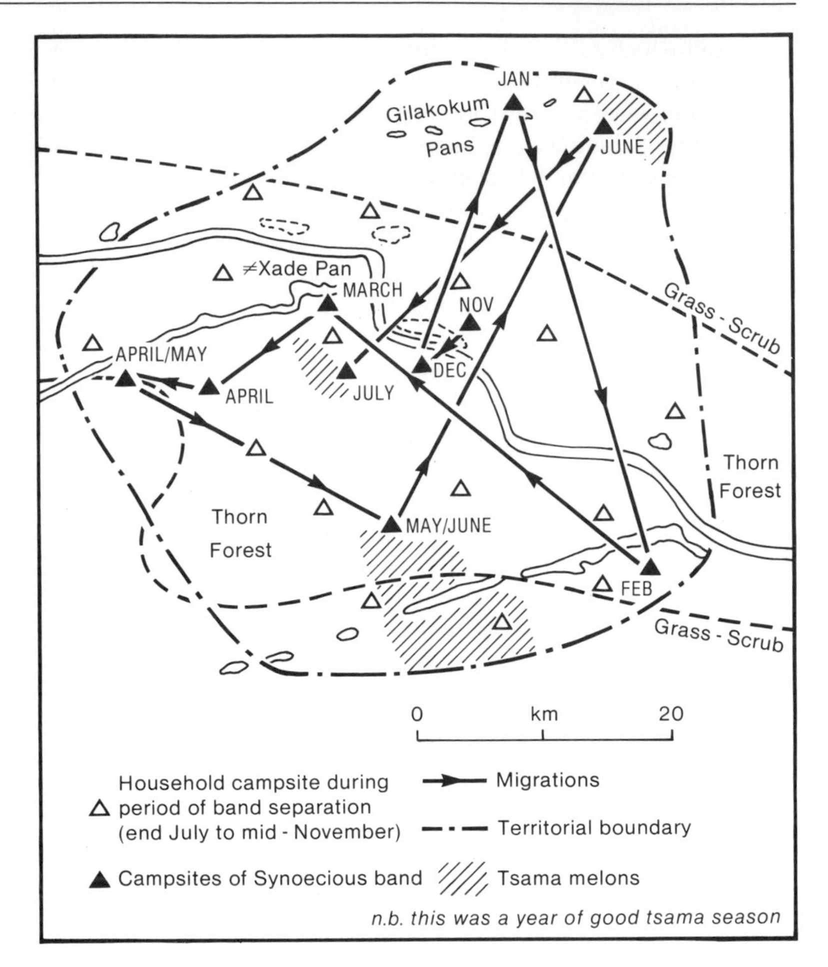

Figure 3.8 The yearly movement of a band of G/wi bushmen, in a time of good availability of the important *Tsama* melon. Note that as in fig. 4.6, the group splits up during the year.
Source: G. B. Silberbauer, 'The G/wi bushmen', in M. G. Bicchiere (ed.) *Hunters and Gatherers Today*. New York: Holt, Rinehart and Winston, 1972, 271–326, at p. 297.

a lot more calories than it needs for food. Thus where the energy ratios of hunters have been measured (the sample of such studies is necessarily very small because of the recency of the approach and the scarcity of traditional hunter groups), the combination of food-collecting skills and environment seemed normally to provide an adequate surplus of output to input to feed not only the collectors but those dependent upon them. The !Kung food systems, for example, appeared to provide an adequate diet for the whole group on 2.4 work-days per week with indeed some more surplus still which is devoted to feeding the hunting dogs, to accumulating physiological fat against the lean season, and sometimes to consuming energy in a dusk-to-dawn dance which involves the entire adult community. Estimates for the mongongo nut suggests that the returns to the !Kung are of the order of 10 500 kcal/man-day; the Batak of the forested interior of Palawan Island in the Philippines can get 13 912 kcal/man-day from digging a wild yam (*Dioscorea hispida*), which means enough food for the collector and 4.6 others, for the 9 months of the year when the yam is available. The Anbarra women who gather cycad kernels in northern Australia accumulate 1000–1400 kcal/hr and so feed themselves with two hours' work per day.

Where a hunting group starts to move away from pure subsistence to an exchange economy which brings in a different (and probably agriculturally produced) food, then it seems that more calories have to be expended in gathering the product to be exchanged. Such a change is reported for the Batak groups just mentioned. In their self-contained subsistence economy, they collected wild yams whereas they have now moved towards collecting a tree resin (Manila copal), selling it, and buying milled rice. The returns to yam digging are 13,912 kcal/man-day whereas to getting and selling Manila copal, the returns are 8096 kcal/man-day, so each person thus engaged has suffered a 40 per cent decline in the returns to his labour, and both undernourishment and nutritional deficiency have been noted among the Batak. But technology may work the other way: when steel tools became available to the Heve people in New Guinea in 1966–67, it became possible to cut timber 4.4 times as fast as with the traditional stone adze, presumably freeing time for other occupations. Here as elsewhere, a negative energy balance may be accepted for some hunts if they bring in valuable fur as well as some meat.

The energy used in processing and preparing food by hunter–gatherers is not usually very great since the food is either eaten quite quickly after it is collected or else it is subjected to a preservation method which involves little expenditure of human effort: for example smoking uses the energy in wood, and drying that of the sun. The Amerind's custom of making pemmican by pounding surplus meat to a paste involved the use of surplus energy at a time of plenty which was seen as an investment in a portable and enduring food supply during the winter. Some energy may go into very careful butchering of a carcass according to traditional apportionment but it is doubtful if this exceeded by any great amount any other sort of butchery. Pit roasting of animals such as pigs must increase input energy but no doubt everybody thinks the result is worth the effort. The energy used in preparing certain species of yams, for example, by washing, slicing and leaching them with hot water is not optional since an alkaloid poison has to be removed; the virtue of the mongongo nut, which has only to be cracked open, is readily apparent. The daily energy costs of transporting or storing food are usually low since hunter–gatherers do not travel very far on hunting trips (many stay within 5 km of their camp although they move it at relatively frequent intervals) and indeed may eat some of the food at the site of the procurement, so that less has to be carried home.

As subsistence hunters move away from traditional modes, as when marketing meat in exchange for other energy sources, then the caloric input into this process must increase, since a maximum amount of meat must be brought to the exchange site and kept there in good enough condition to be attractive to the traders. This may in effect mean that hunting is carried out only at certain times (in the case of the Mbuti Pygmies it is when they are visited in camp by

Figure 3.9 A modern energy flow diagram for an Inuit people from Baffin Island in the 1960s. On the left-hand side are stacked the modern inputs like store food, fuels and ammunition for guns. *Source*: W. B. Kemp, 'The flow of energy in a hunting society', *Scientific American* 224 (3), 1971, 104–115.

commercial meat traders), with uncertain effects upon the animal resource.

Energy flow in a modern hunting group

Various remarks in the preceding section will have prepared us for the fact that the type of study of a traditional society which would be most useful at this point does not exist. A neatly quantitative study of the energetics of an ecosystem, including the hunter's place within it, is not to be found. In more recent times, human groups which still subsist predominantly by collecting food have usually been in contact with industrial economies and have thus altered their behaviour. In some cases this has meant adding some food production to their repertoire, in others exchanging the products of food collection for those of food production elsewhere, and in yet others the enhancement of hunting success by the acquisition of modern aids such as guns and vehicles. The energy flows of one such group of Inuit (Eskimo) for Baffin Island in northern Canada were studied by W. B. Kemp in 1967–8 and the contrasts with the traditional flows noted. The Inuit of this somewhat barren and inhospitable environment exploited all the marine and terrestrial food webs for food, harvesting at least 20 species of animals but concentrating especially on the common and bearded seals. When studied the group had gained access to modern technology largely in the form of outboard motors, snowmobiles, kerosene stoves, rifles and imported food, all of which represent imported energy rather than recently fixed solar energy (plate 3.3). The subsequent flows as measured by Kemp are given in figure 3.9. A number of consequences were noted which include the virtual absence of hunger, due to increased hunting success and government payments, and sedentarization in the sense that while some men may make long hunting trips, the village remains in the

Plate 3.3 An Inuit hunter from Igloolik, Canada. Still basically a hunter, the influence on him of industrialization is evident, principally in the form of the petrol-driven snowmobile. This piece of technology transforms the energy relations of hunting. (Bryan and Cherry Alexander Photography)

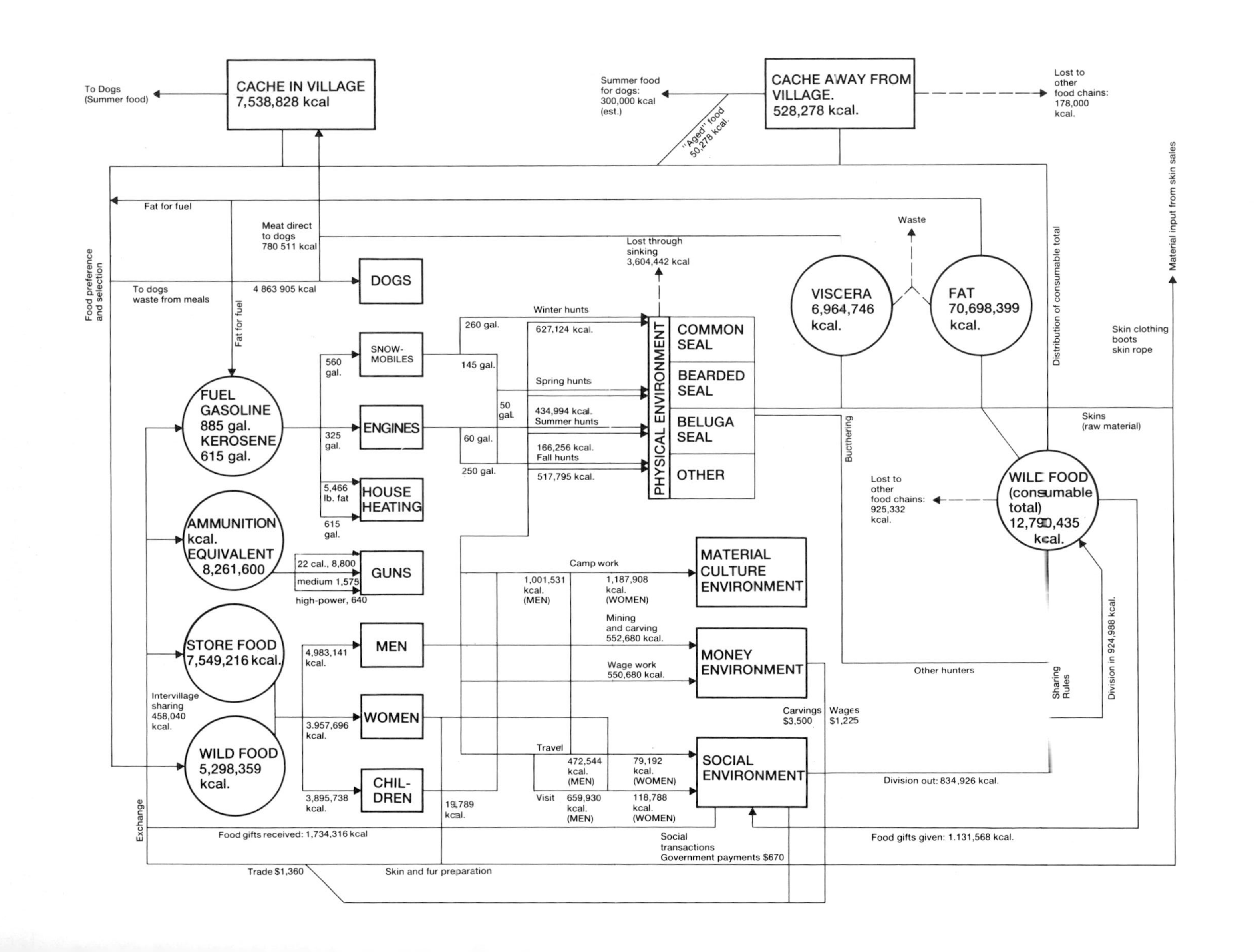

To Dogs
(Summer food)
CACHE IN VILLAGE
7,538,828 kcal
Summer food
for dogs:
300,000 kcal
(est.)
CACHE AWAY FROM
VILLAGE.
528,278 kcal.
Lost to
other
food chains:
178,000
kcal.
"Aged" food
50,278 kcal.
Material input from skin sales
Fat for fuel
Meat direct
to dogs
780 511 kcal
Waste
Lost through
sinking
3,604,442 kcal
Food preference
and selection
To dogs
waste from meals
4 863 905 kcal
DOGS
VISCERA
6,964,746
kcal.
FAT
70,698,399
kcal.
Distribution of consumable total
Skin clothing
boots
skin rope
Fat for fuel
Winter hunts
260 gal.
627,124 kcal.
PHYSICAL ENVIRONMENT
COMMON
SEAL
SNOW-
MOBILES
560
gal.
145 gal.
Spring hunts
BEARDED
SEAL
FUEL
GASOLINE
885 gal.
KEROSENE
615 gal.
50
gal.
434,994 kcal.
Summer hunts
BELUGA
SEAL
Skins
(raw material)
325
gal.
ENGINES
60 gal.
Buchering
166,256 kcal.
Fall hunts
250 gal.
517,795 kcal.
OTHER
Lost to
other
food chains:
925,332
kcal.
WILD FOOD
(consumable
total)
12,790,435
kcal.
5,466
lb. fat
HOUSE
HEATING
615
gal.
AMMUNITION
kcal.
EQUIVALENT
8,261,600
22 cal., 8,800
medium 1,575
high-power, 640
GUNS
Camp work
1,001,531
kcal.
(MEN)
1,187,908
kcal.
(WOMEN)
MATERIAL
CULTURE
ENVIRONMENT
Mining
and carving
552,680 kcal.
STORE FOOD
7,549,216 kcal.
4,983,141
kcal.
MEN
MONEY
ENVIRONMENT
Wage work
550,680 kcal.
Other hunters
Sharing
Rules
Division in 924,988 kcal.
Intervillage
sharing
458,040
kcal.
3,957,696
kcal.
WOMEN
Carvings
$3,500
Wages
$1,225
Travel
472,544
kcal.
(MEN)
79,192
kcal.
(WOMEN)
SOCIAL
ENVIRONMENT
WILD FOOD
5,298,359
kcal.
3,895,738
kcal.
CHIL-
DREN
19,789
kcal.
Visit
659,930
kcal.
(MEN)
118,788
kcal.
(WOMEN)
Division out: 834,926 kcal.
Exchange
Food gifts received: 1,734,316 kcal
Social
transactions
Government payments $670
Food gifts given: 1.131,568 kcal.
Trade $1,360
Skin and fur preparation

same place. Settling down also means that the men may at times be available for wage labour, which results in more food coming from the store rather than from hunting. Environmentally, the consequences do not seem to have been an 'overkill' of animal resources. The rifle and the outboard motor make hunting success easier to attain but at the same time the seals are more wary and less likely to come close to the boat or the ice edge. Similarly, the mobility conferred by outboard motors and snowmobiles makes it easy to visit distant kinsmen, an important form of social life, and missionary activity has promulgated the idea of Sunday as a day of rest so relieving animals of some hunting pressure. As long as a certain number of Inuit men now become members of a cash nexus, engaged in wage labour or services which bring in money which then buys store food, then even with modern technology the remnant hunters are unlikely to break the linkages in the Arctic ecosystems which have sustained them and their ancestors for several hundred years.

This brief look at the energetics of hunters yields few generalizations. One is that the form in which NPP manifests itself is clearly important. Where a great deal of it is edible and collectable as in the Kalahari, then animal biomass is not so crucial for nutrition as plant foods. In the Arctic, NPP is neither easily collectable (and not at all in winter) nor largely edible, and so the stored energy of animals becomes all-important in traditional subsistence patterns. We expect therefore that population densities of the Arctic should be lower than in the Kalahari, but this is now difficult to test because of modern circumstances and differing patterns of seasonal movement. Tropical forests, too, are limited in their available crop at NPP level and require the harvesting of a good deal of animal material; this in turn is difficult because much of the NPP is at canopy level and so are most of the animals. The bow and the blowpipe can be seen as necessary pieces of technology to tap these energy flows, but there is a constant pressure to harvest high-calorie plants such as wild yams or to move to some form of plant food production (plate 3.4). The long vanished hunters of temperate forests must have faced similar problems, except that much of the edible animal biomass would be at ground level rather than in the tree canopy. But although we have no accurate measurements of a modern kind, the proportions of NPP gathered by hunter–gatherers must everywhere have been less than 1 per cent and evidence of food preferences suggests that it may well have been much less.

Impact of management

Spatial and seasonal patterns

Regular movement of groups of people is characteristic of the hunting way of life. Some of the food collectors of the Cape York peninsula of Australia manage with a range of only a few hundred metres; the

Plate 3.4 In the remaining tropical forests of the world (as with these Auca people from Peru), hunting technology may still be largely unchanged: the blow-pipe directs the diffuse energy of face and chest and directs it down the narrow channel of the dart's flight. In this way animals of the forest canopy such as monkeys can be added to the diet. (The Hutchison Library)

Ainu sent out groups of men to kill deer at particular seasons; the !Kung all move up to 3500 km/yr on foot. This looks like a simple ecological adaptation to environment, determined by resource availability and resilience but social factors may also be involved. Getting away from the women (Hare Indians, Northwest Territories of Canada), death (Hadza, Tanzania), socializing (Inuit) and the search for a marriage partner are all documented and may, of course, reinforce the ecological interpretation. Movement has a demographic effect since effective child spacing will ensure that a woman has to carry only one infant when shifting camp, and combined with a low level of material possessions also means that archaeological visibility of past cultures is always likely to be poor.

Equilibria or extinction?

The fact that hunter–gatherer economies survived for such long periods is testimony to a man–nature relationship of a sustainable, though not necessarily static, kind. The question to be asked at this stage is: did their persistence derive from a care not to overexploit their biotic resources (whether this was a conscious perception or a behavioural trait clothed in ideological, ceremonial, magical or other folkways), or did it just happen, in the sense that although no particular 'tenderness' was involved, nevertheless the environment could withstand such pressure as was put upon it? The ethnographic and other cultural-ecological literature is rather spotty on this particular topic so it does not seem possible to give a complete picture for all groups even for near-recent times, let alone the past, though one summary attempt is shown in table 3.1. Some examples can be

Table 3.1 Conservation practices and resource variability in some hunter–gatherers

Groups ranked by resource diversity	*Resource diversity*	*Seasonality*	*Conservation practices*
Western Desert Aborigines	very low	pronounced	Consistent use of peripheral areas whenever possible; use of major waterholes last
G/wi Bushmen	very low	pronounced	Cannot kill more or gather more than needed without angering Supreme Being
Great Basin Numa	very low (13–38cm precipitation)	pronounced	Dislike using fire drives because brush requires many years to regenerate; antelope drives held only every 12 years to permit animals time to reproduce
Montagnais	very low	pronounced	Randomization of kill areas via divination practices, preventing overexploitation
Aranda	low–moderate	pronounced	Sacred nature of major sites precluded hunting within 1-mile radius and provided game refuges in time of environmental stress, thus preventing overexploitation; totemic prohibitions may have also provided differential refuge areas for particular species; compassion for animals
!Kung Bushmen	moderate	pronounced	Use of peripheral areas whenever rainfall permits; conservation of water by using seasonal sources rather than permanent ones when possible
Hadza	moderate–high (56cm precipitation)	pronounced	No vegetable or animal conservation; kill as much as possible
Tanaina	high (ecotone)	pronounced	Pregnant animals not killed, to conserve population levels
Tlingit	very high (ccotone) (10–200 cm precipitation)	pronounced	Competition over resources
Birhor	high	moderate	Allow areas to lie 'fallow' for 1–4 years; prefer not to join groups in order to conserve limited game; conscious of needing conservation for long-term productivity
Mbuti	high	negligible	Conscious effort to use every part of animal; never kill more than necessary for the day

Source: B. Hayden, 'Subsistence and ecological adaptations of modern hunter–gatherers'. In R. S. O. Harding and G. Teleki (eds), *Omnivorous Primates*. NY: Columbia UP, 1981, 344–421.

found of both tendencies: as an example of the 'awareness' mode, we may cite work on the Waswanipi Cree of the Boreal forest zone of Quebec in Canada in the 1960s. H. E. Teit suggests that the culture contains the thought that men have the means to kill too many animals but that it is irresponsible to do so. Beaver, moose and fish comprise the normal consumption in the winter hunting season and it is clear that there is a recognition that beaver and moose can easily be overhunted. This leads to rotational hunting: of 22 territories hunted in 1968–69 or in 1969–70 there were only six cases where the same territories were hunted over in both years, and the Cree recognized that catches were usually lower on a territory which had been hunted over in recent years. An efficient method of harvesting was achieved by reducing the animal population of a territory significantly and then culling the creatures characteristic of a fast-growing replacement population. The impact of northern hunters may also be reduced by preferences for individual animals on the basis of sex and age, and in cultural sharing practices which ensure that no meat is wasted from any one kill, diminishing the need to kill again soon. But the hunters do not aspire to control the reproductive or other habits of their prey-species.

In contrast, T. Ingold suggests that 'the rationality of conservation is totally alien to a predatory subsistence economy, which rests on the fundamental premise that holds one responsible for the existence of Man, rather than men . . . for the perpetuation of the herds.' This he amplifies by reference to traditional reindeer hunting in North Europe, where the principal hunting techniques are non-selective and indeed make for a maximum kill on encounter even if this results in considerable wastage during periods of animal abundance and when the meat is difficult to preserve in the summer months. A degree of similarity with some of the buffalo hunters of the Great Plains may be seen, though some of the kills there were to some extent selective because of the season at which they occurred, while others seemed to be totally unselective. For the Great Basin Shoshone, it is recorded that it took 8–10 years for the antelope populations to recover from one cooperative hunt. Perhaps the predictability of the success of a given hunt and the possibilities for storage are also factors in determining whether a 'conservationist' attitude is taken.

If we accept that groups of both kinds have existed in different places and at different times, then the resolution of the two attitudes in terms of the persistence of hunting and gathering is probably to be found in low population densities which even if 'non-conservationist' were unlikely to render an animal population extinct. Alternatively, perhaps the non-conservationists themselves became extinct from place to place as either they hunted out a prey species or that species cycled down in the normal course of events, causing starvation as postulated for the Noallian phase of the Upper Palaeolithic in south-west France. If men were absent from a region for a time then a new cycle of man–animal relationships would begin again and there might be little trace in the archaeological record of the blank phase.

One process which in recent times has clearly interfered with the environmental equilibria of hunter–gatherers is the ending of their relative isolation and the development of contacts with agricultural and industrial economies. At its extreme this process leads to the disappearance of hunters, but short of that it is clear from a number of examples that changed environmental relations result from contact with more energy-intensive systems. In North America, for example, Wishart has shown that contact between Indians and fur traders led to the devastation of certain animal populations: by 1831 the beaver was extirpated on the Great Plains, and it became extinct in the central Rockies in the 1820s and 1830s. By 1844 there were no mountain bison west of the Three Forks of the Missouri or on the plains of Yellowstone, since they were a major source of food for trappers and were killed by both Europeans and Indians for that purpose. A similar story is reported for the Ojibway of northern Ontario in the years 1810–29. In his study of fur trading policies pursued by the Hudson's Bay Company in western Canada, A. J. Ray records that by 1821, years of overhunting and overtrapping 'had left the trade of many areas on a precarious footing'. The Company then tried to introduce conservation measures as well as expansion into new areas. The former depended to some extent on moving the trading posts around the territory but also on shifting the burden of hunting from beaver to other species, including moratoria on beaver hunting in some places. The use of steel traps was also forbidden. Neither the Indians nor the trading post operators were happy with the attempts at conservation and Ray asserts that any idea of husbanding resources against the future was alien to the Indians of western Canada, particularly since they believed that they were powerless to affect animal numbers: these were determined supernaturally. Land too was a free good and any unexploited resources could be claimed on a first-come basis, so recovery period policies naturally failed. In 1843, the rules were changed, for the felt hat was replaced in fashionable circles by one of silk.

Even if selective killing within a species is not practical then it is possible to practise 'rational' management by adopting 'fallow' periods for areas of hunting territory: 'nursing' as the Hudson's Bay Company called it. Near-recent examples include Boreal forest people like the Waswapini Cree, the Nasakapi and the Ojibway. Another very good example is provided by the Koyukon Indians of the western Subarctic. Although they use modern technology, they have long seen that overharvesting of sedentary, localized biotic resources can reduce them to very low levels. Thus the Koyukon avoid waste and adopt sustained-yield practices; these appear to be grounded in a world-view in which the human and natural spheres are, according to R. K. Nelson's detailed and delightful account, 'tightly interwoven by threads of spiritual power.' But these examples are rarer than those in which severe depletion of indigenous resources has been recorded.

In the late 1950s the Mbuti of Zaire began to have increasing contact with commercial meat traders. These latter people promote intensified net hunting and monopolize the Mbuti trade in meat: in a period during 1974, twenty-two hunts yielded 398 kg of butchered meat, of which the traders acquired 47 per cent; on other days of rather poor hunting, 50–76 per cent of the meat went to the traders. One result of this new orientation seems to be a decline in certain prey species. Net hunting selects for adult rather than young antelope, for instance, and of 383 female duikers (*Cephalophus* spp.) captured during the study, 65 per cent were adults and 74 per cent of these were gravid. Overall hunting success decreased by 32 per cent over 3 years but there was the complicating factor of increasing agricultural settlement in the region. In more remote forest areas, the Mbuti reported that a day's hunting did not yield above 17 antelope, whereas in former times 25 was regarded as a minimum yield. The Zaire government has now imposed a ban on meat trading during the rainy season to protect the principal game animals during their breeding season. In a similar fashion, Neitschmann records the inroads made upon turtle populations by Miskito Indians in Nicaragua once a commercial demand became evident, and numerous other examples can be found.

In general, it appears that the ending of isolation for hunters meant that either their resources were greatly reduced in quantity or the effort to gain them greatly increased, or both, and that this led to demise or to the adaptations of agriculture or horticulture, or wage labour with consequent shifts in diet (figure 3.10). Stable though hunting and gathering can be, it does not seem to possess enough resilience to allow it to survive when in contact with modern energy-intensive economies.

Further examples of ecological impact

If we start with animals and animal populations, then we have to consider selective killing since this seems at once to have the potential of causing environmental impact (e.g. by concentrating on one species) and diminishing it by killing selectively within a species (e.g. by avoiding young animals and gravid females). The former circumstances are the most likely to lead to overhunting, but there is the intermediate condition of concentrating upon a particular species and drastically reducing its numbers rather than reducing the hunting presence so that the species itself responds as if it were colonizing the habitat afresh: there are likely to be large numbers of young animals, for example. Selectivity within species seems to be a relatively rare practice though not unknown: the buffalo kills were selective by virtue of their season, i.e. mature males were generally not killed. At Star Carr, an early Mesolithic site in northern England, the deer antlers were predominantly of males, suggesting a cull of males surplus to the breeding population. The Mbuti do not often kill young deer, because they slip through the mesh of the nets. But in general, the apparently rational practice of not killing the young of a

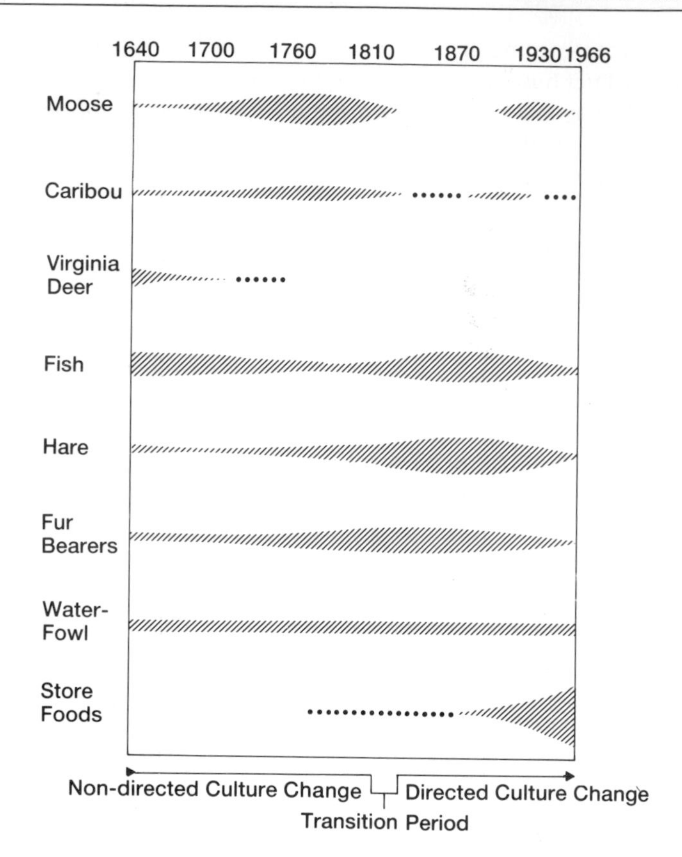

Figure 3.10 The food consumption pattern of a group of Indians in the northern forests of Canada over 326 years. Note first the recent changes brought about by the impact of the southern economy, and then the gradual shifts in earlier times, brought about by some combination of natural abundance and hunting pressure; the fur-bearers become more attractive to eat when their pelts can be sold as well. Mentally translate these patterns into impact upon the animal populations. *Source*: W. A. Bishop, *The Northern Ojibwa and the Fur Trade: an Historical and Ecological Study*. Toronto: Holt, Rhinehart and Winston, 1974.

species or gravid females has not been widely followed and foetuses are often a prized delicacy. The heavy use of large mammal herbivores may interact detrimentally with the apparently natural cycles of abundance of such populations. Animals like caribou and reindeer have natural cycles of numbers and presumably when the numbers are down then the chance of extirpating at least the regional population of that species is quite high. If there are alternative foods, then the human population may survive; if not then it must either migrate to more favourable areas or die of starvation. We may perhaps conclude that the heavy investment of time and effort (and possibly pride) which goes into hunting militates against selective killing and 'rational' management.

Considerable impacts may be exerted if hunters select a 'key species' as their prey and then diminish its numbers greatly. A key species is one whose presence determines the character of the rest of the ecosystem and an example can be quoted from the Aleutian Islands of Alaska in the period 500 BC–AD 1200. The species in this case was the sea-otter, hunted to near-extinction in some places but not, it appears from excavations, in others. Where the sea-otter continued to exist, the ecosystem on the left of figure 3.11 was found; where it was removed, the different system of the right of the diagram can be reconstructed. (The centre shows a food chain apparently unaffected by these changes.) Among the changes, the diminution of

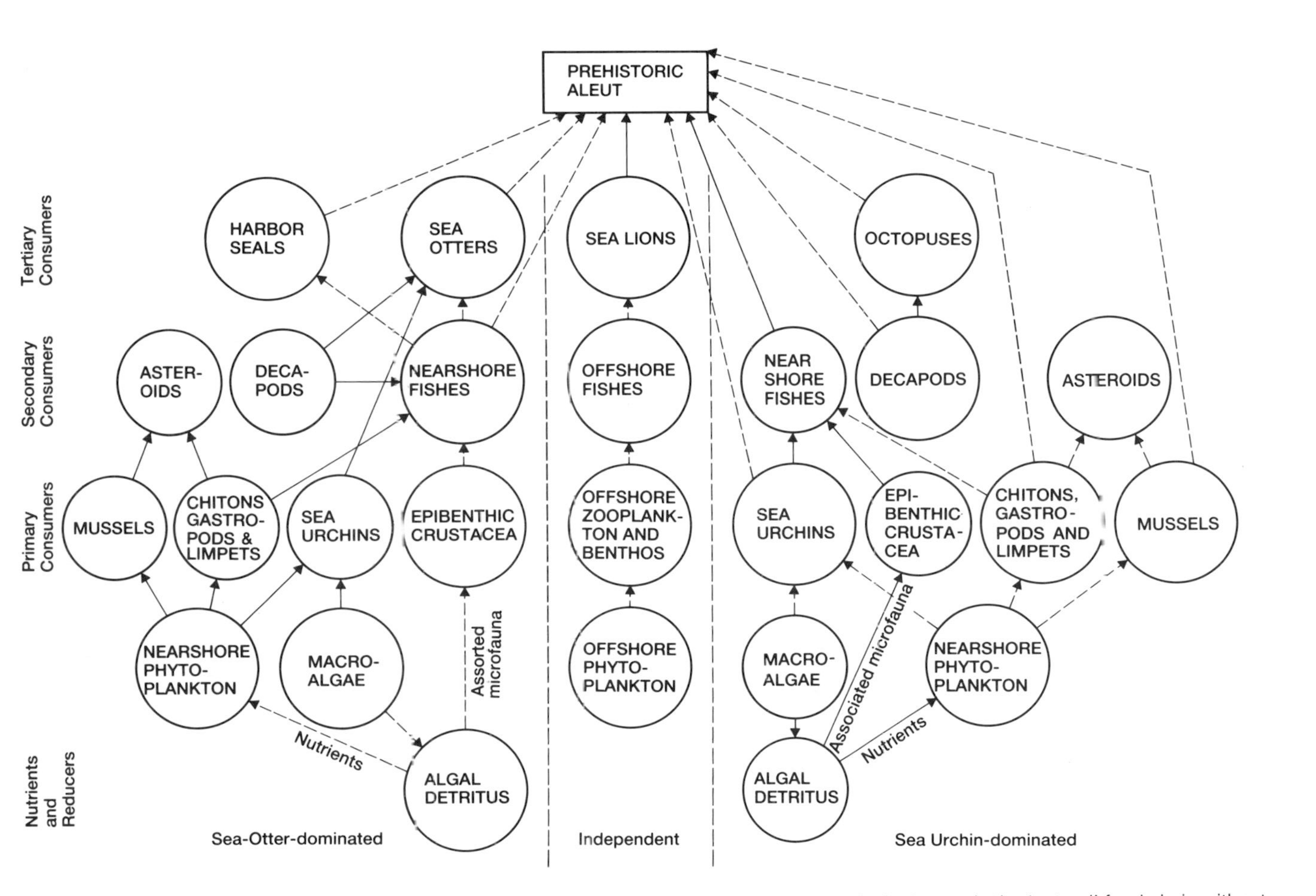

Figure 3.11 The effect of the Aleuts on the principal components of the offshore ecosystem. In the centre is the 'natural' food chain without humans; to the left the relative biomass (size of circles) of a community with sea-otters in it and on the right the effect of hunting down the sea-otter population. In this case the sea-urchin springs into abundance. Heavy lines indicate the principal interactions which vary between the two major states of the system.
Source: C. A. Simenstad, J. A. Estes and K. W. Kenyon, 'Aleuts, sea otters, and alternate stable-state communities', *Science* 200, 1978, 408–11.

harbor seals and fish paralleled by increases in octopus, mussels and sea-urchins reflects the key role of the sea-otter which was a predator of herbivores like the sea-urchin. Dense populations of sea-otters keep sea-urchin populations sparse. This in turn allows the growth of vigorous beds of kelp with associated fish, bird and mammal faunas. If the sea-otters go, then sea-urchins flourish, eat a lot of kelp and diminish the quantity of associated fauna, while allowing more limpets and mussels to survive on exposed rocks.

For the Pleistocene, the outstanding example of environmental impact in terms of animal populations is the phenomenon known as 'Pleistocene overkill'. This term is applied to the history of large (adult weight 50 kg or more) warm-blooded animals (usually mammals but with some birds as well) at the end of the Pleistocene, when some 200 genera of them (the so-called 'megafauna') became extinct (table 3.2). The phenomenon is perhaps most closely observable in North America: here, two-thirds of the large mammal fauna present at the end of the Pleistocene (*c.* 13 000 BC) then disappeared, including 3 genera of elephants, 6 genera of giant edentates (armadillos, ant-eaters, pangolins and sloths), 15 genera of ungulates, and various giant rodents and carnivores. There is no evidence of such extinctions in earlier periods which might have been expected if Pleistocene climatic changes had been the proximate cause, nor any firm evidence of survival of these genera past the 500-year period during which the extinctions appear to have taken place. These data have led to the hypothesis that it was the introduction of man as a predator against which these mammals had no genetically implanted defence behaviour which was the cause of extinctions. In North America, the date of introduction of man via the Bering Strait land bridge is usually taken as being about 12 000BP, and a simulation of the extinction pattern supposes a wave-front advance by hunters whose population periodically exploded because of the new and favourable habitat (figure 3.12). At the advancing front, a human population growth rate of 2.4 per cent p.a. may have given a density of 0.4 persons/km^2 and a forward movement (due to hunting-out of the large mammals) of 16 km/yr. Thus the front could sweep from Canada to Mexico in 350 years and extirpate the large mammals

Table 3.2 Late Pleistocene extinct and living genera of terrestrial megafauna: four continents

	Extinct in last 100000 yr	*Living*	*Total*	*extinct (%)*
Africa	7	42	49	14
North America	33	12	45	73
South America	46	12	58	80
Australia	19	3	22	86

Source: P. S. Martin, 'Prehistoric overkill: the global model'. In P. S. Martin and R. G. Klein (eds), *Quaternary Extinctions: A Prehistoric Revolution*. Tucson: University of Arizona Press, 1984, 354–403.

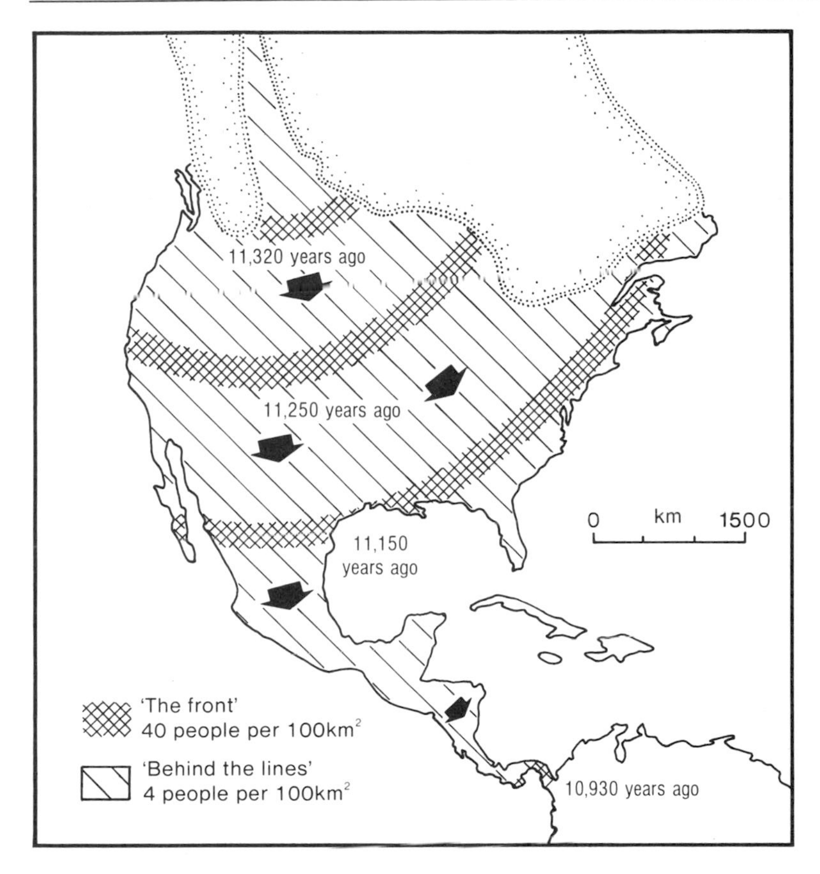

Figure 3.12 A hypothesis for 'Pleistocene overkill' sweeping through the North American continent, with the initial colonization occurring from the north. *Source*: P. S. Martin, 'The discovery of America', *Science* 179, 1973, 969–74.

through superior predation techniques. Some support is given to this hypothesis by the effects of the arrival of man into other hitherto unpeopled places. The disappearance of the moa bird from New Zealand occurred within a few hundred years of the human occupance of the islands; the megafauna (including a large terrestrial bird *Aepyornis*, and a pygmy hippopotamus) of Madagascar disappeared within a similar period after the first human occupation in AD 1000; in Java and Sulawesi, populations of dwarf elephants likewise did not long survive the coming of man. It would appear, therefore, that when introduced into a new habitat man the hunter is capable of rapidly exterminating flightless birds and large mammals.

The end of the Pleistocene in the Old World presents a somewhat different picture. In Eurasia, only a total of 9 species were made extinct, compared with the New World total of at least 24 genera. The Eurasian losses comprised the woolly mammoth (*Mammuthus primigenius*), woolly rhinoceros (*Coelodonta antiquitatis*), giant Irish elk (*Megaloceros giganteus*), musk-ox (*Ovibos moschatus*), steppe-bison (*Bison priscus*), a buffalo of northern Africa (*Homoioceros antiquus*) and three species of associated carnivores. All except the buffalo were animals of the cold steppe which was widespread outside the ice-sheets of Europe during the last glacial phase, and all the herbivores, especially the mammoth and bison, had been hunted for tens of thousands of years yet survived in large numbers until about 12 000–10 800 BP. But this time-horizon does not mark the introduction of man to Eurasia, nor even a particularly great cultural change, although in very general terms it is the period of the Upper Palaeolithic which saw the culmination of the cultural and technological 'explosion' in the Magdalenian. It appears much more likely that climatic change leading to forest growth of birch and pine was the main cause of the animals' disappearance. Of course, continued hunting when the population was under environmental stress would have hastened their demise but the change in habitat appears to be the primary reason. At any rate, when their habitats returned briefly in the cold period of 10 800–10 150 BP, the animals did not. Some other species of the cold steppe proved more adaptable: the reindeer persisted on the tundra, and both the saiga antelope (*Saiga tartarica*) and wild horse (*Equus przewalskii*) took to the grassy steppes of central Eurasia. Comparable with Eurasia is a site in New South Wales, Australia, which appears to show 7000 years of coexistence of man and megafauna without any evidence that the human societies were the cause of the eventual extinction of the other animals.

It is also necessary to consider the possibility that the extinction of megafauna was a secondary rather than a direct effect of human activity. If hunters enter an area and specialize on one species of herbivore, then the local carnivores which also concentrate on the same prey will be starved out during winter competition and become extinct. The prey species then expands in numbers and territory and puts pressure upon the numbers of other herbivore species, in turn causing other carnivores to become extinct. This kind of idea, coupled with the failure to disprove the primacy of climatic change as the primary cause of the extinctions, means that acceptance of human activity as the most important cause of these Pleistocene extinctions is far from complete. Nevertheless, the sudden onset of extinctions in areas like North America, Australia, New Zealand (plate 3.5) and Madagascar, the different times of the onset of the phenomenon between areas, and its correlation with either the advent of man or a change in technology (figure 3.13) keep the idea very much in play. The slow loss of fauna from Africa could be explained by the gradual evolution of humans on that continent rather than their sudden arrival and explosive spread.

Modern accounts of the relationship of hunters with animal populations do not feature attempts to feed the animals at, for example, bad times or in order to bring them nearer the settlements so as to diminish hunting effort, but in European prehistory we have the evidence of large quantities of ivy pollen being found at occupation sites during the Mesolithic. Ivy flowers during the late autumn and the pollen records have been interpreted as the gathering of ivy for red deer in order to provide supplementary feeding at a particular place. Whether the deer were wild or whether they were to some extent a 'herded' population, we do not know. Such a practice, though, does presuppose 'ownership' of animals or territory, and both were no

Plate 3.5 The Polynesian inhabitants of New Zealand are held to be responsible for over-hunting the Moa bird and thus leading to its extinction. Large flightless birds are clearly vulnerable to such a fate: elsewhere the Dodo and Great Auk, for example, were also extirpated. (The Mansell Collection)

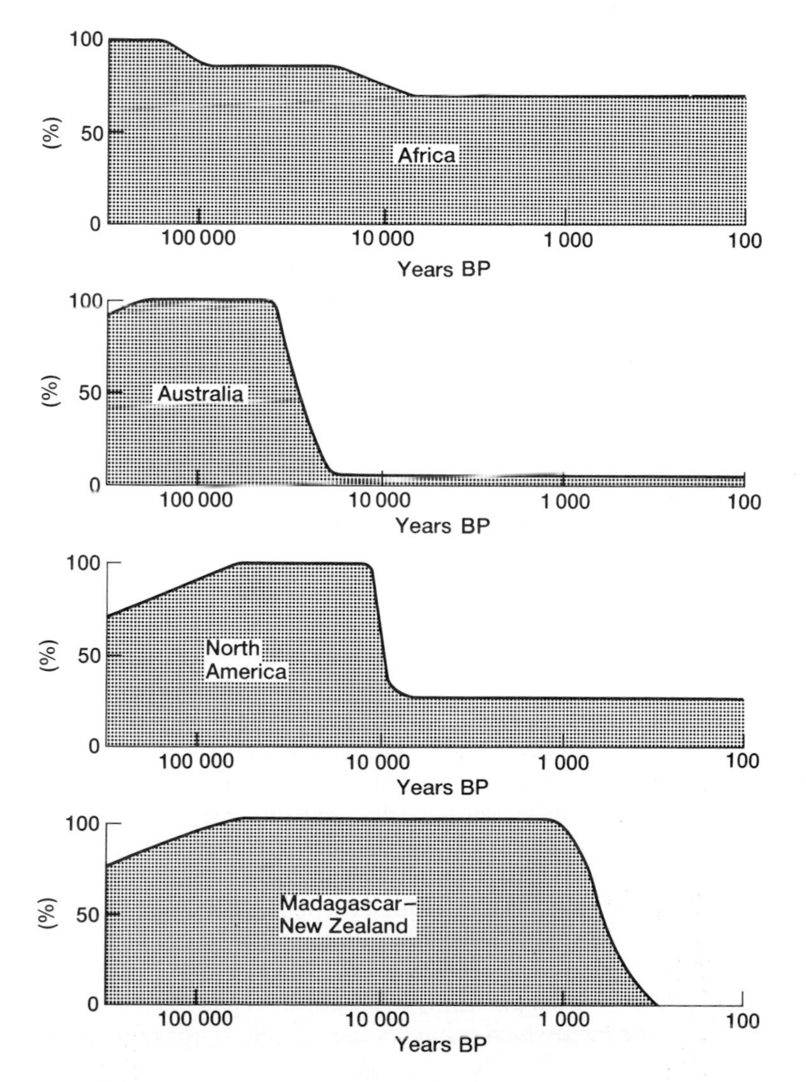

Figure 3.13 Percentage survival of large animals in three continents and two large islands during the later Pleistocene and the Holocene; except in Africa the sharp loss follows cultural development, especially the advent of the species *Homo sapiens*.
Source: P. S. Martin, 'Historic extinctions: a Rosetta stone for understanding prehistoric extinctions', in P. S. Martin and R. G. Klein (eds) *Quaternary Extinctions. A Prehistoric Model*. Tucson: University of Arizona Press, 1984, 354–403.

doubt important social elements in any progress to domestication, though this never happened to red deer. A further step in this direction seems to be provided by the osteological evidence for crib-biting in horses' teeth from the Middle and Upper Palaeolithic of France. Crib-biting is known today as a 'stable vice' resulting from boredom and confinement, and it wears down the incisor teeth in a particular way. Its presence in an archaeological context may constitute evidence of tethering, and if so then presumably selected horses were being removed from the wild herds or perhaps foals were brought up in captivity.

In parallel with impact upon animal populations, selective gathering of plants must be considered although examples of how the life history of a particular plant species or the local ecosystem might be affected are not easy to find. The location of a favoured plant may be changed if it is found possible to grow it close to a settlement or in some place safe from rival herbivores, and we can imagine the sort of replacement of natural selection which would eventually lead to domestication: wild yams are planted on offshore islands in the Cape York peninsula of Australia because fewer animals are likely to disturb them there. If gathering practices spill over into collection beyond the reproductive capacity of the species then the plant may locally become extinct, though probably not so on any wider scale. If a species is easily extirpated in one place then perhaps it is unlikely to have been an important element in the human diet and, as with animals, it is perhaps most probably eliminated at a time of a population trough. On the positive side, tending wild plant crops is easy to imagine, though recorded instances of it seem rather rare: the Kwakitul Indians of the north-west coast of North America apparently looked after beds of clover and cinquefoil. Ethnobotanists tend to think of the weeding of a plant crop or the diversion of a small channel to ensure a water supply as being husbandry which stops short of the genetic selection which constitutes domestication.

It is also possible that some species of wild grasses with useful seed heads would be favoured if their competitors were eliminated by fire. The history of early man–fire relationships and the ecological effects of fire have been discussed in chapter 2 and that account will not be repeated here. But especially relevant here is the realization that fire has been used in, or is a component of, most of the major world biomes. Only those ecosystems which remain damp throughout the year are likely to be fire-proof to humans: habitats which have any dry season (e.g. the margins of the tropical forests, deciduous temperate woodland at its margins in a dry spell during spring or autumn, wetland communities which seasonally dry out) are likely to experience fire at the hand of man, sometimes no doubt accidentally. As far as hunters are concerned, especially in wooded landscapes, fire has the advantage that it raises the value and nutritional quality of food for ungulates by a factor of ten, since it increases the quality of browse as well as improves the protein content of leaves (via the mechanism of mineralizing litter which enhances the nutrient status of the soils), so opening the way for an increased density of animals either on an absolute basis or as a concentration at a chosen place. The ability to affect the movement of animals by attracting them to food sources or by making them flee in particular directions (as with some of the Plains buffalo hunts) must have been a major positive contribution to the energetics of hunting, since it reduces by many times the amount of effort that has to go into harvesting: human energy is replaced by the release of energy stored in dead or otherwise combustible plants. It seems likely, too, that the use of fire reduces uncertainty in the hunt: used as a catalyst of improved forage, this is obvious. In other contexts, it appears that burning to reduce the shrub layer of a forest improves sight-lines and hence kill-frequency, and if a large area is burned, something edible is almost bound to flee the fire in the direction of a party of hunters.

Near-recent examples of the utility of fire to hunters and gatherers are plentiful but perhaps the most notable concentration is in the Antipodes. In the Cape York peninsula of Australia, D. R. Harris reports the usefulness of large stands of cycad trees (*Cycas media*) which yield some 131 kcal/m^2/yr from edible kernels, equal to some cultivated crops. He suggests that these stands are a consequence of firing the area repeatedly, thus encouraging the fire-resistant cycad at the expense of other trees. The cycad *Macrozamia*, if fired, produces all its seeds simultaneously and at 7–8 times the unfired quantity. Further into the interior, the indigenous population managed fire for hunting, land clearance and communications, with a total of 5000 bush fires per year being estimated for the years of early European occupance. They said that fire was used to 'clean up' a landscape (rather as we might – some of us – tidy up a room) and it created a patchy array of habitats with greater floral and faunal diversity than had previously existed. Many of the biota thus both encouraged could be eaten and some of the animals easily flushed out by the flames: long-necked turtle and goanna lizard are still hunted in this way. At the southern extremity of Australia, the climax forest of western Tasmania is *Nothofagus* (southern beech) forest but a high fire frequency converts this to sclerophyll forest and thence (if fire is continually applied) via scrub to fern heath or tussock grassland (figure 3.14). The fire enhances the growth and spread of edible plants like the bracken fern *Pteridium esculentum* which colonized newly burned areas and became a carbohydrate staple; the patchy environment encouraged wallabies, bandicoots and possums, all of which were edible. In New Zealand, the bracken fern was the starchy staple of the Maori and firing was used regularly to encourage its growth and spread. In the period AD 1000–1400, such burning was extended to the podocarp forests which covered most of the eastern half of South Island and much of this forest land became grassland and scrub. Along with the forests and with the first arrival of man, many animal species became extinct: six genera containing 20 species of moa bird became extinct within 500 years of man's arrival, along with 20 other species of bird, and the extinction of

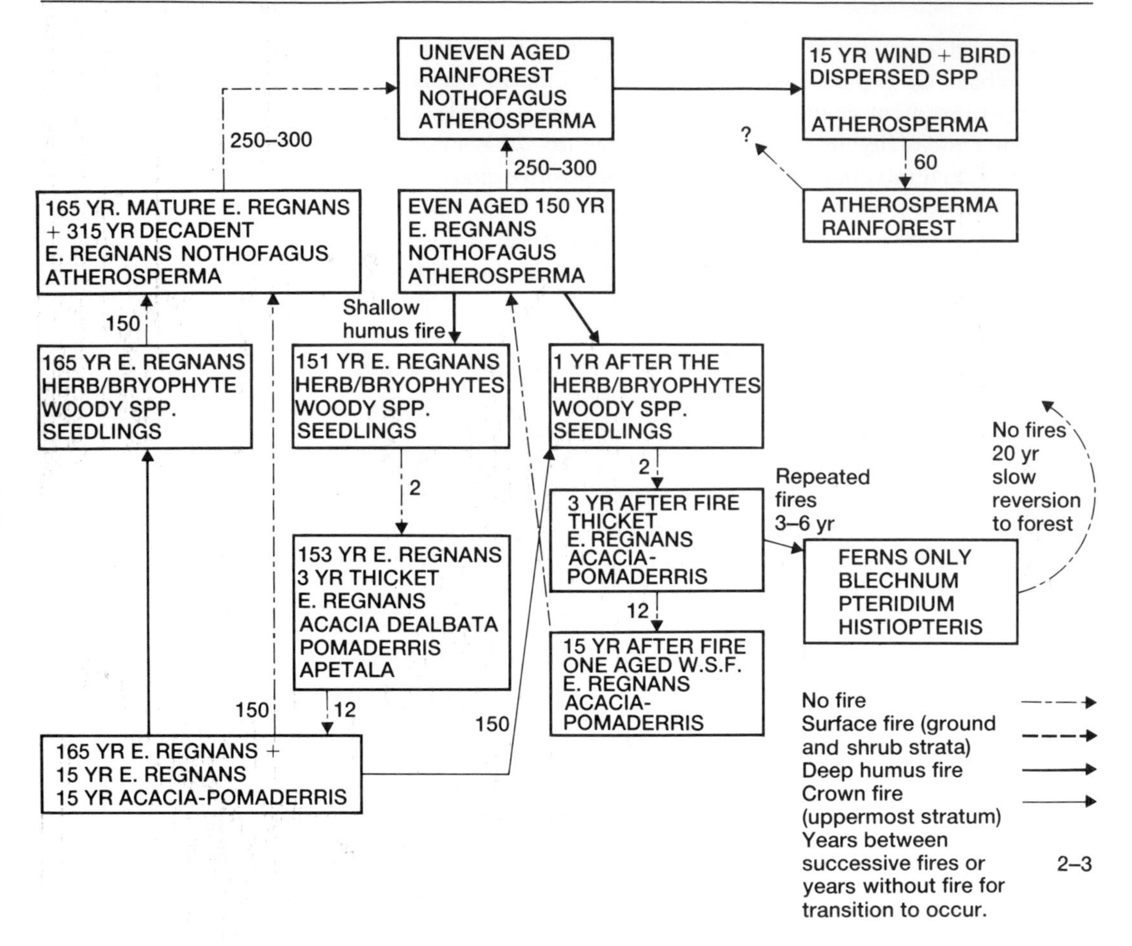

Figure 3.14 Fire in ecosystems is not a simple influence. This diagram charts some of the vegetation pathways following fires of different frequencies and types (surface, humus, crown) in Southern Tasmania. *Source*: A. M. Gill *et al*. (eds) *Fire and the Australian Biota*. Canberra: Australian Academy of Science, 1981.

elephant and fur seals except in remote areas. So predation and environmental change brought about this analogy to 'Pleistocene overkill' in remote New Zealand.

Similar examples of fire's producing habitat change and eventually demises in animal species can be produced for most places outside the wet tropics. Historical examples are postulated from many palaeological investigations, so from the second interglacial to the coming of Captain Cook (and after), we can find examples of hunter–gatherers using fire to alter their environment and we can have little doubt that it was in using this tool more than any other that they exerted any impact upon their surroundings. Indeed, the antiquity of the relationship between men, fire and biota is such that it is possible that some major biomes may be the product of co-evolutionary relationships between these components. Thus the physiognomy and trophic relations (although not, probably, the genetic evolution) of places like the savannas of Africa, or the High Plains of North America, might be the result of thousands of years of occupance by fire-brandishing men.

As for the manipulation of water, we have evidence from south-west Victoria during the period 40 000–7000 BP of artificial drainage of swamps, with channels about 3 km long. These appear to have been management systems for the procurement of eels. The channels were more than harvesting devices since they kept water in swamps during drought and hence kept up the supply of eels in what was in fact a marginal part of their range. Hence, a high density of semi-sedentary hunter–gatherer people, equal to that attained under shifting agriculture in New Guinea, was achieved by means of a broad-based subsistence base supplemented with the products of water control.

If vegetation is affected by subsistence hunters, then we would expect soils likewise to be changed. The evidence for this is generally found in connection with prehistoric rather than near-recent cultures. Following repeated fire and disforestation, we might expect the loss of soil and perhaps regolith from sloping areas and its downhill and downstream aggregation as colluvial and alluvial deposits. It is very difficult to tie such deposits to hunter–gatherer populations, but we may note in passing recent dating of sediments in river valleys in some parts of England which suggests that they were the product of openings in the largely closed deciduous forests of the Mesolithic period of those regions. Aboriginal burning in the mountainous areas of Tasmania and Eastern Australia is thought to have led to soil loss, and it has been further postulated that in the interior of Australia the Aborigines' firing practices may have allowed the remobilization of sand dunes which had had a phase of being fixed by vegetation. It has also been suggested that Late Mesolithic (7000–5000 BC) forest burning in the English uplands allowed blanket peat to accumulate on areas of low slope. When forest was present, the trees acted as water pumps; when they had gone, the soils became waterlogged and eventually up to several metres of peat accumulated.

The population, resources and environment of hunter–gatherers

The first point to be emphasized, as elsewhere in this chapter, is that the population density of hunter–gatherers has always been very low, with only a sprinkling of exceptions in either very rich environments or where some manipulation of the animal population was practised. Calculations suggest that over the world, an average of 26 km^2 has been needed to support each hunter and that the population density of the ecumene in Lower Palaeolithic times was 0.025 persons/km^2 (reflecting a total population of 8.0×10^6 people), rising to 0.1 pers/km^2 in the Upper Palaeolithic (total population 6×10^6 people) and 0.115 pers/km^2 in the Mesolithic just before the onset of agriculture ($0.8–9.0 \times 10^6$ population). That this low density was one of the main causes of the lack of intensive environmental manipulation cannot be doubted, but we cannot assume that it renders unnecessary a 'conservationist' attitude to environmental resources.

It is generally thought that hunter–gatherers lived at low densities because the carrying capacity of their various environments was limited. Carrying capacity is often defined as the maximum ability of an environment continuously to provide subsistence at the level of culture practised by the inhabitants. It is a difficult concept to apply, especially in a historical context, because the terms are difficult to define accurately and both environment and culture are liable to change through time. Yet there is an orderly relationship between territorial area and the maximum number of people which can be supported at a certain level of food-collecting efficiency. Most hunter–gatherer populations are observed to exist in numbers below their objective carrying capacity, maintaining their size at 20–60 per cent of the maximum size possible. This appears to be a successful response to periodic and unpredictable fluctuations in the yield of the utilized resources, since a population of optimum rather than maximum size can offset the occasional periods of stress by exploiting new resource areas, expanding the food base to include species or items hitherto ignored or unselected, or by intensifying the collection of food. Such periods of stress may include the rapid fluctuations which are characteristic of small populations, which may show a four-fold difference between maximum and minimum sizes over a 200 year period. Also relevant is 'environmental stochasticity' which happens simultaneously on several time scales, as with seasonal variations within each year, and to which must be added natural catastrophies which may come about only once every few generations. Newly discovered interactions may also complicate the picture: Calvin Martin explores the possibility that zoonotic diseases (i.e. those endemic to wild animal populations but transmissible to man), such as sylvatic plague and tularemia, were be passed to Amerindian populations and produced in some places a 40 per cent mortality among those who contracted them.

A key element in the low density and low impact of so many hunter–gatherers seems to have been population control by the human groups. As well as the need to confine the demand for resources and to keep a wide margin against times of scarcity, there appears to be the social fact of it being disadvantageous to societies which often moved camp for the women to have to carry more than one offspring. Convenience therefore dictated that one child should be able to walk a day's march before any more were born. There is also the extra workload imposed on a mother by each of her children, which is especially heavy in their first few years, and the relative unavailability to many hunting people of the right kind of soft foods to enable the infants to be weaned at less than two or three years. Taken all together, it looks as if the normal birth interval among now-sedentary hunter–gatherers is around four years, but not all scholars would subscribe to the idea of population control of a conscious kind. Nevertheless, using inferences from recent hunter–gatherers as well as archaeological evidence, W. T. Divale has postulated that a system of population control had first evolved by the Middle Palaeolithic, and that by the Upper Palaeolithic (*c.* 35 Kyr) it was effective. The means used are not always entirely clear since no doubt it has never been a topic to be too fully discussed with visiting anthropologists, let alone missionaries, but infanticide (especially female), abortion, sexual abstinence and prolonged lactation all seem to have been used from time to time and place to place: J. B. Birdsell has argued that systematic infanticide involved 15–50 per cent of the total births. Yet there would always have been a good potential for population expansion if this was needed after a disease epidemic or starvation: a net reproduction rate of 2.0 could probably be achieved, doubling the population of a group in a single generation. But the growth rate in the human population in its first 1–2 million years of existence seems to have been very low: 0.0007 to 0.0015 per cent per annum seems to cover the ranges experienced in the period up to about the end of the Pleistocene. Some authors have suggested that there was a sudden increase in population growth rates and densities at that time; J. L. Angel has gone so far as to talk of a 'population avalanche' at the end of the Upper Palaeolithic, with the great herds of bison, horse and reindeer as the prime food resource. Other writers are more cautious: F. A. Hassan, for example, does not think 'that there was a rapid population increase during the terminal Pleistocene leading to abnormally high densities.' It is common ground, however, that the end of the Pleistocene (i.e. late-glacial and early post-glacial times in central and northern Eurasia and in North America) was a time of rapid oscillations and other changes of climate, leading to considerable fluctuations in the availability of food resources, and thus perhaps sparking off further population control, migration and the different cultural adaptations characteristic of the Upper Palaeolithic. Such adaptations may well have included the re-valuation of wild grasses which led eventually to cereal-based agriculture. If we accept that the Upper Palaeolithic could have been a time when population was pushing hard against its resource base then a number of adaptations could have been made, evidence for all of which can be found in the Solutrean culture of the European Upper Palaeolithic. Examples quoted by P. E. L. Smith include the search for food further away from base camps, more mobile camps, smaller groups of people with fewer large seasonal clusterings, more emphasis on fish, molluscs, seeds, technological aids, the further development of techniques of food preservation and storage, and more intergroup conflict.

If we combine environmental and demographic stochasticity, then it is not remarkable that hunters could maintain (as M. D. Sahlins put it) 'affluence without abundance', provided that they kept moving and that they under used the objective economic potential of their environment. In order to achieve a sustainable way of life they had to be good energy managers and we have seen how some technological development has helped. The concentration of muscular energy through the bow, slingstick and blowpipe are good examples, and the use of the dog as an easy-to-feed but wide-ranging sniffer and tracker is another. The use of fire is clearly of great significance: it is widespread spatially and temporally and has so many virtues both in the camp (including the preservation of food against lean periods) and in the wider landscape that to call its possession and control 'revolutionary' is no exaggeration, as it represents the addition of cultural to metabolic energy. So although hunters were never dense on the ground for very long, and although their technology was by subsequent standards limited, they were able to change their surroundings. Some of this change may have been the product of periods of stress rather than of long episodes of stability. The outcome, however, includes the alteration of many ecological parameters, from time to time and place to place (table 3.3). Domestication (apart from

Table 3.3 Major ecosystem characteristics as affected by hunter–gatherers: a summary

Characteristic	*Effect of hunter–gatherers*
Energy flow	Little deflected except where manipulation severe, i.e. only a few instances as when (1) fire used regularly or (2) animal populations severely reduced or extirpated. Proportion of solar energy as net human productivity very low in all systems
Nutrient cycling	Rarely affected unless dominant plants changed (e.g. by some form of forest clearance), or unless fire used regularly. In latter case some nutrients (esp. N_2) lost in smoke; in former, loss to runoff may be increased. But vast majority of hunter–gatherer–fishers through time did not affect cycling to any degree permanently, though some nutrient stores created in the form of middens near settlements
Biological productivity	Affected by regular use of fire: growth rates of selected plants may become higher (nutrient release from ash, less competition) and this may knock-on into animal populations. Otherwise, influence of man seems to be rare except in very limited circumstances
Population dynamics	Some evidence of selectivity of cull in order to maintain reproductive capacity of animal population; some evidence of limitation of cropping of animals and plants so as to maintain viable populations. Some but disputed evidence of complete extermination of genera in terminal Pleistocene, presumably of population under climatic stress; good evidence for megafaunal extinction by first human populations in e.g. New Zealand, Madagascar
Successional stages	Fire as a tool in promoting retention of earlier successional stages, but eventually adaptation of biota produces permanently deflected succession ('fire climax'). Some plants of successional stages thus held may provide positive feedback in form of allelopathic reactions
Biotic diversity	See also population dynamics, above. Some new habitats created at settlement sites, middens etc. which may allow species such as ruderals to colonize from other unstable areas. Such places often temporary, though longer where HGFs maintain a more or less permanent settlement (e.g. West Coast Indians; Ainu). No proper domestication except of dog, where breeds presumably begin to differentiate
Stability	Recorded instances of inability of ecosystem to recover from human impact very few: if ecosystem destabilized then human population likely to suffer some fate. Is possible that repeated fire changed biotic composition of ecosystems during periods of allochthonous change, e.g. after the Pleistocene. Fire may also change vegetation/soil irreversibly
Degree of modification	Remarkably spotty: overall picture through time and space is one of relatively little modification of ecosystems on anything other than a short-term (*c.* 2 years, perhaps) basis, partly through conscious 'conservationist' attitudes but mostly because of low population densities. In certain instances, evidence of long-term modification of ecosystems as a whole (and not just the plant or animal components) is found, but they seem remarkably few. Is this a real phenomenon or an artefact of the type of research done on near-recent groups and of the archaeological visibility of those of the past?

the dog) is by definition missing from the changes brought about but even that may have sometimes been present in an incipient form. Obliteration is still unlikely beyond a settlement site but system simplification and diversification in their turn must have been common occurrences as habitats were modified. So although hunting was sustained for many millennia it was not part of a mythical golden age when humans took the usufruct of nature without altering the natural world at all. Such a time probably never existed after the evolution of *Homo erectus* and certainly not after the Upper Palaeolithic.

CHAPTER FOUR

Agriculture and its Impact

Girls planting paddy –
Only their song
Clear of the mud
haiku

Introduction

The aim of this chapter is to trace the environmental impact of human activities during the phase of history when the economic base was agricultural. The evolution and impact of agriculture itself are therefore crucial but we must not ignore related changes in forests, at the margins of the sea, and for the purposes of pleasure. Pre-industrial agriculture has a cropping system which relies on recently fixed solar energy, but society as a whole has access to older energy in the form of wood or peat and in this phase wind and water power are added to the technological repertoire. These were used in the service of agriculture as well as that of industry (Plate 4.1): the windmill, the water-mill (and even the tide mill) are most familiar as rural symbols. Agriculture especially, but most of the related systems of production

Plate 4.1 Prehistoric flint mines at Grimes Graves in Norfolk, England. The sites of shafts and the heaps of refuse indicate the heavy environmental impact of mining even in pre-industrial times; it would have been more obvious still when the heaps of debris were fresh since they were largely of white chalk. (Cambridge University Collection of Air Photographs)

as well, meant the replacement of natural or near-natural (or apparently natural since the impacts of hunters were not always recognized) ecosystems with man-directed systems. Most of the measurable parameters, like energy flow, mineral cycling, biological productivity, population dynamics of component species, species diversity and stability, are altered under these regimes, and when an area is transformed to cultivation the term 'agroecosystem' is appropriate (table 4.1). In energetic terms, many food-producing systems

Table 4.1 Major ecosystem characteristics as affected by cultivators: a summary

Characteristic	*Effect of cultivators: Permanent (P) Shifting (S)*
Energy flow	*S* cultivators interrupt normal flows for finite period then allow successional flows to resume; but long-term vegetational change (e.g. forest to savanna) may produce permanent changes in flows. *P* cultivators eliminate many of natural flows and direct new ones to themselves. Quantity of energy going to animals is small in any system, so vegetarians get more nutrition per unit area: many traditional cultivators are more or les vegetarian with animal protein likely to come from fish
Nutrient cycling	Natural cycles often held 'tight' by vegetation–soil complex; agriculture usually increases potential for loss through runofff, soil loss and (to a lesser extent) through cropping. Most systems aim at replacing the nutrients, e.g. by forest fallow or bush fallow (*S*), fallowing as part of a rotative field system (*P*), manuring and/or use of night soil (*P*), folding of animals onto fields for manure (*P*), leading of silty water onto fields (*P*), and *padi* construction with regulated water supply containing nutrients (*P*)
Biological productivity	Global average for agriculture is 650 $g/m^2/yr$ (current figure, so traditional agriculture presumably much less), which is same as a natural temperate grassland. Thus NPP is generally lower than that of natural vegetation which it replaces: smaller plants, growth season shorter, more space between plants are some of reasons. But culturally acceptable crop is usually higher from same patch of land
Population dynamics (Non-human populations)	Selective culling/extirpation of culturally undesirable species, labelled 'weeds' and 'pests', e.g. non-edible plants for predators upon crop animal species. Crop species encouraged nutritionally and genetics gradually altered by selective breeding. Crop may provide favourable expansion for hitherto insignificant species to expand to 'weed' or 'pest' status
Successional stages	*P* agriculture is analogous to early successional stage with much bare ground, and tillage normally aims at keeping it that way. *S* agriculture may (especially in wet tropics) mimic tropical forest and aim to cover ground with growing plants or a 'mini-forest'. Abandonment of plots of either allows natural secondary vegetation to develop; this may colonize more or less to original vegetation if left long enough

have through history produced a surplus which allowed activities not related to subsistence, such as the growth of cities and of industry, and the pursuit of war by professional warriors. These and many other occupational specialisms derived from the surpluses of energy provided by the shift from food collecting to food production.

There are still parts of the world where the effects of industrialization are few and where the ecology and economy are still solar-based. Such areas are sometimes labelled 'traditional' or 'pre-industrial' and studies of them can be used, with caution, to illuminate the past. Essentially, though, some areas of developing countries and a few enclaves in the developed are still 'agriculturalist' rather than 'industrialist' in terms of a four-phase division of history, so the present tense can occasionally be used in this chapter.

The evolution of agriculture

The Palaeolithic legacy

Because of the nature of the evidence, agriculture tends to show up in the prehistoric record when it is well developed, as when distinctive remains of fully domesticated varieties of plants and animals can be distinguished in archaeological contexts. Unless agriculture was virtually an 'overnight' invention, then we must look for foreshadowings of it among the hunter–gatherers.

Possibly the outstanding example is the collection of wild grains in western Asia during the terminal Pleistocene and early Holocene by hunter–gatherers. It seems likely that the wild ancestors of, for instance, wheat and barley immigrated into that region in about 11 000 BP, and that they were subject to systematic exploitation by human groups who gathered the seed heads of the plants. These same people may have diverted streams to bring water to plants, and could have undertaken some elimination of competitive plants and perhaps even scared away herbivorous animals. But the people (who were in any case nomadic) went to the plants rather than brought the crops to the settlements and these forms of elementary husbandry probably did not alter the genetics of the plant population. What is postulated for wheat and barley in south-west Asia may also be true for rice in east Asia where it may have been part of an analogous grass flora placed between the very arid core of Asia and wetter forest land further south.

The equivalent for animals may have been the practice of feeding wild individuals during periods of difficult weather and thus getting them accustomed to human presence. There may also have been a long period in south-west Asia in which populations of sheep, goats and cattle still bred in the wild but were loosely controlled through the provision of salt. Another move towards taming would have been the adoption of the young of wild animals as pets, bringing them up within the seasonal round of hunter–gatherers. Agriculturalists had in front of them the example of the dog, which was domesticated from various species of wolf, and its value must have set the tone for a new relationship with animals. Not all such interactions will have led to similar results and the Upper Palaeolithic can be regarded as a time of sifting. Husbandry of a number of plants and animals took place, and out of this process a few emerged as candidates for full-scale adoption by human societies, with all the genetic and ecosystemic changes which then followed.

The emergence of domesticates

How do agriculture and domestication differ from mere husbandry? Domestication implies that human societies have altered the genetic composition of plants and animals by influencing their breeding so as to make them better fitted to the needs and demands of human societies than they were under natural conditions; eventually the biota are adapted exclusively to man-made surroundings and would not survive in the wild. From another viewpoint, agriculture is the result of the transition from food collection to food production, where the food is no longer gathered from wild ecosystems but is produced from deliberately managed systems. In the new system, the annual offtake of energy as edible food per unit area is much increased. This more concentrated production in space is usually accompanied by a seasonal peaking of production, so techniques of storage become essential. If successful, then an energy surplus is produced which can feed more, who need not themselves be food-producers, and the stage is set for the development of societies whose individuals may specialize, knowing that they can 'trade' their functions for food.

The general picture of early domestications is well known, since a few regions contributed a major share of the most important cultivated species. The main foci were south-west Asia, south-east Asia and Meso-America, and the important initial surges of domestication seem to have been *c.* 7000 BC in south-west Asia, *c.* 6000 BC in south-east Asia, and *c.* 5000 BC in Meso-America, though new discoveries are constantly changing our view of this chronology. A summary of time, place, and species group in the early domestications is given in figure 4.1 and table 4.2.

Domestication involves the replacement of natural selection by man-directed selection and the processes of plant cultivation work toward a product whose edible parts ripen simultaneously. The adaptation syndromes resulting from the planting by people of harvested and selected seed often bring about a doubling or further multiplication of the chromosome numbers of the plant, which produces an increase in size and robustness: the fruit of the domestic red pepper is some 500 times the volume of its wild ancestor, and maize cobs are now seven times the length of the early domesticates. Bitterness, toxicity and thorns, which are a defence against grazing in wild plants, tend to be lost. The plant may lose as well the ability to

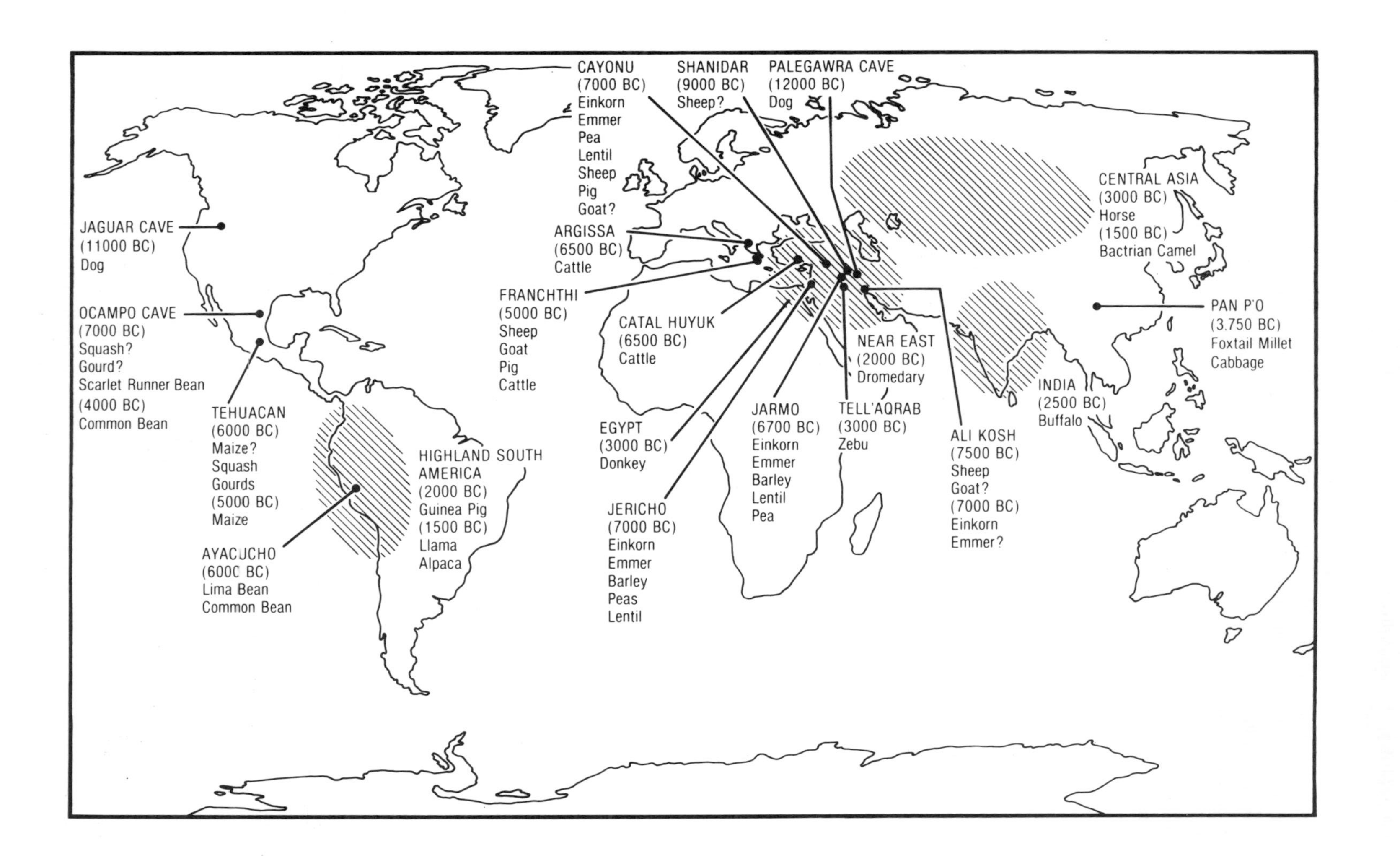

JAGUAR CAVE
(11000 BC)
Dog
OCAMPO CAVE
(7000 BC)
Squash?
Gourd?
Scarlet Runner Bean
(4000 BC)
Common Bean
TEHUACAN
(6000 BC)
Maize?
Squash
Gourds
(5000 BC)
Maize
AYACUCHO
(6000 BC)
Lima Bean
Common Bean
HIGHLAND SOUTH
AMERICA
(2000 BC)
Guinea Pig
(1500 BC)
Llama
Alpaca
FRANCHTHI
(5000 BC)
Sheep
Goat
Pig
Cattle
ARGISSA
(6500 BC)
Cattle
CAYONU
(7000 BC)
Einkorn
Emmer
Pea
Lentil
Sheep
Pig
Goat?
SHANIDAR
(9000 BC)
Sheep?
PALEGAWRA CAVE
(12000 BC)
Dog
CATAL HUYUK
(6500 BC)
Cattle
EGYPT
(3000 BC)
Donkey
JERICHO
(7000 BC)
Einkorn
Emmer
Barley
Peas
Lentil
JARMO
(6700 BC)
Einkorn
Emmer
Barley
Lentil
Pea
NEAR EAST
(2000 BC)
Dromedary
TELL'AQRAB
(3000 BC)
Zebu
ALI KOSH
(7500 BC)
Sheep
Goat?
(7000 BC)
Einkorn
Emmer?
INDIA
(2500 BC)
Buffalo
CENTRAL ASIA
(3000 BC)
Horse
(1500 BC)
Bactrian Camel
PAN P'O
(3.750 BC)
Foxtail Millet
Cabbage

Table 4.2 Early domesticates: some examples

Type	*Examples*	*Localities*
Grass and herb seeds (the basis of cereal agriculture or 'granoculture')	Maize	Meso-America
	Amaranth	Meso-America
	Sorghum	W. Africa
	Millet	W. Africa
	Rice, African rice	SE Asia, W. Africa
	Wheat	SW Asia
	Barley	SW Asia
	Oats	SW Asia
Roots and tubers (the basis of 'vegeculture')	Manioc	Lowland S. America
	Yams	Of diverse tropical origins
	Taro	SE Asia
	Potato	Andean America
	Sweet Potato	Lowland S. America
Pulses	Beans (Lima, Haricot, Runner)	Tropical America
	Cowpeas	W. Africa
	Lentil, Broadbean	SW Asia
	Pea, Chickpea	SW Asia
	Soybean	N. China
	Adzuki and Mung beans	E. and S. Asia
Tree and shrub crops	Avocado	Meso-America
	Chili pepper	Meso-America
	Guava	Tropical America
	Cacao	N. and S. America
	Oil palm, shea butter tree	W. Africa
	Locust bean	Tropical Africa
	Peach, apricot	W. China
	Breadfruit	E. Indonesia/W. Pacific
	Orange and other citrus	SE Asia
	Date	SW Asia/Mediterranean
	Fig	SW Asia
	Grape	SW Asia
	Olive	SW Asia (western part)

NB This is not an exhaustive list.
Source: adapted from D. R. Harris, 'The prehistory of human subsistence: a speculative outline'. In D. N. Walcher and N. Kretchmer (eds), *Food, Nutrition and Evolution*. NY: Masson, 1981, 15–35.

disseminate itself: if its wild relatives were perennial then it becomes an annual. Wild rye for example is a perennial, whereas cultivated rye is an annual.

Changes in the genetics of animals can also be inferred from the archaeozoological record. A large category of animals is recorded as tending to produce smaller individuals under domestication (e.g. dog, cattle, sheep, goat, pig), although some are much the same size (e.g. llama, camel), and a few rather bigger, like the horse. It appears that domestication selects for the retention of juvenile characteristics in the adult, a process called neotony; this is apparent in the skull shapes of domestic dogs, in the short faces and small horns of some domestic cattle, in curly rather than straight hair and in pied coats rather than monotone colours. Size difference between male and female is also likely to be reduced.

Figure 4.1 The major regions and sites (with approximate dates) of some of the most important domestications in the first waves of that process. It should not be inferred that no other regions were at all significant (cotton is not plotted here for example) but the main food plants and animals are shown. The picture changes as new archaeological information builds up. Compiled from a number of sources.

The local environment of the foci of domestication must have undergone changes, although there is normally much less evidence than for the domesticates themselves. Vegeculture is particularly vulnerable in this respect, for there are no hard parts to be sedimented into archaeological deposits. Most investigators are agreed that early cereal farming in south-west Asia was accompanied by sedentarism, so that new niches for plant and animal life are created in and around the settlement itself. Increased opportunities for scavengers such as pigs and for plants of disturbed and broken ground were created: grasses such as *Aegilops* and *Lolium* might be colonists, since they could have been brought in with uncleaned seed corn. Early agriculture would probably also have needed fallow periods and the plants most effective in colonizing such ground would be different from those of permanent grassland or steppe, so that species of *Galium* and *Plantago* became more common in cultivated areas.

No doubt the use of fire in hunting continued, but H. T. Lewis has made a case for deliberate burning as an accompaniment to the emergence of agriculture in south-west Asia. His reconstruction involves the burning of the oak–pistachio woodland, which would favour its replacement with grassland, which would in turn produce an increase in the number and range of wild sheep and goats and possibly wild cattle and gazelle as well. These enhanced stocks could then be either hunted or herded, the latter very probably leading to domestication. Further than this, the large-seeded grasses which were ancestral to the main cereals may have benefited greatly in terms of range and growth habit by the opening up of the oak–pistachio woodlands by fire. The same could have been true of the small trees and shrubs which are the wild ancestors of fruits like the vine, olive and fig, though their domestication came in the fourth millennium BC, after the major cereals.

Agricultural systems

Food-producing groups extend their environmental impact from the biotic constituents only (as with most hunters) to the largely abiotic components of the ecosystem such as soils. Changes in soil humus content, water relations, crumb structure and chemistry would have followed any agriculture which used the same patch of land constantly, and even with shifting agriculture, the effects of cultivation would not have vanished quickly from the soils. Of greater obvious importance is the fact that whenever the native vegetation is cleared from hill and mountain lands, the danger of accelerated soil erosion is present, and there must have been many cultivated places which at some stage lost all or some of their soil to wind and rain. The introduction of the plough in south-west Asia in the fourth millennium BC would have exacerbated the potential for soil loss, although the efficiency in preparing the soil for planting is increased by about four times. Here is an early example of an intensification of produc-

Figure 4.2 Land reclamation in ancient Egypt. Bushes and trees are being cut down on uneven land which is then hacked by hoes and afterwards ploughed and sown.
Source: T. K. Derry and T. I. Williams, *A Short History of Technology*. Oxford: Clarendon Press, 1960, at p. 540.

tion being accompanied by increased environmental impact, even though we are dealing with relatively small-scale rainfed agriculture.

Early grain agriculture might have been improved by a rudimentary irrigation system which involved the spreading of water over alluvial fans by simple ditching networks. From such beginnings, there emerged in due course one of the most outstanding environmental transformations in ancient times, which was the development of irrigated agriculture in the valleys of the Nile and the Tigris–Euphrates. Hitherto, these lowlands were seasonally inundated by their rivers and consisted of great tracts of both seasonal and permanent wetland with papyrus, sedge and reeds, grassland and *Acacia* savanna flats, and forests. All contained wildfowl and big game.

This land was converted to irrigation agriculture: one illustration from Egypt (figure 4.2) shows the process of land reclamation, with the clearance of bushes and trees from the uneven ground surface, which was then broken up with hoes and ploughed by oxen. Water-lifting technology such as the *šādūf* and the water-wheel then brought the necessary moisture to the fields. In Mesopotamia, effective irrigation of the wide plain necessitated the leading of water from the main rivers via canal systems (figure 4.3). Thus embankments were used to control the rivers and naturally meandering distributaries were diked and straightened, so that at any rate during periods

Figure 4.3 The density of irrigation in ancient Mesopotamia as revealed by air photography. Here a whole landscape is transformed from semi-arid steppe to fertile cropland by this early civilization.
Source: R. McC. Adams and H. J. Nissen, *The Uruk Countryside. The Natural Setting of Urban Societies*. Chicago and London: University of Chicago Press, 1976.

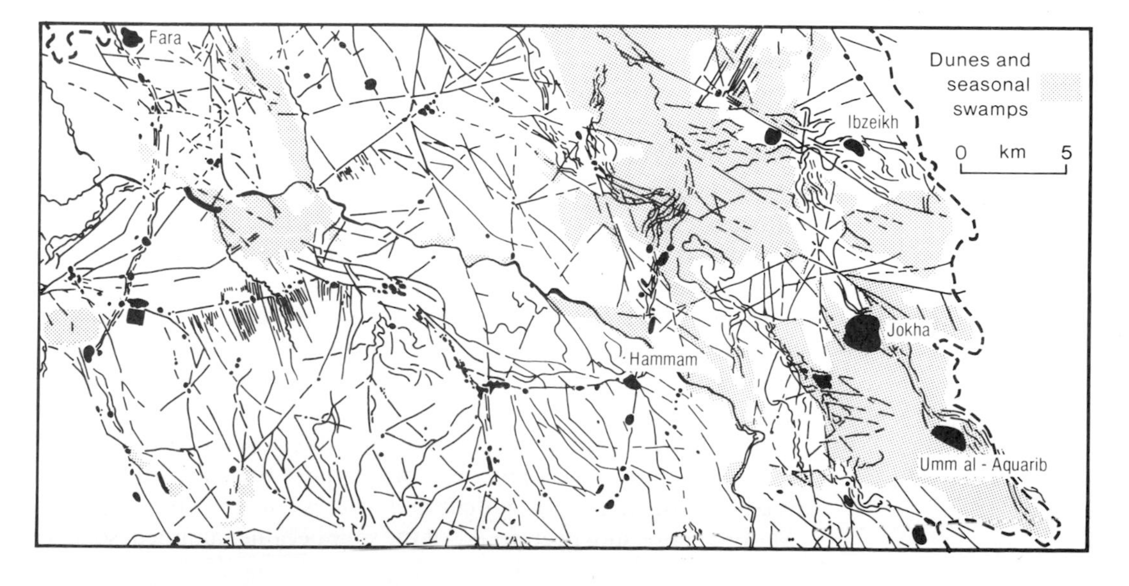

of political stability like the Third Dynasty of Ur (*c.* 2300 BC) and the Parthian and Sassanian periods (AD 1–700) the whole country was intersected with dikes, leading to fields of crops of wheat, barley, sesame and millet, and to groves of date palms. Travelling in the fourth century BC, Herodotus remarks that one queen of Assyria had caused the excavation of a basin for a lake as large as 75 km in circumference (apparently to store water for the canal systems); that rivers had been straightened and that the whole country was intersected by dikes from which irrigation water was drawn by hand-pumps. In Egypt, the Faiyum had its level regulated so that some marginal land could be colonized as early as the Middle Kingdom (2040–1750 BC), and then the Ptolemies (323–30 BC) reduced it to a smaller size but established a system of radial irrigation so that the area of cultivated land near it was enlarged three times.

In Mesopotamia (much less so in Egypt), the agriculture had unintended environmental consequences. The water and mineral balances of the soils were changed and in the late third and early second millennia BC, the decline of crop yields, a move towards more salt-tolerant crops and records of ever-larger areas removed from cultivation, all suggest that salinization was widespread. The grazing of the lands surrounding the plains, together with the faster erosion of soil from cultivated land, brought their own difficulties in the form of heavy silt loads in the rivers and irrigation channels. The systems developed in these irrigated empires may have produced a positive feedback loop. Depictions of harnessed cattle in fourth millennium Mesopotamia suggest that an integrated livestock–crop system (with barley being an especially useful animal feed) was developed. The probable use of milk and milk products may have led to the introduction of higher fat levels into human diets. In turn this could have produced an earlier menarche in women and coupled with cereal grain as an infant food might have allowed rapid population growth, in turn creating more pressure for change in ecosystems.

In the basin of Mexico in prehispanic times there is abundant evidence of irrigation agriculture. At Teotihuacan the inception of a system of canals fed by dams across major streams probably dates from the 100 BC–AD 100 period along with the emergence of the city itself. In addition, spring-head swamps were drained and their emergent streams diverted to feed irrigated systems. This region provides early examples of the drainage of wetlands for agriculture, for in areas of lowland swamp it was the practice to create 'islands' of mud (*chinampa*) upon which to grow crops.

Farming did not remain confined to its centres of origin: by *c.* AD 1, cultivation was well spread in the world (figure 4.4) and a number of distinct types adapted to regional conditions had developed. Each of these, too, had particular patterns of environmental impact, summarized in table 4.3. Within this general pattern, political events or ecological processes might cause the reversion of cultivated lands. Areas of agricultural success might turn, as with Mesopotamia, into near-deserts if the soils became salinified, so there

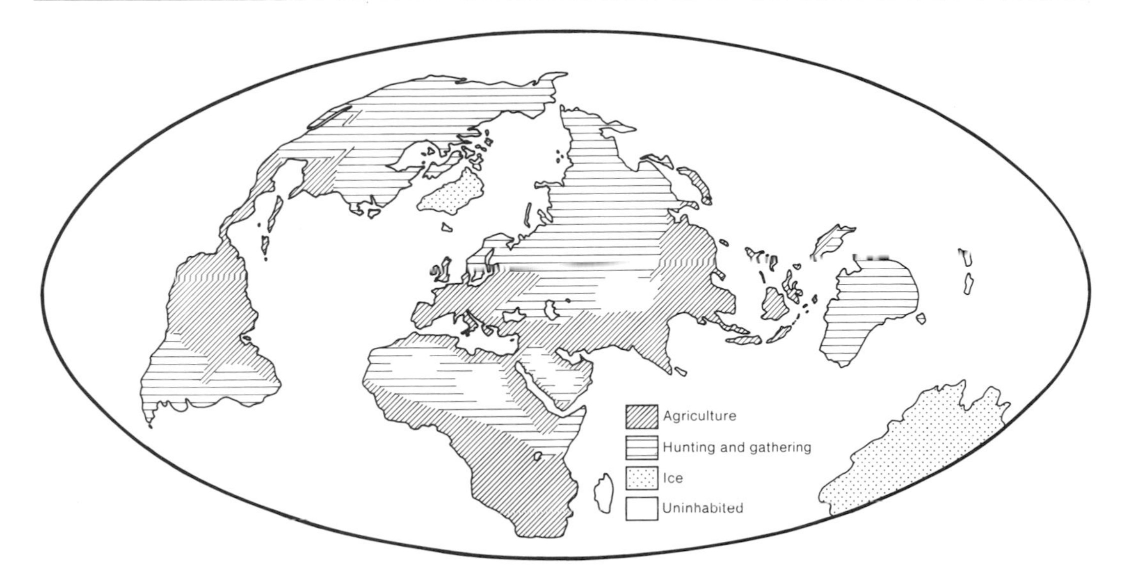

Figure 4.4 The distribution of agriculture in AD 1, with its implications for environmental change where it had replaced hunter-gatherer economies. A great deal of it outside the temperate zone was probably shifting cultivation, producing continually altering mosaics of ecosystem type.
Source: A. Sherratt (ed.) *The Cambridge Encyclopedia of Archaeology*. CUP, 1980.

is evidence for the abandonment of irrigated areas in river valleys and the reversion of the cropland to steppe and scrub. In the Middle East, the Islamic conquerors found that the irrigation systems were largely in ruins.

Such was the complexity and hierarchy of the societies built upon the energy surpluses of irrigation agriculture in the period to AD 1 that their environmental impact spread beyond the limits of their immediate subsistence base. The demand for meat which seems to accompany high social status almost everywhere no doubt stimulated increased production by pastoralists beyond the irrigated zone, and this in turn caused vegetation loss leading to silt inwash to the irrigation channels. Societies with high priestly and aristocratic castes like those of Assyria and ancient Egypt must also have had a considerable impact upon the wild animal communities of surrounding terrain, if we are to believe the many depictions of hunting scenes found among the relics of these civilizations. Given the overall population density of humans beyond the irrigated lands it may seem unlikely that animal populations could have been permanently affected, but remembering that some of these outer lands were used by pastoralists as well, we ought not to be surprised if animals whose trophic level already made them relatively scarce became considerably reduced both by hunting for sport by aristocrats and from pre-emptive extirpation by pastoralists. The lion in south-west Asia was clearly hunted and killed in large numbers (even allowing for flattery in artistic depictions) and other carnivores like the leopard may have been subjected to a similar depletion of range and abundance.

Receiving its crop plants and animals by dispersal from south-west Asia, the temperate forest zone of Europe was host initially to a form of shifting cultivation which was rainfed. Abundant evidence from central, north and western Europe suggests that the first agriculturalists adapted to the conditions by felling trees with polished stone

Table 4.3 Agricultural systems and their impact to c. AD 1

Type	*Distribution in 0 BC/AD*	*Major domesticates*	*Environmental impact*
1 South-west Asian and derivatives	Nile, Tigris-Euphrates, Indus, Turkestan	Wheat, barley, cattle (and rice and coffee in India), millet, sesame	Clearance of savanna, swamps, and fringing forests. Changes in water regime with irrigation by embankment dams, canals, water-wheel, bucket systems. Canal-side ecosystems established. Accessory hunting, incl. predators for sport. Salinization of soils, silting of canals known to happen
(a) Irrigated farming in major valleys			
(b) Dry farming	Original type: hill-lands of SW Asia, N. Africa	Wheat, barley, pulses, sheep, cattle, goats	Clearance of oak–pistachio woods for fields. Fallowing and cross-ploughing to preserve moisture. Soil changes and soil erosion yielding silt to valleys. Animals (domestic) change vegetation beyond fields. Accessory hunting e.g. of gazelle
(c) Mediterranean agriculture	Mediterranean littoral	Wheat, barley, olive, grape, fig, pigs	Clearance of forests and scrub for fields. Overcultivation produces soil loss esp. on limestones, so much soil erosion, silt yield to rivers. Some more or less bare limestone, or semi-arid steppe (e.g. N. Africa) may result
(d) Permanent temperate agriculture	Northern Europe, Ukraine	Wheat, barley, oats, rye, pigs, cattle, sheep	Clearance of deciduous forest for fields. Stable system develops with manure input to fields, as well as fallowing. Soil loss not absent but at relatively low level because of climate and slope factors
2 South-east Asian and derivatives			
(a) Tropical vegeculture	Mainland and insular SE Asia	Taro, yam, bananas, coconut, pigs, poultry	Axe and digging-stick cultivation: shifting cultivation replaces forest with polyculture of crops. Forest perh. does not regenerate to original spp. composition but nutrient cycles restored in fallow period. Accessory hunting

(b) Wet-rice	Widespread, though not in present padi form, in mainland SE Asia	Wet rice, pigs (tended to replace vegeculture)	Grown originally in wet areas beside streams, then in small irrigated plots. Terraces and large-scale padi come rather later
3 Northern China			
Dry farming of cereals and pulses	Not much spread beyond original area	Millet, soy bean, mulberry, pigs; later imports from SW Asia	A form of dry agriculture. Plough in use by 500 BC
4 Africa			
(a) Tropical vegeculture	Shifting cultivation type of humid tropics	Yams	Usual form of shifting agriculture, probably without domestic animals; plough not adopted. Accessory hunting
(b) Cereal cultivation	Shifting cultivation and bush fallow type of savanna zone	Sorghum and millets, cattle	Bush fallow rather than tone shifting cultivation. Latosols prone to erosion and falling yield in savanna environments. Much 'cultivation' of trees and wild plants generally. Accessory hunting
5 Americas			
(a) Root cultivation	Shifting agriculture in various environments	Manioc, yams, potato	Neither plough nor domestic animals as livestock in agriculture system
(b) Irrigation	Confined to a few 'high' cultures in lowland areas	Maize, beans, squash, tomato, peanut, cotton	Irrigation systems channelled spring water to canal systems round largely saline lakes; soils not very prone to salinification(?).
(c) Drainage	Just developing in Maya, Aztec areas		Where lowland swamps occurred a system of 'islands' of mud (*chinampa*) was developed, though major florescence later than this period. Accessory hunting

The basic typology is derived largely from D. B. Grigg, *The Agricultural Systems of the World*. London: CUP, 1974. Other material from numerous sources, see references (pp. 413–14) to text. Pastoralism is omitted from this table.

axes, a trade-mark of the Neolithic in Europe which emerges in the fourth millennium BC all the way from Ireland to Russia. The cultivators then piled up the combustible material and burned it, and planted cereals in the ash-enriched soil. The cropping was continued until the yield started to drop, a likely feature of this system if no nutrients are returned to the soil, or until the effort of weeding became too much for the cultivators. The plot was then abandoned and a secondary succession developed: in Western Europe, grasses, weeds and bracken fern took over the bare ground and were in turn replaced by brambles and light-demanding small trees such as birch and hazel, which then yielded to the shade of the forest dominants like oak and lime. Because of the surrounding forest, soil loss seems to have been low. As prehistory progressed, this system was replaced with a form of agriculture in which permanent fields are found. Their fertility was kept up, the limited evidence suggests, by the periodic spreading of the settlement's midden-heap on the land and by the depasturage of the domesticated animals upon the fields. Any manure from winter-confined beasts would have been used this way as well, so linkage between crops and domesticated beasts was a central part of the system. Fallowing was also practised, and the growing of leguminous crops such as pulses helped to keep up soil nitrogen levels. When the plough became available in western Europe (at first as a simple wooden scratch-plough, later as a mould-board implement with an iron coulter), it helped productivity by keeping down weeds, aerating the soil and possibly also by bringing up nutrients. In turn it changed the nutrient composition of the soils, the crumb structure and particle size distribution. Permanent fields meant also that a larger area of soil was exposed to erosive agencies, and from the time of the beginning of Bronze Age agriculture (1800 BC) in the West Midlands of England, we find deposits of sediment 1–2 m thick in the terraces of the River Severn, so a great deal of soil must have been lost from the fields; the ploughing of grasslands on hillslopes in Britain is estimated to have increased soil movement by 400 times.

Environmental impact spread beyond the fields: if domesticated animals were allowed to graze in surrounding woods and grasslands then shifts in plant species composition became inevitable. In woodlands the seedlings of trees would have been grazed by domestic beasts like cattle, and pigs would root out and eat acorns, chestnuts and beech mast, so that the regeneration capacity of the forest would have been impaired except perhaps where thickets of brambles or holly gave protection to seedlings. Even if woodlands were not felled to allow the creation of new fields, then the exercise of pasturage will have resulted in slow attrition of their area. It would be remarkable if deer populations did not decline as the areas of shelter diminished (Plate 4.2), and certainly the wild ox and the European bison became almost extinct during prehistoric times. No doubt their chief predator, the wolf, suffered decline as well, although it survived well into historic times even in relatively densely settled parts of Europe.

Plate 4.2 Paolo Uccello's oil painting *Hunt in the Forest* (15th century). It depicts a highly managed woodland: the uniformity of the oaks suggests a plantation and the stems have been shredded to remove side branches. (Ashmolean Museum, Oxford)

The practice of cultivating a patch of cleared land and then abandoning it to the natural succession of the original ecosystem is most often associated with the humid tropics and the adjacent savanna zones where it has persisted through to the present. Traditional shifting cultivation in the moist tropics is not difficult in essence. After clearance and burning, phosphorus and calcium build up to values higher than those in the original forest, whereas nitrogen and potassium fall. The normal 2–3-year cropping period therefore takes advantage of the temporarily high levels (figure 4.5) of nutrients in the soil. Falling nutrient levels and the encroachment of weeds cause abandonment of the plot and thereafter nutrient levels rise, but very slowly; in one study, the concentrations of nutrients in

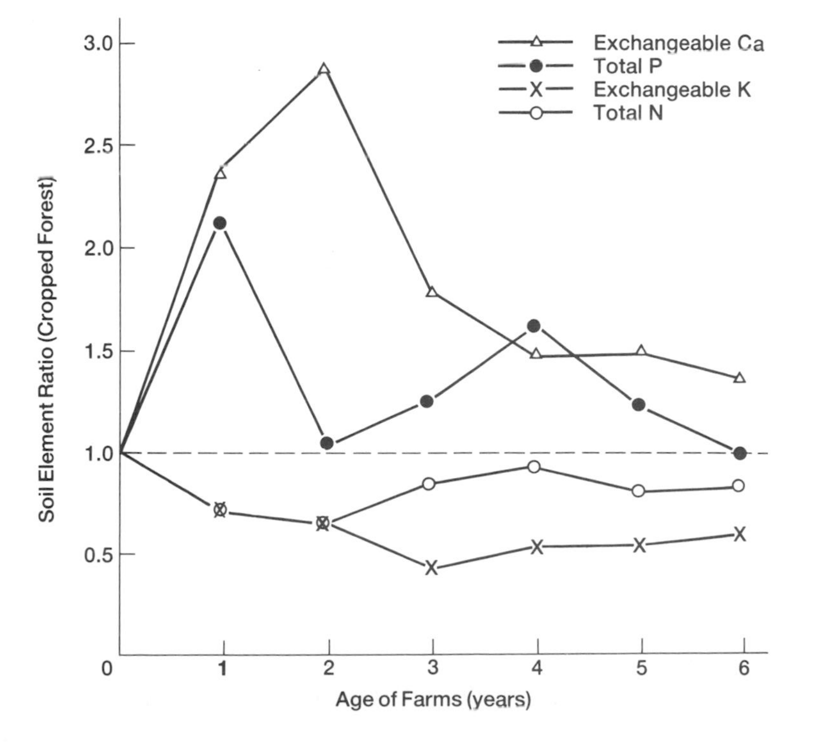

Figure 4.5 The time-course of soil fertility under shifting cultivation in the tropics: an example of the quantities of various nutrients expressed as ratios of the pre-clearance levels.
Source: F. O. Adedeji, 'Nutrient cycles and successional changes following shifting cultivation practice in moist semi-deciduous forests in Nigeria', *Forest Ecology and Management* 9, 1984, 87–99.

secondary woodland were still well below the original values after six years. The nutrient conservation mechanisms of the regrowth are not so efficient as in mature forest and there are considerable implications for the re-use of secondary growth areas should human population growth cause a shortening of the fallow cycle, for if the period of forest regrowth is decreased beyond certain levels, land degradation (e.g. by soil erosion) may often follow. The introduction of terracing into slash-and-burn areas may slow down soil and nutrient loss but even so, in northern India, such plots were abandoned after about six years, only a one-year advantage over the conventional cycle in recent years in similar areas.

As table 4.3 and figure 4.4 show, a differentiated set of agroecosystems, each with distinctive environmental linkages and impacts, was in place by AD 1, supporting the core area of the world's population distribution. Peripheral to them were firstly the remaining hunter–gatherers (then numerous) and secondly another way of life, based on domesticated biota but this time dominated by animals, i.e. pastoralism.

Pastoralist ecosystems

Pastoralism focuses upon a domesticated or semi-domesticated animal as a source of food and other materials, and usually also of symbolic wealth. Both hunting and cultivation are marginalized. It was once postulated as a cultural stage between hunting and agriculture but most scholars would now place its evolution later than that of sedentary cereal-based agriculture. (K. Butzer suggests that the first secure example of pastoralism is to be found in the western and central Sahara in the period 4500–3500 BC where a mixture of cattle and sheep were used.) Pastoralism tended to occupy the lands marginal to agriculture: any extension of farming would likely be at the expense of the pastoralists, and any loss of control of pastoral flocks might release the animals into standing crops. Hence tension as well as trade has attended the relations between sendentary farmer and nomadic pastoralist and has even been held to account for the development of the pork-avoidance taboo since the pig is so preeminently an animal of the sedentary cultivator and not of the mobile men of the steppes and mountains. The versatile pig is not suitable for pastoralism, but there is no shortage of other candidates. Cattle are often desirable though not always the most suitable because of their need for large quantities of water every day, but suitable breeds of sheep and goats, yaks, llamas, camels, horses, water buffalo and the semi-domesticated reindeer all came to be the basis of pastoral economies. By AD 1 the distribution of pastoralists (figure 4.4) was dominated by the semi-arid lands of the Old World, the highlands of South America and the forest–tundra ecotone of Eurasia.

The ecology of pastoralism centres on the ability of the domesticate to turn rather unpromising forms of cellulose into human nourishment like meat, milk and blood, and to provide other

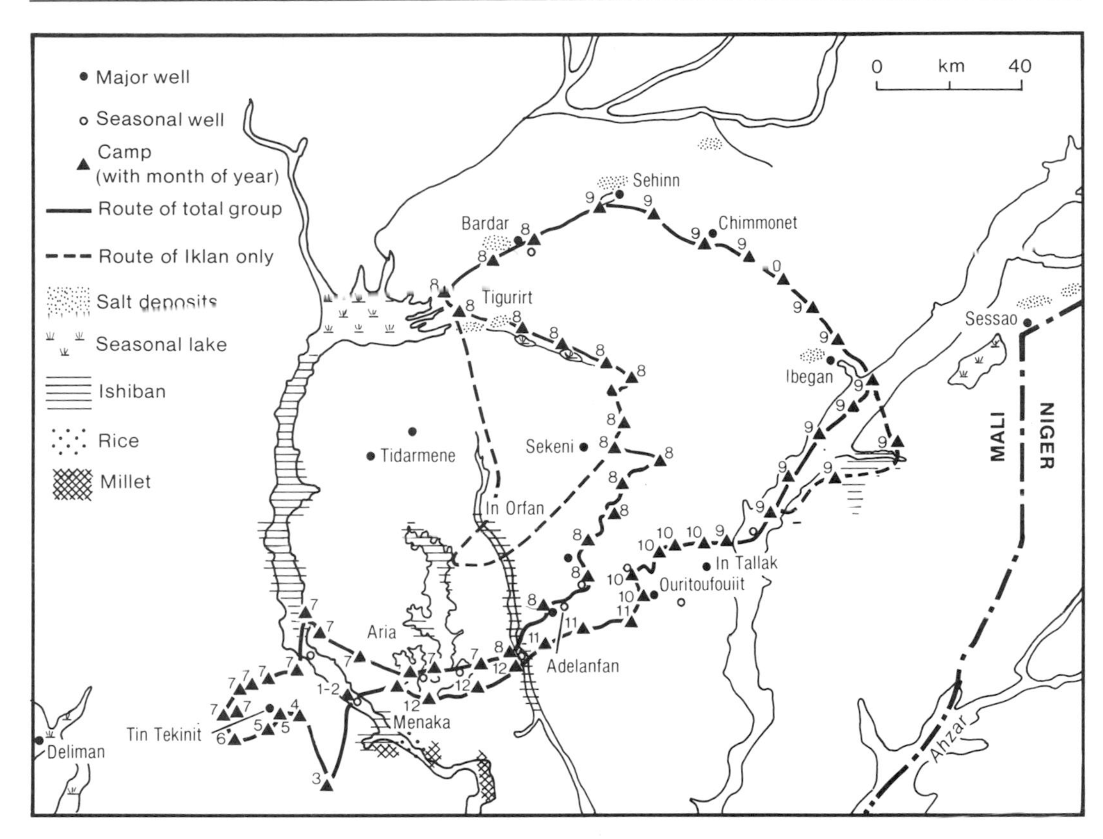

materials such as dung, horn and hide. Ecology and culture combine to make migration a feature of pastoralism (figure 4.6). The need for movement is obvious, for a herd of beasts will drink a great deal of water and they can quickly eat out all the nearby forage and render the area useless to another group or to the same group in another year. So pastoralists must constantly move in order not to exhaust their resources. However, if a prolonged halt at any one place is for some reason possible, some pastoralists will add some accessory agriculture to their economy, using camels, for example, to plough an area of soil which can be cultivated long enough to produce a cereal crop.

Figure 4.6 The annual round of a group of pastoralists in the Sahel zone of Mali. Note that they move little during January–March (consider the impact consequences), and that (cf. fig. 3.9) they split up for part of the cycle. Characteristically, some crops are grown to supplement animal foods. *Source*: S. E. Smith, 'The environmental adaptation of nomads in the West African Sahel: a key to understanding prehistoric pastoralists', in W. Weissleder (ed.) *The Nomadic Alternative*. The Hague: Mouton, 1978, 75–96.

The environmental impact of pastoralism (table 4.4) starts with breeding. Selection of a strain of animals most suited to the terrain is essential, but no great genetic or behavioural changes may be absolutely necessary. Both camel and reindeer, for example, are sufficiently close to their wild ancestors that they can still interbreed with them, although most other species are more firmly tamed. An ecological feature of the system is that even light grazing pressure by domesticates will change the species composition of an ecosystem, especially if repeated over a number of years. An inexorable shift towards inedible plants is therefore likely to have happened, with dominance becoming the preserve of plants which possess features like totally unchewable leaves, spines or toxins. If the grazing

Table 4.4 Major ecosystem characteristics as affected by pastoralists: a summary

Characteristic	*Effect of pastoralists*
Energy flow	Relatively little changed: likely that domestic animals replace a wild ancestor at same tropic level. Culling of rival predators unlikely to affect system very much. Only if system very unstable in terms of 'desertification' are flows likely to alter a great deal
Nutrient cycling	Normal systems relatively little affected; perhaps some concentration of dung at places where flocks have stayed some time: this likely to be used as fuel or possibly as manure if stay long enough to take a catch crop. Otherwise actual processes relied upon to keep up nutritional standard of the vegetation
Biological productivity	Is generally low by global standards: areas of vegetation used by nomadic pastoralists (NPs) are not normally very high in NPP (e.g. Steppe, tundra, mountain grasslands) and then usage by mammal herbivores keeps secondary production low. SNP is usually low enough that meat is not greatly eaten but milk and milk products (and in some places blood) are the staples
Population dynamics (non-human populations) Successional stages Biotic diversity	Unless grazing by the domesticated animals is very heavy then these change only slowly
Stability	If too many animals are depastured on too small an area of land then (a) selective grazing occurs and palatable species begin to disappear permanently, and (b) all vegetation may start to disappear and be replaced with bare soil. Latter known as desertification, probably not very common under traditional pastoralism except at times of interaction with climatic desiccation. Large flocks regarded as increase against drought and other hazards because of poor breeding success under many traditional pastoral systems; interaction with cultivators (exchange animal products for carbohydrate) may also affect survival
Degree of modification	'Once over lightly' seems recipe for long-term survival of traditional pastoralist systems, but many factors – cultural,political, climatic – may militate against this, so NPs often forced into smaller areas or shorter cycles, cf shifting cultivators. Availability of water is crucial and E.I. always much greater around sources of water: little vegetation for animals, or firewood (dung becomes very important), and there is puddling of soils which increases gullying because of reduced infiltration

pressure is heavy then the vegetation will become sparser and much more open ground will result. With low quantities of organic matter on the soil surface, precipitation will run off or evaporate quickly and so the ecosystem will gradually become more desert-like. In this state, the soils are open to accelerated erosion from wind and rain so the syndrome of desertification is established. Just how much of the riven and ruined landscape of south-west Asia's hillsides was thus established by AD 1 is not possible to say, but it seems very likely that at least 2000 years of pastoralism had wrought considerable change upon the ecosystems of the arid and semi-arid lands of the Old World. Nor did the impact necessarily end on those lands, for eroded soil has to go somewhere and much of it ended up as aggradative river terraces in the lower reaches of river basins, adding to their flood liabilities on the way and eventually to their deltas. Some of this alluvial material was, however, trapped in river valleys where it formed the basis of a very fertile soil for use by cultivators. Some indication of the potential extent of the environmental impact of pastoralism is given by the estimate that even at the present day some 47 per cent of the world's land surface is suitable only for grazing by wild or domestic stock.

End of the beginning

By AD 1 a relatively advanced agriculture of a sedentary type, often supporting cities and socially stratified states, was established along a belt from the Pillars of Hercules to the Tsukinoki river and in Andean and Meso-America. Less socially developed forms of agriculture such as vegeculture and shifting cultivation occupied the Amazon Basin, the south-eastern corner of the present USA, western and northern Europe, much of savanna Africa, and both peninsular and insular south-east Asia. Pastoralism was dominant in the Saharan fringes, in the great core of mountain Eurasia and inside the coastal agricultural zone of Arabia. Elsewhere, as in the southern third of South America and sub-Saharan Africa, all of Australia and most of North America, and in the boreal forest belt of Eurasia, hunting and gathering prevailed. A few places were still uninhabited by humans: Madagascar, New Zealand and parts of Polynesia and Micronesia, the Himalayan mountains and Saharan interior, and all except the southern fringe of Greenland. But most of the land surfaces of the world were effectively occupied by human societies and many of them were changing their surroundings, though on a variety of scales of intensity.

The spread and development of agriculture

A world-wide practice

AD 1 as a marker point has no special significance in agricultural history. The dispersals of plants, animals and agroecosystems continued, though with two unusually intensive phases, the first from AD 1 until the drawing apart of the first great political empires of the world round about AD 600, and the second coincident with world-wide trade and colonization from AD 1500–1800. Processes described for those 1800 years, however, look both backwards and forwards for their origins and fates. The achievement of agricultural development was to feed a human population which rose from perhaps 170 million in AD 1, through 200 million in 600 and 425 million in AD 1500, to 900 million in AD 1800.

The dispersal of plants and animals and the transfer of agroecosystems are not necessarily the same, for biota can be transported a great distance and merely be substituted for an existing species in an extant agroecosystem: sheep can be substituted for llamas, for example, without altering the fundamental characteristics of the system, though many details might change; equally, maize might be substituted for millet and still keep the basic functional nature of the cropped ecosystem. Transference of an agroecosystem *en bloc*, however, is symptomatic of much greater environmental change. A category of land transformation wrought by pre-industrial agriculture but not confined to one region is terracing. (Plate 4.3) This is a composite phenomenon, involving both dry- and wet-field products, and altering both hillsides and valley floors. The world distribution of terracing is given in figure 4.7, and table 4.5 sets out a simple typology of terrace types. J. E. Spencer and G. A. Hale postulated an origin for the practice in the ancient Near East (*c.* 9000–5000 BP) in the same environments as cereal-based agriculture, with an eastward spread that eventually met the Aryan wet-rice tradition (perhaps in northern Indochina) and then spread in all directions. Many of the 'traditional' terraces of today are, however, relatively recent: in China there was a big burst of construction in the eighteenth century, in Java in the nineteenth and in peninsular India in the eighteenth and nineteenth centuries. Given its spread in time and place, there can be little contravention of the view that 'agricultural terracing constitutes one

Figure 4.7 The origins and distribution of agricultural terracing. The outer boundary is probably the most important distribution since it shows how widespread the technique has been and the wide variety of environments in which it is found. The data on origins and spread are not agreed by all scholars.
Source: J. E. Spencer and G. A. Hale, 'The origin, nature and distribution of agricultural terracing', *Pacific Viewpoint* 2, 1961, 1–40.

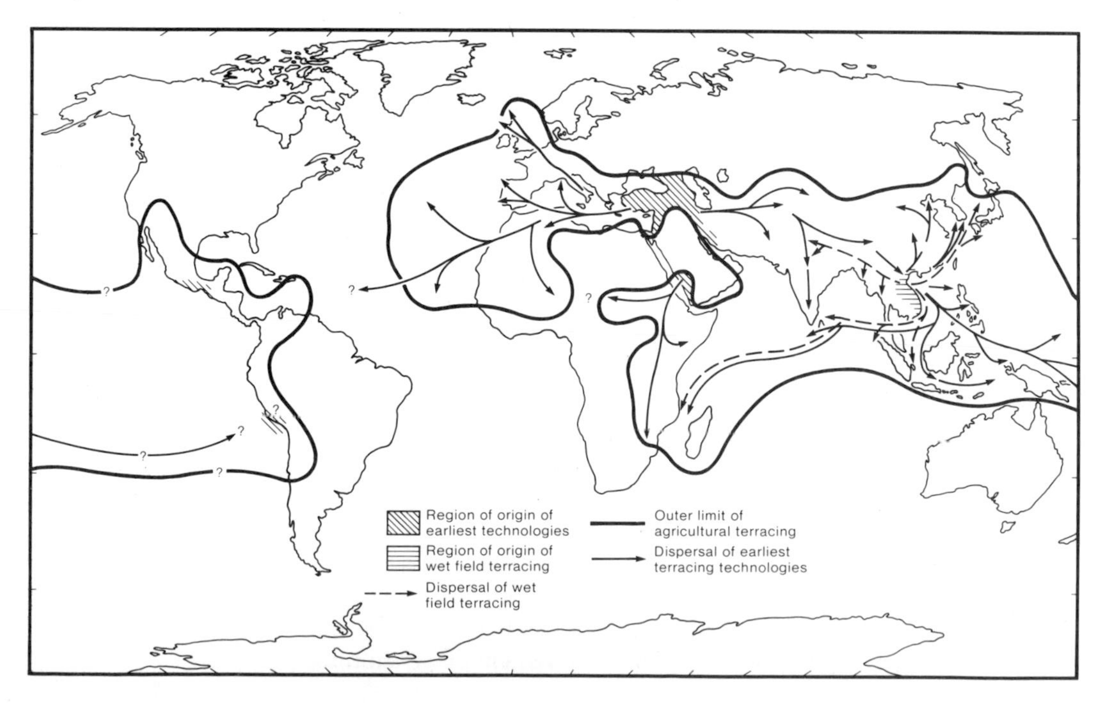

Plate 4.3 Terraced fields near Pisac, Peru. The lower slopes of this mountainous environment have been transformed into a series of irrigable fields with a low slope that retains both soil and water. Compare with Pl. 4.4. (Jenifer Roberts/Panos Pictures)

very important category of phenomena stamping the permanent imprint of man upon the surface of the earth.'

The 'medieval' phase

In terms of the dispersal of domesticated biota, perhaps the most noticeable feature was a 'common market' of transfers of crops between the Old World centres of trade and empire. Thus various

Table 4.5 Types of agricultural terracing

	Type	*Remarks*
1	Weir in non-entrenched drainage channel	New soil formed from trapped silt. Small–moderate fan-shaped fields. Near-East; very arid zones
2	High terrace in narrow drainage channel	Massive stone barrages trap silt and water in areas with few flat surfaces, e.g. the Negev
3	Rock-embanked non-contour terrace with sloping field	Field not horizontal but flatter than original surface: digging and filling employed. Field crops, shrubs or trees grown: wide distribution in world where irrigation not needed
4	Isolated field terrace with rock embankment	At random over slope, slightly flatter than original slope. W. China, Himalayas, SE Asia
5	Contour terrace with rock embankment and diverted water supply	Only slight gradients, often begins near a water source, and may also have drains for surplus water. Semi-arid Old World, Andes
6	Lateral buttress terrace with no attempt to create a slope	Often for trees and shrubs: walls may be irregular or semi-terraced. Middle East and Mediterranean
7	Mud-walled terrace on flat-floored stream bed, impounding water	Flat fields produced by silting rather than soil-shifting. Mud blends near areas of stream flow. Many areas probably transformed to type 8. Asia
8	Lateral contour terrace, impounding water from up-slope	Front embankment higher than fill – is a dam and must be leakproof. Services etc. necessary. Field must be levelled and water control efficient. Asia
9	Field-pond constructed to supply water to fields above or below	Shallow reservoir of water, often cropped for plants, crustacea and fish. Asia
10	Excavated pit field dug to create wet-soil conditions closer to water-table	Taro pit is best example: perhaps not a true terrace. SE Asia, E. Indies, Pacific

Source: J. E. Spencer and G. A. Hale, 'The origin, nature and distribution of agricultural terracing', *Pacific Viewpoint* 2, 1961, 1–40.

plants of early domestication were moved with success within the zone of south-west Asia, the Mediterranean, Europe, North Africa, Turkestan and north-west India. The vegeculture and hill-rice complex of south-east Asia went into China and Malaysia, and the source area of northern China influenced Turkestan and south-west Asia. India stood at a crossroads in many of these movements. On a world scale, the development of rice cultivation is of considerable significance. About 5000–3500 BC vegeculture was reinforced with dry or hill rice growing in mainland south-east Asia, and then *c.* 500 BC wet rice began to gain in importance, spreading during the next thousand years from its heartland in south-east Asia into southern China, Korea, Japan, North Vietnam, India and Java. At the same time rice grown more extensively, without the radical transformation to padi, or in shifting cultivation systems occupied most of the rest of south-east Asia except Java. In India *c.* AD 75–100, padi cultivation involved an ecocultural package of the clearing of forest, the construction of irrigation works, and the establishment of a settlement together with a temple. Quite firm dates tie the cultivation of padi rice to Indonesia and the Philippines in the second millennium BC, to Sri Lanka in 400 BC and to Egypt in 375 BC. Thereafter, to anticipate our story, the spread included Sicily in Roman times, northern Italy in the fifteenth century AD, and export by European colonizers to such places as Brazil, the Spanish Americas, South Carolina in 1685 and Louisiana in 1718, Hawaii in 1853 and by German influence to New Guinea and by French carriage to New Caledonia.

The agriculture of the temperate zone of north and west Europe had to make do with fewer domesticated plant species than the southern neighbours from whose systems it was derived, though the impact of Rome did provide a better scythe, introduced the chicken, and facilitated the spread of the wine grape as far as the Moselle by the fourth century BC. By the end of the first century AD oats were established as a major crop in maritime Europe, where it was well able to tolerate the cool and wet conditions. Rye performed a rather similar role in northern and eastern Europe. Rome itself was the driving force behind the spreading of grain cultivation in the Empire and much 'marginal' land, whether of forest to the north or low steppe to the south, was converted to corn land: Rome alone imported 17×10^6 bushels (620×10^6 litres) of corn a year from Egypt, Africa and Sicily. 'Corn', said Tacitus, 'is the only produce required from the earth.' In the Mediterranean littoral, the vine–olive–fig part of the complex was also extended under Roman influence, at least until the fifth century AD.

Further north a major phase of agricultural development had by AD 1300 taken on the characteristics it was largely to manifest until industrialized. It was, too, the kind of agriculture that was exported to the eastern seaboard of North America. The ecosystems of the deciduous woodland were cleared quasi-permanently and replaced with tilled fields, some of which were fallowed in rotation. Tillage was effected with a wheeled plough which featured a coulter and mould board. Stock had therefore to be kept for use as draught animals, although the growth of cereals along with some pulses was the main purpose of the system. Most of the stock grazed upon the wildlands or upon grasslands in situations where cultivation was impossible, for example near to a river which flooded frequently. So the woodlands provided browse for cattle and 'grazing' for pigs (known as pannage), as well as wood for fuel and construction. Areas of heath or moor might provide fuel in the form of peat or pared turves as well as grazing for sheep.

The success of such a system is judged by its yields of cereals, without which there would have been famine. Thus the agroecosystems of the fields had to be manipulated so as to shed excess moisture, especially on clay soils. Ploughing was used to throw up ridges within the field which were well drained so that at least some of the crop would survive a wet year. Equally important was the resting of fields by fallowing them: in the traditional three-field system, one field was left fallow in each year. The benefits, although essential, were probably modest. Nitrogen-fixing bacteria might flourish during the fallow period and help raise the levels of nitrogen which had been depleted by cereal cropping. Fallow also allowed the processes of weathering to proceed uninterruptedly, but the release of phosphorus, potassium and trace elements was unlikely to have been fast enough to replace those taken away in cropping. Deep tilling by the mould-board plough increased the effective root zone and hence the supply of minerals, and promoted the effective incorporation of the organic wastes which were the source of vital nitrogen.

Ecologically, therefore, the key to the system lay in its nutrient flows. Cereals depend on nutrient uptakes which may be as high as in the forest they replaced, yet arable soils show significant declines in their populations of free-living nitrogen-fixing bacteria. In addition, the alterations in soil structure produced by repeated tilling invite the loss of other mineral nutrients. Hence every possible source of nutrients had to be tapped and brought onto the fields: animals, for example, could be largely fed in woods or on heather and then be put onto fields for a while to manure them, and all animals would be penned at night for easy collection of their faeces. If available, soft chalky rocks might be quarried to be put on the fields ('marling' was noted by Roman visitors to Britain), but the manure heap was the keystone of the system. It was inefficient: since it was neither floored nor covered, some 50 per cent of the nitrogen and 75 per cent of the potassium would be returned to the fields. Should the input of manure decline (if common grazing areas shrink because of population growth, for example, or if the animals die from disease or are taken as taxes) then grain output becomes uncertain. One judgement of the open-field system suggested that it 'offered a means for sustaining a mediocre level of productivity at the price of a judiciously slow depletion of the nutrient reserves of the arable's hinterlands', and that by the late thirteenth century there was considerable pressure of population on the system, which was not easily capable of intensification, so a fresh burst of woodland clearance followed in order to tap the nutrient reserves of hitherto virgin soils.

Two further major dispersals should be noted: in the Americas, the maize–beans–squash complex spread from Meso-America into the south-east quadrant of North America and then in irrigated form into the valleys of the south-west such as the Rio Grande and the Colorado. It appears, too, that maize spread into the Pamana (Venezuela) region of the Orinoco basin between 800 BC and AD 400, more or less replacing the earlier diet of manioc and hunted animals. In central America the remains have been found of a complex urban society to be ranked with those of highland South America and of Ancient Mesopotamia. Recent work has excavated not only the cities and ceremonial centres of the culture but has thrown considerable light on its subsistence base and its environmental relations. The Mayans inhabited what is now part of Guatemala, Mexico (Yucatan), Belize and the fringe of Honduras, mostly in the lowlands but with some extension of their ecumene into the upland areas that interrupt the plains. The chronology usually adopted has a 'Classic Maya' phase from AD 250 to *c.* AD 800, followed by the Collapse of AD 800–900 (with a sudden decline of population and the desertion of cities); thereafter the Postclassic Maya continued at a lower level of complexity until the advent of the Spaniards in AD 1520 completed the disintegration of the ancient culture.

The original vegetation of the area had a number of elements including mesic and monsoon forests, savannas and grasslands, swamps with buttressed trees and small water-filled depressions. The Classic Maya were underpinned economically by a suite of subsistence techniques. Of these, the most important spatially was probably maize grown by shifting cultivation (swidden), with sweet potato and manioc as other important crops. A variation of this technique was the intensive *milpa* practised in areas of year-round rainfall. This involved multi-cropping with a high diversity of crops and high ground coverage to reduce the hazard of soil loss. Even though fallow periods were short, this system was apparently stable. Another important element in the diet may have been the nuts of the *ramon* tree (*Brosimium alicastrum*), which would have been harvested either from a monocultural orchard system or from a more mixed arboriculture with *guayo* and *aguacate* trees. At any rate, the possible yields of 1245kg/ha/yr of nuts would have been a valuable addition to the subsistence base.

Large-scale alteration of the land surface was within the capability of the Maya, including the creation of raised fields in areas of swamp and river flood plain. These raised fields (*chinampas*) are especially characteristic of the valley of Mexico but some 21 per cent of the land surface of the Peten area seems to have been devoted to them. Canals are dug, and the mud dredged up forms a rich soil while the water harbours fish and turtles. The crops grown included maize, cotton and cacao. The act of keeping free the canals added more rich silt to the fields, and some of the waterways had raised sills at each end to act as traps for silt or fish, or both.

On the upland areas which interrupt the lowlands, large tracts were terraced, including one tract of 10×10^3 km^2 in the Rio Bec region. These increased the available agricultural land while avoiding erosion and leaching, and there is evidence that cultivators carried soil upslope within the terrace systems and perhaps even from inundated lowland areas. Such terrace systems (figure 4.8) were supplemented by devices such as rows of stones set into the slope, and by silt trap terraces and weirs set in the valleys where they might accumulate over 50 cm of soil.

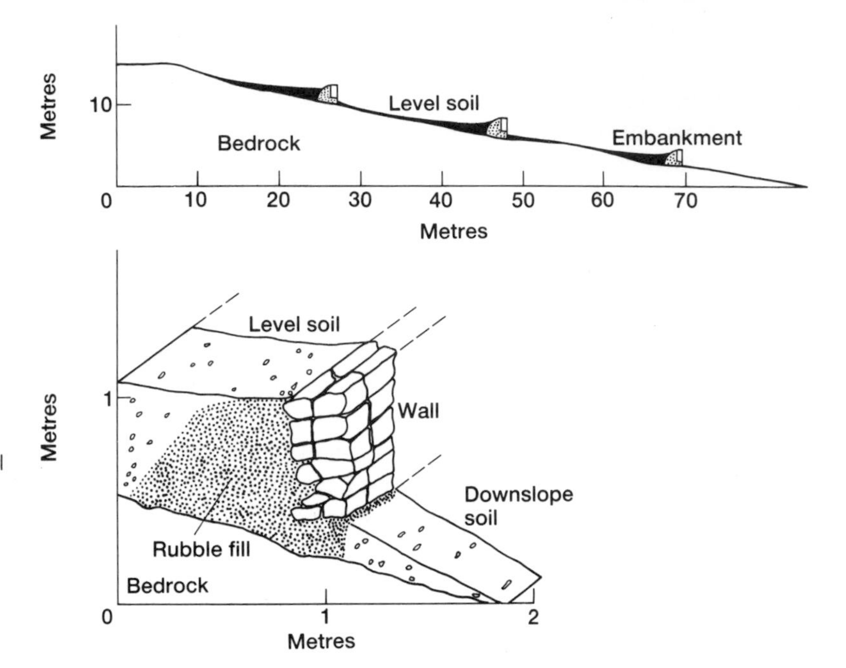

Figure 4.8 Many prehistoric societies (and current ones as well) use terracing with a wall to reduce soil loss. The detail of the construction, which could come from numerous places, is complemented by a section of the hillside showing the approximate spacing of the walls.

The presence of such large areas of terracing argues for a more or less total land use since all the available lowland arable would have been in use before the labour of terracing would be undertaken. What other land uses besides agriculture might have been necessary? One other important environmental resource was wood. Each family would need 15–20 cords/year, with 1 cord of wood equivalent to 120 trees. Since short fallow cycles would yield insufficient wood of the right size from their regeneration phases, woodlot plantations were probably necessary, in the form of either swidden areas with a long fallow for its wood harvest or isolated wood-yielding 'plantations' grown within other agricultural systems. Lastly, we may note that the Classic Maya practised water management: they collected water from limestone caves in Yucatan, lined pools with stone, constructed systems of reservoirs, cut chambers into solid rock with stone 'catchments' leading to their mouths, dug wells, and fed moated fortresses by canals as well as erecting water storage systems for cities as great as 17 km^2 in area (figure 4.9).

It seems virtually certain therefore that the Maya transformed much of the landscape they inhabited. An airborne observer at the time, were such possible, would have seen a largely cultural, rather than natural, landscape. No very sophisticated technology (no plough for example) was needed for a numerous population to transform a set of natural tropical ecosystems into an equally diverse suite of man-directed ones.

In Oceania an agroecosystem based on taro and yams borne by voyagers spread quite quickly: the Marquesas seem to have been focal in the spread of this agroecosystem to Polynesia at one period,

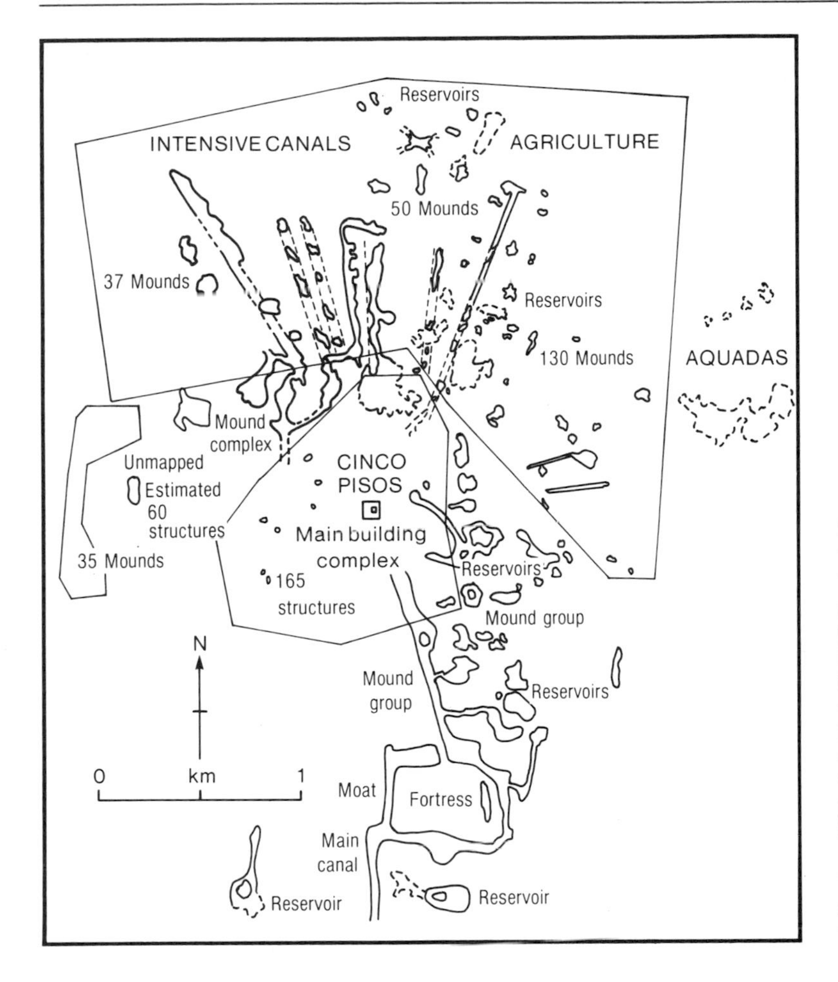

Figure 4.9 At Edzna, in the Maya lowlands, fieldwork and air photography have made possible the reconstruction of the hydraulic features. These comprise canals and reservoirs and clearly supplied the city with water for all purposes, probably including irrigation.
Source: R. T. Matheney, 'Maya lowland hydraulic systems', *Science* 193, 1976, 639–46.

but the actual sequence of spread after about 1500 BC, from a core horticultural area probably located in Papua New Guinea into an Oceania probably totally occupied by AD 1000, has yet to be worked out in detail. Australia, however, remained immune to agriculture of any form until the arrival of the Europeans.

After the fall of the last Roman emperor in AD 476 the great empires of the world tended to draw apart and exchange fewer biota, but the Arabs and other Muslims then became a force. One of the great regional transformations was the rehabilitation of defunct irrigation systems, using both improved technology and new crops. The tapping, diversion and lifting of water was brought to new levels of competence. 'In many regions,' says A. M. Watson of the period AD 900–1000, 'there was hardly a river, stream, oasis, spring, known aquifer or predictable flood that was not fully exploited.' The control of irrigation meant that a new agricultural season (the summer) had been created: rice, cotton, sugar cane, eggplant, water melon and sorghum were introduced, along with systems of rotation and cropping that did away with the fallow for which there is, it is said, no word in classical Arabic. By the eighth century AD the Arabs had introduced the orange, lime and lemon from south China and south-east Asia to the eastern Mediterranean and thereafter to their western lands: the orange came to Seville in the tenth century AD. In the eighth century also, the Arabs introduced rice, citrus fruit, the coconut, cucumbers and mango into east Africa from south-east and east Asia via Mesopotamia.

'Early modern' times

Two linked developments are the driving forces behind the transfers of biota and agroecosystems in this period: the ocean-going vessels pioneered by the Portuguese which meant that Europeans could penetrate to most parts of the world's littorals, and the desire of European states to extend trade and hegemony to the lands thus 'discovered'. Before Columbus, the Portuguese had taken crops to west Africa and the Atlantic islands: in the first half of the fifteenth century, for example, they introduced maize, sugar cane, bananas and the grape to those places, and by 1513 they were trading plants in the spice islands of the Moluccas, between Sulawesi and Timor. Such trickles quickly became a braided flood which cannot be chronicled in detail here but in which major channels would be the introduction of the potato and maize from the Americas to Europe soon after Columbus and the spread of maize onwards to Egypt and the Levant. Europe also received crops like lucerne from Persia and was stimulated to begin to breed legumes, roots crops and grasses from its native varieties: the wild *Beta vulgaris* of the sixteenth century yielded the beetroot and mangel-wurzel and, by the eighteenth century, the sugar beet. In other continents, similar volumes of transfer occurred, such as the introduction of tobacco as a plantation crop into south-east Asia after 1600; the colonization of the hilly lands of south China above the rice limit by groundnut and sweet potato growers in the seventeenth and eighteenth centuries; and the clearance of forest in North America to grow tobacco, sugar and coffee. Some semi-arid zones were transferred from nomadic pastoralism to ranching (and others indeed from hunting to ranching) by the introduction of European cattle or sheep, though much of that type of development belongs more to the nineteenth century with the ascendancy of the merino sheep.

Even the apparently stable open-field and livestock agriculture of temperate Europe underwent considerable changes as the common fields were enclosed into holdings belonging to individuals; such lands might then be used more intensively if an individual landowner was an innovator. The promise of individual ownership made the prospect of the reclamation of heath, moor, fen and salting more attractive and a large estate owner might even take the land out of food production and change it into woodland or parkland. This preindustrial 'modernization' of European agriculture began as early as the sixteenth and seventeenth centuries with a number of processes:

1 The reduction of the fallow, along with the growing of turnips as winter feed for cattle, the use of a long ley with legumes such as clover in a convertible husbandry, and the use of night soil from towns. In this process the Dutch were often in the vanguard and the English early adopters: west of the North Sea the classic four-course rotation was established in the early eighteenth century.
2 The extinction of common rights and the consolidation of strips. This started slowly in the thirteenth but accelerated in the seventeenth and eighteenth centuries, especially in England, Denmark, and Sweden. Elsewhere in Europe, fragmentation remained more common.
3 An increase in the importance of livestock generally, and in the eighteenth century the production of specialized animals: cattle bred for meat or milk, sheep bred for wool or mutton.

All these changes represented an intensification of the land use and between 1750 and 1850 crop productivity in England doubled; between 1700 and 1770 in France there was even a 60 per cent increase. Yet in places like the central Russian black earth region, intensification meant gulley erosion, mentioned as early as 1578 near larger settlements, and by the nineteenth century some 2 per cent of the land was thus classified, with reports of gulleys growing as fast as 200 m/yr.

Thus the period to 1800 saw the establishment of what we now generally regard as 'traditional' agricultural practices and ecosystems: those operating under pre-industrial conditions even if not free of European influence. Many other changes were adopted that derived from a 'scientific' agriculture which preceded the nineteenth century, but these for convenience will be dealt with in chapter 5. On the eve of the European outreach, in *c.* 1500, the pattern of agroecosystems and consequent environmental impact (figure 4.10) show no great changes from that of AD 500, except for the shrinkage of hunting and gathering zones, the firmer hold of settled agricultural systems in much of Africa, and the organization by the Mongols of the economy of central Asia. Between 1500 and 1800 the pattern was to change yet again at its peripheries with the occupation by European agriculturalists of part of North America, Argentina, and Australia and the impact of that style of hegemony in much of Africa, Latin-America and south Asia. But the traditional impacts and the pre-industrial European influences were small compared with what was to follow in the nineteenth century, even given the rise in world population from 425 to 900 million between 1500 and 1800.

Four thousand years of agriculture: examples of impact

One of the characteristics of looking backwards to times which we label as 'traditional' and 'ancient' is that we tend to underestimate their impact on their environment, preferring often to think of a

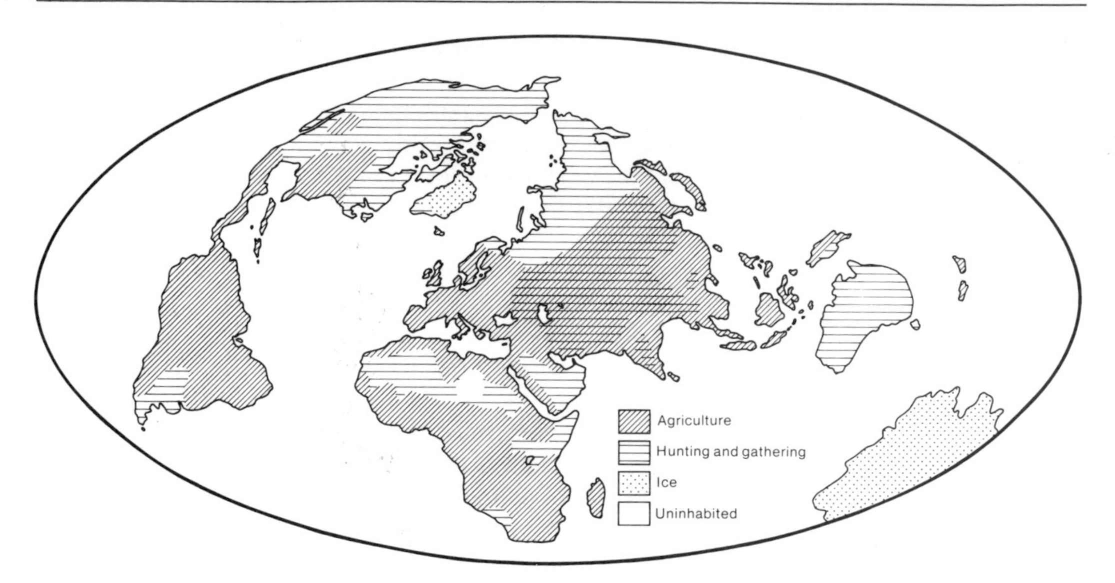

Figure 4.10 Compared with fig. 4.4, by AD 1500 agriculture had made some more inroads into hunter-gatherer territory (though in places there was more co-existence than this map implies), with consequent environmental changes. *Source*: A. Sherratt (ed.), *The Cambridge Encyclopedia of Archaeology*. CUP, 1980.

golden age of man–land harmony when the land yielded its usufruct to the honest toil of a contented peasant in a smiling sort of way. Looking at some examples of the management practices and consequent impacts of the period between the evolution of agriculture and AD 1800, however, it becomes apparent that even in pre-industrial times the manipulation of nature involved in various ecosystems might be quite intensive (see e.g., tables 4.6 to 4.8). We need, though, to beware of regarding these examples as an unbiased sample for naturally enough the more intensive cases are the ones with highest visibility in the literal, literary and archaeological senses: an assessment of the total picture does not yet appear to be possible.

Western culture, for example, celebrates its Greek roots. Yet that particular flowering of intellect took place against the background of very severe environmental change in the rural areas. The human-induced modification of Mediterranean ecosystems probably began in the Late Pleistocene, when people deliberately used fire to modify the dense forest into more open multi-layered woodland interspread with grassland and herbaceous communities. These folk were succeeded by the agricultural cultures which used domesticated grasses as the basis of their economy. The elaboration and development of agriculture and pastoralism in the period from the local Neolithic (*c.* 11 000 BP) until the relatively recent import of industrialization (a few decades ago) thereafter exerted a major impact on the ecosystems of the lands around the Mediterranean and indeed upon the sea itself (figure 4.11).

The farming itself was generally ecologically stable: contour ploughing and terracing were practised so that soil loss was minimized and wherever possible hillsides were reserved for orchards and vineyards. Irrigation on a small scale, using wells and springs, was practised (although dry farming preponderated) and the farmers were aware of the importance of maintaining nutrient cycles. In 400 BC it

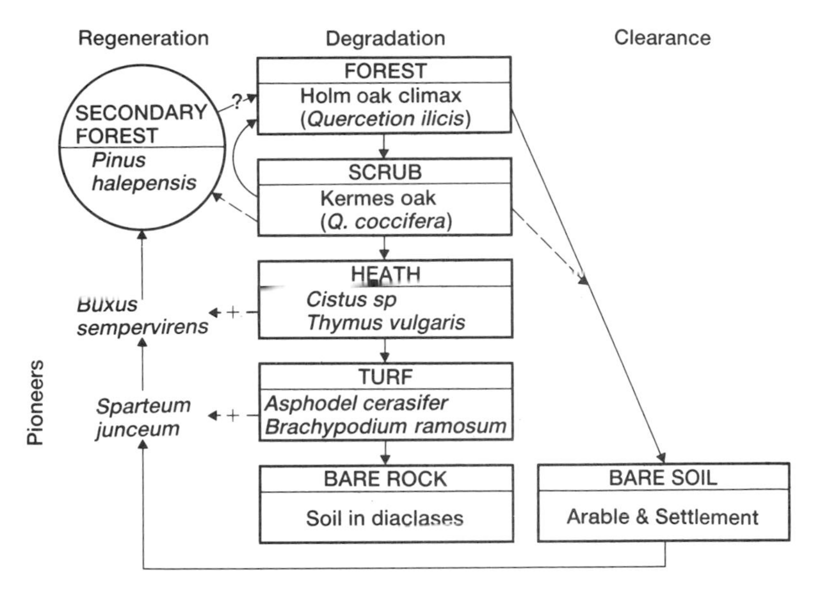

Figure 4.11 The vegetation sequence in the Mediterranean, following human activity along either the pathway of clearance for cropping or that of 'degradation' with grazing and fire. *Source*: C. Delano Smith, *Western Mediterranean Europe*. London: Academic Press, 1979.

was recorded that green crops were ploughed in, lime and marl were spread on the land, and legumes were grown in rotations which also included fallow periods. Yet animal manure, the most obvious source of soil organic matter, was available only for half the year because of the practice of pastoral transhumance whereby the animals (mostly goats and sheep) were taken to the hill lands for summer grazing. (By way of minor compensation, bird dung might be available: the temple at Delos exported pigeon guano.) Since true nomadism was precluded by the nature of the mountains and the political boundaries, the herds were pastured within restricted areas and overgrazing resulted. These grazing lands had been mostly created at the expense of forests which were sometimes deliberately fired by shepherds, who also ring-barked trees to ensure the trees' early death. Writing between 500 BC and AD 25, Herodotus, Xenophon, Aristotle, Theophrastus and Strabo all remark on the rapidity with which forests were replaced with fields or pastures. Deforestation was also carried out for shipbuilding, for construction timber, and for conversion to charcoal for fuel. The twin processes of deforestation and heavy grazing in these climatic and topographic conditions meant that soil erosion was virtually unimpeded from former forest lands, and the ensuing silt was largely deposited in river deltas along the Mediterranean. These features brought about the formation of low-lying swamps which were a paradise for malarial mosquitoes.

The pastoral use of the hills also encouraged the extirpation of wildlife: by 200 BC the lion and leopard were gone from Greece and coastal Asia Minor; wolves and jackals were confined to the mountains. Aristocratic hunting of predators and wild boar contributed to the decline of wild animals, as did the general trapping of fur-bearers like the beavers of northern Greece. Forests suffered in times of war, since they might cover the movements of opposing enemies, and deliberate destruction of crops and vineyards was also a military tactic.

Thus the agroecosystems of Ancient Greece in supplying meat, leather, wool, grain, fruit and wine, caused a considerable impact of a destabilizing kind, and the need for more supplies as the Greek empire expanded led to similar changes throughout the eastern Mediterranean; even the better agricultural land sometimes became exhausted because of double cropping. In southern areas like Crete triple cropping was demanded. 'Nothing in excess' was inscribed on the great shrine at Delphi, but was not, it seems, applied to the use of the environment.

As might be expected, the culture of the Romans exerted a broadly similar effect on the environment. The pre-existing cover of Italy was forest which largely disappeared during the flourishing of the Roman Empire. This loss of woodland was the most widespread effect of Roman occupation of the land, proceeding more slowly in the north and west: the Ciminian forest in Etruria was reputed to be thick, dangerous and impassable. As in Greece, the forest receded because the wood was needed for shipbuilding, for construction and above all for fuel, and because the land was used for agriculture and for herding mixed flocks of sheep and goats. The same processes occurred as in Greece: bare hillsides, rivers choked with silt, floods, and coastal marshlands infested with malaria mosquitoes. This syndrome was not of course confined to Italy: the Empire spread its practices along with its politics, and deposits 10 metres thick of the products of the erosion which followed deforestation occupy some valleys in the former Roman provinces of Syria and north Africa. Although the outlying zones of the Empire might contribute grain, olive oil and wine, any fresh meat required had to be produced nearer at hand (and the Romans were heavier meat-eaters than the Greeks), so the hillsides became heavily grazed, with firing of forest and scrub adding to the potential for erosion. The mixture of sheep and goats meant that almost any plant would be eaten from ground level up to the canopies of the trees into which the goats might climb. Pigs were herded in the oak, beech and chestnut forests and provided the most common meat for Roman tables. The growing of crops in the core areas of the Empire followed the normal Mediterranean pattern: terraces were built with stone walls and soil was even carried back up slopes from which it had been washed down. Fallowing was practised, as was a crop rotation of grain, legumes and fallow; intercultivation also ensured a high productivity per unit area since olive trees might support vines and have grain planted between the rows of trees. But institutional conditions favoured specialization, hence rotations and fallow were restricted in their use; manures (animal, human and green) were, however, used when available, as were vegetable refuse, ashes, marl, lime and burned ground cover after fallow or harvest.

Further, the Romans were skilled at seed selection and at the technology of irrigation and drainage, although they appear to have been unaware of the probability of salinification in irrigated lands.

Plate 4.4 The Mediterranean region sees the growth of many olive trees: part of the foundation economy of wheat, vine and olive oil. The plantations clearly have a vital role in soil retention on this hillside in southern Spain. The foreground is the kind of scrub which will harbour some grazing animals for milk and meat. 1980s EEC policies will likely see the olive groves taken out, with considerable environmental consequences. (David Goulston/Bruce Coleman Ltd)

But many lakes and marshes were drained to extend the arable area, which had placed on it the additional demand of raising horses for war and racing. Alfalfa was a major feed for them, having been brought to Italy in the second century BC from Greece, which in turn had imported it from Persia three centuries earlier. The firmness of Rome's imprint, at home and abroad, included the regular layout of square fields called centuriation, which divided the land from the Po valley to North Africa into squares of *c.* 50 ha in a way not to be encountered again until the nineteenth century homestead laws of the USA. The intensity of manipulation of the environment is said by authors such as J. D. Hughes to have contributed to the fall of this great Empire. But in some parts of the Mediterranean, forests must have been either preserved or replanted for in 1880 an English traveller found Upper and Lower Galilee as well wooded as Britain, which is scarcely to say there were great tracts of dense forest. In the Mediterranean generally there seems to have been some equilibrium between clearance, burning, grazing and coppicing in the region at least until the downfall of the Ottoman Empire, and the conventional picture of unrelieved degradation is probably exaggerated (Plate 4.4).

It is perhaps no surprise that societies apparently as advanced as Greece and Rome exerted such impacts on their surroundings. More impressive in some ways is the list compiled by J. L. Beyer of many of the practices of traditional agriculturalists in Africa. Although many in themselves were small in scale, together they provide an impressive recital of evidence for a very fine-grained appreciation and manipulation of the environment. If the widespread use of fire is added, then the potential for environmental change in most zones outside the moist forests can be readily perceived (table 4.6). A similar pattern of soil management is reported from pre-colonial south Asia (table 4.7) in which the fertility and tillage practices, not in themselves environ-

Table 4.6 Traditional management and environment-changing practices in Africa

Management practice	*Examples*	*Management practice*	*Examples*
Soil fertility and texture control		Water management	
Irrigation	Widespread*	Aqueducts	Kenya*
		Channels	Tanzania
Kitchen gardens	Widespread*	Drainage ditches and diversion dams	Ethiopia*
Mulch from grass, household carpet	Tanzania*	Small dam and diversion system	Zimbabwe* (Rhodesia)
Ash for fertilizer	Widespread*	Seasonal flood padis	Zambia*
Grass dug into mounds and fired[a]	Cameroon, Zambia	Ponding water on hillsides	Madagascar
Composting	Widespread*	Flood water, padi cultivation; cilili gardens	Zambia*
Broadcast-sowing – cattle used to tramp in seeds and fertilize	Ethiopia*	Wells for permanent irrigation	Ghana*
		Drainage canals in peat soils (mataba)	Zambia*
Re-use of soil from abandoned villages (umufundo)	Zambia*	Irrigation of terraced fields	Togo*
Grass mounds evolved into contour ridges (vikuse)	Zambia*		
Rotation of cattle pens on dry litongo soils	Zambia*		
Stumps and stones left in ploughed fields as erosion control	Ethiopia*	Terracing	
		Stone or earth walled	Ethiopia, Cameroon, Togo*
		Ruins	East Africa
Roots of tree stumps and pollarded trees left against erosion	Rhodesia	Contour ridging, grassed	Uganda*
		Contour banks and pits	S. Tanzania*
Recognition and classification of soils	Ethiopia, Tanzania*		
Stall feeding for manure	Ghana, Tanzania*	Fire as a tool	
Grazing animals in fallowed fields	Ethiopia, Lesotho	General effects	Widespread and common*
Straw from cattle pits	Madagascar*	Sub-climax and drier woodland	East Africa*
Ikpezi composting[b]	Nigeria*	Controlled burning	Sierra Leone*
Ant hills used as fertile sites	Zambia*	Game control and clearing	Cameroon
Litter to mulch banana groves	Tanzania*	Brush control after harvest by sedentary farmers	Ethiopia*
Field ridges and strip cropping	Nigeria*		
Mounding with hoe[c]	Widespread*		
Night soil and composting	Tanzania*		

* indicates pre-European but still present
[a] planting takes place on the ash heaps
[b] a rubbish heap on a frequently used path, with fallen leaves and household garbage
[c] these mounds may be 0.3–2.0 m in diameter and up to 1 m high

Source: J. L. Beyer, 'Africa'. In G. A. Klee (ed.), *World Systems of Traditional Resource Management*. London: Edward Arnold, 1980, 5–37.

Table 4.7 Environmentally significant agricultural management practices in south Asia before colonial rule

Form	Land type[a]	Resource complex-crop[b]	Location[c]
Soil fertility			
Animal manure			
Cattle dung	I	P-R, W, M	Everywhere
Sheep dung	I, R	P-R, W, M	Everywhere
Goat dung	I, R	P-R, W, M	Everywhere
Horse dung	I	P-W	P, Ba
Camel dung	I	P-W	P, Ba
Donkey dung	I	P-W	P, Ba
Bat dung	K	P-M	Ka
Human manure	I	P-R, W, M	Small amounts everywhere
Green manure			
Ploughed crops	I	P-R, M	Ka, G, M, WB, BD, O
Leaves branches	I	P-R, M, W	T, Ka, M, Ke, WB, BD
Village debris	I	P-M, W	Ka, P, S
Household refuse	I	P-R, W, M	Everywhere
Ash	I	P-R, W, M, T-SC	M, Ka, Ke, T, AP, O, MP, A, BD, N, MR, S, SL
Oil seed waste	I	P-R, W, M	G, UP, M, P, WB, BD, T
Fallowing	I, R	P-R, W, M, T-SC	Known everywhere
Texture improvement and soil transfer	I	P-R, W, M	M, Ka, Ke, T, G, MP, O, WB, BD, P
Crop mixing	I, R	P-W, M, T-SC	Everywhere
Leguminous crops	I, R	P-R, W, M, T-SC	Everywhere
Crop rotation	I, R	P-R, W, M	Known everywhere but was restricted
Natural silting	I	P-R, W	A, BD, WB, Bi, UP, S, O
Fish	I		Coastal areas of G and S
Flushing soil	I		O, AP
Tillage			
Ploughing			
Light ploughs	I, R	P-R, W, M	Everywhere
Heavy ploughs	R	P-M	Ma, Ka, AP
Multiple ploughing	I, R	P-R, W, M	Everywhere
Wet ploughing	I	P-R, W	Everywhere
Dry ploughing	I, R	P-R, W, M	
Hoeing	R	P-R, W, M, T-SC	Ma, Ka, Ke, T, AP, O, WB, BD, A, Bi, MP, MR
Harrowing	I, R	P-R, W, M	Everywhere, but very important in Ma, Ka, AP
Clod breaking	I, R	P-R, W, M	Everywhere
Puddling	I	P-R	Everywhere
Heating soil	R	T-SC	Ma, Ka, Ke, T, AP, O, WB, BD, A, Bi, MP
Field Forms			
Embankments against salt water	I	P-R	Ke, WB, BD, S
Bunding	I, R	P-R, W, M	Everywhere, but esp. important in Ba, MP, Ma, MB, Bi
Levelling	I	P-R, W, M	Everywhere
Mounding	I	P-R, W, M	Ke

contd overleaf

Table 4.7 (*cont.*)

Form	*Land type*[a]	*Resource complex-crop*[b]	*Location*[c]
Slope management			
Terracing	I	P-R, W, M	SL, T, Ke, Ka, Ma, MP, WB, BD, A, MR, Na, Ne, Bi, K
Field levelling	I	P-R, W, M	Everywhere
Silt-trap dams	I	P-W, M	Ka, Ba
Planted vegetation	R	P-M, W	T, Ka, G, NW

[a] I - irrigated; R - rainfed
[b] M - millet-sorghum; P - peasant; R - rice; SC - shifting cultivation; T - tribal; W - wheat
[c] A - Assam, AP - Andhra Pradesh, Ba - Baluchistan, BD - Bangladesh, Bi - Bihar, HP - Himachal Pradesh, K - Kashmir, Ka - Karnatka, Ke - Kerala, Ma - Maharashtra, ME - Meghalaya, MP - Madhya Pradesh, MR - Manipur, Na - Nagaland, Ne - Nepal, NW - North West Frontier, O - Orissa, P - Punjab, R - Rajasthan, S - Sind, SL - Sri Lanka, T - Tamil Nadu, UP - Uttar Pradesh, WB - West Bengal.

Source: B. J. Murton, 'South Asia' in G. A. Klee (ed.), *World Systems of Traditional Resource Management*. London: Edward Arnold, 1980, 67–99.

mentally significant on a large scale, take place against widespread and landscape-creating practices of making low earth walls (bunds), levelling and mounding. Levelling for terracing is perhaps the best known form of slope control in south Asia, and often operated in combination with a variety of water management practices which include tanks for water storage and bunds for controlling the flow of water through the fields.

Irrigation is even more characteristic of east Asia especially where applied to the hillsides which yield water and nutrients channelled into the fields used for padi rice. (Plate 4.5) The Malayan word *padi*

Plate 4.5 Terraced fields in the New Territories of Hong Kong. Taken one step further than the examples in Pl. 4.3, these are like the aquaria of Geertz's description. Further water management can be seen in the dug-out pond, used for rearing ducks and fish. (Photo: Author)

meaning 'rice on the straw' has been Anglicized to refer to the growth of rice in conditions which require it to be submerged beneath 100–150 mm of slowly moving water for three-quarters of its growing period. Most of it is thus found in fields created on flat lands near rivers but in some places there has been an elaborate terracing of slopes which represent a considerable feat of environmental management. The low nutrient status of tropical soils is circumvented through the water-borne nutrients brought onto the terrace, the fixation of nitrogen by the blue-green algae which proliferate in the warm water, and the release of minerals by the decay of stubble and of other organic matter used to fertilize the fields. Additionally, the working of the waterlogged soil produces 'puddling' so that the soil becomes impervious and minimizes the loss of nutrients by downward-moving water. So the padi is a fertile environment: a farmer may add 10–15 cartloads/ha of manure from pigs or water buffalo, plus human excreta, ashes, soot, straw, alluvium and perhaps green compost from leguminous crops. Further, there may be 16 t/ha fresh weight of algae which may accumulate 100 kg/ha of nitrogen, and other nitrogen-fixing organisms also exist in the root zone of the rice plants. Given that the flooded padi may also yield animal protein in the form of fish or crustacea then its riches as a food-producing system are convincing. In his classic description of pre-industrial padi, C. Geertz says of its environmental relations:

> wet-rice cultivation is essentially an ingenious device for the agricultural exploitation of a habitat in which heavy reliance on soil processes is impossible and where other means for converting natural energy into food are therefore necessary . . . here we have . . . the fabrication of an aquarium.

In order to achieve this floodable environment and to keep its fertility high, a number of management practices are used. B. J. Murton lists 26 soil management techniques which are applied in wet-rice cultivation in south Asia and we may take it that most if not all of them are common in other regions. Animal manure, human manure, oil seed waste, green manure and leaves from trees, wet and dry ploughing, bunding and even soil transfer are prominent in the list. Water management is equally complex: the hydrological cycle is manipulated at several points in order to provide the right quantity of water to the fields at the right time, with a correspondingly complex social framework to ensure the equitable distribution of the resource. Again, Murton lists 23 forms of water management, including storage wells and tanks, lifting devices, delivery systems, application practices, and ways of preventing the flooding of padi by estuarine or river waters in low lying areas, especially in the deltas of the major river systems. The greatest environmental management, however, comes through slope control, with terrace construction, the levelling of fields so that they can be evenly flooded, and the construction of dams in runoff streams to trap any valuable silt which might otherwise escape. Thus at some stage in their history, whole landscapes have been realigned so that flows of energy, moisture and nutrients were channelled, year in and year out, into a relatively small area of cultivated land. Where major nutrient leaks to urban centres took place in the padi system of east Asia, then the cycle was restored by the use of human excreta, and it is clear that the cycling of nitrogen is crucial to this whole food-producing system.

In summary, we can perhaps say that padi rice is not such a radical transformation of an entire landscape as is perhaps commonly thought: it exists in a matrix of less-manipulated ecosystems except in particularly suitable places like riverine plains. But its potential for environmental change extends outwards from the padi fields themselves into dryland ecosystems which may be sources of nutrients, and onto surrounding hillsides which supply the vital water (often with important nutrients in solution or suspension) without which the crop could not be grown. The net effect was the realignment of landscapes to accommodate the rice fields, leaving the hillsides covered in forest areas which might be the home of shifting cultivators or of herders who would produce their customary changes in the ecosystems.

Although both absolute numbers and densities of populations in Oceania are low, the impact impressed on a number of ecosystems was by no means negligible under traditional systems of management (table 4.8). In New Guinea, for example, food production from horticulture was present as early as the period 10 000–9000 BP. So many of the shrublands and grasslands of highland valleys and basins of Papua New Guinea are a result of agricultural and hunting practices: forests and forest regrowth were repeatedly burned to produce grassland, and grassland itself burned to drive out game, ward off enemies, and prepare plots for planting. In wet areas, it appears that by 4000 BP ditches were dug to control the water levels in swamps so that a new crop like taro suited to wet conditions could be grown: according to some interpreters, this was a response to the degradation of soil on slopes and the increase of population on the land. By 4000 BP there were considerable areas of cleared land throughout the Highlands, and the swamplands were used for farming as well as the mountain slopes. By 2000 BP further pressure on the forests saw the expansion of grassland and further incursions into the swamps; by 400 years ago the extent of forest was reduced even more and raised plots were introduced which enabled frost-intolerant crops to be grown at higher altitudes. All this allowed population growth to the point where further intensification took the form of the introduction of the sweet potato (some time less than 400 years ago) which supported high densities of both people and pigs.

In Polynesia, taro and yam were highly important cultivated staples, and numerous practices (table 4.8) both ensured good crops and changed the landscapes. In swampy areas on coasts a reticulation of drainage channels would leave garden islands for planting with taro; on atolls, pits were dug down to a freshwater lens so that the taro could be grown in damp but freshwater conditions. On drier areas, arboriculture for breadfruit, coconut and pandanus added a

Table 4.8 Traditional soil and water resource management in Oceania

Form	Examples of recognized conservation practice Ancient	Present
Terracing	Fiji	New Hebrides, New Guinea
Mulching	Pacific	New Guinea
	Pacific	East Caroline Islands
Leaving tree seedlings after clearing	Pacific	New Guinea
Planting trees within gardens		New Guinea
Planting in isolated holes, pits or mounds	Pacific	Pacific, Gilbert
		Ellice
		Caroline, Marshalls
		French Oceania
		Fanning
		Garden
		Hull
		Washington
		New Guinea
Arranging felled trees or rubbish in transverse rows on hillside	Pacific	New Guinea
Restricting digging of large plots to lowlands	Pacific	
Field rotation	Samoa	Yap
	New Zealand	Guam
Fertilizer uses	Polynesia	Gilbert Islands
	Micronesia	East Caroline Islands
Using pigs for rooting		New Guinea
Burning to release nutrients into soil	New Zealand	New Guinea
Using human waste		New Guinea
Alteration of soil to improve texture	New Zealand	
Using silt to change soil texture	Rapa, Hawaii	
Shelterbelts	New Zealand	New Guinea
Diff. crop intercultivation		New Guinea
Knowledge of soil types	New Zealand	New Guinea
Restrictive taboos	New Zealand	

Source: G. A. Klee, 'Oceania' in G. A. Klee (ed.), *World Systems of Traditional Resource Management*. London: Edward Arnold, 1980, 245–81.

new layer of biological productivity to cultivated systems. In the coastal zones of the larger islands and throughout the atolls of the Pacific, marine resources were always of considerable importance: fish dominated, but invertebrates, dugong, crocodile and turtles were all eaten at some time. These resources were not randomly harvested, but actively managed by a variety of cultural and technological means which apparently ensured that overuse did not occur (table 4.9), practices which were reinforced by quite rigorous controls on the numbers of the human population.

Islands are usually thought of in terms of ecological and biotic instability, often brought about by human activity. By AD 600, to quote one instance, Polynesians had colonized the Hawaiian islands, and between that time and the arrival of Europeans, it is postulated that at least 39 species of land birds became extinct from predation and habitat change. In New Zealand, the use of fire by the Maori to

Table 4.9 Traditional marine resource management in Oceania

Form	*Examples (ancient)*
Traditional fisheries ecologists[a]	Samoa, Lau (Fiji), Lamotrek, Society
Marine tenure systems	
Fishing rights to specific areas[b]	Marshalls, Papeete, Society, Guam
Specific species regulation[a]	Samoa, Hawaii, Mangareva, Society
Optimum fishing seasons according to traditional time reckoning[a]	Society, Mangareva, Tokelau, Niue, Kapingamarangi, Fiji, Trobriand, Palau
Closed seasons[a]	New Zealand, Hawaii, Society, Mangareva, Pukapuka
Food avoidances[c]	Society, Hawaii
Conservation of sea food[a]	Pacific
Traps and ponds	Pukapuka, Mangareva
Preservation	Pukapuka, Tokelau
Magico-religious taboos[a]	Samoa, New Zealand, Society, Tokelau, Hawaii, Marquesas, Mangareva, Pukapuka, Truk, Ponape, Palau
Fines and punishment[a]	Marshalls, Hawaii, Mangareva, Pukapuka

[a] Recognized conservation practice
[b] Recognized conservation practice in some cultural groups and inadvertent in others
[c] Inadvertent conservation practice

Source: G. A. Klee, 'Oceania', In G. A. Klee (ed.), *World Systems of Traditional Resource Management*. London: Edward Arnold, 1980, 245–81.

clear land and hunt the moa (which they made extinct) converted some areas of beech and kauri forests to tussock grasslands. The destruction of the forests of the Canary Islands began in the fifteenth century; they were replaced with sugar plantation, subsistence farming and grazing animals. This was followed later by the taking of naval stores, pine heartwood for building, and even pine needles, and changed the landscapes and environments a great deal. An assessment of the impact of Europeans on Barbados is given by S. W. Tam, who measured the hydrological regime of various plots and showed that with the clearance of forest for sugar cane, runoff was much increased. In turn this accelerated erosion along stream channels, spawned new erosion cells and aided the regrading of slopes through deep-seated slides. In the case of Barbados, deforestation was accelerated by the introduction of domesticates like the goat, which became feral so that non-agricultural areas lost their forest cover in the pre-industrial (though colonial) period. Australia provides an example of an island continent where the advent of Europeans changed the ecology very rapidly. This is summarized in figure 4.12, which shows that directly by various means, and indirectly (e.g. by the displacement of the aboriginal population), considerable changes in vegetation and animal communities has been brought about, and some more details are set out in table 4.10.

Ecologically, mountains have certain features in common with islands in terms of low diversity of species. Equally, they can be rapidly changed by colonial but pre-industrial regimes: in the mountains of South America, overgrazing and deterioration of the pastures

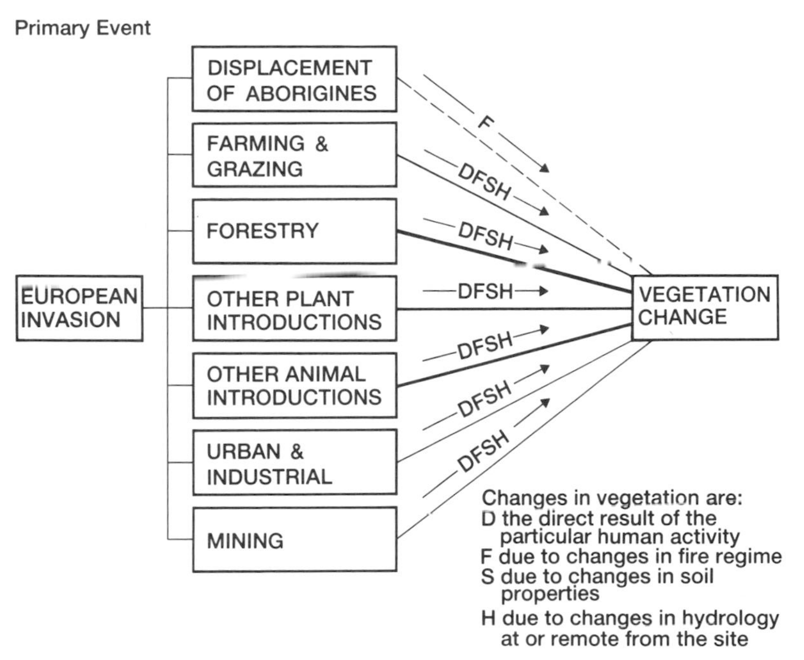

Figure 4.12 A conspectus of the changes in Australian vegetation since the European colonization. The thickness of the connecting bands is roughly proportional to the magnitude of change associated with that factor. *Source*: D. A. Adamson and M. D. Fox, 'Changes in Australasian vegetation since European settlement', in J. M. B. Smith (ed.) *A History of Australasian Vegetation*. Sydney: McGraw Hill, 1982, 109–49.

of the puña are thought to have happened in pre-Conquest times and indeed the puña itself is probably a semi-natural plant community. After the Conquest the introduction of sheep led not only to further erosion but also to the invasion of weedy species together with coarse and unpalatable bunch grasses. Alpine vegetation recovers very slowly if at all from heavy grazing and erosion is inevitable on devegetated slopes. The ecology of the receiving areas is also affected, sometimes positively since more fertile valley soils can result, especially if the silt is held up by terracing. Soil erosion may also expose veins of valuable minerals providing a positive feedback loop for further environmental change. That the use of mountain lands may be stable and non-destructive is shown in the European Alps where hay, potatoes and vines occupied the valleys. The upper slopes where cattle would spend the summer consisted of pasture. But strict regulation of animal numbers was an integral feature of the culture, combined with the spreading of manure on the communally held unfenced grassland. Similarly, the woodland zone was held in common and its use carefully regulated. In some high mountain areas, animals are a key element in subsistence strategies, turning tree branches, chaff, stalks and straw into food and manure. If deforestation occurs then the animals' dung is likely to be used as fuel rather than as fertilizer, with consequent losses in the productivity of the soil: a situation exemplified in the semi-arid mountains of the Middle East, where domesticated sheep and goat populations increased through historic time, so that perhaps only 10 per cent of the forest present in say 2000 BC is now left in the mountains of the lands from Morocco to Afghanistan. The famous cedars of Lebanon, now reduced to a few remnants, stand as a symbol of all this.

Table 4.10 Effects of European settlement on vegetation in Australia

Consequences of European settlement	*Effects and comments*
Displacement of Aborigines	Fire regimes that operated over the previous few to many thousands of years were replaced by new and variable patterns of burning
Cultivation (and crop plants)	
(a) Dryland	Vegetation cleared by ringbarking, felling, bulldozing, burning. Establishment of broad-acre farming, particularly cereals. Also intensive agriculture, e.g. sugarcane, horticulture, etc., from tropical to cool temperate zones. Fertilizers, herbicides, insecticides affect biota. Huge number of plant species introduced. Some types of vegetation almost eliminated. Limited by topography, rainfall and to lesser extent soil fertility. Many areas show specialized problems, e.g. wind erosion on sand mallee soils, water erosion in hillslopes, increase in salinity common in wheat areas in Western Australia. Fire regime of surrounding vegetation altered.
(b) Irrigation	Additional effects which influence vegetation including flooding behind dams; irrigation canals with aquatic weed problems; intensified use of fertilizers, herbicides and pesticides; changed river discharge; salinity increase in soils and rivers
Grazing	
(a) Improved pasture (sheep, cattle and introduced pasture plants)	Vegetation cleared as above, often leaving isolated shade trees which become senescent. Fertilization and sowing of a large number of species of introduced legume and grass species in higher rainfall areas. Soil probably affected by trampling by hard-hoofed animals; grazing habits and intensity of sheep and cattle different from those of native animals. Soil erosion widespread. Fire regime severely altered. Altered insect and bird populations which in turn affect remaining eucalypts (one type of dieback)
(b) Rangeland (sheep and cattle)	Semi-arid and arid rangelands: concentration of grazing pressure around watering places with major vegetation changes, trampling and soil erosion. Widespread changes in composition of vegetation, both in species abundance and in structure. Changed fire regime. Increase in some native grazing animals (kangaroos) with extra watering places. Control of shrubs by clearing, poisoning or massive overstocking has been attempted in many areas, often without success. High rainfall mountainous areas: fire used widely to produce palatable feed. Kosciusko National Park and Victorian Alps were, or are, burned as often as annually.

Forestry	
(a) Native forests	Management to produce uniform-aged stands of fast growing eucalypts by clear-felling and burning led to changes in forest composition by disadvantaging other tree species. Elimination of old trees removes animal breeding sites. Previously uneconomic eucalypt forests are being woodchipped. Changed fire regime is deleterious to Western Australian jarrah eucalypt forests. Relict areas of rainforests are opened to weed invasion along logging tracks. Old trees (to over 2000 years old) containing palaeoclimatic information in tree rings are being logged or lost below dam waters in the temperate rainforests of Tasmania. An argument can be made for a policy of preservation without further logging for remaining rainforest
(b) Exotic forests	Vegetation cleared as for crops by bulldozing, windrowing and burning (pine plantations are seldom established on derelict grazing or agricultural land)
Plant introductions (agricultural species, micro-organisms omitted)	An unknown but very large number of plants has been introduced and many hundreds of species have become naturalized. In the Sydney region, for example, about one fifth (400 species) of the total flora is introduced, and this figure is typical for south-east Australia. Overall about 10 per cent of the Australian flora has been introduced since European settlement. Many naturalized introductions were originally accidental contaminants of seed, etc., or imported as ornamental plants. Eradication of any well-established new plant is not practicable: a new flora is being assembled
Animal introductions (other than sheep and cattle)	Introduced as draught animals (horse, donkey, camel), for hunting (rabbit, hare, fox, deer), as pets (dog, cat), accidentally (mouse, rat), or as food (pig, goat, buffalo). All have become naturalized and a new fauna has been assembled. Rabbits are established throughout non-tropical Australia with severe effects on grass, herb and shrub vegetation, on tree regeneration, and on soil. Myxomatosis has reduced but not removed their effect. Goats are common in semi-arid regions and influence shrub vegetation and soils. Pigs occur particularly in swamplands and influence vegetation and soils. Buffalo occur in tropical swamplands. Deer are firmly established in hilly forest land particularly in Victoria. Camels and donkeys are widespread in the arid regions but are probably limited in numbers by availability of water. A large number of other animals has become naturalized including birds, toads and a wide range of invertebrates. Many have significant direct or indirect effects on native vegetation

Source: D. M. Adamson and M. D. Fox, 'Change in Australasian vegetation since European settlement'. In J. M. B. Smith (ed.), *A History of Australasian Vegetation*. Sydney, McGraw Hill, 1982, 109–49.

To summarize these pieces of evidence and place them in the context of all the work which there is no space to mention, we have to ask whether environmental impact is inevitable in traditional agricultural systems, and in those affected by European colonization in pre-industrial times? In the first case, the impact may be very light particularly if there are feedback mechanisms controlling the intensity of use and, very probably, the level of human population. But even traditional systems may exert very strong manipulative effects (most notably in arid and semi-arid areas) of a kind we would now describe as degradational. Not for nothing, we may suppose, did the Torah prohibit working the land or tending the trees in the seventh year, with a flogging for disobedience, though fields might be watered so that they did not become salty.

Agroecosystems as energy traps

Introduction

In keeping with this book's theme of human access to energy sources, it becomes useful to look at the structure and performance of traditional agricultural systems or agroecosystems in terms of their energy flows, as was done in a preliminary fashion for some hunters (p. 60) and will be done for modern agriculture (p. 239).

The main purpose of an agroecosystem is to provide humans with food, skins or fibres. Nutrition is the most important of these, so that the system has to produce more energy per unit area than it consumes and also to provide proteins, minerals and vitamins. A 'traditional' system must also be sustainable in the long term since societies cannot continually produce technological fixes to solve problems. A surplus for trade is also a desirable feature of an agroecosystem: even 'traditional' groups have rarely been totally isolated.

Human-managed and natural systems

The main characteristics of a natural ecosystem have evolved in response to a set of pressures of an evolutionary and ecological kind during a variety of time-scales. With human use, the temporal horizons change, since our demands are usually immediate: a year is a long time in farming given that people can starve in considerably less. Compared with hunting and gathering, agriculture is an intensification of use, representing a cultural desire to take more energy per unit area time than could be cropped from the more natural system. (This is not to say more than was present in the natural system, but more that was culturally acceptable. A grassland created out of deciduous forest and grazed by cattle may yield at most 0.3 per cent of NPP as cattle tissue production, whereas the forest if harvested for wood, wild animals and plant foods might yield as much as 10 per cent of NPP.) In the most general of terms, management of ecosystems involves the replacement of wild species with domestic ones, necessitating the elimination of a lot of NPP which is labelled 'weeds'. Thus the overall NPP of the agroecosystem is lower than the natural system it replaces, and the culturally acceptable crop is a small proportion of the total NPP, getting smaller if animals are kept. Nevertheless some agroecosystems can compare favourably with steady-state natural ecosystems since their productivity is cropped during a phase of rapid growth rather than at a plateau stage of self-maintenance. Comparison between the early successional phase of the natural ecosystem and the cultivated crop would perhaps be more realistic. Looked at in global perspective, however, the mean NPP of cultivated lands today (650 $g/m^2/yr$) is about that of a temperate grassland rather than that of a tropical forest (2200 $g/m^2/yr$) or an algal–coral reef (2500 $g/m^2/yr$).

Solar energy has to be fixed in chemical form if it is to be useful to humans so the supply of nutrients is also critical, and can indeed be limiting. The manipulation of a natural system, particularly one formerly in a steady state, will have some effect on the nutrient cycles. Mature systems generally seem to have evolved a set of conserving mechanisms which retain the nutrient pool at a steady level: losses to runoff are balanced by inputs from soils and the atmosphere, and these flows are generally small in relation to the quantities cycling between soils and plants. The system may colloquially be described as 'tight'. Many of the retention mechanisms are likely to be broken down in an agroecosystem: the soil, for example, will probably be exposed to rainfall for much of the year, enhancing the potential for nutrient loss via runoff. Similarly, higher soil temperatures may result in more rapid release of nitrogen from humus into the soil and this in turn may be quickly lost if not immediately taken up by the crop. Nutrients harvested as agricultural crops may be recycled to the soil pool as animal manure or human excreta (the latter, here expressed in its Latin form, is mostly known in the literature by the delicate euphemism of 'night soil', a convention reluctantly adopted henceforth in deference to readers with an aversion to plain Anglo-Saxon). On the other hand, the nutrients may be traded off to another place and not be replaced other than by the slow mechanisms of input from soil and atmosphere. The balance between input and export is clearly critical in the maintenance of the output of the system through time.

Energy flow and production: some examples

The energy yield of a crop must have deducted from it the cost of its production in energy terms. Lacking the access to hydrocarbon fuels of the industrial-age agriculturalist, a solar-powered cultivator has to use his own energy and that of his family and animals (all of whom have to be fed, though the animals may be able to crop wild plants indigestible to humans) in the manifold tasks of preparing the ground, sowing, weeding, fencing, scaring, harvesting and processing. One of the measures of an agroecosystem, therefore, will be the ratio E_r (input/output), normally expressed in a 1 : n fashion. This is not the whole story for even a high ratio does not necessarily guarantee that there is sufficient protein for a family's nutritional needs, nor that the crop when harvested can be successfully stored until the next one is due. The E_r statistic is thus a guide to the fitness of an agroecosystem, but like all reductionist measures it tells only part of the story.

In this section some examples of studies of energy flow and production in pre-industrial agroecosystems are summarized. They are not uniform in purpose and methodology so exact comparisons are difficult to make, but there are sufficient indications of their structure to make comparisons with both earlier societies and those with access to industrial technology.

(a) *Plant crop systems: shifting cultivation* Although once widespread, this mode of production is now mostly confined to the tropics and studies are from modern agriculture in those regions. Just how such systems at the present day differ from their equivalents before European contact is difficult to say with any accuracy, although most workers would accept the generalization that fallow periods have become much shorter since the nineteenth century, often because of increasing population density. Along with these changes have come greater degrees of land modification, more extensive use of animal rearing to provide protein and manure, the progressive transformation of adjacent wild plant communities and the concomitant elimination of many of the wild fauna.

A society which suffered little from high population densities and land shortages was studied in the upper Rio Negro area of the Amazon Basin (Venezuela) at about 120 m above sea-level, where the people cleared patches of mature tropical forest 0.5–2.0 ha in size and planted yuca (*Manihot esculenta*) as a staple (90 per cent of edible crop production) along with a variety of other roots, fruits and pulses. Even after considerable effort, weeds comprised 5.6 and 12.8 per cent of the total above-ground plant productivity in years one and two respectively of the clearing, and some 17 per cent of the phosphorus in the yuca was removed by species of sucking and chewing insects. The food crop production was 146.5g/m^2 in year one and 119.2 g/m^2 in year two, this decline being common in shifting cultivation systems, and the E_r for year one was 1 : 14.8, and for year two 1 : 12.6, with an average of 1 : 13.9 for the two years during which the plot was maintained. A nearby plot allowed to colonize back to forest had a lower NPP than the cultivation in the first year (532 vs 109 g/m^2) but far exceeded it in the second year (1446 vs 529 g/m^2), showing the difference in NPP between cultivated and natural vegetation.

Recent investigations of shifting cultivation have also been carried out in the highlands of Papua New Guinea where progressive intensification and sedentarization of field systems can be seen. At the 'extensive' end of the regional spectrum are the Miyanmin who live in montane rainforest between 450 and 1400 m above sea-level. Their staple crop is taro but they also grow a wide variety of other crops, with a practice of only partially clearing the forest and taking only a single crop before abandoning the plot. A certain level of pig-rearing is carried on, though wild boars are relied upon to fertilize the domestic sows; fields are little fenced against either. Hunting of wild pigs and native mammals is also important. The E_r of their cultivation is 1 : 17.5, of hunting and plant food collection 1 : 8.9, of pig-herding 1 : 4 to 1 : 6 so the net efficiency of their subsistence is 1 : 7.8 where the inputs include only labour put into food production and collecting. If the net food gain is divided by total labour, including travelling, inter-group politics and warfare, then 1 : 3.8 is a more reliable level. The partially quantified basis of the energy flow through the Miyanmin system is shown in figure 4.13.

Other groups of New Guinea highlanders provide us with a linkage to rainfed but permanent-plot agricultural systems. The Tsembaga Maring, for example, live up to 1800 m a.s.i. in a rainforest which is interspersed with savanna grassland. The Tsembaga are more intensive cultivators than the Miyanmin. They fence their fields and keep them in use for several years, expend energy on the prevention of soil erosion, and control the species succession of the fallow. Their energy flow is shown in figure 4.14; it results in a cultivation efficiency of 1 : 18. They have more pigs per head than the previous group so the efficiency of their pig herding is 1 : 15, bringing the overall efficiency down to 1 : 10.2 but delivering 349g/head/day of protein from the pigs. Thus some energy is sacrificed to provide protein in meat form (and to support a ceremonial system based on pig slaughter), showing again that E_r measurements are not a total measure of how 'good' a system is. The channelling of plant energy into animals is taken even further by the Raiapu Enga who live between 1500 and 2000 m above sea-level in a savanna-grassland-dominated valley without access to forests. All their pigs are domestic (there is no contact with wild boars) and are fed on some of the produce of continuously cultivated open fields devoted largely to the cultivation of the sweet potato, along with a variety of subsidiary crops. As much as two thirds of the garden produce goes to the pigs, which are kept in a ratio of up to 2.3 pigs/head, compared with 0.8 pigs/head for the Tsembaga Maring and 0.1 pigs/ head for the Minyanmin. Ritual exchange and ceremonial involving pig slaughter distribute these animals' products over an area of 8000 km^2. Now the Enga also grow coffee and western vegetables which they exchange for canned food, rice and other store goods (figure 4.15).

One footnote should be added to studies of E_r in shifting cultivation systems: for an accurate accounting, the energy value of the burned-off vegetation should be deducted from the output. In savannas this is relatively small, but in forests, one estimate for a forest stand suggests that 1720×10^6 kcal/ha should be levied against the calorific value of the ensuing crops. This modified an E_r of 1 : 13.9 to one of 1 : 0.005 and suggests that, contrary to received wisdom, shifting cultivation is an inefficient agricultural system. The corollary that the longer the trees are on the site, the higher the efficiency, has implications for current use of the tropical moist forest biome.

Figure 4.13 (p. 132) The first of three diagrams showing different energy flow patterns among montane people of Papua New Guinea, and especially their use of the pig. Here in the Miyanmin people, pig management is extensive rather than intensive: much of the pork consumed is from the wild and most of the village pigs are in fact tamed wild piglets; people shift their settlements in accordance with wild pig activity. Units are 10^6 kcal/ hamlet/yr.
Source: G. E. B. Morren, 'From hunting to herding: pigs and the control of energy in montane New Guinea' in T. Bayliss-Smith and R. Feachem (eds) *Subsistence and Survival. Rural Ecology in the Pacific*. London: Academic Press, 1977, 273–315.

Figure 4.14 (p. 133) The Maring people manage pigs more intensively. Most pork is from domesticated animals and most pigs are from domestic sows; land is brought into cultivation in order to grow crops, about one quarter of which is fed to pigs. Units are 10^6 kcal/mi^2/yr.
Source: G. E. B. Morran, 'From Hunting to Herding'.

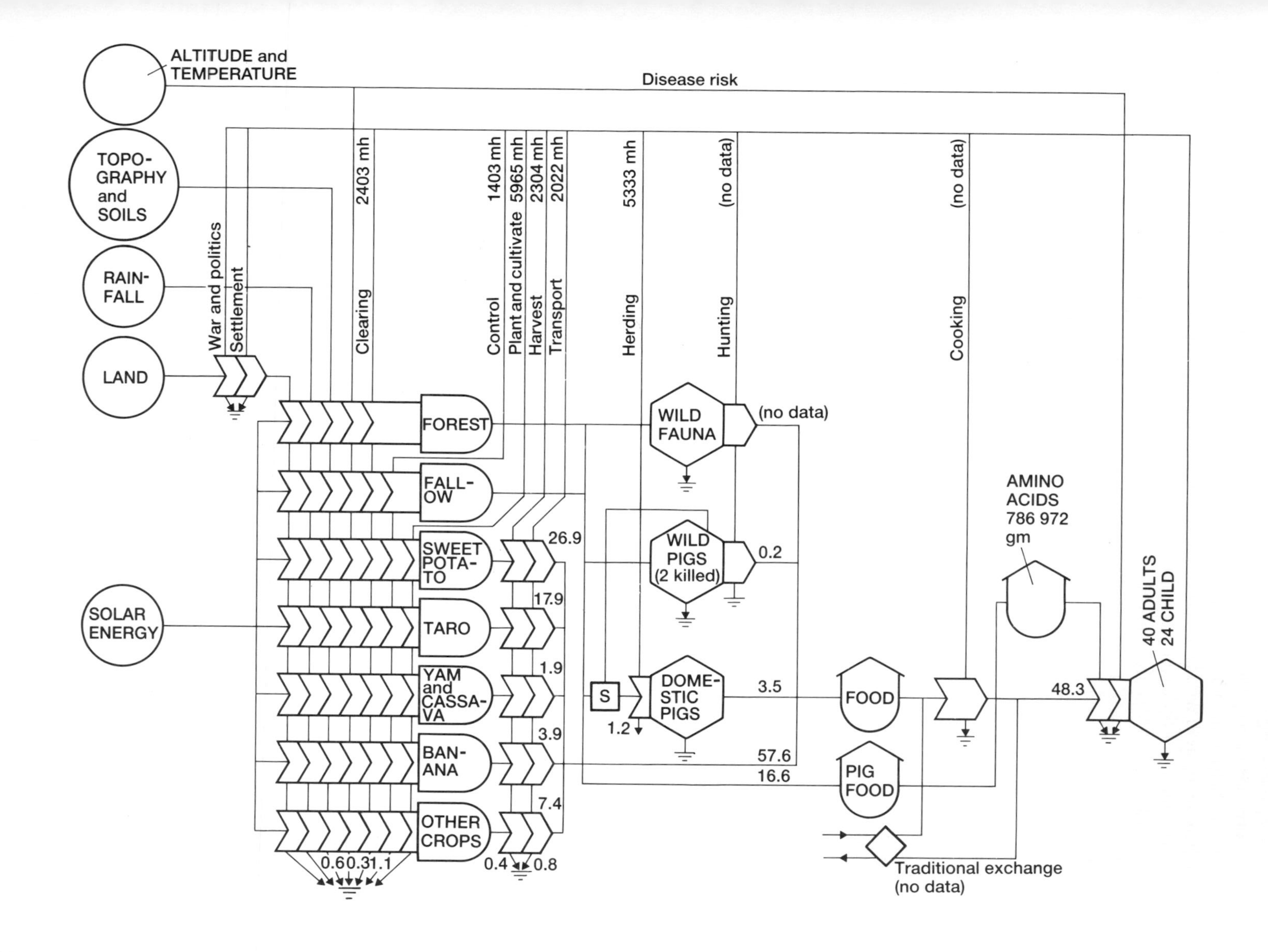

ALTITUDE and TEMPERATURE
Disease risk
TOPO-GRAPHY and SOILS
RAIN-FALL
LAND
War and politics
Settlement
Clearing
2403 mh
Control
1403 mh
Plant and cultivate
5965 mh
Harvest
2304 mh
Transport
2022 mh
Herding
5333 mh
Hunting
(no data)
Cooking
(no data)
FOREST
FALL-OW
SWEET POTA-TO
TARO
YAM and CASSA-VA
BAN-ANA
OTHER CROPS
SOLAR ENERGY
0.6
0.3
1.1
0.4
0.8
26.9
17.9
1.9
3.9
7.4
WILD FAUNA
(no data)
WILD PIGS (2 killed)
0.2
S
1.2
DOME-STIC PIGS
3.5
57.6
16.6
FOOD
PIG FOOD
AMINO ACIDS 786 972 gm
48.3
40 ADULTS 24 CHILD
Traditional exchange (no data)

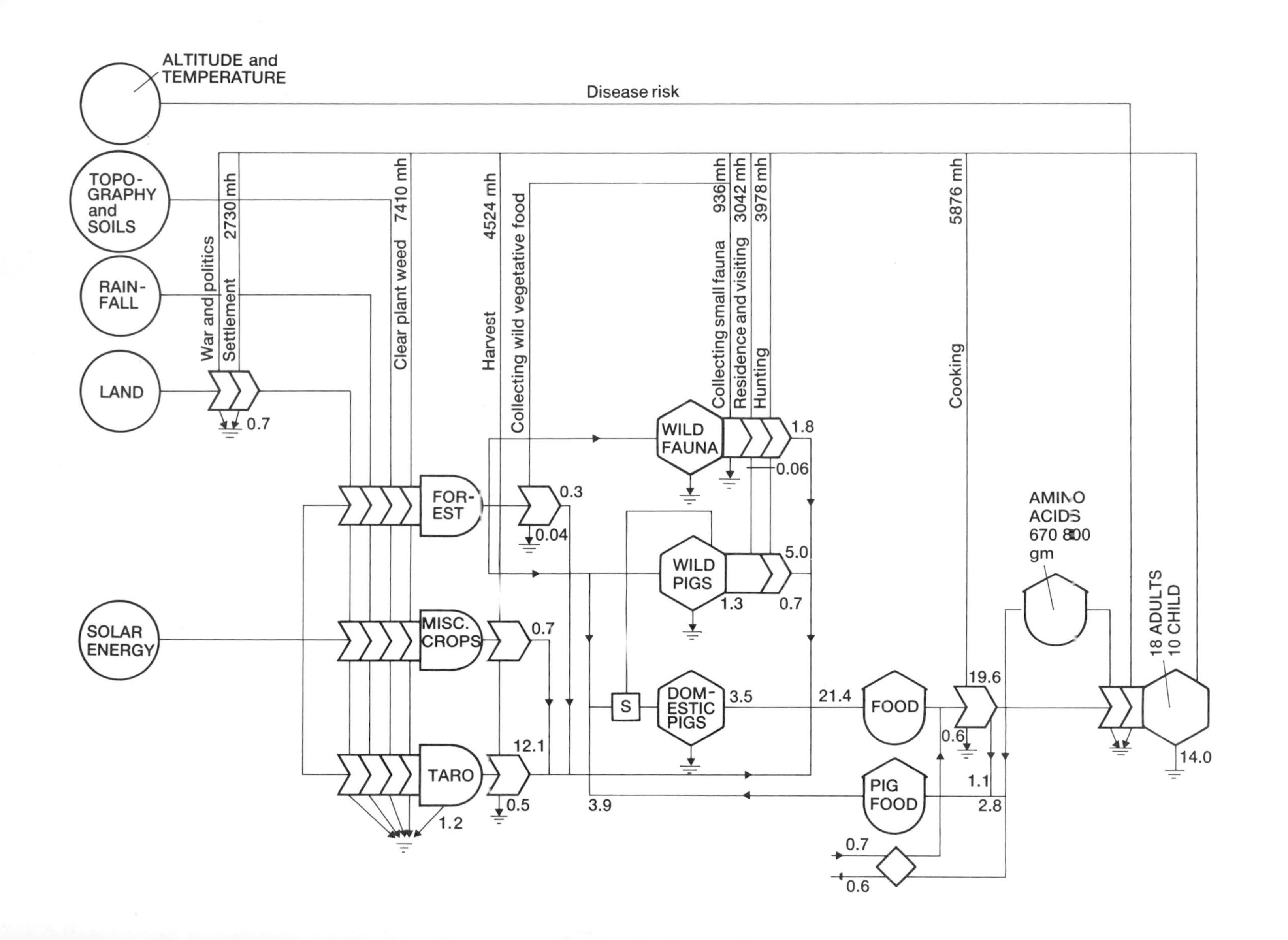

ALTITUDE and TEMPERATURE
Disease risk
TOPO-GRAPHY and SOILS
RAIN-FALL
LAND
SOLAR ENERGY
War and politics
Settlement 2730 mh
Clear plant weed 7410 mh
Harvest 4524 mh
Collecting wild vegetative food
Collecting small fauna 936 mh
Residence and visiting 3042 mh
Hunting 3978 mh
Cooking 5876 mh
0.7
FOR-EST
MISC. CROPS
TARO
1.2
0.3
0.04
0.7
12.1
0.5
WILD FAUNA
1.8
0.06
WILD PIGS
5.0
1.3
0.7
S
DOM-ESTIC PIGS
3.5
3.9
21.4
FOOD
PIG FOOD
0.7
0.6
19.6
0.6
1.1
2.8
AMINO ACIDS 670 800 gm
18 ADULTS 10 CHILD
14.0

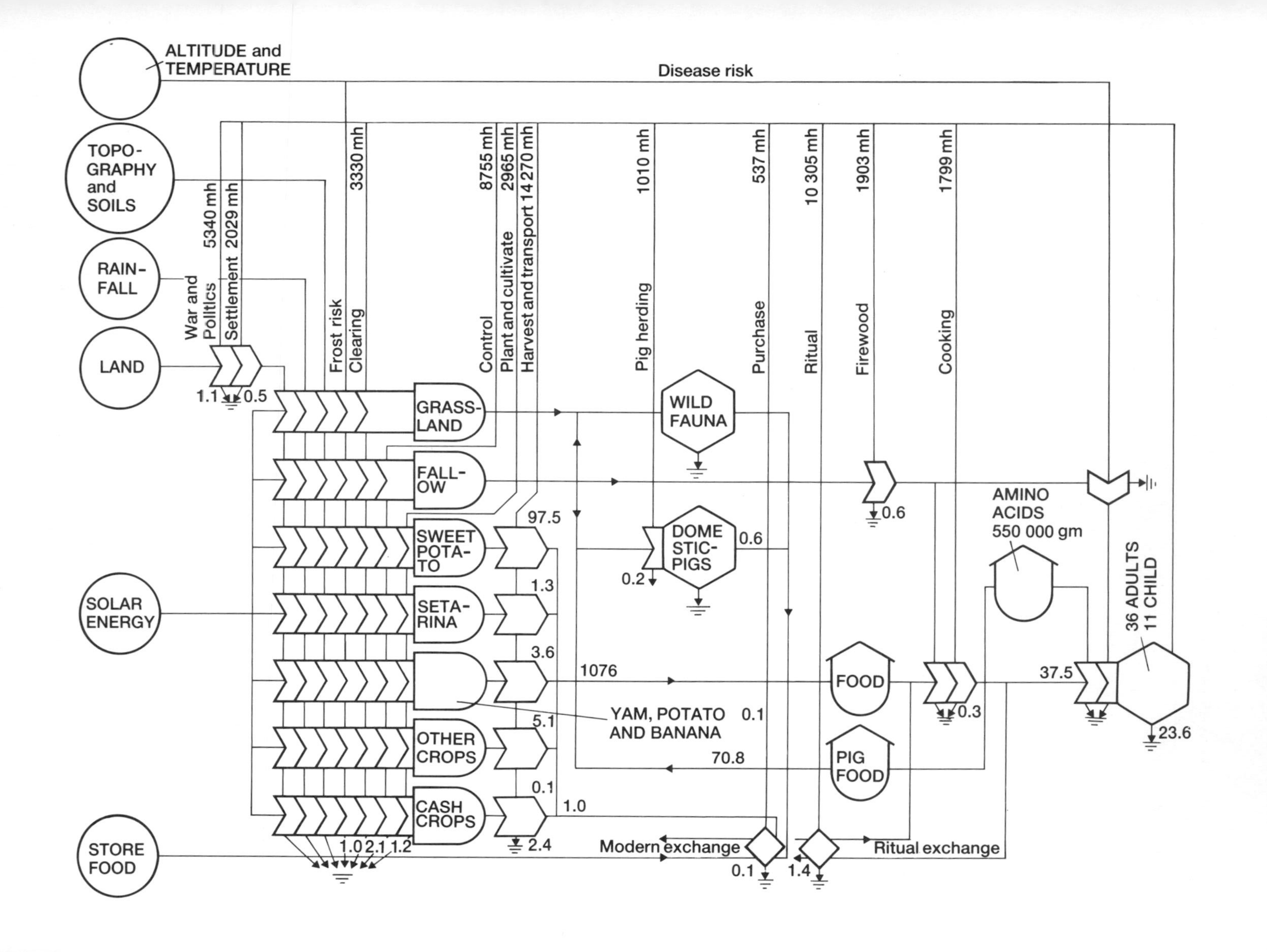
ALTITUDE and TEMPERATURE
TOPO-GRAPHY and SOILS
RAIN-FALL
LAND
SOLAR ENERGY
STORE FOOD
Disease risk
War and Politics 5340 mh
Settlement 2029 mh
Frost risk
Clearing 3330 mh
Control 8755 mh
Plant and cultivate 2965 mh
Harvest and transport 14 270 mh
Pig herding 1010 mh
Purchase 537 mh
Ritual 10 305 mh
Firewood 1903 mh
Cooking 1799 mh
1.1
0.5
GRASS-LAND
FALL-OW
SWEET POTA-TO
SETA-RINA
OTHER CROPS
CASH CROPS
1.0
2.1
1.2
97.5
1.3
3.6
5.1
0.1
2.4
1.0
WILD FAUNA
DOME STIC-PIGS
0.2
0.6
1076
YAM, POTATO AND BANANA
0.1
70.8
FOOD
PIG FOOD
0.6
AMINO ACIDS 550 000 gm
0.3
37.5
36 ADULTS 11 CHILD
23.6
Modern exchange
0.1
1.4
Ritual exchange

(b) *Plant crop systems: sedentary agriculture* An example of non-industrial agricultural production with only a little irrigation is that of broad beans in China during the period 1929–33, written up by K. Newcombe. Both yield and energy input are low, with heavy applications of night soil (25–50 t/crop ha) and some animal labour; the E_r is 1 : 0.64, showing that the energetics of the crop were negative so it is possible that the main role of the beans was protein rather than energy. Maize in Mexico using only manpower raised an E_r of 12.5, whereas in north-east China in the 1950s, maize was grown with no fossil-fuel-based inputs except tools and horse-drawn implements. This system yielded an E_r of 2.44. Growing grain sorghum in Nigeria using draft animals produced an E_r of 0.90 and bullock-pulled wheat cultivation in Uttar Pradesh an E_r of 0.96. Root crops may give more organic matter: sugar beet in Germany (horse-powered) gave 2.03. The congruence of the grain-crop figures suggests their validity and also the general precariousness of any nutrition based solely upon such crops, especially if post-harvest losses (through pests or taxation) are significant.

Rice is known for its high yields where there is gravity-fed irrigation: in experimental conditions in the Philippines, a traditional system gave an E_r of 9.8, but a modern high-energy regime only 7.9, and yields were not significantly different. The introduction of irrigation needs higher levels of energy input: one study cites a 145 per cent increase in total human energy expenditure when irrigation is carried out, and clearly the E_r declines when animal power supplements human energy input. The E_r figures for wet rice grown in rainfed conditions in Thailand, by gravity irrigation in Taiwan and water-lift irrigation in Madras were shown to be 12.5, 18.8 and 1.7 respectively. Thus there is a considerable fall-off in efficiency when energy has to be expended in lifting water. What may be critical in such systems is whether the animals can feed from forage gathered with little or no human effort or whether in fact a great deal of crop energy is expended in feeding the animals.

A great advantage is conferred upon a society with a reliable gravity-feed irrigation system, such as the Nile valley. Reconstructions of its agriculture at the end of the eighteenth century show that year-round cropping was made possible by the retention of Nile floodwaters in small ponds, and that salinity was not a problem. Even using a roughly equal number of animal and human workdays, the Nile valley input of 9.61 GJ/ha/yr[1] was rewarded with 19.25 GJ/ha/yr of edible energy for humans, together with 62 animal feed days of clover and 122 feed days of straw. A human population density of 4.5 per ha was supported. These data suggest that yields were four to ten times greater than for English wheat at the time and more than twice those of France. If such relationships held good in earlier times then the availability of wheat for export to the Roman Empire would have been very great, and given that 2500 kcal/day plus 85 g of plant

[1] The prefix 'G' = Giga (× 10^9).

Figure 4.15 (*left*) The Raipu Enga groups extend the paterns of the Marings. All pigs are domesticated, settlement is sedentary, and more than half the produce grown is fed to pigs. The food chain leading to pigs has become dominant at the expense of the other food chains, a sequence that can be seen throughout these three diagrams. Units are 10^6 kcal/community/yr.
Source: G. E. B. Morren, 'From Hunting to Herding'.

protein could be supplied to each person in AD 1800 by 2 h/day of adult male labour, then a labour surplus could have been diverted to public building works in the Old Kingdom (2686–2160 BC), for instance, without hazarding food supplies. A pre-industrial farm in England, using the best knowledge of the time, could, however, manage an E_r of 1 : 40 (data for a farm in Wiltshire in 1826) with only 5 per cent of the energy input being attributable to fossil fuels, mostly in the form of tools. Such a farm produced bread, beer and bacon for its own subsistence, but 98 per cent of the crops of wheat, barley and sheep were sent away from the farm (figure 4.16).

Contemporary China represents a national agroecosystem on the brink of becoming an industrial rather than a pre-industrial system (Plate 4.6). The data (figure 4.17) show the continuing importance of energy from people and animals and the reliance on plant fuels rather than fossil fuel, and the way in which the lack of surplus leads to a diet dominated by plant material. Not that the agriculture is 'poor': output/ha is 5 times that of the USA, and 3.5 times that of France, though 20 per cent below the Dutch and 40 per cent below the Egyptian systems of the present day.

(c) *Animals in food systems* A generalized model of the energy flow for savanna pastoralists is shown in figure 4.18. A skeletal energy flow diagram for a pastoralist group (the Karimojong cattle herders of northern Uganda) has been interpreted to suggest that the animal products (blood, milk and meat) would supply about 20–35 per cent of the people's energy requirements, the rest coming from traded sources. About 0.07 per cent of NPP is converted to net secondary production in the Karimojong system, a figure which can be pushed up to 0.47 per cent for cattle grazing on the High Plains of the USA,

Plate 4.6 The rice harvest at Yangshuo in Guangxi Province, China. Even without much mechanization, this form of agriculture demonstrates how completely the available surface is transformed: right up to the foot of the precipitous peaks. (MarkEdwards – Still Pictures/Panos Pictures)

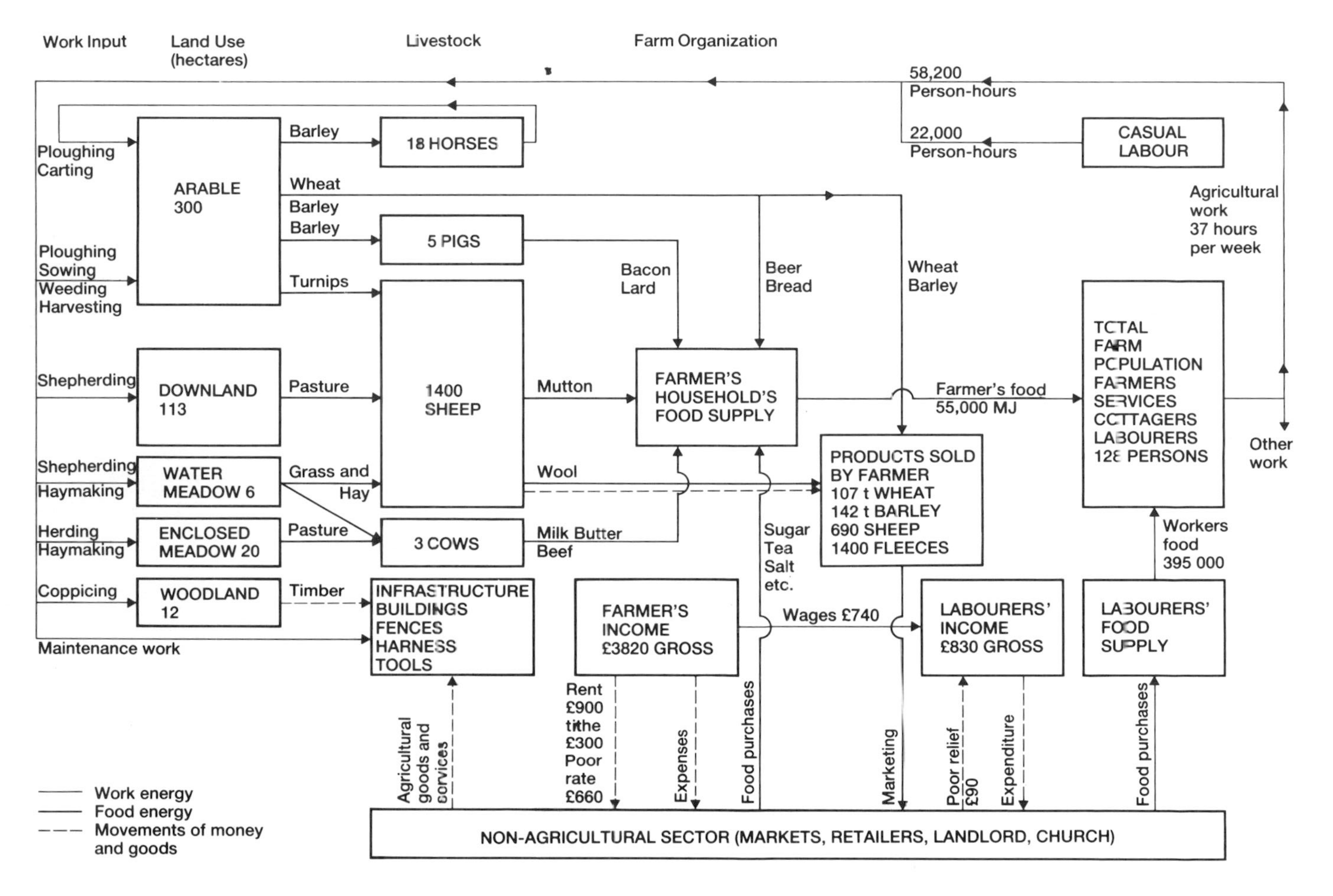

Figure 4.16 The flow of energy and materials on a largely pre-industrial Wiltshire (England) farm in 1826. Note that the interchange with the non-agricultural sector involves no purchase of energy.
Source: T. Bayliss-Smith, *The Ecology of Agricultural Systems*. CUP, 1982.

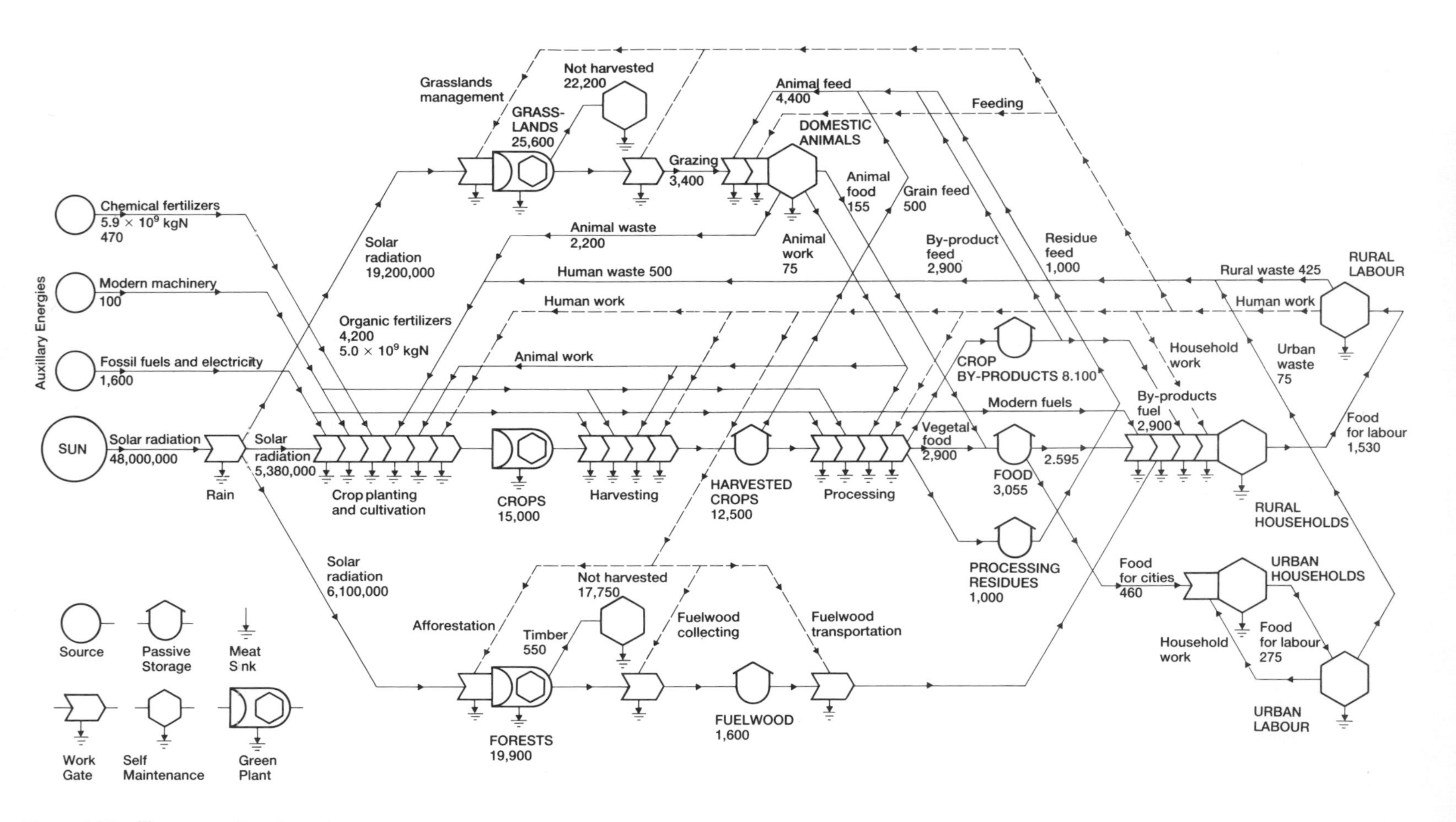

Figure 4.17 The energy flow through rural China in the 1970s, before the present surge of modernization. Fossil fuels and electricity are not absent but their magnitude compared with the solar input is very small. The dashed lines represent control loops in the form of human work. Units are 10^{12}kcal/yr.
Source: V. Smil, 'China's agro-ecosystem', *Agro-Ecosystems* 7, 1981, 27–46.

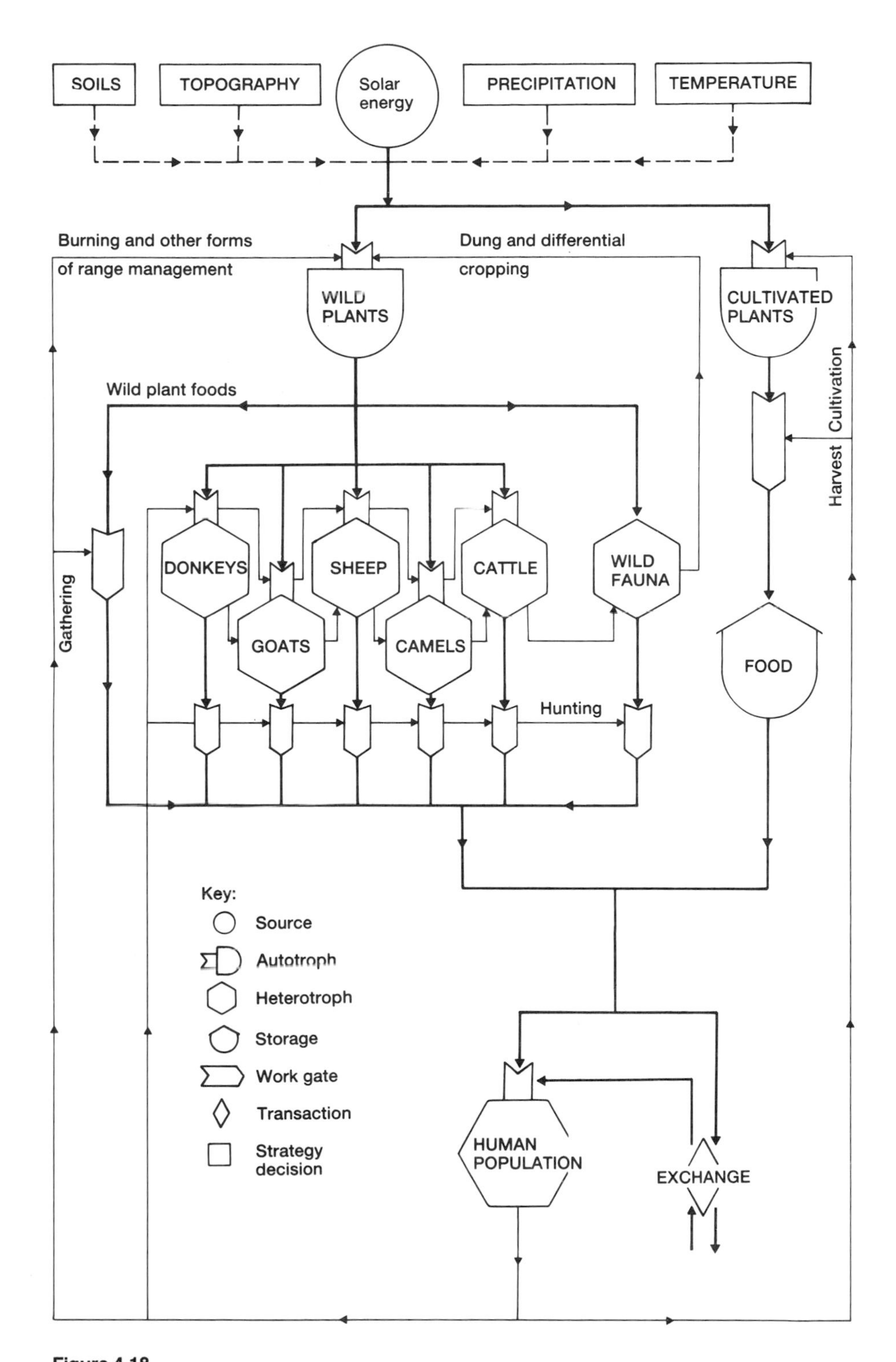

Figure 4.18
Energy flow for savanna pastoralists: a general model. The flow of solar energy is from top to bottom; control loops from human work run the opposite way and are in lighter lines.
Source: D. R. Harris (ed.), *Human Ecology in Savanna Environments*. London: Academic Press, 1979.

using the optimal six months of the year and a good deal of scientific knowledge.

Animals which form part of an agricultural system often have two distinct roles as sources of food. One is to furnish protein in the form of meat and milk (less often as blood), and the other is to be sold outside the immediate locality to provide money for, variously, luxuries, or more animals, or foodstuffs not supplied locally. A modern example of this is seen in the margins of the Sahara in Tunisia, where sheep, goats and camels provide only 3 per cent of the caloric intake. This of course puts pressure on the cultivated parts of the regional ecosystems. Thus if the plant crop yield declines, even more pressure is put on the semi-arid grasslands, with the consequent risk of desertification.

The apotheosis (somewhat literally) of grazing animals comes in the shape of cattle in India, which have a gross energetic efficiency of 17 per cent and are seen by some to be a drag on the nutrition of the human population, especially in rural areas. Other interpretations, however, stress that they are not in competition for food supplies and that the cattle convert material inedible to people into work, milk, and dung. This last product is potentially a great source of fertilizer although one study in Bengal showed that around 70 per cent of it was used as fuel. So 3770 cattle at a density of 252 per km^2 produced 16 341 GJ/yr of dung, of which 10 894 GJ was used as fuel, together with 2346 GJ/yr of work from bullocks, and 752 GJ/yr of milk. In fact, there are further products, because older female cattle are permitted to be sold to Muslims for slaughter.

The integration of the Indian cattle into the rural economy was to some extent paralleled by the similar incorporation of oxen and horses into rural pre-industrial Europe. The work of T. Bayliss-Smith on agricultural ecology provides a more closely documented example of an English pre-industrial farm of 460 ha in Wiltshire. It is, however, for the 1820s when a good deal of post-medieval social, economic and technical change had taken place. Nevertheless, in his diagram the role of sheep in a crop-raising economy (figure 4.16) emphasizes an element of continuity: there was very little fallow and so nutrient levels for the crops of barley and wheat were kept up by means of a flock of sheep which fed successively on downland, arable land planted with swedes, and water meadow. When the sheep were feeding on the down or the meadow, then they were folded onto arable areas at night, when they dropped most of their manure. So two crops out of three were cereals and there was wool to be sold off the farm. A large farm could thus support 32 employed persons and 98 dependants without an energy subsidy, and without imported fertilizer a good income was provided for the farmer together with a net energy production of 7390 MJ/ha.

In central Pennsylvania today there live within a similar environment modern American farmers and two types of deliberately conservative, religiously motivated, Amish communities who have turned their backs on modern energy-consumptive agriculture while still absorbing modern scientific knowledge. The energy input/output ratio (E_r) for modern farmers was 0.553, i.e. every unit of food production at the farm gate required two units of energy input apart from the sun. For the Old Order Amish the ratio was 1.009, with much of the produce being in the form of milk. The American farm thus required about 83 per cent more energy to produce the same quantity of milk as the Amish, and the yields per hectare are much the same. The most conservative Amish, known as the Nebraska Amish, produce an E_r of 1.058 but their products are deemed to be of a lower quality: their milk for example cannot be cooled to the requirements set by governmental regulations for potable milk. Their yield was 7155 MJ/ha/yr, whereas the other Amish and the Americans achieved outputs of the order of just over 12 500 MJ/ha/yr. Thus it appears that relatively labour-intensive farming, without the environmental impacts associated with high-energy modern farming (see pp. 00) can compete in terms of productivity.

Another interesting example of the role of animals in a society is given by studies of the Indians of the Nuñoa area of the Peruvian altiplano, at about 4000 m above sea-level (figure 4.19). Here the energy of the people comes partly from crops such as potatoes and other tubers, and the native grains *quinoa* and *canihua* which are Chenopods. But the Nuñoa people also herd llamas, alpacas and sheep on the *puña* grasslands and these animals play a vital role in the energetics of the region. By contrast with cultivation ($E_r = 11.5$), the efficiency of herding is relatively low ($E_r = 2.0$), but the people engage in trade with lower ecological zones and exchange the animal products (meat, wool and hides) for high-calorie plant foods (maize, wheat and barley) as well as material goods. Thus the net energetic efficiency of herding rises to 7.5 and in fact herding then produces altogether 83 per cent of the energy of the Nuñoa population. This does not include the value of the dung which is produced by the livestock and used as fertilizer for the crops. The decision as to how much dung is used as fertilizer and how much for fuel is an important one: there is no access to unlimited forests as in medieval Europe.

Following this theme, an instance of the role of forests in agriculture can be found in the hill lands of the central Himalaya where combinations of wheat/soybean and potato/cabbage crops are grown. The vegetable systems yield 1.7 times the energy of the grain crops but require 25 times the fertilizer and 151 times the manure of the wheat and soybeans. The E_r of wheat crops (including by-products) was 14, and of vegetables 0.46. Much of the input comes from the surrounding forests in the form of fodder for the bullocks which are the main source of power in the system: some 76 per cent of their fodder requirement is thus satisfied, compared with 22 per cent from the croplands. The forest also provides fuel for the village so that dung can be used as fertilizer (figure 4.20). The hill-lands of north-east India provide an opportunity to compare the E_r values of shifting agriculture, terrace cultivation and valley rice, with respective values of 3.17 (for an average 5-year cycle), 1.7 and 16.22

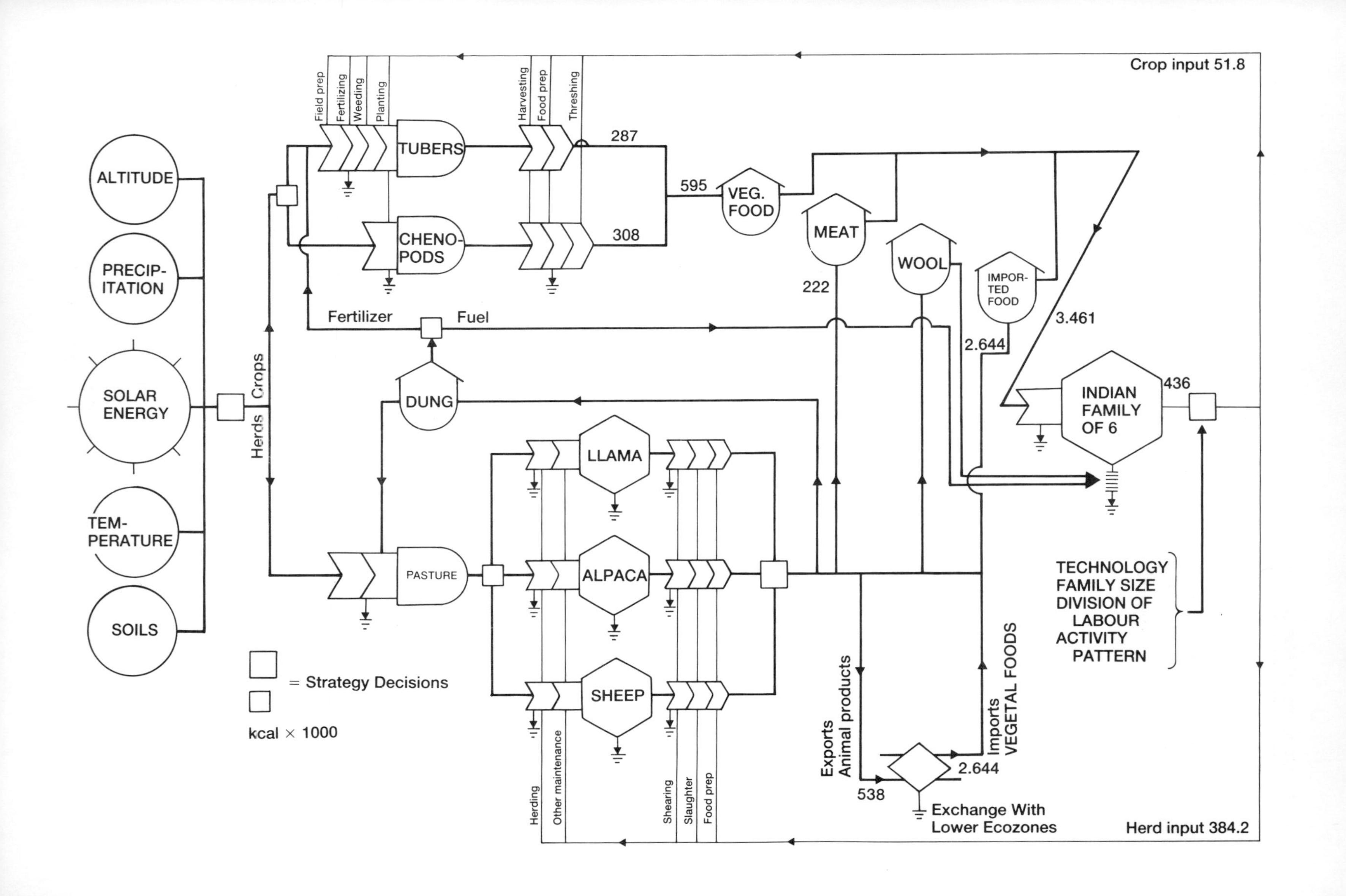
Crop input 51.8
ALTITUDE
PRECIP-ITATION
SOLAR ENERGY
TEM-PERATURE
SOILS
Crops
Herds
Field prep
Fertilizing
Weeding
Planting
TUBERS
CHENO-PODS
Harvesting
Food prep
Threshing
287
308
595
VEG. FOOD
MEAT
WOOL
IMPOR-TED FOOD
222
3.461
2.644
Fertilizer
Fuel
DUNG
INDIAN FAMILY OF 6
436
LLAMA
ALPACA
SHEEP
PASTURE
TECHNOLOGY
FAMILY SIZE
DIVISION OF LABOUR
ACTIVITY PATTERN
= Strategy Decisions
kcal × 1000
Exports
Animal products
Imports
VEGETAL FOODS
538
2.644
Exchange With
Lower Ecozones
Herding
Other maintenance
Shearing
Slaughter
Food prep
Herd input 384.2

respectively. Where there is space for a 15-year shifting cycle, the value is 25.6. Thus the effort needed to construct and maintain terraces (for mixed cropping in this case) appears to make it inefficient energetically since it does not add anything to the mix of crops, and requires heavy inputs of fertilizer which are not needed in the other two systems. Even so, the soil appears to deteriorate after 6–8 years. Terracing is not part of the indigenous repertoire of the region, and was introduced by the Soil Conservation Department in the 1970s. It seems as if the ancients got it right.

Perhaps the most closely integrated agroecosystem of pre-industrial times based on animals was the city. A study of Paris based on sources of 1858 and 1869 suggests that urban cultivation provided a significant proportion of the city's fresh food, gained a valuable export market and transformed a pollution problem into a fertile asset. The basis was the total of about 96 000 horses used to power the city's transport system. The manure from these animals was applied to 7800 ha of gardens at a rate of 30 cm/yr and the heat of fermentation plus cloches, straw mats and 2 m high walls made possible the production of 3–6 crops/yr of salads. Although the E_r was low (1 : 0.25) because of the labour-intensive nature of the work, 110,000 t/yr of produce provided 1.5 per cent of the city's calories and 2.4 per cent of its protein (and incidentally added 6 per cent to the solar flux: no wonder Paris was such a desirable place in the naughty nineties). Outside its limits at that period Paris had an irrigated agricultural system based on the city's sewage.

Efficiency and stability

Looking over the range of energy input/output ratios in these pre-industrial agroecosystems, it is apparent that those based on plant foods produce ratios such as 1 : 10–20 where rainfed shifting cultivation or semi-intensive cropping systems are used. Even higher ratios can be obtained from non-cereal crops, notably tubers, though often at the cost of a lower protein content. Irrigation increases the gross output, but is noticeably labour-intensive and human input may double compared with rainfed systems; and the human-plus-animal input may rise by an order of magnitude if water has to be lifted, so that gravity-irrigated rice may return an E_r of 1 : 19 but water-lift irrigation using both human and animal labour yields an E_r of 1 : 2. But such statistics can be misleading: firstly, the ratios calculated for E_r are usually those between the agricultural workers and the output, not between all those dependent on the farm products such as the very young, the old and the infirm. Further, even the workers are expending energy all through the year though they may not be engaged in farming. So an E_r of 1 : 20 for the workers might more realistically be 1 : 4 or even 1 : 2 for a whole family for a whole year. In addition, crop yields vary widely from season to season in such low-technology systems (especially in rainfed systems and semi-arid areas), and food storage, processing and preservation methods are

Figure 4.19 An energy flow diagram for the Nuñoa people of the Andean highlands, in 10^3 kcal/yr. Like the savanna pastoralists in fig. 4.18, some crop-growing is important, as is exchange of surpluses with another ecological zone. An important strategy decision, therefore is whether to use the animal dung as fuel or fertiliser: there are few alternatives for either.
Source: R. B. Thomas, 'Human adaptation to a High Andean energy flow system', *Penn State Occasional papers in Anthropology* 7, 1973.

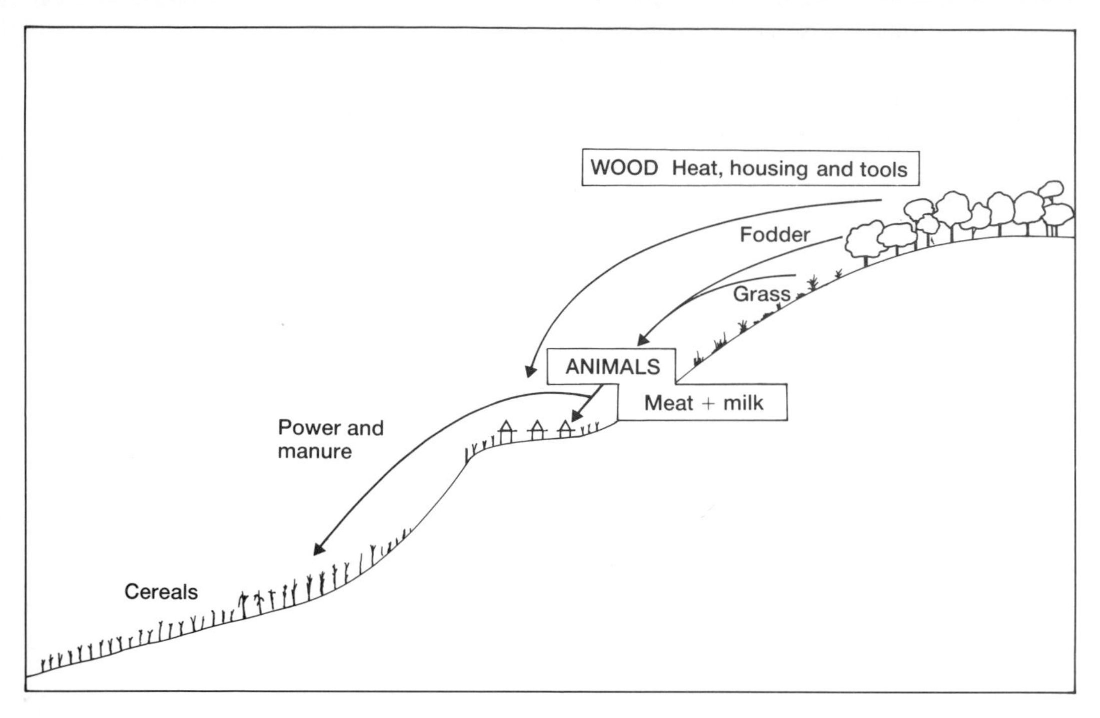

Figure 4.20 The flow of resources (and hence the sites of environmental manipulation) down a hillside in Thak, Nepal. In a series of steps, organic matter finds its way onto the cereals; note however the likely concentration of impact on the forests which are at the head of steep slopes.
Source: A. Macfarlane, *Resources and Population: a Study of the* Gurungs of Nepal. CUP, 1976.

seldom adequate for major carryovers of surpluses from one season to the next. On the other hand, minor sources of energy may buffer a household against such fluctuations and the importance of minor field crops, a backyard plot, and wild food gathering must not be overlooked. Nevertheless, it can perhaps be said of many traditional agroecosystems both of the past and of less-developed countries (LDCs) today that conventionally expressed E_r figures of 1 : 10–20 do not necessarily represent a state of nutritional security, and we must use care in comparing them with the apparently poorer ratios obtained under mechanized and industrially-based agriculture. But clearly the E_r must have been adequate overall for traditional non-industrial agriculture in the sense that from (very roughly) 5000 BC to AD 1800 it supported the bulk of a global population which increased from perhaps 100 million to 720 million, with a steadily rising curve for every continent; except the Americas, whose lands still retained a large hunting and gathering population until quite late in that time-span and their native populations suffered widely from death by disease and conquest in the fifteenth to eighteenth centuries AD.

Endword

It causes no surprise to think of agriculture itself as the major vehicle of environmental modification between the periods of its invention and AD 1800. But as we have seen, there were a number of peripheral systems, notably grazing, hunting and forestry but also urban–industrial systems, wildlife conservation and others which were not negligible, especially in regions which were technologically advanced such as ancient Rome, post-Dark Ages Europe, and ancient China. If the people involved in all these enterprises did not develop theoretical science to underpin their work, in no way did they lack the empirical knowledge and appropriate technology of forestry, agriculture, drainage or deep ocean fishing which enabled them to make large scale and lasting (in some cases virtually irreversible) changes in the natural environment. If, as the late eighteenth-century philosopher Herder thought, 'Da wir hier die Erde als einen Schauplatz der Menschengesshicte beträchten' ('the earth is therefore the theatre of the history of man') then the players in this case have themselves constructed the scenery as the play progressed (Plate 4.7).

Related systems of production and protection

Pastoral nomadism

As discussed earlier, this form of economy developed first in the fourth millennium BC and enabled people to occupy environments marginal to agriculture. Herds of mammals form the biological base, set in the context of a territory which allows movement between fresh sources of water and forage. As well as the familiar examples of goat, sheep, cattle, llama, reindeer and camel, the horse occupies a position of some importance since it made possible the occupation of new

Plate 4.7 The 19th-century advent of iron as an architectural technology hastened the movement of plants around the world since bigger glasshouses could be built. The Palm House at Kew Gardens, London. (A–Z Collection)

lands such as the Tien Shan and the Altai in the second millennium BC and gave birth to a new form of warfare which could overrun sedentary agriculturalists. In more peaceful settings, the way of life was rarely isolated: trade with agriculturalists was common and if sufficient time was spent at any one place then a crop of grain or onions might be grown by the herdsmen.

In *c.* AD 1500 the distribution of pastoral economies was apparently limited, though figure 4.4 does not easily show areas like the high grasslands of the Andes and those places where it was closely interspersed with cultivation as in savanna West Africa. But the essential marginality in spatial and ecological terms is borne out; particularly, pastoralism provided a way of tapping the biological resources of arid and semi-arid lands, provided that the animals can constantly be moved to fresh sources of water and food. Even camels have to drink from time to time but with cattle the dependence is acute: the Dodo of Uganda depend on only 18 water holes for 75 000 cattle in the dry season. Given all the constraints, it is not surprising that traditional pastoralism led to few energy surpluses of the kind enjoyed by some sedentary cultivators. Cities, for example, do not arise out of movement and material simplicity, though they might be founded in pastoral regions as places of trade and secure water supply.

This way of life has acquired a reputation for high environmental impact. Only the very lightest of grazing pressures by domesticates would leave a natural grassland unaffected since the preferential foraging of the animals will give a competitive advantage to plants with thorns, leathery leaves or toxic exudates. In addition, many pastoralists will attempt to manipulate the grazing lands with fire to encourage certain species or perhaps to flush out predators like the wolf. To add to this, the normal strategy of herd management tends to bring about a heavy grazing pressure since it is in the interests of the herdsman to keep a large flock as an insurance against very bad years when a large proportion of his stock may die. (Plate 4.8) Provided some animals are left then he still has a breeding nucleus. As a result, we see some evidence of soil erosion in the Lake Titicaca area of Peru over 2000 years ago, in pre-Inca times, which is attributed to heavy use of the llama pastures. Recent experiments in Hawaii which have excluded feral goats from semi-arid areas (figure 4.21) showed how grazing provided habitats for many non-native plants and how the recovery of endemic and indigenous species is enhanced by the cessation of grazing and browsing by an introduced domesticated herbivore.

Even given that type of change, traditional pastoralism does not seem to have resulted in the marked environmental degradation of recent times. The pastures were scattered, small-capacity wells were the main water source and animal diseases kept the number of animals from becoming too high. Thus damage to the pasture round the wells was limited and could easily be repaired naturally during the following year. An example is a study of a semi-arid pastoralist group

Plate 4.8 The animals of pastoralist groups are a reservoir of food in bad times but any survivors may take time to regain their condition even when the vegetation is more prolific again. (Mark Edwards – Still Pictures/ Panos Pictures)

in Kenya in the 1980s. This section of the Turkana covered over 1000 km/yr, relocating up to 15 times in that period. The livestock provided some 76 per cent of human food energy (figure 4.22) and some of them were bartered to brining in sorghum, maize and sugar with each person receiving 1.95 GJ/yr of total food energy. Overall, the system yielded about 25 MJ of human food per person per year, whereas modern intensive agriculture can give 31 000–77 000 MJ/cap/yr. But the system is adapted to self-maintenance in a climate where rainfall fails one year in four or five, with the animals themselves forming a mobile reserve of food and wealth. Yet they only crop 7 per cent of the non-woody above-ground biomass. So this type of economy is a highly appropriate adaptation to a landscape of low NPP. An analogous equilibrium was attained in the Mediterranean between Classical times and the downfall of the Ottoman empire. Burning, grazing and coppicing produced a mosaic of scrub, woodland and grassland that supported sheep and goat herds which added milk and meat to the produce of small patches of tilled land and larger areas of terrace. The outcome was a set of sustainable ecologies and societies.

Intensification of the system produces the sort of environmental impact discussed in chapter 5 for modern times. A precursor and

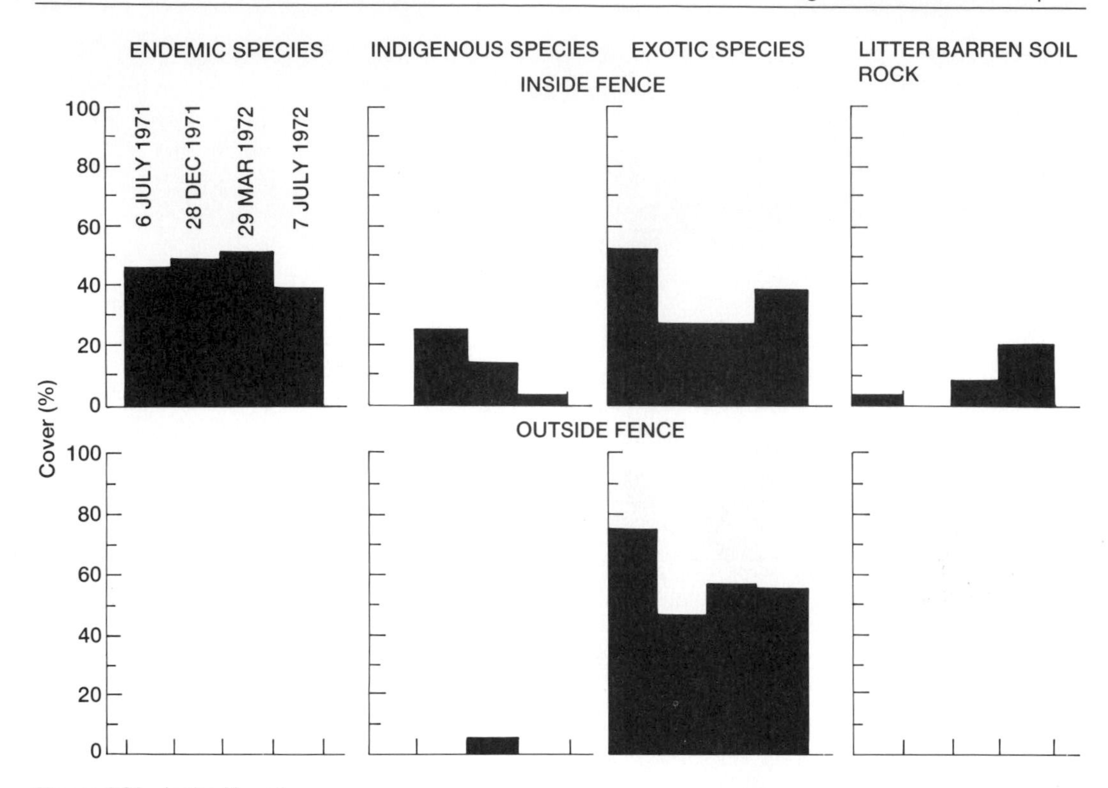

Figure 4.21 In the Hawaii Volcanoes N.P. an exclosure to keep out feral goats was erected for 1 year. The cover changes of various kinds of plant species inside and outside the fence are plotted on this diagram. The resurgence of endemic species is most notable, as is the diminution in barren soil.
Source: D. Mueller-Dombois and G. Spatz, 'Influence of feral goats on the lowland vegetation in Hawaii Volcanoes National Park', *Phytocoenologica* 3, 1975, 1–29.

exemplar of such events is seen in the provision of live sheep and goats to the valley of Mina near Mecca for sacrifice by pilgrims. The number of animals must nearly equal the number of people and they must be ready for the tenth and twelfth days of the month of Dhu-Hijja. Since the calendar is lunar these dates fall 10–12 days earlier in each solar year and it becomes possible for the sacrifices to fall near the end of the dry season. Large numbers of beasts then have to be kept on grassland of low carrying capacity and at the expense of agricultural land and forest, in a climate whose long dry season may end with thunderstorms.

The risks of pastoralism are therefore always present, especially in semi-arid climates. The prophet Joel seems to have known of them when he said (1 : 20) 'The beasts of the field cry also unto thee: for the river of waters is dried up, and the fire hath devoured the pastures of the wilderness.' But if nomadic, traditional pastoralism can probably last thousands of years; the commercial pastoralism of chapter 5 and the interfacing of traditional ways with the industrial world look aged after only decades.

Hunting

We may suppose that a way of life which has been pursued for at least 90 per cent of our own species' evolutionary history is not easily eradicated from either cultural or genetic makeup, thus giving hunters a rather enhanced status in some societies. In *The Canterbury Tales*, Chaucer says of the Monk,

> He yaf not of that text a pulled hen
> That seith that hunters ben not hooly men,

a view which reflects the nature of hunting in agricultural societies from Assyria onwards, where hunting for pleasure and for sacrifice (hence perhaps the 'hooly men') has been carried out whenever an energy surplus allows a diversion of effort to it. But there can be a sharp break in motivation between hunting for sport and what we may call 'accessory hunting', where the meat provided is a necessary addition to an agriculture-based diet, perhaps providing a large proportion of the animal protein for the human group concerned: the pioneer European farmers in North American who shot deer and buffalo for meat come into this category. Into the former we may place the limiting practices of the Pharaohs of Egypt, which were a sign of their divine status, and the pursuit of lions in Assyria (Plate 4.9) which was so much the prerogative of the King that a few beasts were kept in cages ready for a royal hunt at any time.

Where accessory hunting is concerned and where circumstances permit, management of the hunted species and their habitat is virtually absent: the animals are there and they are hunted. If they are gone then some substitution is found: plant products may then be traded for meat from nearby agriculturalists or pastoralists, for example.

Nevertheless, some examples of management may be found. One of these is the provision of winter feed for mammal herbivores in order to keep their population levels high and perhaps also to attract the animals towards the hunters. In Mesolithic Europe (*c.* 5000 BC) there is evidence that may be interpreted as the practice of putting piles of ivy near settlements in early winter to provide nourishment for red deer. Salt-licks are another example of this type of management. A more complex method is the duck decoy of medieval and later Europe, which is a hunting aid that uses a specially shaped pool, nets and a dog to lure duck into a one-way netting system from which they cannot escape save into the neck-breaking hands of the decoy operator. Such devices supplied many of the wild waterfowl which decorated the tables of the better-off of Europe before the nineteenth century. But in none of these do we find any conscious attempts at conservation of the game species' populations, even at the level discussed for some hunter–gatherers in chapter 3.

Ecologically, two types of hunting seem to have developed. The first is that of hunting essentially free populations of wild animals wherever they were found, using techniques like netting, deadfall and pit traps, and falconry. Driving by dogs into the path of the waiting hunters was perhaps the most popular, with a simple chase coming next. In some places there was no management of the population of the hunted species: they were killed in accordance with social rules rather than biological ones and inevitably local or regional extinction

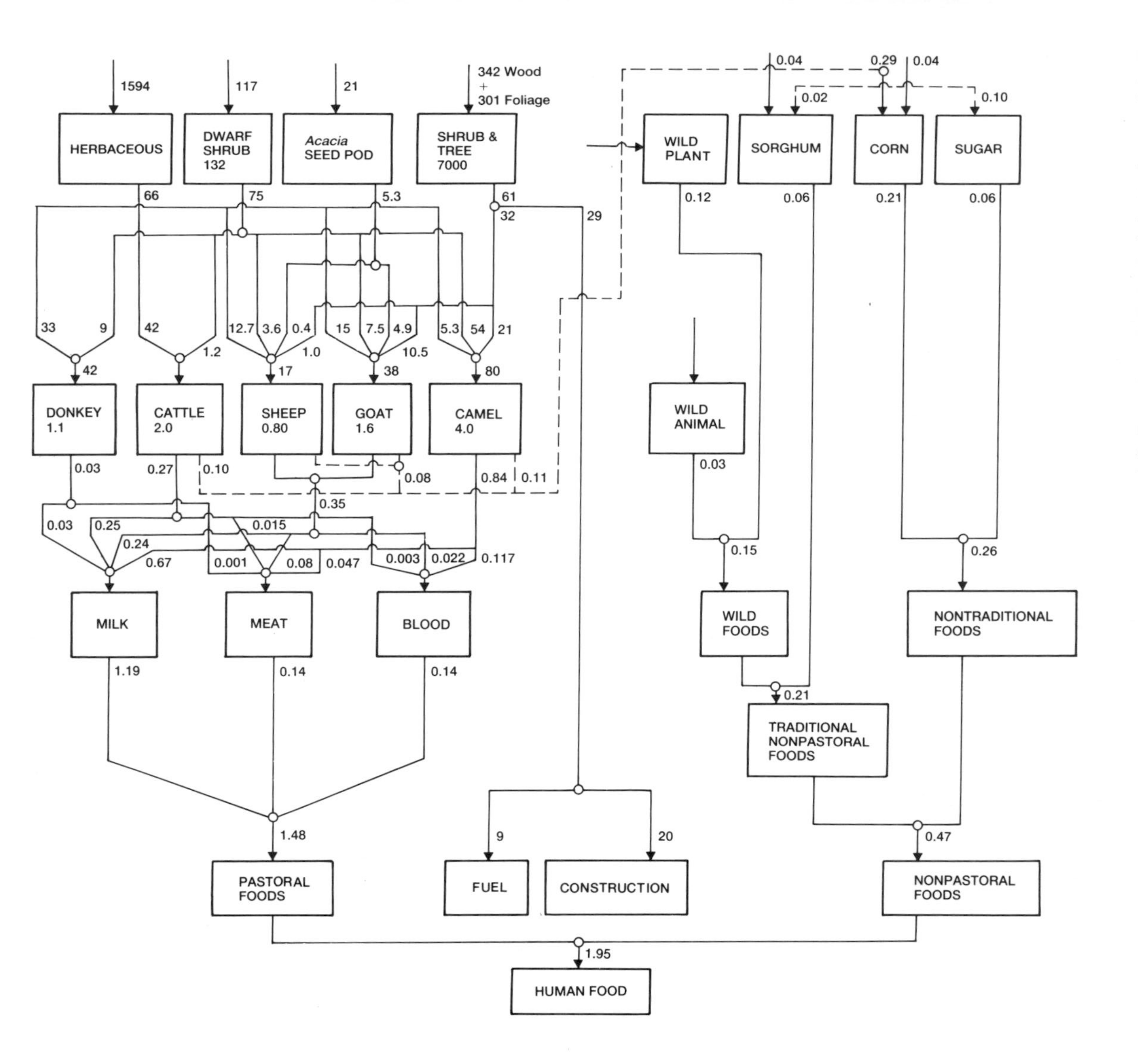

Figure 4.22 Flow of energy (in GJ/yr) through the Turkana people of Kenya, whose economy is mainly pastoralist but which includes, typically, some crops as well (about 32 per cent of the energy value of the milk, meat and blood). The dashed lines indicate the movement of livestock for sale or for barter.
Source: M. B. Coughenour *et al*., 'Energy extraction and use in a nomadic pastoral ecosystem', *Science* 230, 1985, 619–25.

Plate 4.9 King Assurbanipal's lion hunt, *c.*650 BC. Like many others, the kings of Assyria regarded such hunts as part of their royal prerogatives and status. It has been suggested that these reliefs would have been put up even if the King for some reason did not actually engage in hunting lions. (Reproduced by Courtesy of the Trustees of the British Museum)

occurred: a combination of hunting and habitat conversion (of steppe to wheatland) eliminated the lion from North Africa in Roman times. In the Middle East, hunting of birds and mammals using hawks, falcons and the *saluqi* dog seems to have been of this ecological type, whether in medieval Islamic times or recently. Numerous other examples can be found, of which the history of the grizzly bear in North America and the wolf in Europe provide examples of the hunting of animals by a wider range of social classes (a common practice in North America anyway) because the animal was feared. Conversely, the disappearance of open land in Europe, with the enclosure of open fields and the reclamation of heaths and moors, made falconry more difficult and eventually more or less extinct as a hunting method, just as the combination of efficient shooting and conversion of grassland made the great bustard extinct in Britain by 1838.

This type of hunting sometimes involved management of the desired species: the captive breeding of game birds such as pheasants is one example. The eggs are hatched under a domestic hen, reared safely away from predators and released into the wild shortly before the shoot, when such an aritificial density of birds ensures that even the most rheumy-eyed old aristocrat is likely to hit one. (There is quite a lot about pheasant-rearing in D. H. Lawrence's novel *Lady Chatterley's Lover*, though it cannot be said to be the main theme of the story.) Another common method of regulating game populations is by killing their predators: thus the wolf has often been persecuted in areas of deer hunting, and the osprey was eliminated from Britain during the nineteenth century because it was held to be a major predator of river salmon.

Beyond the management of populations is the management of the habitat of the desired species. England provides a good example in the case of the development of fox hunting in enclosed landscapes. The fencing and hedging of the open fields might otherwise have meant the disappearance of the small areas of relict woodland where foxes' earths were usually found, so these coppices were preserved deliberately by landowners. Field boundaries were kept in hedge and post-and-rail even after the availability of wire so that the traditions and rituals of fox-hunting might be kept up. (Though even at its height (1815–70) it had its detractors, culminating in Oscar Wilde's famous epithet of 1893 that fox-hunting was 'the unspeakable in full pursuit of the uneatable.') So here the rural landscape was managed to provide not only the animals but the right setting for the chase.

The second major ecological type of hunting for pleasure took place in enclosures, in England often called 'parks' in this context. Here game populations were maintained largely if not entirely for the hunt and the landowners would provide a staff of people whose job it was to safeguard the animals for that purpose. The earliest examples of such parks were probably in ancient Assyria and Persia, where records suggest that gazelle, boar, deer and even lions and tigers were kept in large enclosures, presumably subdivided. This tradition was followed by the Romans, but not apparently by the Greeks, and independently invented by the high civilizations of Andean America, such as the Aztecs. In about 300 BC in India the Mauryan emperors had laid down that there should be royal hunting forests surrounded by a moat, stocked with tigers, elephants and bison which were to be deprived of their claws and teeth; further, the royal hunters were protected by bodyguards of armed women. Such an unsporting and ungentlemanly set of conditions could never have survived the coming of the British Raj, even if it had lasted that long. One other more or less random example is that of Sultan Firoz (1351–88) who, following Persian tradition, built a royal enclosure some 38.6 km long near Delhi in which to hunt deer, nilgai and jungle fowl. Similar examples of parks can be found in China: in Weichang the Quing dynasty established a hunting park in 1681 (for the chase of bear, deer and tiger) which was opened to peasant cultivation after 1865, with considerable impact of the kind commonly experienced after the disforestation of upland areas.

We see a like tradition emerging in Western Europe under the Norman kings, who kept parks of relatively limited size with enclosed boundaries that held captive populations of deer (Plate 4.10) which were hunted by dogs and foot-followers. The objects of the sport were driven into a central and pre-ordained place where the ensuing slaughter might be watched by those unwilling or unable to have participated in the chase itself. It sounds indiscriminate but we know from the fourteenth century Middle English poem *Sir Gawain and the Green Knight* that this killing could in fact be selective, avoiding gravid females for instance. In terms of areas affected, the great contribution of the Normans in France and England was the Royal Forest. This was a much larger unit than the hunting park and had its own laws which took precedence over the common law. The term

Plate 4.10 Medieval hunting was by no means simply the chasing of animals across country. Here, a predator is lured into an enclosure baited with meat; and dogs and spearmen drive deer into a net stretched across country. (Museé Condé, Chantilly; Photo: Lauros-Giraudon)

foris is a legal expression, not an ecological one; within the forest no animals could be taken without express permission, and the right to cut wood and to plough land was severely restricted. In the New Forest of southern England at least 40 000 ha were subject to forest law in the eleventh century, and the total area in England reached its height by about 1190 declining thereafter by about one-third between 1250 and 1325. In the same spirit as the King had his Royal Forests, lesser nobility were allowed graded privileges: bishops were granted rabbit warrens, for example. In terms of management and impact, these protected areas firstly kept away the encroaching plough and came to represent large islands of semi-wild ecosystems in a sea of arable land and villages. Secondly they preserved within them species which otherwise would not have survived the conversion of their habitat to arable land, notably the red deer which is not now common in England outside enclosed parks. Thirdly, although the royal prerogatives passed from the scene by the seventeenth century, enough of some forests has survived to form an important open-country resource at the present day: the New Forest, and Epping Forest, just north-east of London, are two examples.

Hunting is always discussed in terms of terrestrial systems, but in ecological terms fishing is very close to it, for until the twentieth century, proper domestication of fish can scarcely be said to have taken place. As with land animals, fish used as an accessory source of food have generally come from unmanaged ecosystems in the sense that they have been caught whenever possible but with little selectivity. Rather more environmental impact has been found where ponds have been created for captive fish, whether these be for food or pleasure. The construction of water bodies is an integral part of the environmental making-over that constitutes gardening; in the case of 'working' fish-ponds, the size may range from a small pond near a house, through the set of larger water-bodies that were found near many medieval monasteries in the West, to the large-scale water-level manipulation of hitherto natural lake systems as near Trĕbon in southern Bohemia (Czechoslovakia), which provided carp for urban markets from the fourteenth century onwards. All of these would receive additional management in the form of food inputs for the fish.

Although difficult to quantify, it is clear that throughout the tenure of the earth by agriculturalists, hunting of terrestrial animals has had its impact upon ecosystems. In some places, it may have meant the local or regional extirpation of a hunted species or one of its predators; at other times it may have meant the protection and preservation of species, ecosystems and landscapes which might otherwise have been transformed into agricultural land or some other form of intensively used system. Chaucer's Monk is a reminder that a 'hooly man' may be a somewhat ambiguous character, and perhaps that same judgement can be applied to the effect of hunting down the centuries.

Gardens from ancient times to AD *1800*

Gardens (the word is cognate with 'yard') can be defined as plots of enclosed land near a settlement or dwelling. Gardens may be for use or ornament or especially a combination of both. Their biota therefore is characterized by grass, by flowers, vegetables and fruit and the input of human energy is rewarded by edible material as well as the provision of pleasure. In the time-period being considered, a great range of sizes has been known, from those of a few square metres' size to be found in Japan, to the great landscaped areas of country houses and palaces: Versailles, for example, is 6075 ha. There have always been constraints from the natural environment but the intensive input available from human labour has meant a great deal of manipulation so that very few 'natural' biota are left unless this is consciously desired. Gardens can therefore usually be made over in a particular cultural image. Hence, some gardens simplified wild areas with mostly natural components, such as an area of woodland with the undershrubs rooted out to let spring flowers bloom in their season. Towards the other end of the spectrum, they can be composed almost entirely of highly bred varieties of flower and vegetable, interspersed with hedge, statuary and path so that nature seems a long way off, a trend that finds its extreme in the rock and sand gardens of Japan.

We have practically no archaeological evidence of gardens belonging to hunter–gatherers, although there is one account of a pre-agricultural garden:

> And the Lord God planted a garden eastward in Eden . . .
> (*Gen*. 2: 7)

in which was found

> Every tree that is pleasant to the sight, and good for food
> (*Gen*. 2: 9)

This example set the tone, in that both pleasure and utility were important, and that after the Fall humans were cast out in order to earn their living by agriculture. Perhaps every garden reminds us of this myth and harks back to an early affluent society in which work was not valued all that highly. We have another echo of this theme in the ancient Persian pleasure-parks, which were called *pairidaeza* which became *paradeisos* in Greek.

In the beginnings of plant husbandry, it may well have been difficult to differentiate between a garden as defined above and the first agricultural plots, adjacent to settlements. In ancient Mesopotamia, the elements of the garden as it developed in south-west Asia and the Mediterranean were already present: an enclosure, a rectilinear layout, pools and the pergola. Edible crops had priority and thereafter those with medicinal properties, and there was a statue to the tutelary deity of the garden, the earth or mother-goddess. The pleasure-principle was emphasized in variations on this theme such as those on the *ziggurat* at Babylon (*fl*. 814–810 BC) which became known as the 'hanging gardens' because of their terraced nature; all we know of the contents is that 'lilies' (probably *Lilium candidum*) were probably grown there, and maybe date palms as well.

A developed form of this type of garden (without implying diffusion since independent invention is entirely possible) is seen in ancient Egypt, for which the evidence is much better: here we see the same elements of beauty and utility combined in a rectilinear array of water, herbaceous plants, shade trees, medicinal plants and possibly even a small pavilion (Plate 4.11). Wall paintings make clear the presence in the second millennium BC of the date palm, the doum or gingerbread palm, the sycamore fig especially valued for its shade, the common fig, the pomegranate and the vine. The water storage tanks contained lotus, water fowl and edible fish. Other plants known from Egyptian gardens include red poppies and blue cornflowers and perhaps acacias and tamarisk, both known to attract bees. Monkeys were apparently trained to gather figs, but watering was a constant

Plate 4.11 The useful aspect of gardens is emphasized here in this example from the XVIII Dynasty of the New Kingdom of ancient Egypt. The fish pond contains ducks and fish, while the surrounding trees bear *inter alia* pomegranites and dates. (Reproduced by Courtesy of the Trustees of the British Museum)

task to be provided by human labour, mostly via the counterbalanced bucket or *šādūf*. Specialized gardens appeared to include those of priests who were skilled in medicine and one list suggests 23 species of plant grown in them: the use of castor beans is fairly clear, but just what was cured by cucumber is not quite so obvious. Perfume was also derived from plants, including the morning-glory, the sweet flag, white jasmine and rose. Mortuary rites required the produce of myrrh trees and those yielding incense and henna, and these were often grown in sacred groves.

If some emphasis is given to the gardens of ancient Egypt at the expense of other gardens of antiquity, it is not only because the visual (and written) evidence is so good, but because the basic design, of a rectilinear layout, the presence of water, the inclusion of both useful and beautiful plants, and the construction of buildings dedicated to pleasure, is a theme common to most gardens in the western world until the eighteenth century, when there was a reaction to it. True, there were gardens which had curved lines but the essential formality remained.

Outside the nascent West, linearity of layout was not necessarily so popular. The best documented example is that of China, where the garden paid much more attention to the imitation of nature than did the Egyptian, which was a total transformation of the scrub and reed-swamp of the Nile. The Chinese gardens of the Ch'in and Han dynasties (221 BC–AD 220) were landscaped hunting-parks with religious purposes as well as those of mere pleasure, where indeed Taoist thought can disentangle the two. But the later development of the Chinese garden, especially under the influence of Buddhism from the fourth century AD, parallels the development of the English landscape in the eighteenth century in being much influenced by developments in other arts such as poetry and landscape painting. The Chinese garden developed as a picture, to be contemplated with deliberation and indeed sometimes from a particular viewpoint or angle; it contained what is proper to a natural landscape in terms of trees, shrubs, rocks and water, and evoked the equivalent sensations.

In Europe, the take-off point for the post-Hellenic garden seems to have been eleventh-century Sicily with, as Edward Hyams puts it, 'the pleasure gardens of the Saracen Emirs and of the Norman kings who succeeded them and emulated them, and copied their manners.' The Islamic garden is generally rectilinear and symmetrical, being based especially upon the flow of water which is a feature of the Muslim paradise. The gardens of the Moorish palaces in southern Spain (e.g. the Alhambra in Granada, plate 4.12) and the Shalimar garden in Lahore (*c.* 1634) are good examples; the latter has a canal which discharges into a large marble basin from which rise 450 fountains. Some important evolutions stood a little aside from this type of development: the garden of a western monastery from the Dark Ages to the Medieval period was largely concerned with vegetables and medicinal herbs (in which there was a brisk international trade resulting in the transference of species from their native areas) but retained a rectilinear layout often with irrigation canals. Interestingly, part of the garden might be called a paradise though here it was a grassy area set with flowers where sinners waited to be received back into the church. This presence of flowers would be unlikely before the sixth or seventh century, since before then the Church had a policy of condemning them as heathen. But eventually their use and symbolism was assimilated, with the Madonna lily being taken over from Aphrodite, and the symbolism of the rose (the great gift of Islam to the gardens of the West) being similarly apotheosized. The monasteries also took up the idea of schools in gardens which had originated in classical Greece where the very word *academe* was the name of Plato's garden. Taken to more northerly climes, however, the philosophical disputation and its heirs soon disappeared indoors, whence it may yet again emerge given the parsimony of certain western governments towards education.

In Medieval Europe, the taste for gardens spread from the monasteries to the aristocracy and at their largest, the gardens were most like hunting-parks. Smaller gardens, however, had trees, shrubs, grassy meads and small buildings and were once again both for pleasure and for yield. Some of the grander gardens might also possess a maze, which appears to be a stylized sacred grove symbolic

Plate 4.12 The beautiful aspect of gardens is shown to advantage in the Garden of the Alhambra in Granada, Spain. The fountain is a particularly strong symbol of the control of nature in the form of the hydrological cycle. (Photo: S. Hartley)

of the Old Religion. The larger symbolism of the garden is clear: nature was still all around but in the garden and the park its lineaments could be observed in a tamed and harmless form. In more practical terms, these gardens were responsible for the transfer of plants across considerable distances. Europe gained the citrus fruits in this way, via Spain and Sicily from North Africa, and by 1389 a garden in northern Italy had carnations, originally from Persia. Also at that time the rose came to be the most important flower throughout Europe. So there was an English gardening book in 1400 which names 97 plants, of which 26 were certainly alien introductions.

Between the Renaissance and the eighteenth century, a new formality appeared in the larger gardens of Europe, with Italy setting a fashion which then spread to France, the Low Countries and eventually to Britain. A stiff, regular layout of trees, hedges, grass and water ensured that the gardens were more than ever like rooms and the utilitarian parts were banished behind a wall. Such gardens could become increasingly complex and baroque, and the combination of statuary and waterworks led to great hydrological elaboration, with joke fountains to wet the unwary and runnels down *al fresco* stone dining tables to cool the wine. Such waterworks proclaim the triumph of human ingenuity over the hydrological cycle and some interpreters have seen in the layout of these gardens the formal geometry of a post-Copernican world, other commentators the rigidity of society, and yet others an expression of the Englightenment-inspired logical order inherent in the music of Haydn and Mozart. The high point of this style is almost certainly André Le Nôtre's design for the gardens of the Palace of Versailles, commissioned in 1662 and taking 50 years to complete (plate 4.13). The spirit of Versailles was imitated all over Europe and indeed exported to some of the overseas possessions which European nations were then acquiring. Flows of new species from such territories helped also to stock the European botanical gardens which date mostly from the sixteenth century. Padua claims to be the first, having an area of 20 000 m^2 by 1545. These gardens often cultivated a lot of plants

Plate 4.13 Gardens as conquest: the geometric imposition of paths, woods, lakes and fountains in the great 17th-century park at Versailles near Paris. With the growth of many of the woodlands, today's Versailles looks in places rather more 'natural' than this plan suggests. (Bibliothèque Municipale, Versailles; Photo: Lauros-Giraudon)

new to Europe and acted as reservoirs for their later spread: cyclamen, jasmine and the potato are examples, as are the tulip tree, lilac and magnolia. The garden at Leyden in the Netherlands (where Linnaeus studied at one time) was famous for the importation of plants from the East Indies and the Cape: in the year 1726 it received 1416 different kinds of seed from overseas. Gardens like Kew in England (established in 1730) received plants from all over the world and published catalogues of them. Kew was also instrumental in a profound way for agroecosystem changes since it evaluated commercial plants for use in the British Empire: this garden was crucial in the translocation of rubber trees from their native Brazil to the East Indies and Malaya, for example.

Diversification of garden plants was taking place all the time through more straightforward imports: we hear of the tulip being brought to Augsburg from Constantinople in 1599, and of the mulberry coming to England during the reign of Elizabeth I, and the historian Holinshed wrote of 300–400 novelties in the previous 40 years. The importation of new genetic material led to renewed breeding possibilities among plants already established, so that one English gardening book of 1681 catalogues 20 varieties of apples, 20 of pears, 44 plum, 25 cherry, 35 peach and 5 quince. In all, gardens became places not only of highly manipulated ecosystems, but of highly domesticated genotypes as well.

That there should be a reaction against so much artificiality is no surprise. That the reaction should come from England is perhaps more unusual since she had not been hitherto at the forefront of garden innovation. But associated especially with the names of William Kent (b. 1684), Lancelot ('Capability') Brown (b. 1716), and Humphry Repton (b. 1752) is a whole new style of gardening for the houses of the rich. The style of the 'English landscape garden' sometimes retained an element of continental formality in a garden near the house (even that was sometimes replaced with a simple grass parterre) but beyond that, the view from the house was to be of a 'natural' landscape of trees, grass and water: not really natural, of course, but an idealized or tamed nature which exhibited the kind of harmony made popular in Claude Lorrain's landscape paintings. In one way it appears to be a response to a countryside now tamed to the point where it offered no threat and so did not need overt subjugation; its expression was catalysed by the poet Alexander Pope in particular, who elaborated a poetic–philosophical basis for this form of gardening: 'In all, let nature never be forgot', when trees might be planted, but not pleached, pollarded or clipped; paths were not to be straight but were nevertheless to have a definite destination; water appeared not in elaborate fountains and in canals but as focal waterfalls and in irregularly shaped lakes. To this somewhat Puritan conception might be added fanciful constructions of a classical nature, like belvederes, shepherd's cottages, columned mausoleums, temples of the winds or sun or Flora, and grottoes containing Neptune or a river god; or even a hermit's cell with a wise, bearded old man poring over learned tomes when he was not off-duty down at the village tavern.

The English garden was taken abroad to the colonies and dominions but its spirit even found an echo in France, where several large estates had a wild and irregular *jardin Anglais* and perhaps a picturesque mill or group of cottages forming a *hameau*, a world of make-believe rustic simplicity much influenced by the ideas of Rousseau in his novel *La Nouvelle Héloïse* (1761). The most famous *hameau* of all was that belonging to Marie Antoinette at the Petit Trianon, where the 'landscape style' replaced some of the formal gardens after 1781.

All through this period, we have concentrated on the great gardens of Europe but we must not forget the world scale of the gardens of the peasants and the yeoman farmers, together with the plots and the odd corners cultivated by town dwellers. In the eighteenth century a lot of ground was actually under garden, if we take into account the whole spectrum of land use covered by that term: in neither a visual nor a functional sense could the impact of gardening be said to be unimportant. A garden may be a consciously protected miniature wilderness, or it may range to the total artifice of those of the Aztecs who set up models of wild animals which could not be kept captive: what structuralist interpretation, we may wonder, links that practice and today's garden gnomes? But the energy-intensive nature of gardening and the place of the garden as a recipient of the exotic has nearly always meant that the cultural has triumphed over the natural.

Woods and forests

Societies living in areas surrounded by forest are unlikely to ignore it as a source of materials. Its plants, animals and fungi are another source of food, the wood itself is valuable for construction and for fuel, and the soil beneath the trees may be perceived as a reservoir of agricultural land. For most of these purposes some manipulation of the forest's ecology is inevitable.

One long-term use of woodlands in both temperate and tropical zones of the world has been shifting agriculture (see pp. 99–100) which in one form alternates agriculture and forest, with recovery of the latter's physiognomy, though it is likely that the species composition is never quite the same in the regenerated forest. But with repeated clearances, especially beyond the moist lowlands of the tropics, the forest becomes more of a savanna or even a parkland and the term 'farmed parkland' may then be appropriate with individual trees or groves being preserved for their fruits, wood, forage and shade. In temperate deciduous forests the pasturage of domestic animals was for centuries a way of turning herbage such as leaves and twigs, forest floor grasses, acorns, beech mast and chestnuts into animals such as pigs and cattle and thence into valuable manure. The ecological effect of such grazing and browsing varied according to the species of animal involved and the density at which they grazed. Pigs, for example, may eat most of the beech seeds in a poor mast year and thus apparently inhibit regeneration, but in such a year most of the mast would be eaten anyway by small mammals. Cattle will eat not only grass but the leaves and twigs and seedlings of deciduous trees that are within reach. Thus in many deciduous and mixed forests of the temperate zone, it is not hard to imagine that grazing will inhibit tree regeneration and that the loss of recruitment of new individuals will eventually diminish the area of woodland.

Two main wood products have been taken from forests by agricultural societies. The first of these, to use the traditional English terms, is the 'underwood', the day-to-day needs of an agricultural society, and the second is 'timber', large pieces of wood hewn from mature trees for great construction projects, like medieval cathedrals or eighteenth-century ships of the line.

The principal way of managing the underwood was the coppice rotation. A number of European tree species (e.g. hazel, oak, lime, ash) sprout large numbers of shoots from the stump when cut (see figure 4.26). These shoots are allowed to grow for typically 4–8 years after which they are harvested and the cycle begins again. Among the coppiced trees, some individual trees would be allowed to grow for 12–20 years and some to full height ('standards') and so the resulting woodland is called 'coppice with standards'. This coppice produced a steady supply of logs, charcoal, faggots, rods, hurdles and fencing material, and the standards provided for local constructional needs. Where grazing was common then pollarding (see figure 4.27) might be used to keep the young shoots out of reach of cattle; in turn the animals might benefit from the continuing cutting of all side branches for fodder, a practice known as shredding in English but most common in continental Europe. It is seen in plate 4.2.

Another function of the underwood was to provide resource for the industry which preceded the coal-based Industrial Revolution, of which two important examples were the supply of oak bark for tanning, and charcoal for iron smelting. (Plate 4.14) It would be in the interest of every tanner and iron-master to ensure a continuous local supply of these products rather than see woods cleared. In fuel terms, therefore, the forests could provide a renewable source of power, as distinct from the coal and oil which were to succeed them.

Trees grown in woodlands tend to be tall and straight, so that the great use of trees to produce shipbuilding or constructional timber was as concentrated on hedgerows and other open locations as on woodlands themselves. So, according to O. Rackham, shipbuilding may have been less of an influence on European woodland then, for example, tanneries, except in periods of very high demand such as 1780–1850 in Britain. In the open, trees such as oak developed the spreading habits which gave rise to wood out of which might be cut the variously shaped pieces of timber needed for shipbuilding (figure 4.23). In France the competition between the navy and industry for timber, especially oak and fir, was severe. In many *départements* between 1660 and 1789, the naval authorities sought

Plate 4.14 Woodland management in temperate lands has often produced low but rapid growth so as to maximize pole or bark production. Here coppicing is yielding bark for tanning in the traditional way in Devonshire, England. (Countryside Commission)

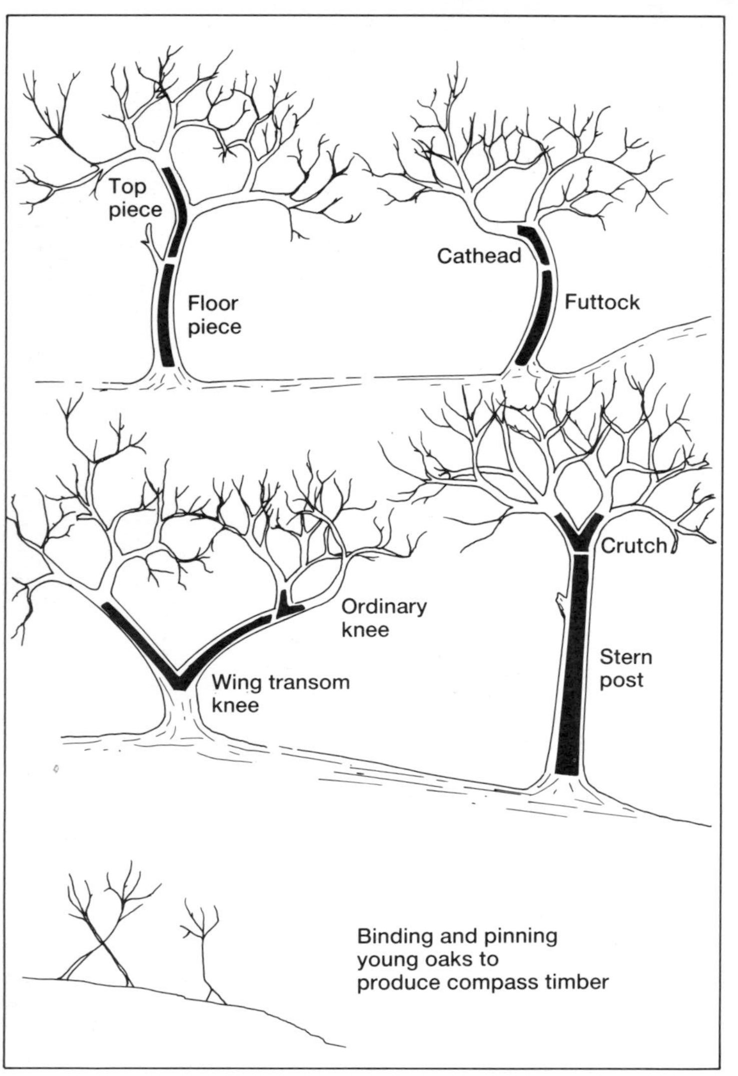

Top piece
Floor piece
Cathead
Futtock
Crutch
Stern post
Ordinary knee
Wing transom knee
Binding and pinning young oaks to produce compass timber

(a)

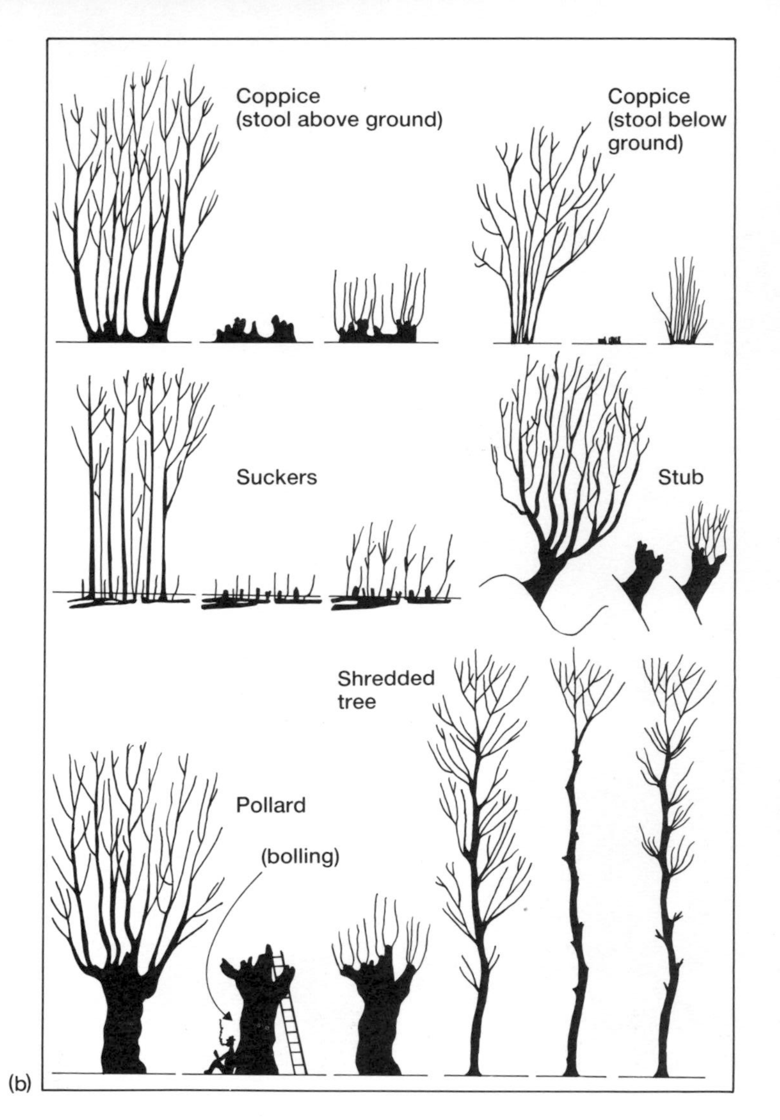

Coppice (stool above ground)
Coppice (stool below ground)
Suckers
Stub
Shredded tree
Pollard
(bolling)

(b)

out all the usable oaks and firs and felled them without regard for the condition in which the forest was left: often so few shade trees were left that birch regeneration was very dense and so prevented the regeneration of the timber species. Cut-over woodland was then appropriated for pasture and fuel by the peasants and very often turned into cultivated land, an age-old theme in the history of forests.

Steady depletion of the availability of the products of native trees through the sixteenth to the eighteenth centuries in Europe led to shifts in species to be used in planting up forests. Conifers were often favoured since they were relatively rapidly growing. Thus about 1800, German forests began to change from being predominantly oak and beech to pine and spruce; attempts to seed pine dated back to the 1680–1720 period and some larch and spruce were brought from the east of the German lands in the second half of the eighteenth century. North American conifers were planted as early as the eighteenth century, but in most places it was the nineteenth century which saw the massive deployment of trees like the Sitka spruce, Douglas fir and Lodgepole pine.

The age-old tension between the utility of the forest itself and the demand for the land on which it stands is illustrated in the numerous studies of forest removal, past and present, from the many areas of the globe which have borne forests under natural conditions. The clearance of forest by human agency is a longstanding process on most continents from before the beginnings of agriculture. Even prehistoric industrial development will have added to the area of deforested land. Since no coherent account on a global basis of the history of forests has been found, we shall have to be content with the citation of a number of examples of the process from various times and places, ending with a rough estimate of the overall picture, continent by continent, to about AD 1800.

In the Neolithic of western Europe, much of the agriculture of the forest zone took the form of shifting cultivation so the forest regenerated, but in the Bronze Age (*c.* 1800–400 BC) we have evidence, preserved for example in upland regions of the British Isles, of apparently permanent field systems and even of 'estates' with definite boundaries. Both of these suppose the permanent clearance of woodland and some form of tillage or improved grassland on the cleared areas. In Bronze Age Wales, it has been demonstrated that areas of soil clad with the lime tree *Tilia* were preferentially cleared for agriculture, presumably since the tolerances of *Tilia* indicated that there were reasonably deep soils beneath, with good moisture and nutrient contents.

We presume also that prehistoric metalworking led to a demand for fuel which was met from forests although we do not know whether sustained yield management was practised. In Classical times forests certainly disappeared under 'industrial' impact (see p. 181); however, this was only part of the process which saw amongst other things the clearance of trees from lowland parts of Greece for agriculture. The removal of forest tended to denude

Figure 4.23 Pre-industrial woodland management. (a) the cuts of oak needed for the various parts of major ships ('great' and 'compass' timber); (b) cutting for agricultural usage: compare the middle shredded tree with the oaks in pl. 4.2. *Sources*: (a) R. G. Albion, *Forests and Sea Power*. Harvard University Press, 1926; (b) O. Rackham, *Trees and Woodlands in the English Landscape*. London: Dent, 1976.

limestone rocks of any cover whatsoever: combined with grazing of small trees and bushes, agriculture soon produced a skeletal landscape, except where special practices were applied to protect the soil. In the *Critias*, Plato says of Attica that in the space of two generations 'What now remains . . . is like the skeleton of a sick man, all the fat and soft earth having wasted away, and only the bare framework of the land being left', and Lucretius said of Italy that 'the forests . . . recede daily higher up the mountainside and yield the ground below to agriculture'. The observations of Classical authors were put together with his own literary gleanings and personal observations by George Perkins Marsh to compile the chapter 'The Woods' which forms the heart of his famous book *Man and Nature, or Physical Geography as modified by Human Action* (1864). This chapter is perhaps still one of the most comprehensive attempts to compile an inventory of the impact of forest clearance on the rest of an ecosystem, though some of its scientific bases have not been confirmed, and it lacks spatially systematic treatment.

In more northern and central parts of Europe, the removal of forests in early historic times has been outlined by H. C. Darby, who emphasizes especially the period AD 1050–1250, and the role therein of the eastwards expansion of Germanic peoples. His two maps (figure 4.24) show something of this process, though by AD 1900 some reafforestation had taken place. But that a landscape revolution had taken place cannot be disputed, even though it left central Europe a heavily wooded area by the standards of the Mediterranean, southwest Asia or the British Isles. In medieval Europe, the clearing of the forest and its replacement by agriculture could be regarded as a religious duty, since the forest in its visage of wilderness might represent the old pre-Christian order and therefore to clear it was to extend God's domain. In their colonization of 'waste' places, the Cistercian monks had a special terminology for the clearance of woodlands: one group (the *incisores*) cut down the trees, a second (*extirpatores*) took out the trunks, and a third (*incensores*) burned up the roots, boughs and undergrowth. It is fair to point out that there are counterexamples from rather more secular societies. The Venetian Republic, for example, was active in the fifteenth century in managing the forests of the Adriatic, keeping a semi-natural type of forest with oaks, beeches, firs and spruces. In some contrast, their Florentine equivalents established intensively used, even-aged stands of silver fir in the mountains of Tuscany.

The effects of forest clearance are especially marked in mountain regions where not only do the slopes suffer soil erosion at an enhanced pace, but the valleys quickly aggrade, with deposits of highly fertile silt as a compensation for more frequent flooding. Channelling of rivers as a response to flood leads in turn to downcutting, so that the whole basin ecosystem and hydrology is altered as a consequence of deforestation. Soil loss from open areas like cropland and landslides is 2.3 times that from forests in the Hima-

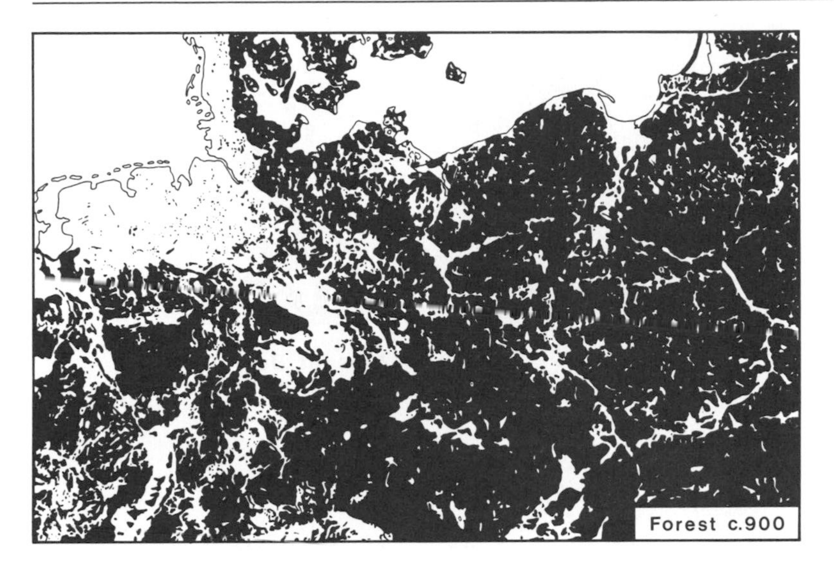

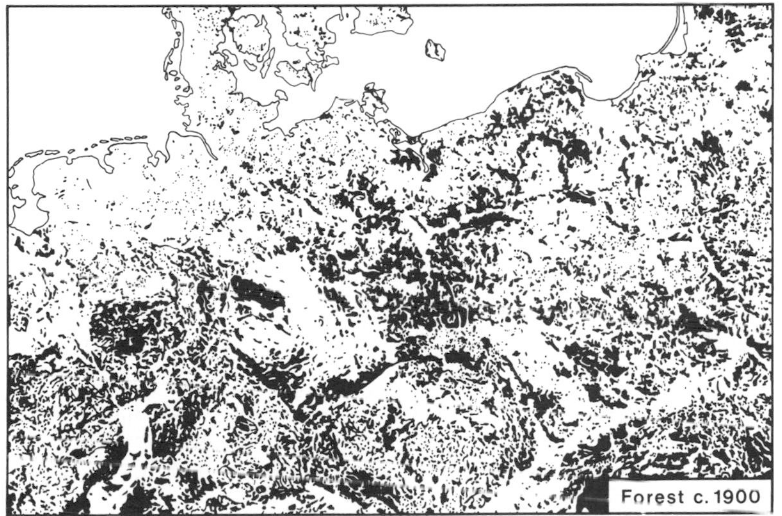

Figure 4.24 The clearance of forest from central Europe AD 900–1900. Compare the speed and intensity of the change with fig. 4.26. *Source*: H. C. Darby, 'The clearing of the woodland in Europe', in W. L. Thomas (ed.) *Man's Role in Changing the Face of the Earth*. Chicago: The University Press, 1956, 183–216.

laya: cropped areas may lose 64 kg/ha/yr whereas dense forest yielded only 25 kg/ha/yr.

The extension of the European experience to the vast forests of eastern North America took place in a similar cultural framework, in which the replacement of the forest (in this case along with its indigenous inhabitants) was regarded as a sacred duty in the Puritan areas and a commercial one in the more mercantile colonies. The Indians had themselves burned the forest and created gaps, glades, hilltop openings (the 'balds' of Appalachia) and the oak-openings of the mid-west; the European pattern was to remove all the forest where possible and replace it with settled agriculture (figure 4.25). Before European intervention, one estimate suggests, the USA had 370×10^6 ha under forest, of which now only 17.6×10^6 ha remain as little-modified forests; some 288×10^6 ha are still under trees but mostly as secondary woodlands or highly managed forests.

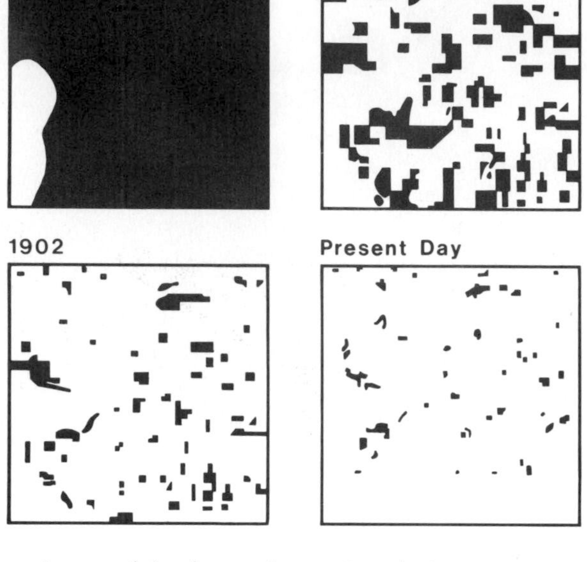

Figure 4.25 Changes in the wooded area of Cadiz Township, Wisconsin, USA, during the period of European settlement. The straight edges result from the methods of land allottment, i.e. after surveying had taken place. What took millenia in Europe was completed in 120 years here. The area of each rectangle is about 90 km^2.
Source: W. M. S. Russell, *Man, Nature and History*. London: Aldus Books, 1967.

A clear picture of the forests in pre-Revolutionary Russia is given by R. A. French. He notes that most of the possible uses of forests took place in European Russia, including temporary clearance for a form of shifting agriculture as well as permanent removal of the trees. By 1800, 60–75 per cent of the northern region of the forest-steppe was under the plough, and some 36 per cent of the southern zone also. Non-agricultural demands were severe: shipping took a lot of timber, as did construction and early industry. Overcutting was noticed in the seventeenth and eighteenth centuries: by 1716 the Petrazadvosk ironworks was using 3131 t/yr of charcoal, and at the beginning of the eighteenth century the Ural region ironworks (then the leading world producers) were consuming 0.5 million m^3/yr of wood, a figure which trebled by the end of the century. Potash was also an important industry; 1 kg of potash needs 3 m^3 of wood and by 1663, Russian production was 3000 t/yr. Making salt by evaporation meant depletion of much woodland and between 1700 and 1750, the Kama works got wood from over 300 km away. Near the Baltic, such pressure was subject to incremental demands for timber for export and British companies played a notable, if deplorable, role in depleting forests. All these processes culminated in the decline of forest cover of European Russia from 53 per cent of the land surface in 1725, through 45 per cent in 1796, down to 35 per cent in 1914.

In Africa, the centuries have seen some tropical forests converted to savanna by cultivation and fire, although the historical geography of the change is not closely documented. In the savanna zone, the preservation of individual trees or clumps of trees for their produce or their shade had led to the notion of 'farmed parkland': the dominant land use is agriculture but the physiognomy of the landscape is characterized by the presence of many trees. The relation of other old civilizations to forests is demonstrated for example in China by the calculation that at the end of the Palaeolithic period (*c.* 8000 BC), some 70 per cent of the land was covered with forest, the present figure (in spite of massive reafforestation programmes since 1949) being 8 per cent. Some of the cleared areas have become grasslands, some put under cultivation, and some have reverted to semi-arid steppe and even near-desert. For India, it is said that the desert of Thar in the Punjab consists of 150 000 km^2 of arid land which 2000 years ago was an impenetrable jungle. Texts dating from between the fifth and tenth centuries AD enjoin kings to encourage forestry. 'the king . . . should grow vast tracts of land . . . with walled enclosures, trees and plants like the following . . . he should have great forests, radiant with groves of trees . . . well guarded by brave soldiers and brave fighters for the benefit of the people.' But the advent of Muslim control meant increased removal of forests since they recognized no sacred trees and had no religious scruples against their destruction. Further, some of the indigenous population fled to hill-lands and there practised shifting cultivation, exacerbating the environmental impact. In later but pre-colonial times, the forests of south Asia yielded many important products (e.g. lumber and firewood, cattle grazing in the dry season, gum, honey, fragrant flowers and woods), but the gathering was licensed by local overlords and the contractors frequently completely gutted forests to make a quick profit, leaving only the 'royal trees' designated by the rulers of territories. In the equatorial rainforests (Plate 4.15), clearance for agriculture has gone on at least since 1000 BC in Africa, 5000 BC in South and Central America, and 7000 BC or earlier in New Guinea, though none of these impacts equal the current loss of these forests (see pp. 000). The secondary rainforest which often resulted is lower in height than its virgin predecessor, poorer in species and often uniform in age: the tangle of climbers and young trees make the word 'jungle' perhaps properly applicable here.

The differentiation of the forests (and other ecosystems) of tropical South American mountains, described by H. Ellenberg, is summarized in figure 4.26. Here, combinations of shifting cultivation, herding of livestock of both native and introduced species, and the collection of wood for fuel have, through centuries of occupation by simple native cultures, the Inca and the Spanish conquest, produced many changes in vegetation physiognomy and species composition, none of which worked in favour of the retention of forests. If the systems were unstable then an occasional burning, cutting, or browsing by goats sufficed to degrade such ecosystems quickly, producing savanna-like mosaics of woodland remnants, patches of scrub, annual grassland and bare ground, with some of the last undergoing deep gullying. Big candalabriform cacti are found near villages (and may be regarded as pasture weeds) but are quickly overshadowed by bushes and trees if a fallow period extends to 20–30 years, suggesting that forest is the 'natural' vegetation of the area.

As G. P. Marsh noted in 1864, the removal of forest and its

replacement with other rural economic systems exerts many secondary effects. His list included some very speculative ideas on atmospheric chemistry and heat conduction but there is no doubt that deforestation results in accelerated soil erosion (figure 4.27) with enhanced nutrient and silt loads in the runoff, probable flooding by these rivers, and the rapid growth of river deltas which in the past have been particularly effective habitats of malaria mosquitoes. Marsh cites Lower Provence as an example of a region where the clearing of forests in the fifteenth to the seventeenth centuries allowed torrents to inundate valleys, the erosion of some good agricultural land, and permitted landslides to cover other parts of the more fertile valleys; the effects in some communes were compared to those of an earthquake. In regions of high relief it is clear that not just soil is spread over the flood plain of a river but larger size cobbles, rock and even boulders. Another documented example is that of the Po delta: Ravenna was built in a lagoon, like Venice, but in 1900 years became separated from the sea by 6.5 km of deltaic deposits with an eastward expansion of 60 m/year according to Marsh's source.

The point need not be laboured: the role of forests as stores of water and stabilizers of soils is as unquestionable as the demonstrated consequences of their removal. More debatable is the climatic impact of the removal of large tracts of forest. Current opinion is presented by H. H. Lamb when he suggests that deforestation on a local scale as within a single river catchment is unlikely to have any significant effect on climate save within the area of any remaining forest. On a larger scale (as with the removal of the temperate forests of the northern hemisphere, reckoned by Lamb to be 'the greatest change in the terrestrial environment so far produced by man'), the prevailing wind strengths have been increased but no effect upon heat absorption is suggested. By contrast, removal of low latitude forest may have more far-reaching effects: the increase in surface albedo would be to reduce heat absorption and hence reduce convection and rainfall, perhaps by 10 per cent, though the predictive certainty of such modelling studies is not yet established.

Though clearly not efficacious at a world scale, there were various influences working in favour of protecting certain areas of forest in the period to *c.* AD 1800, including hunting and wildlife conservation. Others were religious as in the sacred groves of Classical Greece (and in the monastic lands of Mount Athos at the present day) and the clumps of trees related to ancestral graves in China. In one region of Guatemala the activity of furniture-making ensured the protection of the forests of white pine (*Pinus ayacahuite*) which was the raw material; continuity of administrative arrangements after independence has ensured that the forests are still preserved. In Oceania, some traditional island cultures included a 'conservation officer' who watched over garden crops and trees. He made ecological decisions such as the need to plant mixtures of tree species to provide a diversity of crops, of habitats and of predators against any native or imported agricultural pests, or to protect mangroves which

Plate 4.15 The process of woodland removal, though often associated with today's inroads into tropical forests (as here in Amazonia), has been through time the single largest ecosystem shift produced by humankind. (Photo: Andrew Young/ World Wildlife Fund)

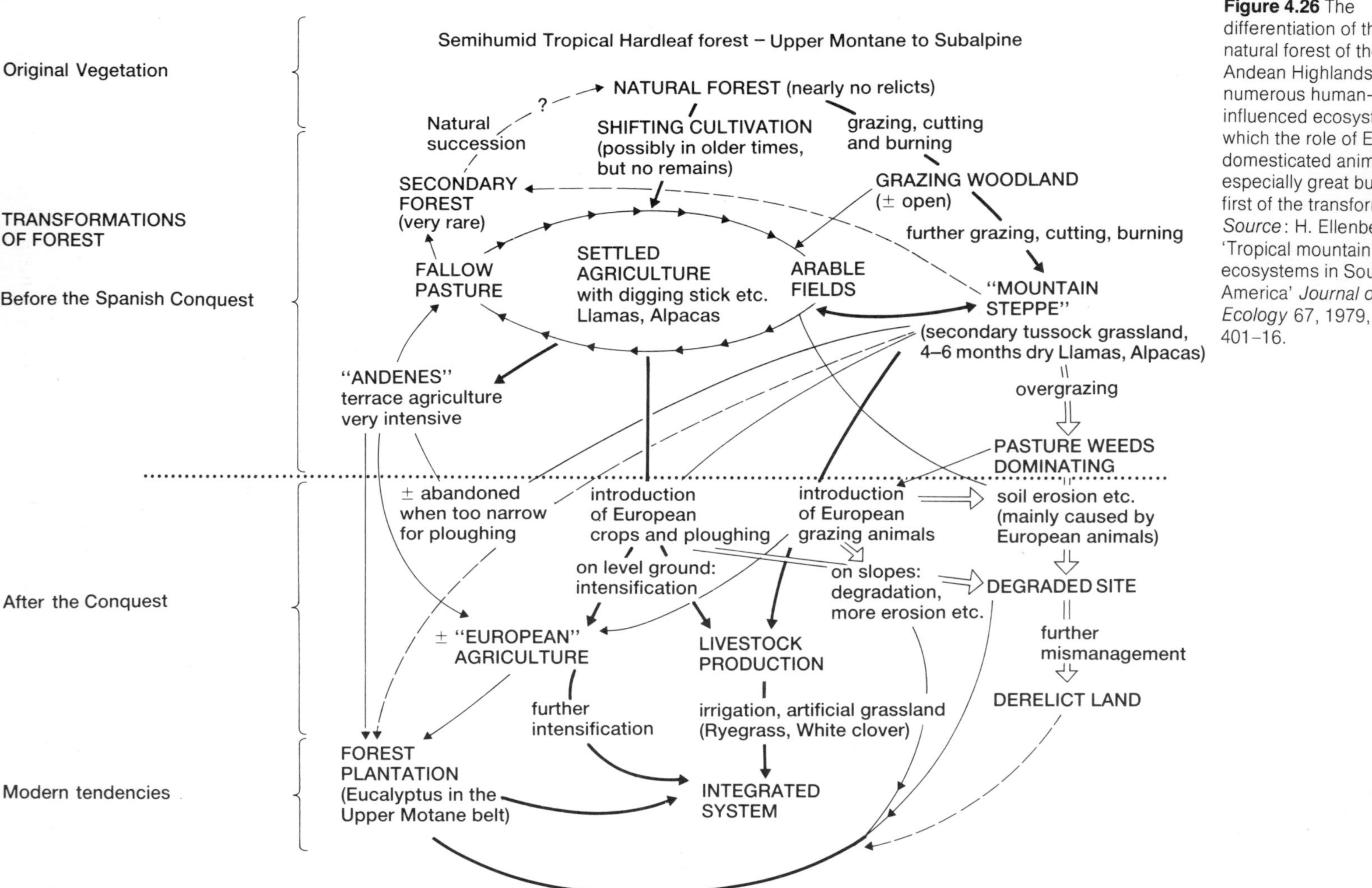

Figure 4.26 The differentiation of the natural forest of the Andean Highlands to numerous human-influenced ecosystems in which the role of European domesticated animals was especially great but not the first of the transformations. *Source*: H. Ellenberg, 'Tropical mountain ecosystems in South America' *Journal of Ecology* 67, 1979, 401–16.

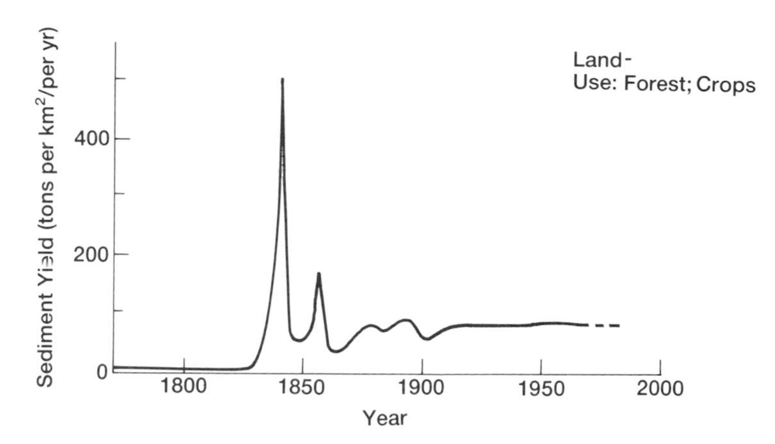

Figure 4.27 The overall history of soil erosion in a lake basin in Southern Michigan, showing the low rates beneath forest being accelerated by its initial conversion to cropland, with a steadying therafter. *Source*: M. B. Davis, 'Erosion rates and Land-use history in Southern Michigan', *Environmental Conservation* 3, 1976, 139–48.

themselves were 'fasteners of the shore' against storms. The Tsembaga of New Guinea look to the growth of nearby trees and shrubs as clocks to remind them of when it is time to abandon their plots and start afresh elsewhere. And as elsewhere, sacred groves acted as agents of forest protection: examples are reported from Kwajalein and New Guinea.

It would be good to end this section with a quantitative, chronologically exact summary of the progress of forest removal from each continent in the period between the introduction of agriculture and AD 1800. The data for such a synopsis do not appear to be available, however, and table 4.11 must suffice. For all its deficiencies it chronicles one of the most enduring and significant impacts which societies have produced in their tenure of the earth and which, although its destabilizing effects are now well-recognized (Plate 4.16), continues unabated through this later, industrial, phase of the human economy.

Nature conservation

It is perhaps a truism that when wild nature is everywhere, no human societies have shown much interest in an altruistic protection of it. Pragmatic reasons such as the perpetuation of a food supply or the maintenance of animals for hunting may have been widespread, but the disinterested preservation of wild plants, animals and ecosystems was rare. Clarence Glacken quotes Emile Male as saying that the European Middle Ages 'gazed at every blade of grass with reverence' but that did not prevent the Europeans from making over much of their continent at that time, and they were not alone in the world; interestingly, Glacken's book indexes none of the keywords of nature conservation. Thus in the agricultural phase of world history there has been a considerable loss of species during the transformation of wildland ecosystems to other types of terrain, whether the results be agricultural land, peat diggings, or a multitude of other land uses. The extension of Polynesians and Europeans into hitherto inaccessible regions of the world made possible extinction of plants and

Table 4.11 Summary in outline of the deforestation of major world regions

Region	*5000 BC*	*O BC/AD*	*AD 1000*	*AD 1500*	*AD 1800*
Europe	Sporadic	Persistent, some permanent	Great clearances for agriculture	Slower rate of clearing	Timber exports N-W Europe Some reafforestation
Mediterranean	Sporadic	Great clearances (agriculture and pasture)	Little reafforestation		
SW Asia	Widespread	Great clearances	Little reafforestation		
S. Asia	Limited clearance	Continuous recession of forests			
E. and SE Asia	Small-scale clearance	Great expansion of cultivated area at expense of forest			
N. America		Limited recession			Great clearances
C. and S. America		Recession for native agriculture and grazing			More land cleared for agriculture and imported stock species
Oceanic Islands		General conservation of forests and trees in indigenous cultures			Deforestation with advent of European domesticated animals

Plate 4.16 The presence of forest in the background reminds us of the regenerative powers of forests if fire, clearing and grazing all stop and the soil is not irreversibly lost. This area of Nepal could become reforested if the inhabitants saw it as a sustainable resource which contributed to their development. (World Wildlife Fund)

animals on remote islands: the dodo is a famous example. This book is not the place for a systematic account of introductions and extinctions, but for balance we should note that some biota inevitably benefited from the human-organized transformations of the agricultural era: malaria would be an example since more deforestation meant more silty deltas with stagnant pools. Rats and fleas doubtless did well out of urban growth, and the common fox in Europe found the increased quantity of woodland edge abutting farmland highly congenial just as more recently the same species has taken well to city life. Overall, however, many species have suffered a diminished range even if not total extinction and in most places this was found acceptable; in the 'development' of Europe from say AD 1000–1800, nobody was very concerned to conserve wild pigs, the European bison and the wolf, in the way we now urge upon the states of east and central Africa, for example: is this progress (Plate 4.17) or a double standard?

Plate 4.17 This modern picture of the capture of a white rhino at Murchison Falls N.P. in Uganda reminds us that wild animals have been captured and managed in parks or zoos for centuries, though the scale of management was never perhaps as great as today. (Photo: John H. Blower/WWF)

While most of the transformation of the wild occurred in the cause of making a living for country and city people, there was one example of enormous inroads upon wild animal populations for what is now regarded as an indefensible purpose: the procurement of wild beasts for the circuses of ancient Rome. Only the larger animals were used since they had to be seen by thousands of spectators at once: the list includes elephants, ostriches, lions, leopards, tigers, hippos, rhinos and crocodiles. Nero, with his talent for the far-fetched, showed polar bears catching seals. The number of animals killed was enormous: Augustus had 3500 animals killed in 26 staged hunts; at the dedication of the Colosseum, 9000 were destroyed in 100 days, and Trajan's conquest of Dacia was celebrated with the slaughter of 11 000 wild animals. Thus the bigger creatures became extinct in any place accessible to the professional hunters and trappers who supplied the circuses. The elephant, rhino and zebra were extirpated in North Africa, the hippo from the lower Nile, and lions from Thessaly, Asia Minor and parts of Syria; tigers disappeared from Hercynia and northern Persia, the nearest source to Rome. Other processes were at work in these areas, but the Roman demand was probably the single most important force in bringing about such extinctions. We could be forgiven for wondering if the same attitudes persist in today's Italian practice of netting and shooting migratory birds, even down to small finches and warblers.

A more mixed picture can be drawn of T'ang China (AD 618–907), where there were attempts to limit or prohibit the slaughter of animals which nevertheless always met severe opposition from the ancient customs of livelihood or from the vested interests of privilege. In the latter instance, kingfishers were killed to provide adornment for ladies of fashion (though this was finally stopped by imperial edict in 1107), as were musk deer to provide scent, and alligators to furnish drumskins. Regulations concerning the slaughter of the wild fauna of China were promulgated in T'ang times: for example in the first half of the eighth century all catching and killing of wild animals was forbidden in the first, fifth and ninth months of each lunar year, and no hunting was permitted at any time within 300 *li* of the two capitals. But it is evident that these rules were differently (and perhaps diffidently) enforced in different periods. So, as J. L. Schaefer concludes, 'although enlightened and humane monarchs issued edicts, conformable to the best morality of their times, these were ignored by their successors . . . And so they were ultimately ineffective.'

Somewhat parallel circumstances are recorded for early India, where King As'oka issued edicts (*c.* 247–242 BC) which gave a list of animals which it was forbidden to kill, 'including all quadrupeds which are neither utilized nor eaten', and also set out that 'forest fire shall not be lit unnecessarily and with a view to killing living beings'; on certain fast days, animals living in certain places (including the 'preserves of the fisher-folk') shall not be killed. In wild places, elephant forests were to be formed in order to keep up a reserve of those most useful animals; there was a Superintendant of Elephants. The dispersal of Buddhism may well have created conditions for the protection of certain species associated with the life and death of Gautama Buddha (567–487 BC), in particular the pippala (*Ficus religiosa*) and the sacred lotus (*Nelumbium nelumbo*).

We get a glance at traditional African practices from ideas like those collected from Matabeleland (now part of Zimbabwe), where it was believed that certain forests were inhabited by the ancestral spirits and were therefore sacrosanct; equally it was customary not to hunt game animals during the breeding season, and when mammals had calves at foot. Totemism meant also that certain species were not eaten by some human groups: the elephant, eland and buffalo are said to have benefited in this way. Just how these practices worked is not recorded, but they scarcely survived the coming of European rule. In Europe, the prime example of nature conservation is the forbidding of hunting of certain animals except by certain social classes (pp. 00), and the creation of deer parks. The park was securely enclosed by a bank and ditch, with the former surmounted by a fence of oak stakes. A park was usually 40–80 hectares in size and a great landowner might possess many, especially in their hey-day of expansion between 1200 and 1350. In England, for example, the Archbishop of Canterbury in the Middle Ages had 21 deer parks and the Bishop of Durham 20; in the country as a whole there were at least 1900. After the Black Death, however, they were not so popular: in the time of Elizabeth I, the number was around 700, and an estimate of 392 was made for 1892. It is a moot point whether the protection of deer to provide sport and meat is really nature conservation. The use of parks also to provide timber and pannage, and as sites for fishponds, perhaps confirms that they are as much part of a nexus of productive ecosystems as of protective ones.

Other pre-industrial modifications

The major modifications of the rural environment during the agricultural time of our past have been described, but there are others which deserve a mention, even though it is not possible to treat them in detail. Most of them are documented from Europe but we should beware of ignoring other parts of the world merely because there is not such a density of evidence.

If we begin with the dry land, then we ought to recall that in some parts of Europe, the result of forest clearance was not necessarily cultivable land but areas of grassland, heath and moor which were mainly used for grazing. These mostly had acid, siliceous soils and at high altitudes the moors vegetated by heather, cotton-sedge and purple moor grass often had peaty soils, or indeed in the wettest places a blanket of peat often 2–6 metres deep. These heaths and moors were a useful source of grazing and of fuel from peat, but were often eyed as potential agricultural land by groups of people who were willing to invest personal and animal energy in transforming them. The heaths and the lower edge of the moorland might be potentially cultivable by squatters, who were dispossessed or entreprenurial individuals seeking to create an agricultural holding. Equally, more established farmers might see the moorland as being a worthwhile pool of agricultural land provided grain prices were high enough, as they were during the Napoleonic Wars in Europe. The boundary between enclosed and cultivated land then moved upward as 'intakes' of land were made; conversely at times of lower prices or harsher climate, the intakes were abandoned and the vegetation slowly reverted to the wild as the stone walls fell and crumbled away. Paring, using a breast spade, and burning was a commonly employed practice in reclamations. Medieval farmers reclaiming upland moors in England had to pare off and burn 8–10 cm of acid peat, as well as clear away stones and construct field boundaries. Burning the peat, followed by ploughing, produced a soil with better drainage, higher pH and lower organic matter levels which was friable and fertile enough for cultivation. The lowland heaths (as found in Germany, the Netherlands and England, for instance) once enclosed were easier to maintain as grasslands or arable, so areas once dominated by them have seen the piecemeal disintegration of large areas of heathland into smaller relict pieces: the gloomy air of the Wessex heaths so strongly pervasive of Thomas Hardy's novels has been dissipated by the conversion of many of them to improved pasture. In the Suffolk sandlings until the early twentieth century the heaths were an integral part of a rural economy of sheep and rabbit warrening. This has now disintegrated, leaving odd patches of heath for military training and for outdoor recreation. A third group of colonizers of such wasteland were medieval religious houses. The Cistercians, for example, were ordered to settle in waste places far from the temptations of the towns and so not only improved lowland agriculture but converted moor to better grassland and doubtless changed the vegetation of the open lands by altering the ratio and density of cattle and sheep. Later events such as consistently unfavourable weather, soil exhaustion and the Black Death would cause retreat from some of these lands and allow them to revert to wild vegetation once more. Another example of land alteration is the stabilization of sand dunes (Plate 4.18) to prevent them blowing inland upon settlements and crops: planting dunes with pine trees for this purpose is recorded from Portugal in 1325.

No exactly similar sets of processes seem to have been chronicled elsewhere in the world, although we may note in passing the colonization of tropical pastures with European weeds, and doubtless the conversion of African savanna into farmland came and went, just as cropland ebbed significantly from North Africa after Roman times, leaving a semi-arid steppe. The Great Hungarian Plain was recovered from the Muslims in 1699 and after functioning for a while as a grazing area, was converted to wheat and maize growing; the Ukraine underwent this particular transformation in the second half

Plate 4.18 Sand dunes are at once a sterile habitat that apparently produces nothing and a barrier against flooding by the sea. The response in Europe (as here in The Netherlands) has been to plant trees to yield wood products and further to stabilize the sand which, if it becomes loose, can blow inland and inundate agricultural land. (Aerocamera – Bart Hofmeester)

of the eighteenth century, foreshadowing what was to happen to the North American prairies in the late nineteenth and early twentieth centuries.

Those who looked for agricultural expansion in Europe saw great potential in wetlands. Intaking from the sea will be considered later; here we mention the areas of inland inundation in the Netherlands, for example, which were embanked, ditched and drained from very early times AD and which we know as polders (Plate 4.19). Control of an energy source was crucial. In this case the wind was harnessed to pump water up and away from the waterlogged lands, allowing the marsh and fen vegetation to be replaced with crops. Dutch engineers were responsible for the drainage of the great English fenland in the seventeenth century, though building upon smaller reclamations that had been going on since at least the twelfth century, which was also the time of the reclamation of the delta of the Po in Italy. These engineers worked not only in Holland, England and Italy, but in Poland, France and Germany as well: in 1753 Frederick the Great boasted that a big drainage project in Prussia had allowed him to conquer a province in peacetime. Not only was fertile agricultural land created this way, but the breeding grounds of malarial mosquitoes were diminished. The control of rivers also had medieval beginnings, with weirs on the Thames in the thirteenth century, and sluices and locks becoming common from the fourteenth century in Europe. Fertilization by silting was carried out in waterside fields by leading flood-waters onto the land outside the growing season, and in the La Breune region of France exhausted soil was inundated by building a barrage at its downstream end; when enough silt had accumulated, the dam was breached and a new field was available to be ploughed and planted.

Plate 4.19 The reclamation of wetlands in pre-industrial times was made possible by pumping the land dry using wind power. In The Netherlands (as here) the resulting land was called a polder and was highly productive provided it was kept dry enough for cultivation. (Photo: D. Cook)

Mining and quarrying

The riches of the surface of the land can often be complemented by those underneath it. With agriculture came specialized hewers of stone resources, and thus between 3500 and 1300 BC flint mining was practised in Europe; certainly in the period 2400–1600 BC the flint mines at Grimes Graves in eastern England were in use, with shafts leading down to galleries which were held up with pillars of unworked rock. Some waste was taken to the surface and so the environment was changed from an undulating heathland to a pock-marked, almost cratered, terrain (figure 4.28). Shaft and adit mines were also used to extract copper ore during the European Bronze Age: in the second millennium BC, large numbers of mines were operating in Austria. The tunnels were timbered and the ore was smelted nearby, so we can make rough calculations of production: in the Salzburg district there were perhaps 500–600 mines and each three-adit mine would use 20 m^3 of timber per day. So a significant environmental impact in terms of changed land form, contaminated waters and atmosphere, and forests cleared or managed for timber and fuel can easily be imagined.

In the Mediterranean, Greek and Roman societies were well known for their use of stone and for access to metals. Herodotus declared that the search for gold on Thasos had thrown a mountain upside-down, and Mt. Pentelius bore a gleaming white scar on its side where the marble had been removed. More lowly quarries supplied potters with clay, and builders with common limestone. For metals, they initially preferred open pit mines to underground shafts, but eventually they went as deep as 152 m in their Etruscan mines

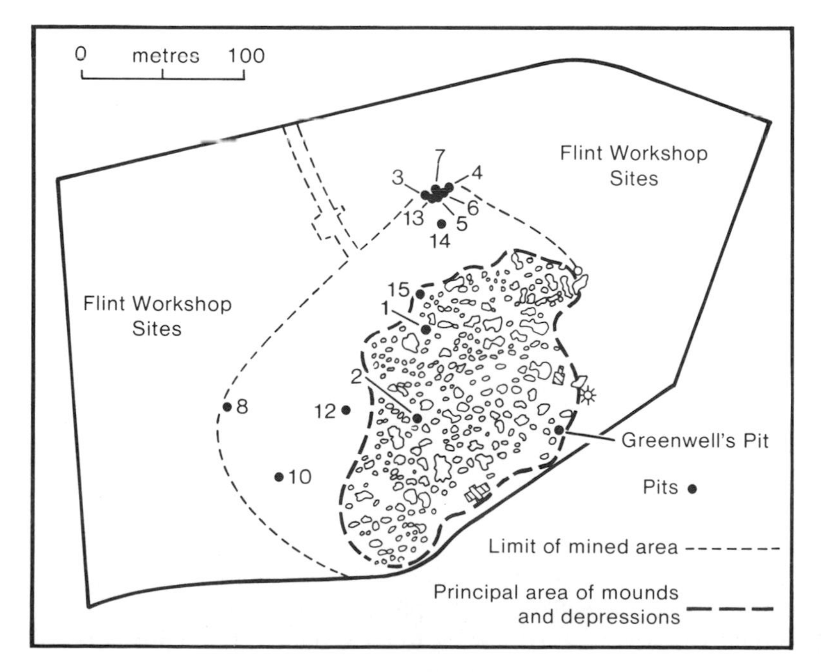

Figure 4.28 The layout of shafts ('Pits') and waste materials at Grimes Graves in eastern England. Greenwell's Pit is the site of the first archaeological investigation of the site, by a Canon Greenwell in 1870.
Source: R. Shepherd, *Prehistoric Mining and Allied Industries*. London: Academic Press, 1980.

and exported this technology to their empire. Mining is known to have clogged streams with debris and to have poisoned both surface and ground water with detritus from the winning of lead, mercury and arsenic. An improvement in technology allowed the Romans to rework old slags for silver and lead, just as their waste heaps were in turn re-used by later groups.

In medieval and later times mining and quarrying spread along with agriculture, providing materials for cultivators, industry, city-dwellers and armies. On a world-wide basis many consequent examples of mining and quarrying can be found, often extending outside the agricultural zone since hill-lands are often easier to mine than they are to cultivate. In most settled areas of the world where agriculture was practised there were quarries, pits and small-scale metal-working; charcoal was the main fuel used in smelting (Plate 4.20).

Sometimes the demand might be great: one Cistercian monastic house took 40,000 cartloads of stone to build. The impact might also be great: in the fourteenth century the digging of stream gravels for tin ore in Devon, England, destroyed farmland at the rate of 120 ha per year. Yet almost inevitably, the story of mining leads to iron as a key material of the developing economies of the world; by the sixteenth century it had become much more expensive since both war and peace now provided a continuous demand for it. There was too a growing scarcity of wood for smelting-charcoal in Europe: by the eighteenth century Sweden was a leading producer of iron largely because of her wood reserves. In England, iron production in the important areas like the Weald, the Forest of Dean and Cleveland was by the late fifteenth century restricted by wood supplies even given the shift of woodland management to an intensive coppicing system. Another instance is nineteenth-century western North America, where mineral bonanzas relied on wood fuel for smelting. In Eureka, Nevada, in 1873 there were thirteen furnaces which had a daily capacity of 595 tons of gold and silver ores. Working at capacity they needed the produce of piñon pine every day. The hills around Eureka were stripped of that tree for a distances of as far as 80 km.

Lastly, we deal with a fuel with a future. The use of coal was known in Europe since Roman times (being burned for warmth on the northern frontier of Britain, and for ritual purposes on the altar of Sulis at Bath) and was exported to London from the Tyne from the thirteenth century onwards, although it never became a fuel which was used on a large, industrial, scale until the eighteenth century. Medieval and early modern coal pits were shallow and bell-shaped and precluded the use of the land for other purposes: a foreshadowing of industrialism to come.

Industry

Much of the non-extractive industry before the nineteenth century was confined to small processing plants scattered through essentially rural areas and in what are by today's standards small towns. In

Plate 4.20 In 1556, Agricola illustrated some of the technology of mining in Europe which underlay a great deal of environmental alteration. The raw tree stumps hint at the quantity of wood always needed for smelting ores and for construction timber. (The Mansell Collection)

addition, the environmental impact of industry before the nineteenth century was confined in the sense that industry itself was stronger in Europe, East Asia and West Africa than on other continents. Its energy base was largely that of wood through the medium of faggots and of charcoal, and in Europe of water harnessed via a wheel. But environmental change was not entirely absent: we have noted, for example, the systems of woodland management needed to produce the charcoal for iron smelting and the bark for tanning. Coal was one fuel in an industry of importance in Europe from very early times: that of producing salt from sea-water. This necessitated boiling sea-water or a mixture of sea-water and silt after the sun had performed part of the concentration process. In medieval England and the Netherlands another possible fuel was peat, and in the Lincolnshire fens as in many parts of the Low Countries, large flooded pits were the result of peat digging to fire the salt pans. Many lakes remain in the Netherlands as a result of digging peat to fuel not only salt-works but many other forms of industry too. Perhaps the best known of all the remnants of extensive peat-digging for both domestic and industrial purposes are the Broads in Norfolk, England: a series of meres whose origin lies in medieval peat-pits flooded by subsequent rises in sea-levels.

Where industry was located, waste products were inevitable. Even a small bloomery produced a slag tip which was hostile to living organisms for a long time; molten lead or lead wastes resulted in toxic gases or contaminated water which would both kill most plants and animals nearby. As early as 1582, Dutch linen-bleachers were dumping lye and milk wastes into the canals and as a consequence were ordered to use separate disposal channels, called *stinkerds*, for effluvia. Other industries in Holland also used canals to carry away wastes and thus polluted the water supply of others; the list of industries producing noisome wastes between 1500 and 1800 includes sugar refining, cotton printing, brewing and distilling, oil-seed processing, papermaking, and tanning using whale-oil. Thus the air, the water and the land could become contaminated even in those centuries, to the detriment of crops, flora, fauna and people, though the extent and intensity could not have been anything like those of the urban–industrial concentrations of the nineteenth century.

Cities

Since the days of the development of the compact city in ancient Egypt, Mesopotamia and India (*c.* 3000 BC onwards), this phenomenon has persisted, though becoming much larger in size and population after the industrial revolution. In environmental terms, the city represents a wholesale transformation of the local environment, although features may remain such as the general shape of the land, the presence of a river, some trees and perhaps some open space which although much altered serves as a reminder of what preceded the urban growth. Pre-industrial cities no less than any others have their own fauna and flora: the latter is one characterized by a limited suite of plants which can flourish in the odd waste places, on walls and roofs, and in gardens; the former includes successful parasites such as fleas and scavengers like rats, feral dogs and cats, and birds of which the red kite in medieval and early modern Europe is an example. The city itself represents a concentration of materials and energy which has very largely to come from outside its boundaries, so it exerts environmental influences at a distance, as in the quarrying of stone for building materials, the digging of coal or cutting of trees for fuel, and where necessary the carriage of water; to say nothing of the satisfaction of the demand for food which will probably determine the agricultural pattern of a pre-industrial city's immediate hinterland, pulling in its surpluses. Water supply was very often a difficulty: Samos and Athens in ancient Greece both had covered or underground aqueducts with settling basins, and the aqueducts of the Romans are justly famous although the use of lead pipes was not perhaps quite so clever. Later cities in many parts of the world rarely reached the standard of water supply of these classical examples, being mostly reliant upon local wells or rivers.

The nucleation of people in the city inevitably brought a concentration of wastes which themselves altered the environment of the city and its surroundings. Noise is an immediate effect, to the point where Julius Caesar prohibited all wheeled vehicles from entering Rome between sunrise and two hours before sunset; this law, however, fell into disuse in the third century AD. Possibly more significant is the contamination of the air caused by the effluents from its energy sources: again in ancient Rome, citizens coming back from the country commented upon the smoke and dust which obscured the sun and caused them to lose their tan. In medieval London, the smoke caused by fires burning coal (known as sea-coal because most of it came by sea from north-east England) high in soot and sulphur residues caused problems as early as the last quarter of the thirteenth century. By the second half of the sixteenth century it became so serious that Queen Elizabeth prohibited the use of sea-coal in brewhouses within a mile of the Court; later, in 1627 the smoke from alum works was said to be the cause of tainting the pastures near the city and even of poisoning the fish in the Thames. Waste water likewise became more contaminated as cities grew. Sewers might have been built from ancient times, but they generally discharged to the nearest watercourse and in time the by-products of all those manufactures which were attracted to the city also ran off into the common drain. In some cities, night soil was collected from pits or directly from privies and sold for fertilizer, but elsewhere the rivers or the sea were the sites of the biological transformation of ordure back to simpler molecules. The solid wastes created by an urban population would require disposal as well but a simple pit outside the built-up area would likely suffice; it could be built upon easily when full, and disjecta per person were no doubt a great deal lower than those of today's western populations and more like the thrifty dwellers of Third World cities. The decomposition of the human body itself might also change land use patterns near the city, since the dead might not be buried within the city either because it was a sacred place or because they were crowded out by more active land developers. At all events, the pre-industrial city was in absolute terms much smaller than those of the nineteenth and twentieth centuries, and its environmental impact was even less simply because each citizen had access to lower quantities of energy than fossil-fuel-using successors.

War

Men seem always to have been willing to find the energy (and the materials) to fight, and threatening the survival of a population by destroying the ecological base of its existence is a long-standing tradition and one probably older than the records of it. Techniques have centred upon poisoning water sources and perhaps above all on destroying crops. Armies throughout time have been noted for their stripping of the land not only to feed themselves but to deny food to the enemy force and its sympathetic civilians. We see an early example of this in *Judges* 15: 3–5 when Samson tied burning torches to the tails of 300 jackals (or foxes, depending on the translation) and sent them into the grainfields, vineyards and olive groves of the Philistines. Similar actions were common during warfare in ancient Greece, and Tacitus says of the Roman conquest of Britain that 'they made a wasteland and called it peace.' The Romans sowed the fields around Carthage with salt so that they might not bear crops, but by way of compensation both Greek and Roman writers remark on the fertilizing effects of bones and blood on crops grown later on the sites of battles. Not only, says Plutarch, did the people of Massalia (scene of the defeat of the Teutons) fence their vineyards with bones, but the soil 'grew so rich and became so full to its depths of the putrefied matter that sank into it, that it produced an exceeding great harvest in after years.'

General environmental destruction was wrought by armies for their own purposes: timber cut down, forests scoured for wild game, and pastures heavily depleted by pack animals so that war in effect depleted a region's energy sources. Forests were especially vulnerable, since they hid enemies and there are examples of burning them to flush out soldiers from seventeenth-century Scotland and the ancient Mediterranean: at the battle of Pylos, all the vegetation of one island was burned so that the Athenians could see the movements of the Spartans. When invading Persians burned forests, Herodotus regarded it as an act of impiety, presumably because of the sacred groves within the woodland.

War has secondary effects too, since it speeds up the demands for more raw materials and may destroy cities which control agricultural systems or water supply networks. In addition to the primary impacts described above, therefore, war has been from early times a local and sporadic but potent manipulative force in the environment. In pre-industrial times, happily, few of these effects were completely irreversible: the salt disappeared from Carthage within a century, and Delilah rather cramped Samson's aggressive capabilities.

Reclamation from the sea

One response to pressures for more agricultural space has been the reclamation of the upper parts of the intertidal zone on low coasts. The natural processes of silt accretion build up the land and when it is sufficiently raised above mean tidal levels, human and animal energy is deployed. A bank is built on the seaward side and the builders hope that subsequent high tides, storms and changing sea-levels will not break it down: sheltered estuaries and bays were thus especially popular sites. The soil has to lose salt and gain organic matter but when ready it is then fertile, stone-free and flat. Perhaps the most spectacular examples of this process have been in the Low Countries of Europe: here the practice started some time after AD 900 but escalated into peaks of activity in the eleventh and twelfth centuries AD (figure 4.29), and in the sixteenth century with another peak in 1600–25. Along the coast of northern Holland a post-Roman rise in sea-level caused the inhabitants to build mounds as refuges for people and stock: these *terpen* were of the order of 3–10 m high and 2–15 ha in area. An analogous set of circumstances across the North Sea produced the phenomenon of reclamation but not the *terpen*. From early medieval times onward, the coast around the Wash in

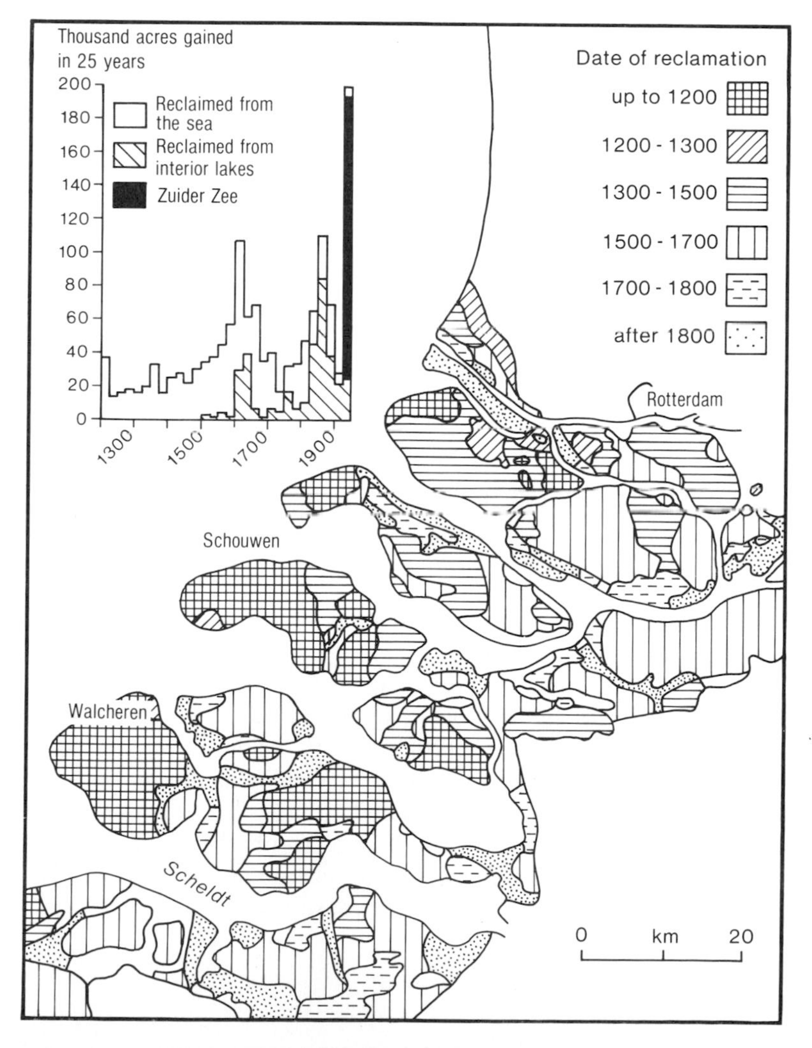

Figure 4.29 The spatial and temporal progress of reclamation in the whole (graph) and part (map) of The Netherlands, though this is now out of date with the completion of the delta plan (see pl. 5.16). The land reclaimed from 'interior lakes' is that which is called *polder*. Note that unlike South Lincolnshire (fig. 4.31) there is not a simple progression from dry land outwards.
Source: C. T. Smith, *An Historical Geography of Western Europe before 1800*. London: Longmans, 1967.

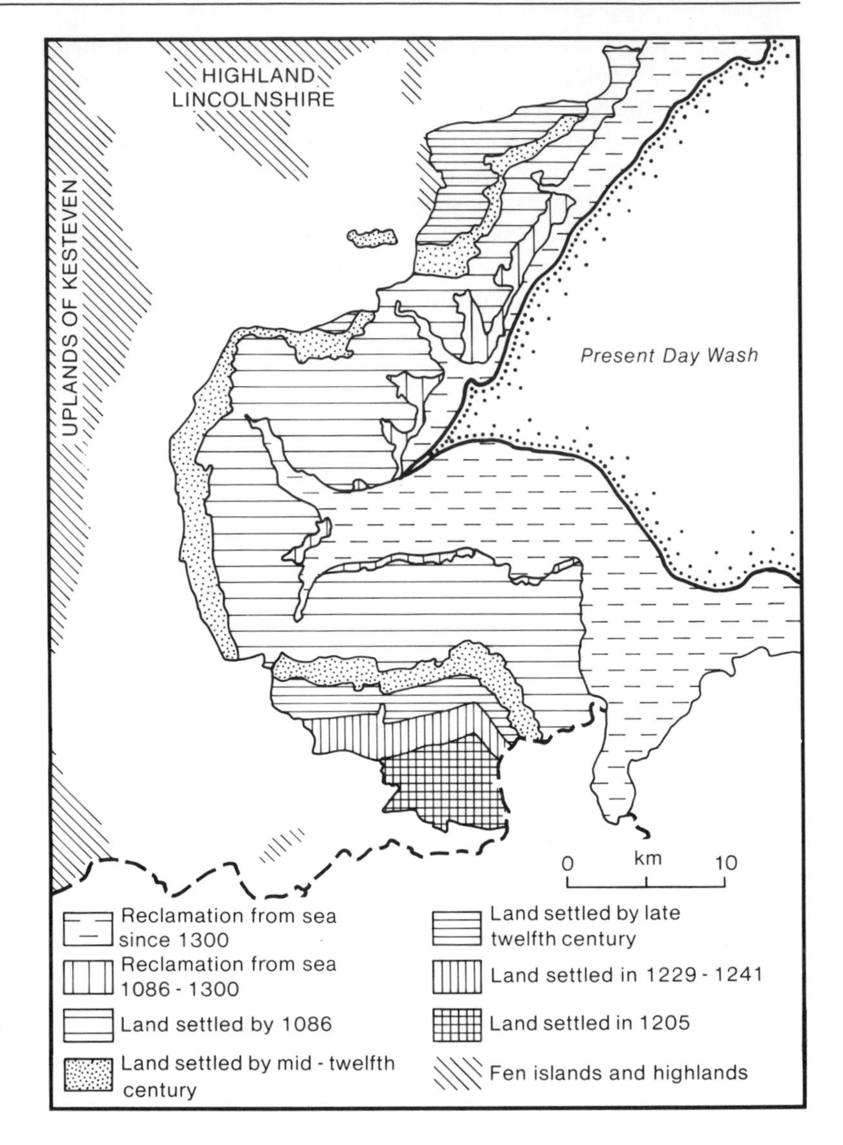

Figure 4.30 The detail of reclamation from the wetlands round the Wash in eastern England. (Land taken in after 1300 is not differentiated by period.) Some of the inland settled areas were on land reclaimed from freshwater fen rather than saltings. *Source*: H. E. Hallam, *Settlement and Society: a Study of the Early Agrarian History of South Lincolnshire*. CUP, 1965.

England was subject to piecemeal intaking, producing a series of banks parallel to the sea, with enclosed fields in between: between 1150 and 1300, some 40 km^2 were reclaimed from the Wash in Lincolnshire alone (figure 4.30). Estuarine lands in south-east Asia were converted into fish ponds, and 'silt fields' were created in twelfth-century China which look very similar to the processes in Europe at the same time. Lagoons and estuaries have also been sites for land reclamation in the form of cities and harbours. Venice and Ravenna are the best examples, though subsequent silting has left the latter some distance from the present coastline.

Impact on the oceans

It may seem absurd to think that oceanic animal stocks could be seriously depleted before AD 1800, but E. L. Jones suggests that this was true of whaling. Whale-oil was an important product in the days before mineral oil ('whaling vessels were the oil tankers of the pre-petroleum world'), especially for pre-industrial lubrication and lighting. Thus, from a time no later than AD 1150 from the shores of the Bay of Biscay, the whales were hunted in the Atlantic and the Arctic (by 1725, a thousand Dutchmen summered in Spitzbergen, at Smeerenburg or Blubbertown), then into the South Atlantic, the Pacific, Tasmania and Antarctica. It so happened, says Jones, that the whalers found new grounds just when the old ones were on the point of being hunted out, so confirming the idea of bringing the pelagic population to economic if not biological extinction; at any rate in 1850 the seas between 80°N and 55°S were thickly populated with whaling grounds but by 1880 most of these had shrunk in size and yield even though the hunting technology was still mostly pre-industrial (Plate 4.21).

The population, resources and environment of agriculturalists

One perspective upon these relationships in agriculturally based societies is that of seeing constraints and buffers. Production is constrained within certain boundaries but human societies use a variety of methods, including environmental manipulation, to try and buffer themselves against effects such as famine and social breakdown. One constraint of obvious importance is that of climate, which is related to environmental impact through second- or third-order effects

Plate 4.21 Pre-industrial whaling was common in most oceans, but the effectiveness was never such as to produce the immense declines in Cetacean populations that accompanied the steamer and associated technology. (From a coloured lithograph in the Mansell Collection)

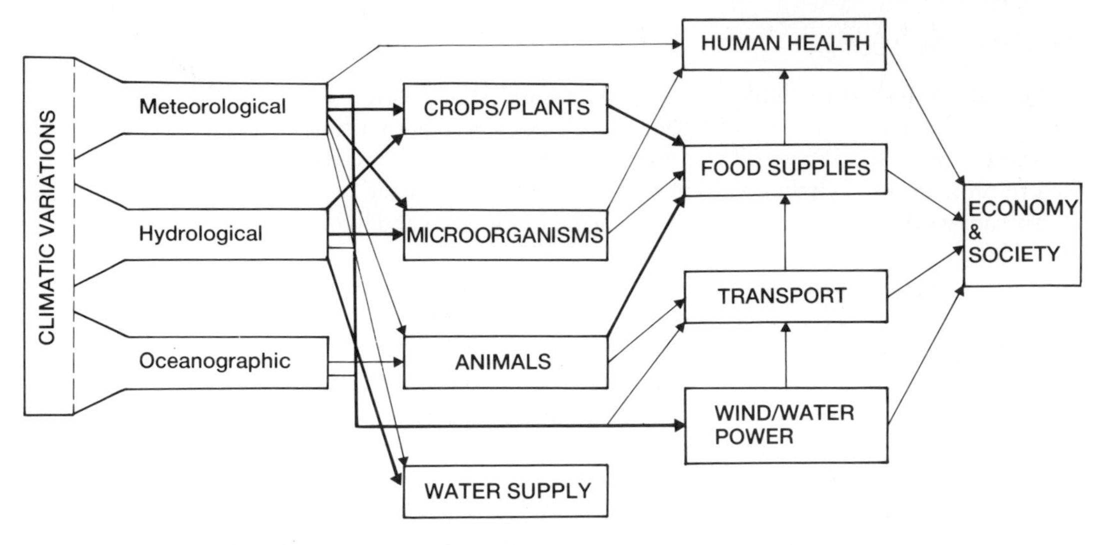

Figure 4.31 A diagram of *1 . . . n* order effects of climate on environment and society. As we move further away from the 'cause', more interactions intervene to disguise and modify the links. Many more arrows could be added to this diagram: a selection only of the more obvious are plotted.

(figure 4.31) and their feedbacks. Some rapid adaptations can of course be made: water storage and drainage, the import of new crops, the improvement of animal strains to cope better with adverse weather, are all first-order buffering devices.

The land itself may provide a second set of constraints if its characteristics are unpropitious for tillage. Here, environmental alterations can be remarkably efficient in producing a better crop: the construction of terraces or the reclamation of some kind of wetland are two outstanding examples. Although an environment may be reasonably conducive to the production of food or another organic crop, there are often minor environmental modifications that can be made to enhance the yield: ridging, fallowing, drainage, and many more of the types compiled by the contributors to R. Klee's book on traditional agricultural practices. So buffering may require the permanent alteration of pre-existing ecosystems either as a deliberate project as with large (even if traditional in their technology) irrigation schemes, or by the piecemeal accumulation of smaller changes.

Compared with hunting and gathering the flows of energy through the human-directed parts of an agricultural society are very large, since the hunter–gatherer generally deflects only a minor part of the fluxes of the ecosystem towards himself whereas the agriculturalist changes whole parts into crops which are of direct value. Much of the human energy in an agricultural society is expended in agriculture itself: in modern (1971) India about 55 per cent of the nation's total expended human energy of 45 000 GJ goes into agriculture. Further, some 82 per cent of total energy use in rural India is directly related to food and a similar situation was probably true of much of the world before AD 1800. There were probably exceptions only where there was such an abundant cropping that less energy needed to be invested; here we may suspect that in many societies the ruling groups kept up the energy inputs via labour in order to bring about large surpluses for the disposal of the largely urban-based kings, priests and chieftains. So our overall picture is of an energy intensity in an agricultural society which is much lower than that of an industrial one: India's 1971 energy-use intensity per hectare has been calculated at 6.49 GJ, whereas in the USA it was 18 GJ, a difference factor of nearly three.

Yet, as table 4.12 shows, quite high population densities can be sustained on agriculture which is basically solar, aided by the labour of men, women, children and domesticated beasts. There is a range of densities since climate plays an important part, as does the growing period of the crop, so that padi rice occupies one end of the density scale, and the simpler forms of shifting cultivation the other. Clearly also some types of agriculture are capable of supporting rapidly growing populations. Some do this by increased conversion of wild ecosystems into agroecosystems; others by intensifying the energy throughput per unit area, that is, by intensification. If the conversion process has definite spatial limits, as with most shifting cultivation systems, then intensification is the normal ecological response to population growth, just as emigration is a customary demographic result. Societies may adapt to change in various ways. Figure 4.32 shows for example different pathways of change for cropping systems in the tropics under conditions of population growth and intensified farming. Further, intensification is not always merely a consequence of indigenous population growth; the introduction of cash crops is also generally accompanied by an intensification of cropping and land use. Where short-term profits and capital accumulation have been sought, then environmental impact has frequently outpaced that of subsistence societies in the same settings. Thus in the Europe of the late middle ages, a stable population did not engender ecological stasis. The high cost of labour encouraged investments in large flocks of sheep, for example, with a high environmental impact. Incentives were provided as well for technological developments like firearms, printing and bigger ships, at least two of which were to result in larger scales of ecological change.

Islands are good laboratories for population–resource studies. A number of studies have suggested that in pre-colonial times, equilibrium populations were achieved which had sustainable agricultural systems. It appears in examples from Pacific atolls and the Caribbean that either population control was practised, or surplus people colonized other islands. Where isolation added to the problems, then some breakdown of society might happen: on Easter Island the population had risen to 20 000 and deforested the island by AD 1680. About that time, the construction of the giant statues ceased, probably because of intergroup conflict and because the large timbers needed to move and erect the statues could no longer be found.

Agricultural ecosystems may experience ecological breakdown. The commonest forms are invasion of the land by inedible vegetation

Table 4.12 The characteristics of the land use stages in Esther Boserup's model of intensification of land use

Type	*Range of population densities (per km^2)*	*Frequency of cropping (%)*	*Fallow length*	*Type of fallow*	*Tools in use*	*Fertilizer*	*Labour needs*	*Productivity*
1 Gathering	0–4	0	–	–	–	–	Low	High
2 Forest fallow	0–4	0–10	One or two crops then twenty-five years fallow	Secondary forest	Axe, fire, digging stick	Ash	Land clearance with axe and fire. No cultivation or weeding	High
3 Bush fallow	4–64	10–40	Two or more crops, eight to ten years fallow	Small trees, bush	Hoe	Ash, vegetation, turfs mixed in soil	Less land clearance but some weeding	Falls as yields fall and extra weeding needed
4 Short fallow	16–64	40–80	One or two crops then one or two years fallow	Wild grasses	Plough	Manure and human waste	Extra preparation of seed-bed, extra weeding and carting manure	Falls as extra cultivation, extra weeding collecting and distributing manure, care of draught animals
5 Annual cropping	64–256	80–100	One crop each year with a few months fallow	Legumes and roots	Plough	Manure, human waste, green manure, marling and silting, composts	Extra cultivation, weeding, terracing, irrigation and water control	Falls as extra cultivation and weeding, collecting and distributing manure, constructing and maintaining irrigation, terracing, water control, etc.
6 Multi-cropping	256–512	200–300	Two or more crops without any fallow					

Source: D. B. Grigg, *Dynamics of Agricultural Change*. London: Hutchinson, 1982.

that cannot be cleared by pre-industrial technology, and accelerated soil erosion. In general, though, it seems likely that both of these processes have occurred much more frequently after a group has been in contact with industrial technology, and so more extended discussion will take place in chapter 5. But there have been instances of soil erosion in the pre-1800 period, just as there have been many famines. The environmental impact of pre-industrial famine deserves perhaps a brief word: at such times almost anything will be eaten: for example, grass, tree leaves, bark (the cambium of elm trees was a known source of emergency protein in Europe as late as the nineteenth century), any animal, and even the soil itself.

In the long perspective, the environmental lesson of the agriculturalist period of human history is one of immense population growth, from an estimated 150 million in 400 BC (not the beginning of agriculture but an early date for which a sensible figure can be given) to *c.* 720 million in AD 1800. This rise is small compared with what was to come but was nevertheless involved with a great deal of environmental change, with new lands being occupied for agriculture on both local and continental scales, as in the European occupation of Australia, for example, where no indigenous agriculture had ever developed. The greater number of people created demands for forest resources which altered the ecology of the supply areas, whether this was in the form of replacement of one species by another more suited to a specific purpose, or by the manipulation of a semi-natural forest ecosystem. Even the sea experienced this impact, with species of whale being regularly hunted out and the whalers moving on to another stock. However, the 'pelagic bounty' fish stock of the Grand Banks of Newfoundland, for example, did apparently manage to sustain itself in the face of the English and French fishing fleets. Occasionally, of course, there was a retreat from environmental manipulation, as when depopulation occurred after a plague or a politically motivated shifting of people, or when a disease of plant or animal meant that cropping or herding underwent an enforced fallow period when natural processes might reassert themselves for a time. We even have the view that environmental breakdown underlain by the ecological fragmentation of the agroecosystems was strongly implicated in the fall of the Mesopotamian and Mayan civilizations, and that the stripping of the forests of the Mediterranean put the skids under Rome.

But clearly the agricultural *genre de vie* has had survival value: winnowed by many centuries of selective pressure both natural and social, it supported a growing population of humans without very many extended breakdowns (Plate 4.22). Constraints were, in time, offset by buffers. As a whole the system exhibited the agroecological conditions for success: (1) multiple cropping wherever possible, and reliable year-upon-year yields where not; (2) crops which where possible provide animal forage as well as human food; (3) a land use system which integrates the intensive use of the fertile land with the production of animal forage or other fertilizer from the marginal

Humid Climates

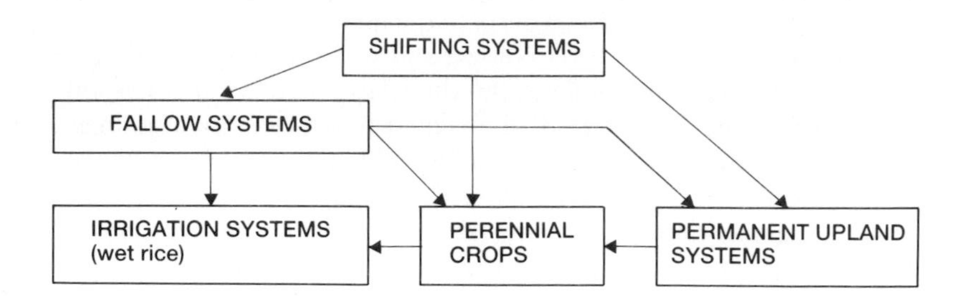

Semihumid Climates

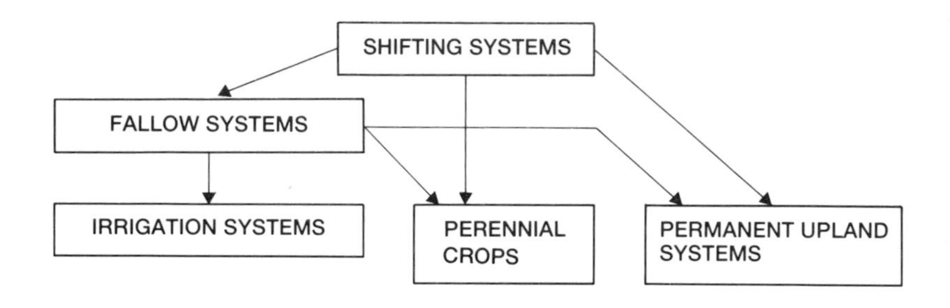

Semiarid Climates

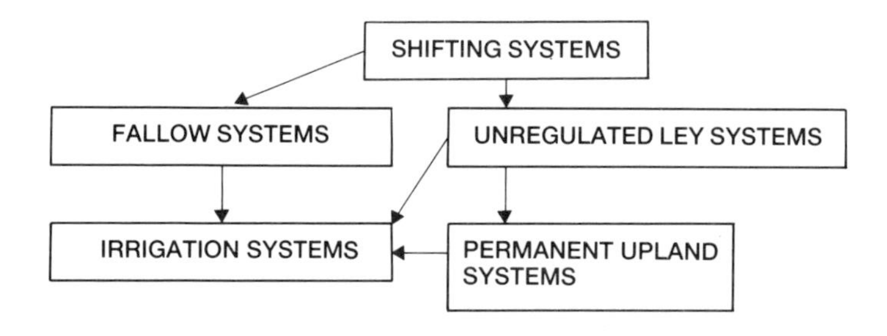

High altitudes

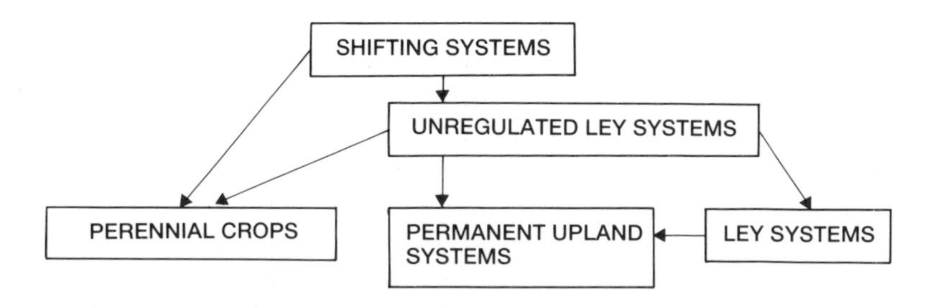

Figure 4.32 Generalized models of the impact of population growth on farming patterns (and hence on environmental impact) in the tropics. In most of them, following the arrows seems to lead sooner or later to irrigation.
Source: H. Ruthenberg, *Farming Systems in the Tropics*. Oxford University Press, 1977.

Plate 4.22 The simple technology of a plough, a beast and water control fed many millions of people for thousands of years before industrialization, and did not threaten any of the global stabilities upon which life depends. A rice field in Indonesia today, cultivated by largely traditional methods. (Paul Harrison/Panos Pictures)

lands, and which is run by a society which somehow limits its family size; and (4) a system which uses most of its manure as fertilizer rather than as fuel. In some places this system still persists in an essentially solar-powered form but in many others it is the recipient of an energy subsidy from fossil fuels. The transformation of agriculture to an 'industrial' base is just as important environmentally as the coming of urban–industrial economies in the conventional sense, and they will both be examined in the next chapter, together with their relationship to the surge in population from *c.* 545 million in AD 1650, when the take-off appears to have begun, to the 5000 million of 1988.

CHAPTER FIVE

Industrialists

Wonders are many and none more wonderful than man . . .
Subtle beyond hope is his power of skilled invention,
and with it he comes now to evil, now to good.

Sophocles, *Antigone*

Introduction

The hydrocarbon resource

Like the control of fire and the invention of agriculture, industrialization represents a turning point in the history of man–nature relations. An essential economic characteristic of the period since AD 1800 has been the manufacture of goods in a factory and by a machine, for sale outside the neighbourhood concerned. As a result, the structure of environmental relations over most parts of the world has been changed by the use of coal, oil and natural gas. In ecological terms, the stored photosynthetic energy of these fossil hydrocarbons has been added to the current energy of the sun, and used by human societies in a variety of ways to gain access to resources and to alter the structure and function of their surroundings. Since 1945, the energy of the atomic nucleus has been added to the repertory of energy sources available to mankind, but this is dealt with in chapter 6. At this stage, we may pick out two distinct features of the fossil energy sources. Firstly, there is always an initial cost in using energy: some has to be invested before there is any surplus, just as agriculturalists must use human or animal energy to prepare the field and tend the crop before there is a yield. But the fossil fuels are so concentrated in their energy content (table 5.1) that their return relative to initial inputs of energy is very high. By analogy, we might think of priming a pump: in traditional agriculture we pour water into the pump and we get more, but not that much more, back; with industrial energy we pour in a lot of water but then the pump produces buckets-full irrespective of the weather (table 5.2). Secondly, the fossil hydrocarbons are finite in quantity on the human time-scale. If indeed the geological conditions of the eras which produced coal, oil and natural gas (especially perhaps the Carboniferous, Jurassic and Tertiary) are being repeated on earth today, then the new deposits are being laid down and transformed at rates far lower than those at which we convert the ancient materials. For

Table 5.1 Energy content of various fuels

1 kg of fuel	*kcal*	*MJ*
Hard coal	7000	29.4
Lignite	4600	19.3
Crude oil	10000	42.0
Natural gas	7700	32.3
Wood	4584	19.2
Sun's daily output	8×10^{28}	3×10^{27}
1 kwh electricity (1-bar fire for 1 hour)	2800	11.76
Uranium-235	1×10^{10}	4.2×10^{5}

The heat released from some fuels is an approximate figure since there is natural variation in all the organic materials.

Table 5.2 Energy return and investment[1] for some fuel supply technologies in the USA

Process	*Energy return*
Oil and gas, 1970s	23.0
Coal (at mine head)	
1950s	80.0
1970s	30.0
Oil shale	0.7–13.3
Ethanol (ex plant residues)	0.7–1.8
Methanol (ex wood)	2.6
Solar flat plate collector	1.9
Hydropower	11.2
Nuclear (LWR)	4.0
Liquid geothermal	4.0

[1] 'The ratio of gross fuel extracted to economic energy required to deliver the fuel to society in a useful form.' The measure presented here excludes, however, the energy costs of refining, transport, labour, government and environmental services.

Source: C. J. Cleveland, R. Costanza, C. A. S. Hall and R. Kaufman, 'Energy and the US economy: a biophysical perspective'. *Science* 225, 1984, 890–97.

practical purposes, therefore, the fossil hydrocarbons are not renewable.

The 'new' energy sources have been allied to metals, chemicals and plastics to elaborate a technology underlain by a scientific (i.e. systematic and predictive) knowledge of the universe in all its aspects from the core of the planet to the limits of the galaxy and beyond. The period since 1800 has seen the unprecedented transformation of energy into knowledge and thence into machines which in turn produce artefacts, some of which facilitate the manipulation of the natural world on a scale hitherto unimaginable, except perhaps by the very few: we might recall some of Michaelangelo's schemes. They also produce environmental consequences of equal degree.

At the basis of these developments lie first of all the fossil fuels themselves, whose whereabouts on the globe are mostly, though not entirely, known (figure 5.1). All of them were used to some degree before 1800, but it is the nineteenth and twentieth centuries which have seen the industrialized nations of the world, and the areas under their hegemony, become dependent upon continuous supplies of hydrocarbons. The energy locked into these materials can eventually be released and transformed in such forms as heat, steam, in the internal combustion engine, and as electricity. In such shapes they facilitate access to other power sources such as hydro-electricity, geothermal energy and, eventually, nuclear energy.

The interaction of energy use with the extraction, refining and use of metals is an important allied theme. Iron is the most important and the history of its rise is that of savings in fuel use as well as the development of new technologies which improved the yield, since a traditional bloomery might extract 15 per cent of the iron from the ore, but a modern blast-furnace 94 per cent. From the late sixteenth century there were developments like those in charcoal production which enhanced yields in western Europe from 26 m^3 of wood per ton of charcoal to 8 m^3 per ton. Critically for industrialization this was followed by the substitution of coke for charcoal, and eventually by specific technologies for fuel-saving. That the nineteenth century was forged in iron is undeniable but its hardened version, steel, together with metals like copper and aluminium, cannot be overlooked in the twentieth century. Energy availability has also been the

Figure 5.1 The basis of industrialization and the western life-style: an estimate in the 1970s of the reserves and distribution of two key fossil fuels: oil and natural gas. These are the owners rather than the users, so to speak.
Source: I. G. Simmons, *The Ecology of Natural Resources*: London: Edward Arnold, 1981, 2nd ed.

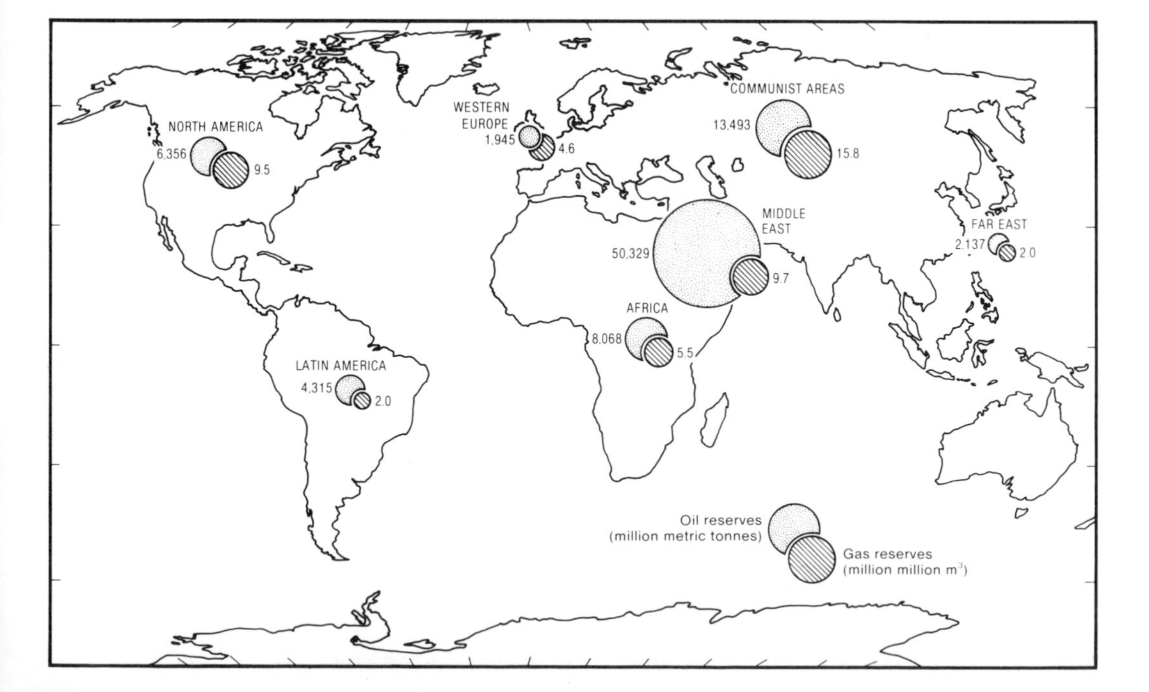

foundation of a chemical industry which produces compounds hitherto unknown in nature, particularly complex organic molecules. In an everyday sense, plastics are the symbols of this progression, although they are not greatly significant in environmental terms except as waste products which are very slow to degrade under most conditions.

Whereas the initial landscape features of the early phases of industrialization, like the canals and railways, were constructed almost entirely by human and animal labour, today's equivalents employ machines like the bulldozer and the dragline. Thus, these last two centuries have in essence seen a hundredfold expansion in the capacity of our species to alter the materials of nature, both *in situ* and in removing them to another place to transform them into artefacts. To come to grips with the seemingly infinite number of processes and changes, we shall start (in a manner parallel to that of chapter 4) by looking at the hearths of industrialization and their environmental impacts, and then go on to examine the major impacts of industrialization. Apart from industry and urbanization themselves, many related systems have been affected, notably agriculture, but also forestry and water resource development, for instance, and whole new activities such as mass tourism have been made possible. There is then a brief discussion of the spatial dissemination of wastes of a concentrated nature. Lastly, the population–environment–resources linkage of the last two hundred years is discussed in terms of being, as W. Zelinsky has put it, the years of 'the great exponentials'. The sustainability of these trends has been questioned, but an inquiry into those views will wait until the final chapter.

Fossil fuels and the early development of industry

To accept AD 1800 as the point when the use of coal becomes a significant factor in the potential for environmental change, does not mean that the fossil fuels had no human use before then. Rather, there was a 'prehistoric' period when they were used in some places but without much lasting environmental effect, and certainly not to produce a whole new way of life. The Chinese apparently mined coal 2000 years ago, and in the UK the Romans certainly took it from the natural outcrops and burned it for heat. The export of coal from the Tyne to London was briefly mentioned in the last chapter and by 1700, England and Wales had a coal output of 3 million tons per year, about five times as much as the rest of the world, and there it was virtually the only fuel used in the manufacture of alum, saltpetre, gunpowder, soap, candles and in brewing and salt extraction from sea-water. Some analogous developments took place in continental Europe: J. S. Bach had shares in a coal mine. Oil likewise had earlier uses: before the tenth century AD the Burmese were hand-drilling wells to tap oil and there was a world trade in it during the fifteenth and sixteenth centuries as a medicinal substance. Aristotle says that the Persian kings had food cooked in caves where leaks of natural gas were set aflame, and in the seventeenth century the Chinese drilled through solid rock to reach natural gas which was then supplied to buildings via a system of bamboo pipes and some towns were lit by its flames.

These examples demonstrate that there was a long period of 'below the horizon' use of hydrocarbons, loosely analogous perhaps with the husbandry of certain wild grasses before full domestication. Their ascent to prominence is in conjunction with two processes which happened in England in the eighteenth and early nineteenth centuries. The first of these was the development of a reliable steam-engine and certain cotton-spinning machines. The second may have been a growing shortage of wood to make charcoal, though since this was generally obtained from coppices on a sustained yield basis it may have been more decisive that the labour costs in making coal or coke available at the blast-furnace were lower than those for charcoal. Technological change had brought about the development of machines in the cotton industry in the eighteenth century: Eli Whitney's cotton gin in 1793 in the USA, and spinning machines in England in the 1760s. These latter were originally powered by water but by the end of the eighteenth century had been converted to steam and needed cheap fuel. A parallel development had been the steam-engine, usually credited to Watt *c*. 1765, though his early engines had no greater horsepower rating than a windmill or watermill. In parallel, Abraham Darby of Coalbrookdale, England, had in 1709 substituted coke for charcoal in the process of iron-smelting (Plate 5.1) to the point where by 1803 the east Shropshire iron industry was

Plate 5.1 The hearth of the industrial revolution, at Coalbrookdale in England. An engraving of the ironworks when charcoal was still an important fuel: it is being produced in mounds by the side of the impoundment whose water power drove the bellows of the furnaces. (The Ironbridge Gorge Museum Trust)

consuming 260,000 tons per year of coal (figure 5.2). The ascent to prominence of coal came with the alliance of demand for it to power machines with the invention of a heat engine which would pump up water from shaft mines. A small amount of coal invested in such an engine is the catalyst for production of energy and materials on an ever-larger scale. By 1800 there were 1200 steam engines in England and Wales, mainly in the cotton industry of the Midlands and North, but soon an 'explosion' of numbers took place, reflecting the fact that the date of 1780 saw the rate of industrial growth in England rise from 1 per cent per annum to 4 per cent, a rate which was then maintained for 100 years.

In these developments, we need to emphasize the role of transport. First the canals and then improved roads and especially railways all made possible cheap bulk transport and facilitated processes like the major shift from charcoal to coal in the iron industry between 1760 and 1820. Railways were initially wooden tracks on which were waggons hauled by horses and then by standing steam-engines (the first wagonway was probably 1605, the first steam-engine 1712, both in the English Midlands), but the great development was the steam locomotive since it carried its own fuel supply. So we might take as symbols of the catalysis which produced the Industrial Revolution such artefacts as the oldest surviving railway bridge in the world (Causey Arch in County Durham, England, 1725, plate 5.2), the successful early steam locomotives such as Locomotion (1825), the first passenger railway in the world, the Stockton to Darlington

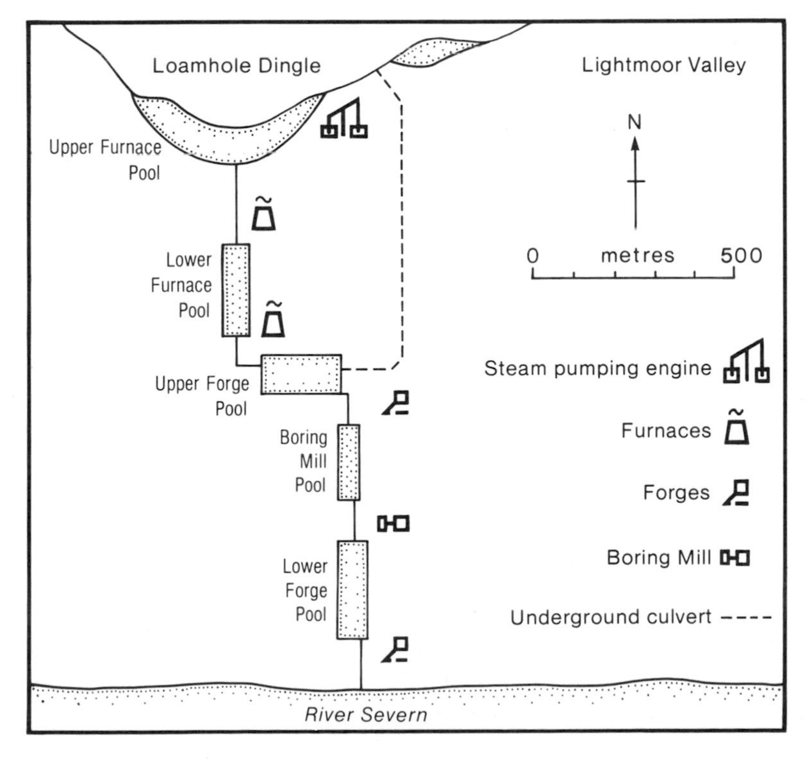

Figure 5.2 Even in very early (before 1800) developments at Ironbridge, control of water was very important and the steam pump was used for recirculation; a series of ponds held up useful heads of the water to power various machines. The crescentic pool is seen in pl. 5.1.
Source: Ironbridge Gorge Museum Trust Museum Guide 2.02, *Coalbrookdale*, 1979.

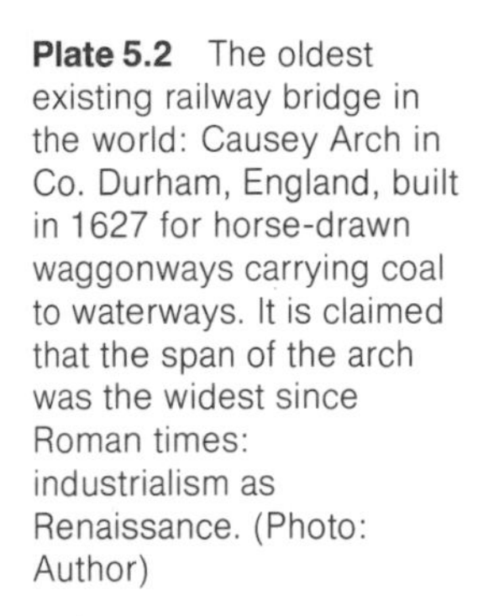

Plate 5.2 The oldest existing railway bridge in the world: Causey Arch in Co. Durham, England, built in 1627 for horse-drawn waggonways carrying coal to waterways. It is claimed that the span of the arch was the widest since Roman times: industrialism as Renaissance. (Photo: Author)

railway (1825), and the great iron passenger liner, the *Great Britain*, designed by Isambard Kingdom Brunel and launched in Bristol in 1843, now being restored in the dock in which she was built (Plate 5.3). (I employ these artefacts as specimens partly to avoid listing in tedious detail every event of those years and partly because they, or replicas of them, are still visible and can with insight be seen not as isolated 'things' for the tourist trade but as symbols of a tremendous change in man–environment relations, especially the passenger-carrying devices since they helped to spread ideas as well as goods.) The railways in particular acted in the early nineteenth century as a positive feedback mechanism: the energy used in them was an instrumental investment in the opening up of yet more sources of coal and of iron.

The early history of petroleum use is rather more dispersed. It springs from a late eighteenth century demand for improved interior lighting, and 'coal oil' was tapped from a spring of it in a Derbyshire coal mine. But the key date and event was the drilling of the first modern oil well (without powered machines, it may be noted) in Pennsylvania in 1859 (Plate 5.4). A refined spin-off product was the fuel for an internal combustion engine which had its origin in N. A. Otto's gas engine of 1876, the oil engines of Dent and Priestman of Hull, patented in 1886, and of Rudolf Diesel, patented in 1892 and made to use petrol (gasoline) by Gottlieb Daimler and Karl Benz in the 1880s. From such beginnings has grown the oil industry of today and its environmental consequences, both in terms of local effects of extraction and refining and the secondary impacts of the use of motor vehicles.

Plate 5.3 The steamship *Great Britain*, a symbol of the outreach of the industrial world, (a) her launching at Bristol in 1843, and (b) under restoration in the same dock. What started as a commercial venture ends as an object for tourism; neither group of users will think much about the environmental changes in which they are participating. ([a] The Mansell Collection; [b] Author)

In the British heartland of industrialization, the second half of the nineteenth century saw a marked increase in the availability of per capita energy: from 1.7 tce/cap[1] in 1850 to just over 4.0 tce/cap in

[1] Note: tce/cap is tonnes of coal equivalent per head.

(a)

(b)

Plate 5.4 The equivalent of Coalbrookdale for oil: the first well at Drake, Pennsylvania, with Drake himself in the well-shaped hat. He is talking to the Titusville druggist, which reminds us that oil is not merely a source of energy but of many materials (including pharmaceuticals) as well. (Drake Well Museum, Pennsylvania)

1919, at which stage it levelled off until after 1950. This energy powered an industrial economy whose index of production in 1860 was 31.7 compared with 100 in 1913, and in which an average family's calorific intake rose from 3240 per person-day in 1860 to 3926 in 1913.

The environmental impact of nineteenth century industrial developments is small by comparison with contemporary industry or even that of say 1914, but it was not negligible. The sites of mines and forges were places of heat and smoke, together with toxic wastes discharged into streams or built up into unvegetated piles. From such centres ran out roads, canals, and railways. Water-courses already altered to provide power for the new machines were further diverted, and as factories became the foci of industrial production, so settlements grew up to house the burgeoning workforces, replacing fields and other open ground (figure 5.3). We can perhaps generalize the developments in terms of a centre where the environmental manipu-

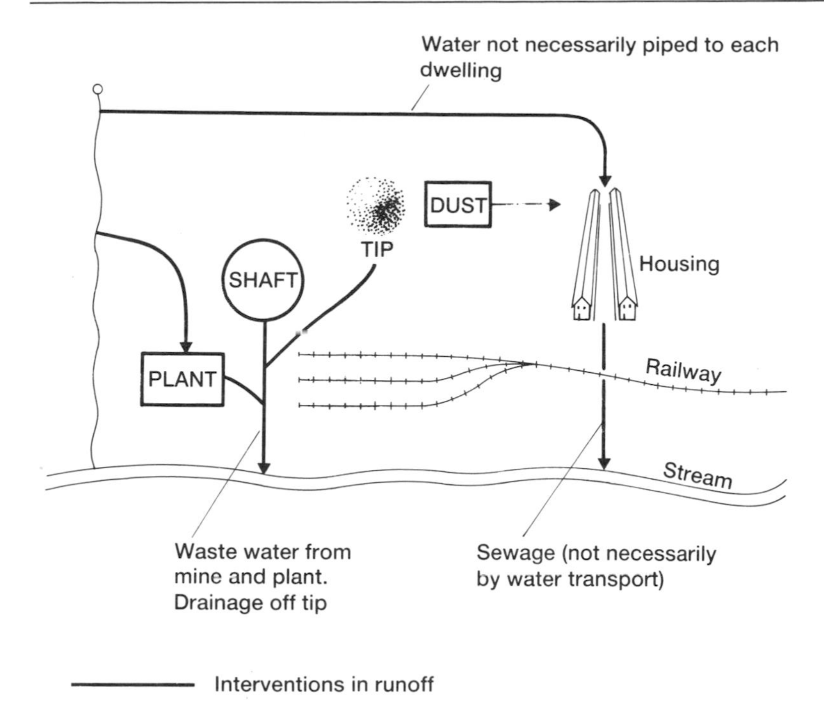

Figure 5.3 The local environmental context of an early shaft mine for coal. In most places this assembly of structures would have replaced agricultural land.

lation was greatest (be it coal mine or ironworks), and a periphery where the effects shaded off into the rural context, though in the late nineteenth century this zone was often squeezed out between coalescing centres of urban–industrial growth. Published in 1835, Alexis de Tocqueville wrote of Manchester in his *Journey to England and Ireland*:

Thirty or forty factories rise on the tops of the hills I have just described. Their six stories tower up; their huge enclosures give notice from afar of the centralisation of industry. The wretched dwellings of the poor are scattered haphazard around them. Round them stretches land uncultivated but without the charm of rustic nature, and still without the amenities of a town. The fetid, muddy waters, stained with a thousand colours by the factories they pass, of one of the streams I mentioned before, wander slowly round this refuge of poverty. They are nowhere kept in place by quays: houses are built haphazard on their banks. Look up and all around this place and you will see the huge palaces of industry. You will hear the noise of furnaces, the whistle of steam. These vast structures keep air and light out of the human habitations which they dominate; they envelope them in perpetual fog; here is the slave, there the master; there is the wealth of some, here the poverty of most.

Unlike the case of earlier economic revolutions, we have a full record of the polemic of those for and those against acting as a feedback loop in the succeeding 200 years. The great changes of the nineteenth century provoked considerable moral reactions: Huskisson wrote in 1824, 'If the steam engine be the most powerful instrument in the hands of man to alter the face of the physical world, it operates at the same time as a powerful lever in forwarding the great cause of civilization.' Rudyard Kipling, on the other hand, was suspicious of the

> . . . heathen heart that puts her trust
> In reeking tube and iron shard.

But in spite of any intellectual misgivings, the new way of life and all that it implied environmentally was rapid in spreading outwards from its hearthlands.

The spread of industrialism

In spatial terms, the new industrialism first became firmly established in Europe and North America. Before 1870, areas like Lancashire, Yorkshire, South Wales, Clydeside, and Ruhr, Saar and Sambre–Meuse regions were outstanding, and by 1890 the USA had a larger rail network than all Europe, including the British Isles and Russia, with the fastest economic growth rate in the world between 1825–1919, mostly concentrated in the north. Until 1870, however, the bulk of the fuel of industrial growth in the USA came from wood, with water power an important second source. Around 1880, Russia began to make big strides, including the development of oil wells, and there was a mass exodus from the countryside in the last 20 years of the nineteenth century to the older industrial regions like Moscow, St. Petersburg and the Urals as well as the fast-growing newer centres such as the Kingdom of Poland, Riga, the Ukraine and the Baku oil area of the Transcaucasus.

In 1814 the largest civil ocean-going vessel was the *Oceanic* of 3800 GRT and capable of 15 knots; in 1914 it was the *Aquitania* of 47 000 GRT and 23 knots and this reflects not only the increased size and power of machinery, but also the role of transport. The steamship and the steam railway played a part in the spread of industrial materials and ideas, just as the ocean-going sailing vessel had spread crops, weeds, diseases and faith after AD 1500. This technology was at the heart of an industrialism in which the development of industrial growth was like an escalator moving upwards at 1.5 per cent p.a. with different nations getting on at different times: 6 of them before 1870, another 15 before 1914, yet another 15 between the wars, and 20 more in 1950–70. Most of the earlier growth took place in what are now the more developed nations, though within them there have been changes in spatial distribution and environmental impact as, for example, minerals have become uneconomic to work in the face of cheaper imports.

The first revolution of coal and iron was supplemented in the second half of the nineteenth century by that of steel and electricity, with chemicals being added after 1870. In the latter years of the nineteenth century there was a core area of industrialism. It consisted largely of Europe, Russia, the USA and Japan, with outliers such as South Africa. There came also to be established a periphery of nations where industrialism was by no means absent but in which it never became a dominant way of life. Thus before 1930 industry in India and China was in enclaves and did not directly touch the lives of most of the inhabitants; most of India's steel production in 1930 was by one firm, and 60 per cent of the foreign-owned firms in China between 1895–1917 were in two provinces only. In 1929, for instance, China had 21 to 30 per cent of its working population in non-agricultural employment, compared with 81 to 90 per cent in the USA; India had 31 to 40 per cent, compared with 61 to 70 per cent in Japan. The introduction of industry into the periphery has been patchy and in many cases was associated with colonial status during the era of European imperialism in the nineteenth century, though nations attaining their independence (e.g. from Spain and Portugal) at that time were often not especially quick to adopt the new ways. The core industrialized countries did not depend upon the periphery for materials in their initial phases of growth, but population growth and rising expectations created a huge demand for materials by the end of the nineteenth century from both temperate and tropical areas beyond the industrial nations. Wheat was a key commodity from the temperate zone, as were meat and wool, and the list of tropical and semi-tropical products is very long, including fibres like rubber, cotton, silk and jute, together with hides, oil seeds, sugar, tea, coffee, and cocoa. In 1913 the role of the tropics in supplying minerals to the core was small (mainly Peru, Mexico, Malaya, Bolivia, N. Rhodesia and Indonesia), but by the 1960s it was very large, with the UK, France, Germany and the USA collectively taking 29 per cent of their supplies from the tropics. Outposts of the core, like Hong Kong and Singapore, and densely populated peripheral places like Egypt, India and Java meant that when cheap transport became available, rice could be moved around outside the core, and so Burma, Thailand and Indochina became major exporters after 1870.

The environmental relations of agricultural development for export from the periphery depended a great deal on the crop and on the economic context in which it was grown. Some plants were cultivated on a small scale as adjuncts to peasant agriculture and with no greater environmental change than that represented by the 'host' system: in Colombia, coffee was interplanted with maize, beans and yucca, and little machinery was required for its processing. In Venezuela, by contrast, the same crop was cultivated on large estates in plantations which monopolized the cultivated land. Similarly in Ceylon (now Sri Lanka), the colonial government seized large areas of 'unclaimed' land and made them available specifically for coffee plantations (Plate 5.5), though in this case they failed after 1870 because of disease, and were succeeded by cinchona (for quinine production), some cocoa, tea, and rubber, mostly planted in areas of former forest. In general, the greatest transformation of natural or semi-natural forest systems to cash crop harvesting took place where there was 'empty' land (which is not to say it was not used by shifting cultivators or perhaps pastoralists) with a good rainfed water supply. Many areas of forest, secondary woodland and moist savanna were

Plate 5.5 A colonial plantation in Ceylon (now Sri Lanka) growing coffee. Such an environment would probably have replaced a mosaic of secondary forest with patches of virgin vegetation, and would have been seen as a great improvement; the natives were probably tidied up as well. (BBC Hulton)

altered in this way with sometimes a succession of crops which responded to overseas influences as much as to local conditions: in southern Nigeria, rice was introduced by Christian missionaries, and then oil-palm, cotton and cocoa successively were grown for export, all by 1902. Where rainfall was deficient, colonial governments might build, or as in Ceylon revive, irrigation systems, with the attendant risk of spreading malaria.

So just as we would expect the core industrialized areas to exhibit a spectrum of environmental manipulation, from an all-pervasive obliteration of nature near the pit-head to the more subtle changes in vegetation of hills subject to sulphur rainout many kilometres away, so in the periphery alternatives ranged from the virtually imperceptible addition of an export crop to peasant shifting agriculture (as with some oil-palms), to the clearance of tropical forest and its total replacement, as with rubber tree groves. Further still, there was metamorphosis of the grassland vegetation of the Pampas, Australia and New Zealand, as great flocks of domesticated beasts were deployed with subsequent ploughing-up of some of these grasslands, most notably on the Prairies and High Plains of North America. In both types of region, the mine and the quarry were developed on an entirely new scale.

Trends established during the period 1870–1930 have not subsequently been greatly affected by either war or decolonialization, to the point where the influence of the core on the material production patterns of the periphery is now called 'neo-colonialist' by some commentators. Most peripheral countries have tried to develop manufacturing industry and some even basic heavy industries, thus adding to environmental impact. These developments have increased the rate of urban growth, but this has not meant less environmental change in the rural areas since industry-linked farming is now common and this brings about more metamorphoses than 'traditional' agriculture, as will be discussed later.

The outreach of industrialization has been very great. it has taken men to the top of Mt. Everest by means of bottled oxygen and to the moon via spacecraft, and more importantly can rapidly emplace people, goods and ideas almost anywhere on the planet's surface. We might wonder if there are any natural environments still unaffected by industrialization and conclude that there are still some places that are nearly pristine: on ocean beds, amongst high mountains, in deserts, tundras and a few forest areas. But such areas are either shrinking in area or receiving greater quantities of industrial fall-out from the atmosphere. Human societies unaffected by industrialization are probably equally absent by definition since if we know about them then they have probably adopted some facet of industrial life. Very far from an industrialized life-style are the Panare of the Venezuelan headwaters of the Orinoco, but they have machetes from Birmingham (England) and beads from Czechoslovakia and Italy, and they also possess steel adzes, nylon line and steel fishing hooks, shotguns and battery-operated torches for night fishing. A 1982 study concluded that as long as any of the Panare could remember they have been dependent upon industrially manufactured goods which they obtained by trade with the non-Indian population.

Impact of early industrialization

An attempt has been made in figure 5.4 to show some of the qualitative effects of an imaginary industrial site in terms of its linkages. The conception is of a late nineteenth century European settlement with coal mining and/or steel industries. This core of activity blots out the original ecosystems on site, as do the heaps of solid wastes that mines and smelting works produce. Workers' housing near to the plant also replaces pre-existing systems, which were most likely agricultural, whereas the lower density housing of the better-off might retain some open land and trees; city centres are very densely built up for administrative and commercial concerns but may contain a park or two in the formal nineteenth-century style.

Beyond its immediate bounds, the towns reach out. Local links include the need for water, met in the early stages by wells and by tapping rivers; and the need for any raw materials not available at the site of the mine or smelter itself, e.g. iron ore and limestone needed for steel-making. Expanding employment opportunities bring people to the town, necessitating expansion of housing, and creating a

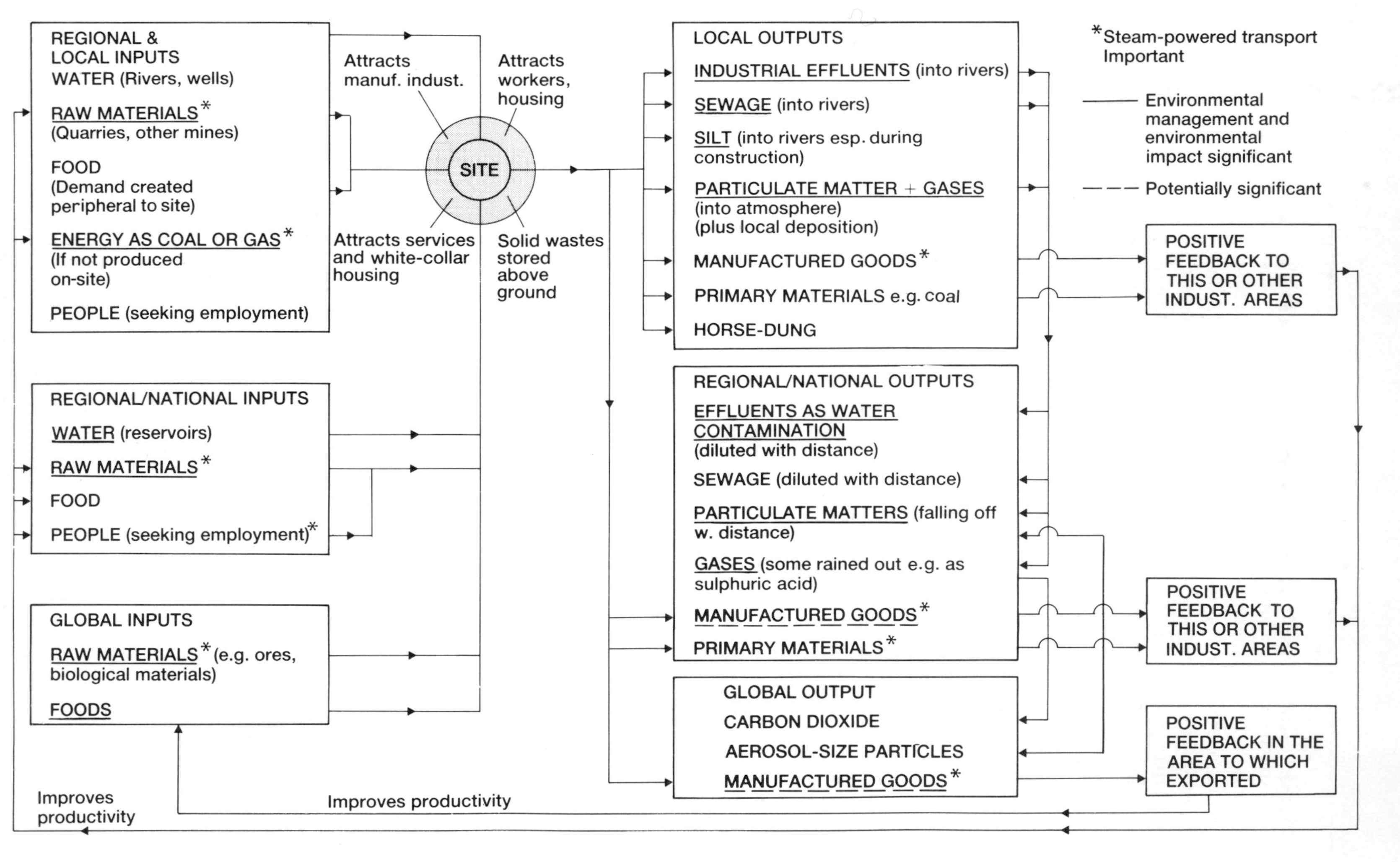

Figure 5.4 Manipulation by industry and to feed industry: a model of environmental impact in the late 19th century. The actual site is in the centre; to the left are the inputs arranged in spatial terms and to the right the outputs similarly arrayed.

demand for food production in the surrounding area. Since this zone is likely to be agricultural, then no major environmental changes are produced by this except perhaps the 'improvement' of some marginal areas as farmers seek to produce more, Some, however, may decide to intensify rather than extend their tilled acreage and make use of the chemical fertilizers which are, after 1870, becoming available; or perhaps to buy horse-dung from the towns as a manure.

Viewed as an open system, the town has outputs as well. The industrial plant and its wastes are responsible for the discharge of effluent to water courses and the atmosphere. Mine drainage and mine-tips are likely to acidify surface waters, and some of the sulphur dioxide from coal burning will immediately rain out as sulphuric acid and attack buildings as well as increase the acidity of freshwater bodies. During phases of construction, a great deal of silt will be washed off into rivers and will be added to the sewage from the town which at least until late in the nineteenth century is unlikely to have received any treatment. The biochemical oxygen demand of the sewage is almost certain to reduce the diversity of life in the river near the outfall, as are the other effluents. Sewage may contaminate drinking water and disease will result, though knowledge of the necessity of a pure water supply increased steadily throughout the nineteenth century. The town will export some of its coal and most of its factory goods, so that more land is devoted to transportation systems, especially railways.

Growth of the town into a city, perhaps coalescing with others to form a conurbation, brings about extensions of these processes. Local water supplies are unlikely to be sufficient so that the city ramifies into rural areas and builds reservoirs which either act as water stores or regulate the downstream discharge so as to prevent long periods of low flow. More raw materials are pulled in as local sources become exhausted or as industry diversifies, and of course more people (both from immigration and from a high rate of natural increase in the newly urbanized population) need more food. The search for building materials is carried out over a wider zone, and the outputs of contaminated air and water invade even larger areas, to the point of running together with those from other industrial areas.

There may then arise a stage where an industrial area becomes a transformer of material brought from overseas. Ores are brought by steamship and railway; biological raw materials like jute or rubber are made into manufactured goods, and new types of effluent may be produced. Similarly the cheap transport may allow the dissemination of tropical and sub-tropical foods beyond the rich: tea and oranges are found not only in the Hall at Shortlands but in Coronation Street as well. Just as the inputs become global in their connections, the outputs begin to achieve a universal distribution on Earth. Aerosol-sized particulate matter from chimney stacks and dust from bared areas of land starts to go up into the global circulation and be deposited in what are now detectable quantities in the Greenland ice-cap, for example. The burning of coal produces more carbon dioxide than can be scavenged out by the world's biota and so its concentration in the atmosphere begins to rise to the levels currently detected. So also the dissemination of manufactured goods to all parts of the world is launched and populations of remote hunter–gatherers have steel knives and machetes made in Pittsburg or Dusseldorf or Birmingham. The taste for these and many other products of number one nations remains even after they have ceased to be imperial masters. 'The Captain and the Kings depart', but nobody forgets the transport systems, the iron-shod ploughs, the movie shows and the medicine (to quote just a few examples) that they brought with them. Add instant global telecommunications (chapter 6), and the internationalization of economy and ecology is well on its way.

Magnitudes of energy use

Energy use per head in an industrial world can usefully be compared with past times. Hunter–gatherers can only tap solar energy and nearly all of this comes in the form of food and fire. Their energy throughput as food is perhaps 2000 kcal per head per day (837×10^3 KJ per head per day). Even without access to fossil fuels, agriculturalists nevertheless use draught animals, construct irrigation channels, use wind and water power and may produce surpluses: they can be responsible for throughputs of 10–20,000 kcal per head per day ($4186–8372 \times 10^5$ KJ). Calculations suggest that in the earlier phases of industrialization, the members of such societies average a throughput of 70,000 kcal per head per day (2936×10^7 KJ) and that members of today's full industrialized societies are at the level of 120,000 kcal per head per day ($50\ 232 \times 10^7$ KJ) (Plate 5.6). The elite selected to engage in space travel are supported, it seems, by 2,740,000 kcal per head per day ($11\ 469 \times 10^9$ KJ). In terms of nation states, the yearly continental averages are displayed in table 5.3. The difference between for example Africa and Europe is telling, as is the contrast between say Canada and Ghana. Some nations have so small a commercial energy consumption that in such classifications they are blanks, i.e. less than 1: Nepal and Kampuchea are instances. On the other hand, each citizen of the USA becomes the possessor of the equivalent of some 73 quite well-fed slaves. Since these calculations exclude food and organic fuels they are misleading about LDCs because wood is still the most important energy resource in most of them and indeed was the world's most important direct energy supplier until the 1960s. The general picture is of a considerable range between the most prolific consumers of fossil energy and those who use very little (or indeed have it used on their behalf, since these are data for total national consumption divided by population and say nothing about regional or class inequalities) and are thus virtually pre-industrial in their economies.

National data for energy throughput ('consumption') are also reflected in statistics such as GDP and GNP, reflecting the fact that energy availability and material wealth are intimately linked, though

Plate 5.6 An industrial society, among other things, can stretch the working and leisure day, with its access to cheap electricity for lighting: Hong Kong at night. It seems likely that rural electricity is an essential ingredient of 'development' in resource-poor societies of the South. (Paul Wakefield/Bruce Coleman Ltd)

as figure 5.5 shows, there is no simple correlation between energy consumption and wealth, since the efficiency of energy use may be an intervening factor. For our present purposes it would be interesting to know how close is the relationship between energy consumption and environmental alteration; however, as the latter term has not been measured in any reliable way which can produce a single numerical value we can only think in the most general way. The fragility of environments and densities of population, to name only two ingredients of a complex process, similarly promote a lack of simple correlations.

The growth of energy use in the world as a whole is shown on figure 5.6. Perhaps the most important lesson for the future of these data is the perception they have implanted, namely that the fossil fuels are physically finite in quantity, even more restricted in terms of economic availability, and therefore not sufficient for all time. This has led to some moves to use energy more efficiently in houses, in industry and in vehicles: industry in the USA consumed one-third less energy in 1982 than trends established before 1973 would have predicted, partly from a reduction in activity but also a result of improved efficiency of use and shifts away from energy-intensive industries. Such changes have helped to slow down depletion rates (and hence the environmental impact of exploitation), as well as

Table 5.3 Commercial energy consumption per capita per year, 1960 and 1984

	1960 (10^9J)	*1984* (10^9J)
By region		
World	38	55
Africa	6	12
C. and N. America	174	179
S. America	16	28
Asia	8	20
Europe	72	124
USSR	81	175
Oceania	76	134
National examples		
USA	236	281
Canada	164	286
Ghana	3	2
China	13	19
India	3	7
Bahrain	13	389
Qatar	44	646
Italy	45	95
UK	126	139
Australia	99	180
New Zealand	67	113

Source: UNEP, *Environmental Data Report.* Oxford: Basil Blackwell, 1987.

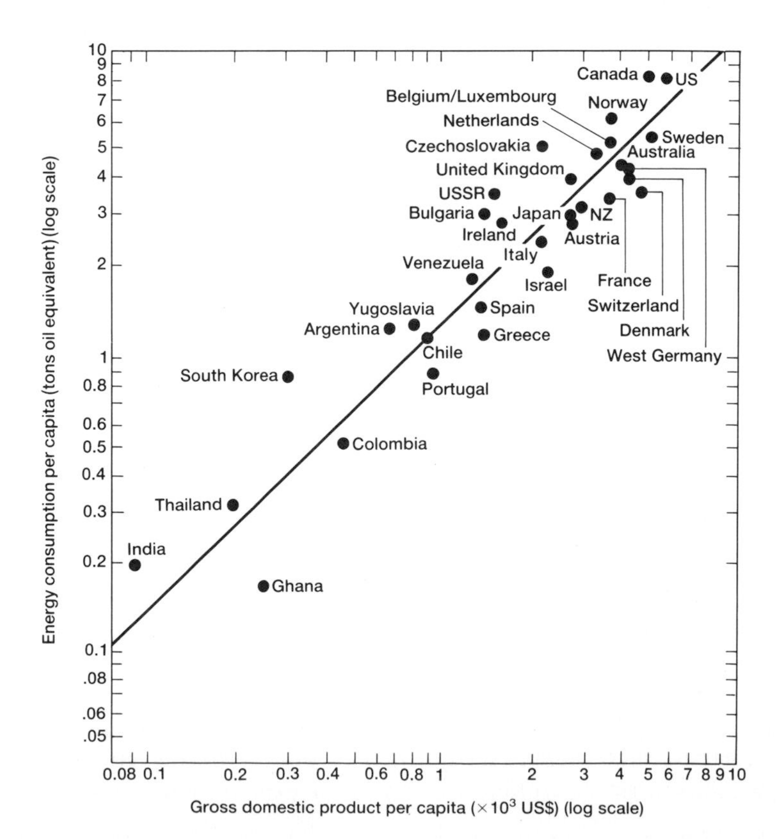

Figure 5.5 The standard graph of the relationship between individual wealth as GDP/cap and commercial energy consumption. Local energy sources such as fuelwood and dung are not included in these calculations. Note the variations in energy efficiency at the top end. *Source*: G. Foley, *The Energy Question*. Harmondsworth: Pelican Books, 1981.

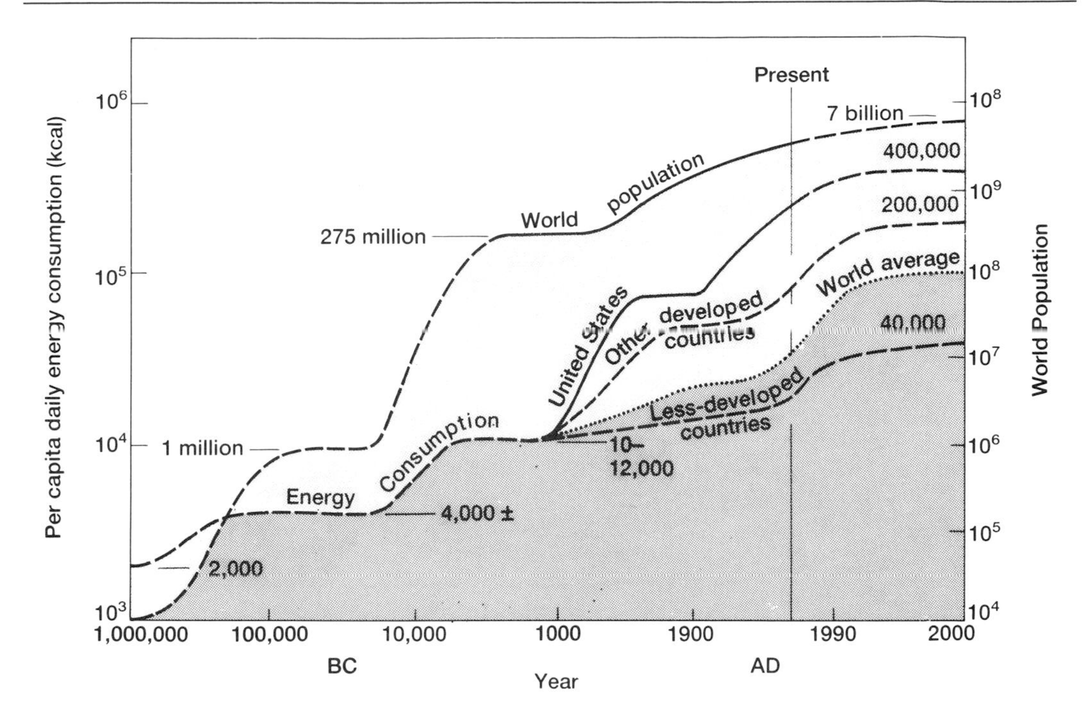

Figure 5.6 Even taking a world average, *per capita* daily energy consumption has risen through time. Added to the growth of the global population, this constitutes a formidable demand for energy and hence a very great potential for environmental change at all parts of the energy-use flow system. *Source*: E. Cook, *Man, Energy, Society*. San Francisco: W. H. Freeman, 1976.

promote the search for further deposits of fossil fuels (and spur the environmental effects of exploration); they have also accelerated attempts to develop the so-called 'alternative' sources of energy, some of which have a lower environmental impact than fossil fuel use, or to move to nuclear energy (chapter 6) which in general has as high an impact as the use of the hydrocarbons.

Given the predominance of energy supply in so much of economic thinking today, it may be apposite to put it into a global perspective. J. F. Alexander has calculated that the global energy flux due to fossil fuels is of the order of 0.11 cal/m^2/day (table 5.4). For comparison, the solar energy absorbed by the earth is 4900 cal/m^2/day and the primary production by plants is 7.8 cal/m^2/day, and volcanoes 0.005 cal/m^2/day. So while exceeding some natural processes, fossil fuel use is still one of the minor fluxes of the planet: compare it with, for example, the weather at 100 cal/m^2/day.

Energy production and transformation

On-site impact of energy and mineral extractions

If enhanced rates of energy flow are the vehicles of industrialization, then we are bound to take an interest in the environmental circumstances of their production, transmission and use. They are used along with technology to effect environmental change, but how is the energy-getting and conversion itself an agent of environmental manipulation?

Table 5.4 Global mean energy flows for various natural and human-induced processes

Process or event	*Energy flow* ($cal/m^2/day$)
Solar energy to earth	7000
Solar energy absorbed by earth	4900
Weather	100
Primary production by plants	7.8
Hurricanes	4
Tides	1.54
Animal respiration	0.65
Cities	0.45
Forest fire	0.3
Fossil fuel	0.11
Urban fire	0.065
War (non-nuclear)	0.05
Floods	0.04
Earthquake	0.001
Volcanoes	0.0005

Source: J. F. Alexander, 'A global systems ecology model'. In R. A. Fazzolare and C. B. Smith (eds), *Changing Energy Use Futures*. NY, Oxford: Pergamon Press, 1979, vol. II, 443–56.

The basis of industrial technology is fossil fuel energy, metals, chemicals and those plastics which are petroleum derivatives. So we ought to start with the environmental manipulation which surrounds their extraction from the earth's crust. In the early Industrial Revolution this process (and the construction of associated structures such as railways) was largely manually powered but progressively more powerful machines have replaced human labour and in turn have created larger upstream and downstream impacts. Once installed, the machine produces a lot of coal and this in turn initiates another set of pervasive environmental alterations. Figure 5.7 shows the flow of coal from its winning out of the earth's crust through combustion for electricity or in other industrial and domestic uses, and presents a simple scheme of qualitative environmental linkages. Even if local in incidence, these may be significant in scale: in the USA in the 1970s there were 262 urban areas that were in jeopardy from mining subsidence: depressions, pits, troughs, cracks, compression bulges, sinkholes and small ponds are all recorded. Where the overburden is less than 15 m, then depressions of 0.30–2.5 m are found, outlining in extent the underground mined areas and perhaps filled with water. The formation of these features may take place more than 50 years after mining has stopped. Table 5.5 describes some of the major environmental impacts of coal use. Reclamation costs are very variable but are perhaps 4 per cent of production costs in free market economies: US$7.5-11k/ha in the Rhineland and US$10k/ha in the UK. There is also likely to be considerable active management of regional water resources in order to produce the large quantities of water needed for dust control and coal washing where open pits are used: a consumption of 17 litres per tonne coal mined is representa-

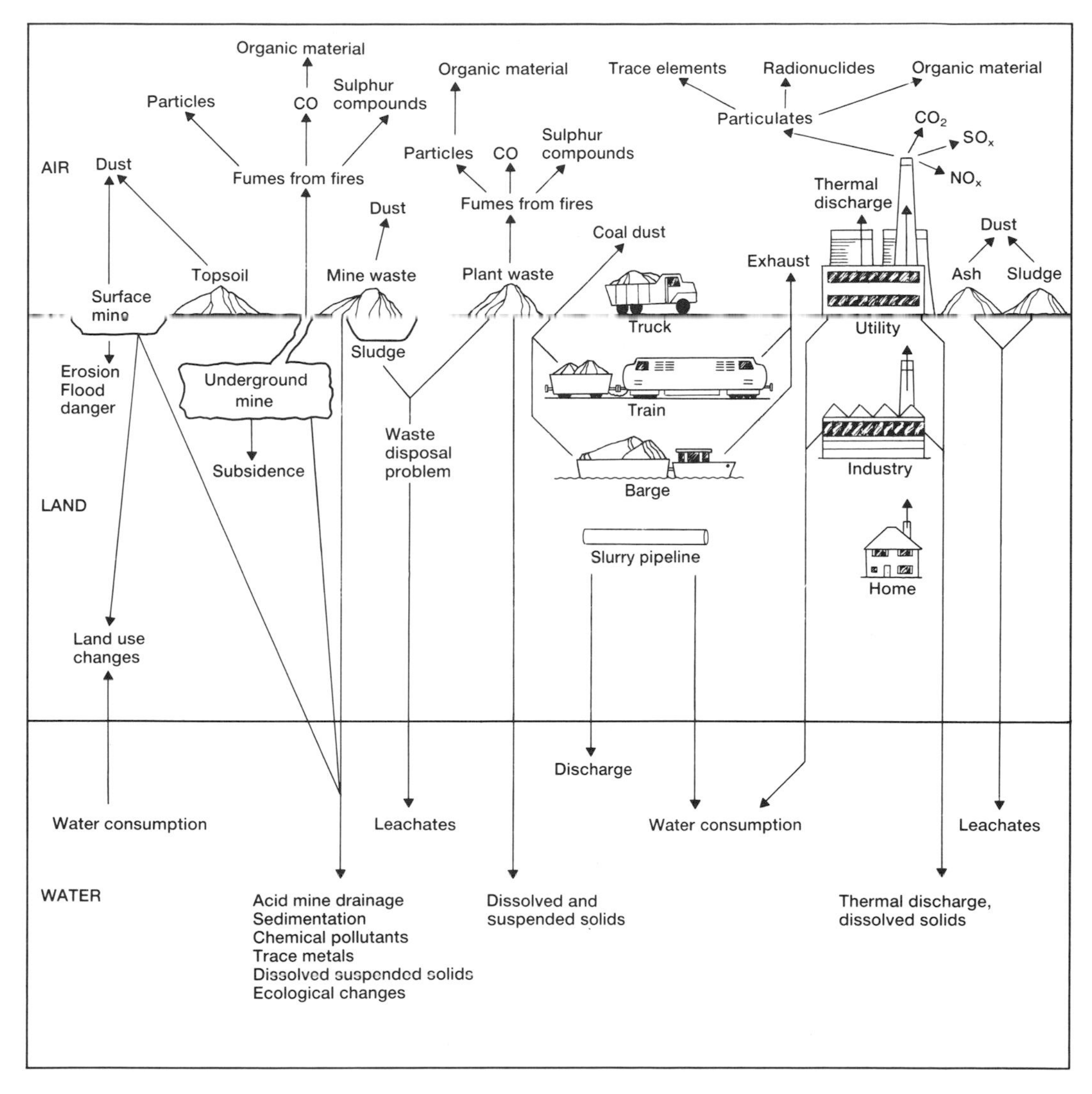

Figure 5.7 Modern coal extraction and use involves a variety of environmental linkages which are summarized in this diagram. All three of land, air and water are affected.
Source: After an original by the US Office of Technology Assessment.

tive for the USA; underground mining uses 63–120 l/tonne, plus 33 l/tonne for surface waste disposal. If coal is gasified then another 6–36 × 10^6 litres of water per million cubic metres of gas are required and a coal-fired power plant may also require 1.86 litres per kWh of electricity generated. Such a plant also has numerous effluents (table 5.6). Even given large quantities of input energy to a mine (table 5.7), the energy ratio is highly favourable; in one example of a United States' open pit, the input:output ratio was 1:6.5–7. Of interest for later comparison are the data that the E_r for corn on the same land would be 1:5.75 and for ethanol distilled from corn, *c*1. 1:1.40.

When oil and natural gas are considered, there is an extra dimension, since exploration and extraction may take place in marine offshore environments as well as on land: in 1978, some 17.2 per

Table 5.5 Major direct environmental impacts of coal use (excluding human health)

Impact	*Characteristics*
Accumulation of CO_2 in atmosphere	Effects potentially global but unpredictable spatially and temporally. Resistant to technical amelioration
Acid precipitation from SO_x and NO_x; acid mine drainage	Regional damage (often across international boundaries) to forest productivity and fresh water fisheries. Emission control of gases feasible
Land use for mining operations	Local to regional scale. May be only temporary if reclamation is practised: technology for this is available
Consumptive water use for conversion plants, open cast mining and reclamation	Regional impacts on agro- and other ecosystems where water is scarce. 'Mining' of aquifers possible. Resistant to technical amelioration

Source: J. P. Holdren, 'Energy and the human environment: the generation and definition of environmental problems'. In G. T. Goodman, L. A. Kristoferson and J. M. Hollander (eds), *The European Transition from Oil. Societal Impacts and Constraints on Energy Policy.* London: Academic Press, 1981, 87–120.

Table 5.6 Effluents from a coal-fired power station

Type of effluent	*tonnes per GW(e) per year*
Airborne effluents	
Particulates	3×10^3
Sulphur oxides	11×10^4
Nitrogen oxides	2.7×10^4
Carbon monoxide	2×10^3
Hydrocarbons	400
Liquid effluents	
Organic material	66.2
Sulphuric acid	82.5
Chlorides	26.3
Phosphates	41.7
Boron	331
Suspended solids	497
Solid waste: ash	3.6×10^5

Source: E. El-Hinnawi, *The Environmental Impacts of Production and Use of Energy.* Dublin: Tycooly International, 1981.

Table 5.7 Energy balance sheet for an open coal mine in Indiana, USA, 1976

Input total (Diesel oil, petrol, electricity, explosives, machinery)	402 MJ/tonne coal produced
Input/ha/yr	6.136×10^6MJ
Yield/ha/yr	390×10^6MJ
Net energy yield/ha/yr	383.8×10^6MJ
Energy input/output ratio	1:63.7

Source: G. W. Land, 'Energy use and production: agriculture and coal mining'. *Landscape Planning* 10, 3–13.

cent of the world production was produced off shore. Such operations are a potential source of oil spills (table 5.8), as are the various stages of oil transport by sea (e.g. terminal losses, tank cleaning and tanker wreck), whose long-term consequence are in general unknown, especially where the impact of oil residues upon fragile environments such as coastal wetlands, tidal marshes and sheltered bays is concerned. The effect of a major tanker spill depends on a number of variables, such as the tanker contents (crude oil or refined products), the meteorological conditions and coastal physiography, the clean-up methods employed and who is to pay the cost of the mopping up.

Oil refineries are major industrial installations (Plate 5.7) and as such may interact strongly with environment, not least in the changes

Table 5.8 Sources of oil in the oceans, 1970s

Source	*Contribution* (tons/yr)	%
Transport (tankers, terminals, bilges, accidents)	2,350,000	34.9
Coastal refineries (municipal and industrial waste)	875,000	13.0
Offshore oil production	87,500	1.3
Rivers and urban runoff	2,100,000	31.2
Atmospheric fallout	660,000	9.8
Natural seepage	660,000	9.8

Source: J. Golden et al., *Environmental Impact Data Book*. Ann Arbor: Ann Arbor Science, 1979.

Plate 5.7 An oil refinery (as seen here at Galeota Point in Trinidad) is not only an environmental transformer itself in terms of land use changes and emissions but a symbol of many more alterations wrought by access to the energy which it makes available. (Philip Wolmuth/ Panos Pictures)

of land use needed to provide sites. Since proximity to deep water is usually required, coastal wetlands are often modified for this purpose. Once in operation, the refinery demands water for cooling and chemical processes, of the order of 195 litres of water per barrel of oil processed, and in turn emits wastes to the air and to the outgoing water. Just as the chemistry of oil refining is complex, so is the battery of wastes, which are set out in tables 5.9 and 5.10, though it should be emphasized that companies are continually reducing these emissions where they are under pressure to do so; others may respond by building refineries where controls are less strict or even non-existent, as in some LDCs.

Oil and gas may be burned to produce electricity, however, in which case there are further demands for cooling water, together with the production of solid and gaseous wastes (table 5.11). Heat is produced and if this is discharged into fresh or offshore water, the local fish population is usually changed; in some places the low-grade heat is used to warm homes. The power station itself, like a refinery, represents a considerable change of land use as well as a source of environmental contaminants so care needs to be exercised over the siting of such a development. In Israel, for example, the evaluation between sites at Zarka and Hadera took into account not only economic costs and benefits, legal–administrative factors and public perceptions, but landscape factors, ecological linkages and probable dispersal of contaminants as well. It was considered also that irreversibility of change to natural systems should be an important criterion and the presence of one of the last pure-water streams in Israel together with rare species of plants and animals in an undeveloped rocky stretch of coastline was in the end decisive.

The extraction of oil from tar sands and oil shales is relatively uncommon but attracts attention when other oil and gas sources dry up or become uneconomic. The mining of tar sands causes widespread environmental disturbance and revegetation of the area mined is not usually possible; there is also a very large output of solids to settling ponds (table 5.12). Oil shale residues can be revegetated but characteristically the refining process requires very large quantities of water: 550 litres/bbl oil is a typical figure. (This is 113 litres per million Btu (1.06×10^9J) energy recovered, compared with 2.3 litres for surface coal mining, and 11.3 litres for oil and gas production, but 890 litres for nuclear power production.) Given the inputs needed, it is not surprising that oil shale production, for instance, should be predicted to have low E_r figures: the range is of the order of 1:2.8–10.0, which is positive, although accurate reclamation energy costs for spent shale dispersal are not yet available.

Environmentally, then, the extraction and conversion of energy cause considerable impact. Obvious features like land use change (figure 5.8) and contamination of the air are complemented by less conspicuous alterations of which the most pervasive is that of the demand for water, with the concomitant manipulation of rates of flow and of quality which are involved.

Table 5.9 Petroleum refineries as sources of wastes

Petroleum refinery products and the wastes generated by their processing

Products	*Wastes*
Aldehydes and alcohols by the Oxo method	Rectification residues with dissolved hydrocarbons and aldehydes
Hydrogen cyanide from natural gas and ammonia	Distillation residues (from the cyanide stripping process) with a small HCN and unreacted hydrocarbon content
Chlorinated derivatives of methane and ethylene	Column effluents containing chlorinated hydrocarbons, salinated to a certain extent with lime sludge
Acetylene from hydrocarbon cracking	Carbon black, hydrocarbons and hydrogen cyanide
Ethylene and propylene by thermal cracking	Wash liquors, oily wastes
Polymerization and alkylation	Alkaline wastes (NaOH), hydrocarbons, benzene derivatives, catalysts (phosphoric acid, aluminium chloride)
Alcohols from olefins by sulphonation and hydrolysis	Large quantities of wastes containing sodium sulphate, polymerized hydrocarbons and butyl or isobutyl alcohol
Aldehydes and alcohols by oxidation of hydrocarbons	Organic acids, formaldehyde, acetaldehyde, acetone, methanol and higher alcohols
Butylene from butane and butadiene from nutylene	Small quantities of wastes with a high hydrocarbon content
Aromatic hydrocarbons by reforming processes	Condensates polluted with catalysts, hydrogen sulfide and ammonia

Potential sources of specific air emissions from petroleum refineries

Emission	*Potential Sources*
Oxides of sulphur	Boilers, process heaters, catalytic cracking unit regenerators, treating units, H_2S flares, decoking operations
Hydrocarbons	Loading facilities, turnarounds, sampling, storage tanks, wastewater separators, blow-downsystems, catalyst regenerators, pumps, valves, blind changing, cooling towers, vacuum jets, barometric condensers, air-blowing, high pressure equipment, handling volatile hydrocarbons, process heaters, boilers, compressor engines
Oxides of nitrogen	Process heaters, boilers, compressor engines, catalyst regenerators, flares
Particulate matter	Catalyst regenerators, boilers, process heaters, decoking operations, incinerators
Aldehydes	Catalyst regenerators
Ammonia	Catalyst regenerators
Odours	Treating units (air-blowing, steam-blowing), drains, tank vents, barometric condenser sumps, wastewater separators
Carbon monoxide	Catalyst regeneration, decoking, compressor engines, incinerators

Source: P. N. Cheremisinoff and A. C. Morresi, *Environment Assessment and Impact Statement Handbook*. Ann Arbor: Ann Arbor Science, 1977.

Table 5.10 Examples of non-product wastes produced by petroleum refineries

Type of Waste	*Sources*	*Description*	*Characteristics (with typical quantities when available)*
Process solids	Crude oil storage, desalter	Basic sediment and water	Iron rust, iron sulphides, clay, sand, water, oil
	Catalytic cracking	Catalyst fines	Inert solids, catalyst particles, carbon
	Coker	Coker fines	Carbon particles, hydrocarbons
	Alkylation	Spent sludges	Calcium fluoride, bauxite, aluminium chloride
	Lubrication oil treatment	Spent clay sludges, press dumps	Clay, acid sludges, oil
	Drying and sweetening	Copper sweetening residues	Copper compounds, sulphides, hydrocarbons
	Storage tanks	Tank bottoms	Oil, water, solids
	Slop oil treatment	Precoat vacuum filter sludges	Oil, diatomaceous earth, solids
Effluent treatment solids	API separator	Separator sludge	Oil sand and various process solids (8320 lb/day)
	Chemical treatment	Flocculant aided precipitates	Aluminium or ferric hydroxides, calcium carbonate (2080 lb/day)
	Air flotation	Scums or froth	Oil, solids, flocculants (if used)
	Biological treatment	Waste sludges	Water, biological solids, inerts (3150 lb/day)
General Waste	Water treatment plant	Water treatment sludges	Calcium carbonate, alumina, ferric oxide, silica (17,260 lb/day)
	Office	Waste paper	Paper, cardboard (1200 lb/day)
	Cafeteria	Food wastes (garbage)	Putrescible matter, paper (60 lb/day)
	Shipping and receiving	Packaging materials, strapping pallets, cartons, returned products, cans, drums	Paper, wood, some metal, wire
	Boiler plant	Ashes, dust	Inert solids
	Laboratory	Used samples, bottles, cans	Glass, metals, waste products
	Plant expansion	Construction and demolition	Dirt, building materials, insulation, scrap metal
	Maintenance	General refuse	Insulation, dirt, scrapped materials – valves, hoses, pipe

Source: P. N. Cheremisinoff and A. C. Morresi, *Environmental Assessment and Impact Statement Handbook*. Ann Arbor: Ann Arbor Science, 1977.

Table 5.11 Effluents from an oil-fired power station

Type of effluent	*tonnes per GW(e) per year*
Airborne effluents	
Particulates	1.2×10^3
Sulphur oxides	37×10^3
Nitrogen oxides	24.8×10^3
Carbon monoxide	710
Aldehydes	240
Hydrocarbons	470
Liquid effluents	
Suspended solids	497
Sulphuric acid	83
Chlorides	26
Phosphates	42
Boron	331
Chromates	2
Organic compounds	66
Solid waste	9190

Source: E. El-Hinnawi, *The Environmental Impacts of the Production and Use of Energy.* Dublin: Tycooly International, 1981.

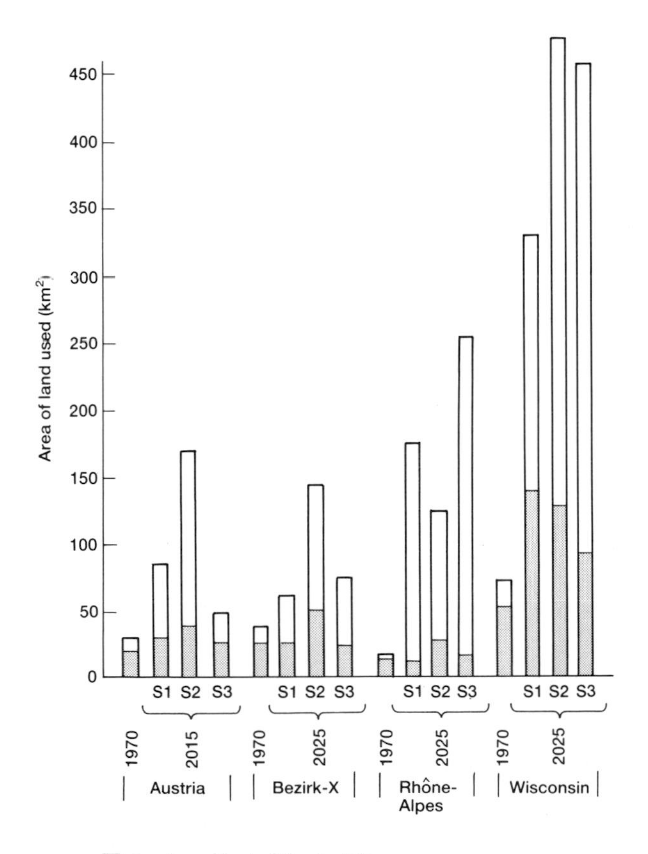

Figure 5.8 A comparison of the actual land use for energy extraction and processing in 1970, with the projected use under various scenarios of demand and development. This *excludes* hydropower, which would add markedly to the totals in Austria and France. In Wisconsin, the S2 scenario would mean that 0.3% of that State would be used for energy production.
Source: W. K. Foell, 'A four-region overview' in W. K. Foell and L. A. Hervey, *National Perspectives on Management of Energy/Environment Systems*. Chichester: Wiley for IIASA, 1983, 203–45.

Table 5.12 Environmental impact of tar sands mining

Operation or source of impact	*Potential impact*							*Changes of ground water*	
	Increased landslide risk	*Destruction of existing vegetation*	*Alteration of habitats*	*Topographic changes*	*Drainage diversion*	*Increased noise*	*Solid waste generated*	*Physical*	*Chemical*
Site preparation		X	X	X	X	X			
Surface clearing (cleared area)		X	X	X	X	X	X		
Stripping		X	X	X			X	X	X
Tar sand extracting (mined area)				X		X		X	X
Haul road transportation (construction)		X	X	X	X	X	X		
Tailings disposal						X	X		
Bitumen in tailings, or low grade tar sand waste		X							X
Solubles or waterborne particles in overburden			X					X	X
New surface									X
Increases in surface slope from waste disposal	X								
Rehandling of material: back-filling, grading and recontouring						X			

Source: E. El-Hinnawi, *The Environmental Impacts of the Production and Use of Energy*. Dublin: Tycooly International, 1981.

The city

One of the most obvious uses of the power made available since the nineteenth century has been the florescence of urban areas. Not that the city originated then, but the last 150 years have seen an unparalleled phase of growth, often in association with machine-intensive industries. Even where industrialization is at a relatively low level, as in LDCs today, cities have grown. Sometimes they came to prominence as centres of colonial administration (power in its other sense), and sometimes because of the influx of people displaced from rural areas, frequently by a more energy-intensive agriculture. Once in the urban area, nearly all city-dwellers share a dependence on fossil fuel power for their maintenance and convenience, and exert a number of environmental impacts at the site and elsewhere (figure 5.9). Thus the city itself represents an almost total transformation of the land surface (figure 5.10), the aquatic and ecological systems, and often noticeable alterations to the lower atmosphere as well.

The city has historically been parasitic upon other areas for its energy. Coal for example dictated certain patterns: in 1915, Chicago handled 23 million tons per year of coal via barge, train and wagon, so that large tracts of the city had to be given over to coal since it made the work of the city possible. The advent of fossil fuels had the effect of disaggregating land uses: it was more economic to deliver

Figure 5.9 The modern city is a kind of energy transformer, and this diagram summarizes the inputs, the immediate products of the energy use and the 'downstream' effects outside the city limits.

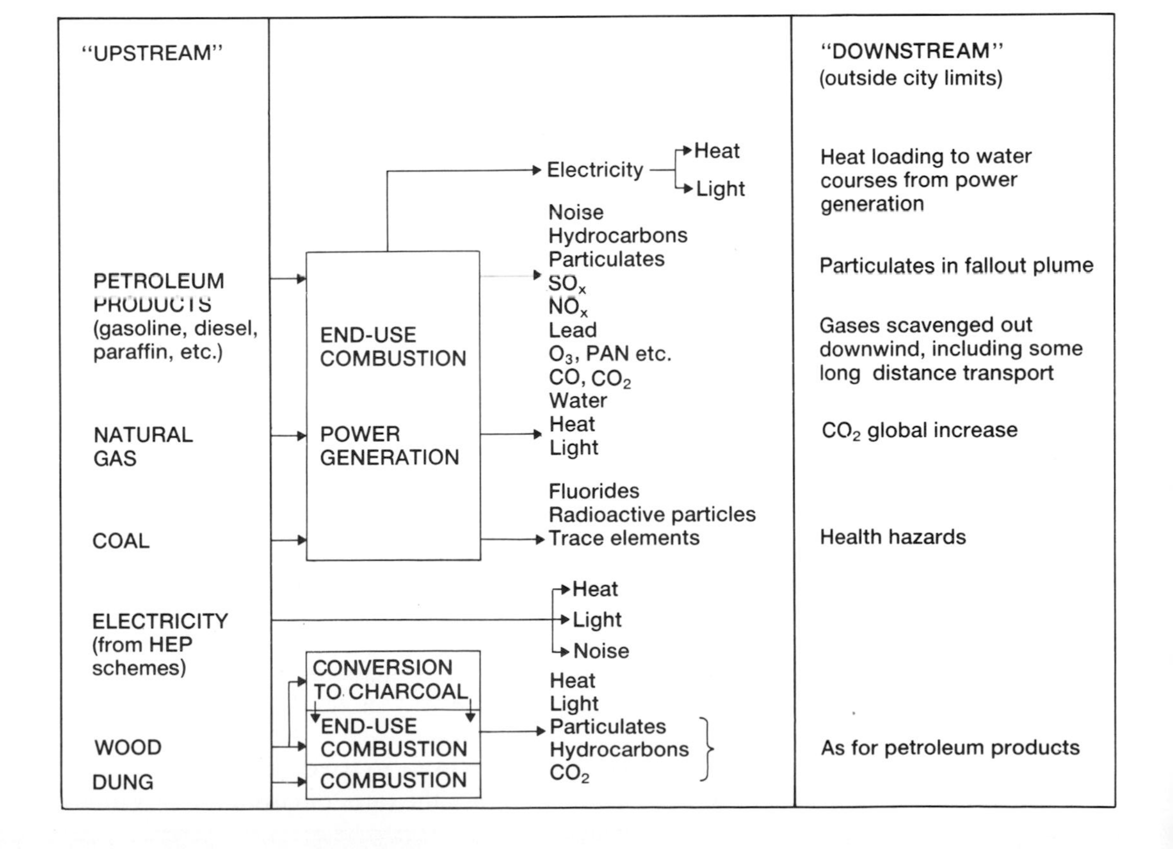

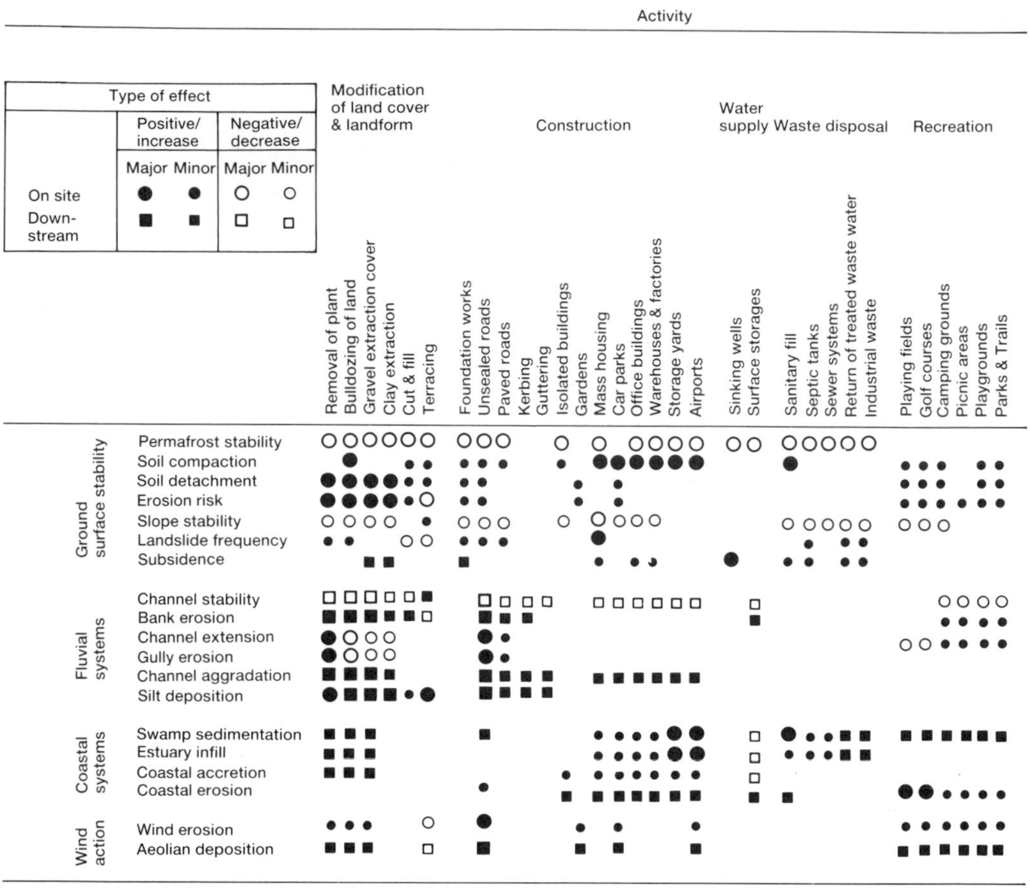

Activity

Type of effect	Positive/ increase Major	Positive/ increase Minor	Negative/ decrease Major	Negative/ decrease Minor
On site	●	•	○	○
Down-stream	■	▪	□	▫

Modification of land cover & landform: Removal of plant; Bulldozing of land; Gravel extraction cover; Clay extraction; Cut & fill; Terracing

Construction: Foundation works; Unsealed roads; Paved roads; Kerbing; Guttering; Isolated buildings; Gardens; Mass housing; Car parks; Office buildings; Warehouses & factories; Storage yards; Airports

Water supply: Sinking wells; Surface storages

Waste disposal: Sanitary fill; Septic tanks; Sewer systems; Return of treated waste water; Industrial waste

Recreation: Playing fields; Golf courses; Camping grounds; Picnic areas; Playgrounds; Parks & Trails

Ground surface stability: Permafrost stability; Soil compaction; Soil detachment; Erosion risk; Slope stability; Landslide frequency; Subsidence

Fluvial systems: Channel stability; Bank erosion; Channel extension; Gully erosion; Channel aggradation; Silt deposition

Coastal systems: Swamp sedimentation; Estuary infill; Coastal accretion; Coastal erosion

Wind action: Wind erosion; Aeolian deposition

Figure 5.10 The very form of the land surface is affected by human activity. This chart summarizes the effects of various processes inherent in urbanization upon the local landforms and 'downstream'.
Source: I. Douglas, *The Urban Environment*. London: Edward Arnold, 1983.

large quantities of coal to a few sites only, and at the same time cheap public transport made it possible for a lot of people to live at a distance from industrial or downtown zones of the city. Petroleum and electricity carried these processes further so that energy was no longer a constraint on land use and both population and work-place could now be diffused over a much larger area: the need for the compact city had gone.

A considerable amount of energy goes into construction, much of it as diesel fuel for machines: the first two years of a residential construction project may consume 25 TJ and 71 TJ per unit built respectively. Once occupied though, the energy consumption is considerably greater, of the order of 106–128 TJ per unit per year. A growing city will suck in materials like sand, gravel, stone and brick; once built and in working order, it affects almost every aspect of the form and function of the systems with which it interacts, and the infinitesimal details of this make adequate description in a concise manner almost impossible. Table 5.13 shows the qualitative fate of the resources imported into the city.

The city under construction is likely to have certain impacts upon soil and water. Table 5.14 shows the rates of erosion from various

Table 5.13 The city as environmental transformer

Input to city	*'Upstream'*	*Transformation within city (maintenance rather than construction)* *'Downstream'*	*Output from city*
Atmosphere	Unmodified unless by other cities upwind	Air velocity changes around buildings Air turbulence increased over roofline of city Loadings increased of – particulates – increase thunderstorm frequency heat radiated from buildings gases, esp. CO, CO_2, SO_x, NO_x PAN + NO_x = photochemical smog as in California, Japan lead, fluorides Precipitation – water is shed quickly from paved and built areas snow melts relatively quickly	Air flow resumes normal characteristics Particulate loadings may increase precipitation immediately downwind Particles fall out downwind Photochemical smog may damage organisms NO_x, SO_x fall out as acid rain
Ground and surface water	River systems modified for flood control and water storage – reservoirs channelization Watershed cover modified for water yield and quality Groundwater tapped for urban use	Evaporation of water from many sources Incorporation of water in organic and inorganic mass stored in city Calefaction of waste water, esp. from power generation Water as waste carrier – in solution / in suspension – affects local water quality Waste water to runoff. May add to flood peaks	Flood peakiness of river enhanced Contaminated rivers affect fauna, flora. Decreases with dilution Contaminants may be toxic to life, affect aesthetic quality of river Flood hazard increases engineering of river, e.g. channel modified for faster flow, cut-off channels Water may be led off for aquifer recharge Water may be purified in sewage treatment works and re-used or put into river for use lower down
Land and surface Forms	Little effect except as consequence of other processes e.g. water management, road construction Sand and gravel, brick clays extracted to built city and adjacent infrastructure. (Also may affect seabed.) Other rocks used for urban–industrial construction	Construction of embankments etc. across flood plains may cause ponding back at times of peak flow Slope stability – infiltration changes may hasten slides; buildings constructed in unsuitable zones e.g. Hong Kong, some Japanese cities high hazard when earthquake risk also present: zoning for land use may mitigate damage. Japan, California Subsidence – extraction of groundwater or removal of e.g. coal or other solids e.g. Mexico City, Venice, Long Beach Surface changes – removal of vegetation increases silt yield and causes local flooding Stream course engineering – to manage and avoid floods, improve navigation Coastal cities – dune-blowing, river channel dredging, storm protection construction, piers, seawalls Waste dump accumulation Coastal landfill for industrial sites, housing, airports e.g. San Francisco Bay	Silt transported away more rapidly to sea Removal or alteration of sediment transport may affect landforms at some distance, esp. increasing coastal erosion hazard Solid wastes must be disposed of: circum–urban fill sites sought e.g. quarries, gravel pits. Contaminants from tips may affect ground water Necessity for sewage treatment unless river or sea volume sufficiently large to make it appear unnecessary Eutrophication of water unless N_2 and P removed by multi-stage treatment
Food	Urban demand may exert strong influence on agricultural patterns Transport network reflects necessity to move food to city	Some food stored, most consumed and transformed to organic wastes	
Plants and Animals	City acts as roosting/nesting zone for rural feeders e.g. starlings, pigeons Fire-prone vegetation may need to be managed to reduce hazard to city	Vegetation more or less all changed: managed vegetation often grass with shrubs and trees ('urban savanna') unmanaged vegetation dominated by weedy species, incl. quick-growing trees and shrubs some species cannot cope with contaminated atmosphere (esp. lichens, mosses) some species adapt to city life e.g. songbirds, fox, badger, escaped pets	Escaped pets may establish themselves in rural areas (e.g. Australian parakeets in S. England)
Energy	Construction of conduits, e.g. pipe and transmission lines, transport networks to import energy; generating plants	Energy resources into power at generating sites (steam, electricity). All forms into wastes (heat, particles, gases)	Gradual dilution of effects of energy transformation and use

Table 5.14 Representative rates of erosion from various land uses in the USA

Land use	*Erosion rate* (t/km²/yr)	*Relative to forest = 1*
Forest, undisturbed	8.5	1
Grassland	85	10
Cropland	1700	200
Harvested forest	4250	500
Active surface mine	17000	2000
Abandoned surface mine	850	100
Construction	17000	2000

Source: PADC Environmental Impact Assessment and Planning Unit (ed.), *Environmental Impact Assessment.* Boston: Martinus Nijhoff for NATO Scientific Affairs Division, 1983.

kinds of land uses: the soil yield to wind and water is 100 times that of the cropland it is most likely to replace. The sediment alters the water quality considerably, and the bare surface sheds water rapidly, so that flood peaks are higher and sooner than before. The channels may experience rapid bank erosion as well as aggradation and even a change from a meandering to a braided state. If we move on from this phase to that of stable urban maintenance, then the potential changes in the hydrological cycle at and downstream from a city can be very marked, as figure 5.11 indicates. If the city is in a tectonically unstable zone then many natural hazards are exacerbated by its growth, as R. U. Cooke has shown for slope erosion, slope failure and channel flows in Los Angeles.

An example of a very thorough study of the material and energy balances of an urban area is that undertaken for the city-state of Hong Kong. In the case of energy (figure 5.12), the flow shows quite clearly the dependence of Hong Kong upon petroleum products for its commercial–industrial base and for the support of some of its food production as well. The nutrient budget can be calculated in detail: each year 2.39×10^6 kg of nutrients enter the Hong Kong food system: some imported, some incorporated locally, and of this, 74 per cent reaches consumers. A key element in this system is phosphorus, of which large amounts are imported (figure 5.13) as food (5434 kg/day), fertilizer (1247 kg/day) and animal feed (3145 kg/day), and much of which (7300 kg/day) is discarded as a pollutant. This is lost to the Hong Kong ecosystem, and adds to the contamination and eutrophication of the harbour and adjacent waters as well. Clearly, a method of recovery and recycling of this phosphorus would be a great benefit all round.

Figure 5.11 The flow of water in the urban environment. Notice (a) the number of human-made channels of movement compared with the 'natural' channels, and (b) the comparative number of channels where flow is increased rather than reduced by urbanization.
Source: W. Pope, 'Impact of man in catchments (ii) roads and urbanization, in A. M. Gower (ed.) *Water Quality and Catchment Ecosystems*. Chichester: Wiley, 1980, 73–112.

As shown by this example, cities are major sources of phosphorus and nitrogen loadings to rivers, though not so intensive as agricultural land in the developed countries (figure 5.14). In areas like mountains where agriculture is not important, then urban areas are

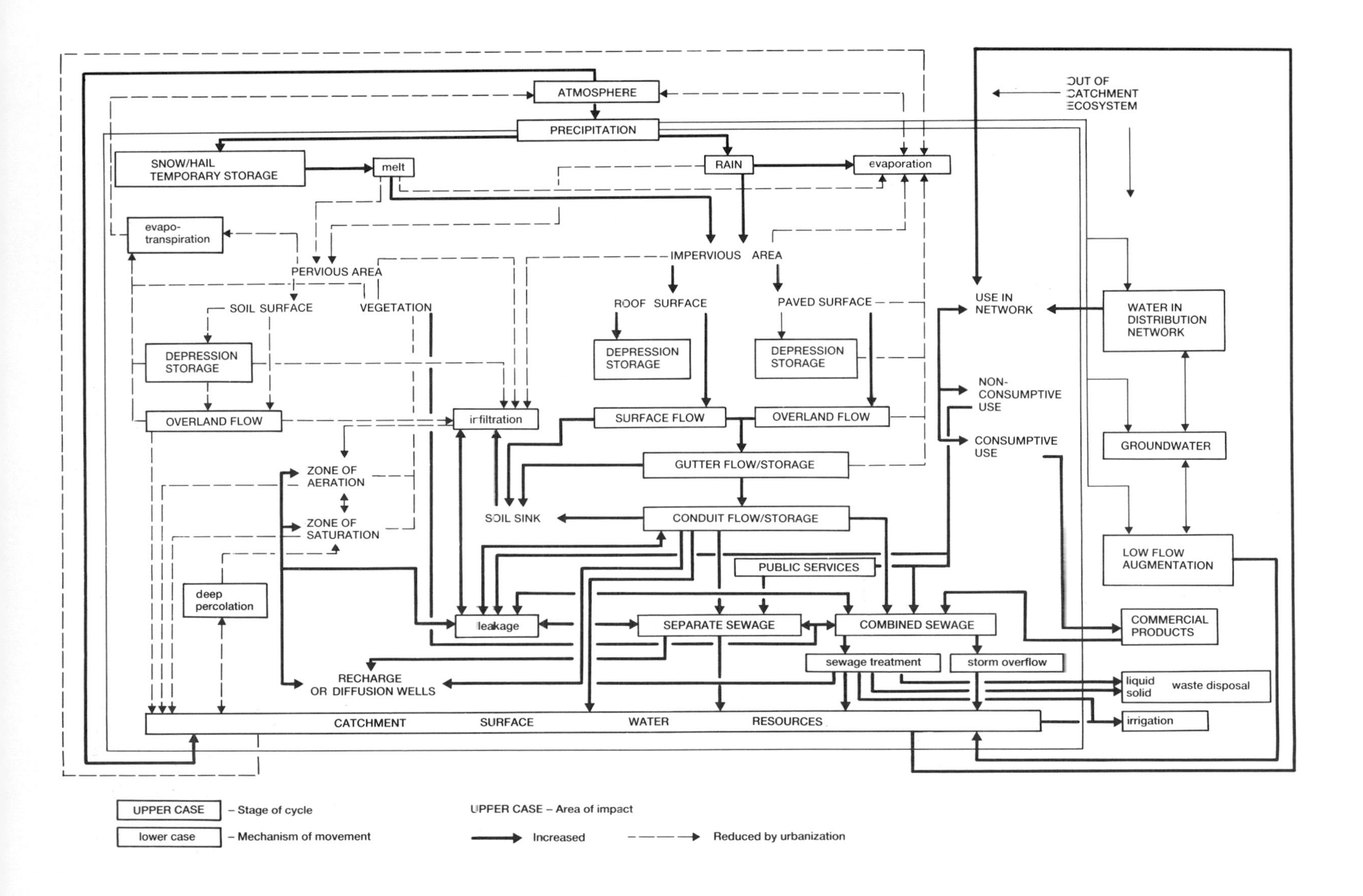
ATMOSPHERE
PRECIPITATION
SNOW/HAIL
TEMPORARY STORAGE
melt
RAIN
evaporation
OUT OF
CATCHMENT
ECOSYSTEM
evapo-
transpiration
PERVIOUS AREA
IMPERVIOUS AREA
SOIL SURFACE
VEGETATION
ROOF SURFACE
PAVED SURFACE
DEPRESSION
STORAGE
DEPRESSION
STORAGE
DEPRESSION
STORAGE
OVERLAND FLOW
infiltration
SURFACE FLOW
OVERLAND FLOW
USE IN
NETWORK
WATER IN
DISTRIBUTION
NETWORK
NON-
CONSUMPTIVE
USE
CONSUMPTIVE
USE
GROUNDWATER
GUTTER FLOW/STORAGE
ZONE OF
AERATION
ZONE OF
SATURATION
SOIL SINK
CONDUIT FLOW/STORAGE
LOW FLOW
AUGMENTATION
PUBLIC SERVICES
deep
percolation
leakage
SEPARATE SEWAGE
COMBINED SEWAGE
COMMERCIAL
PRODUCTS
sewage treatment
storm overflow
RECHARGE
OR DIFFUSION WELLS
liquid
solid
waste disposal
CATCHMENT SURFACE WATER RESOURCES
irrigation
UPPER CASE – Stage of cycle
lower case – Mechanism of movement
UPPER CASE – Area of impact
Increased
Reduced by urbanization

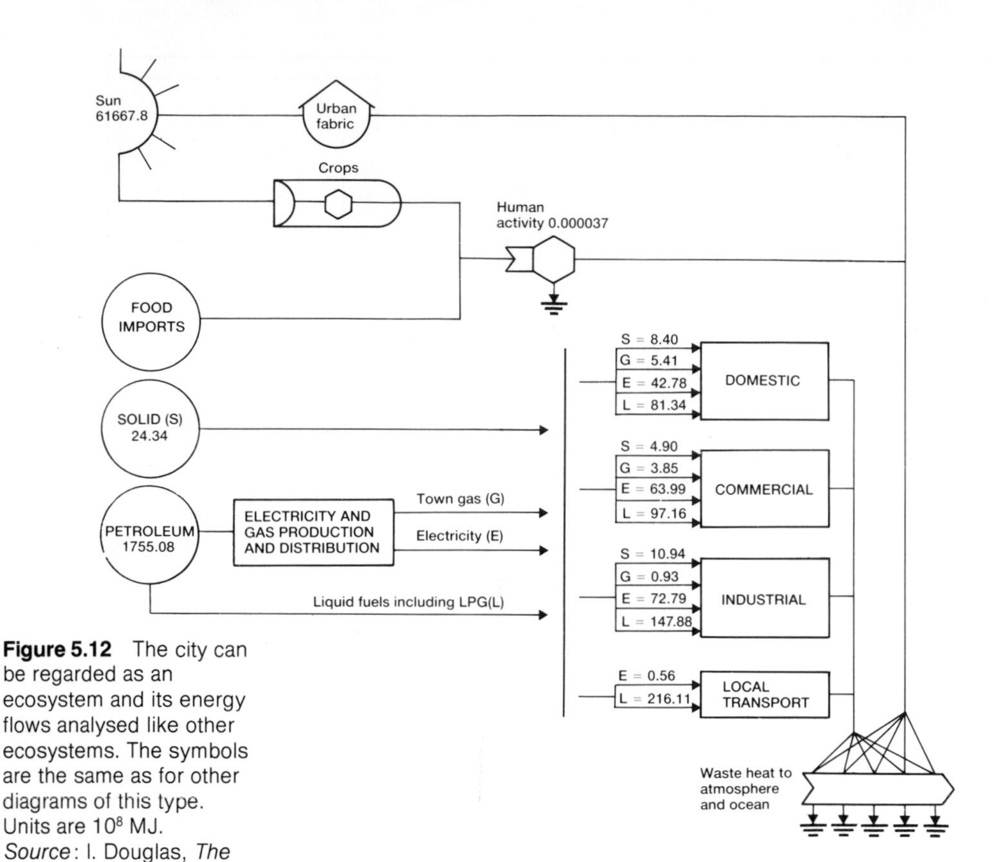

Figure 5.12 The city can be regarded as an ecosystem and its energy flows analysed like other ecosystems. The symbols are the same as for other diagrams of this type. Units are 10^8 MJ. *Source*: I. Douglas, *The Urban Environment*. London: Edward Arnold, 1983.

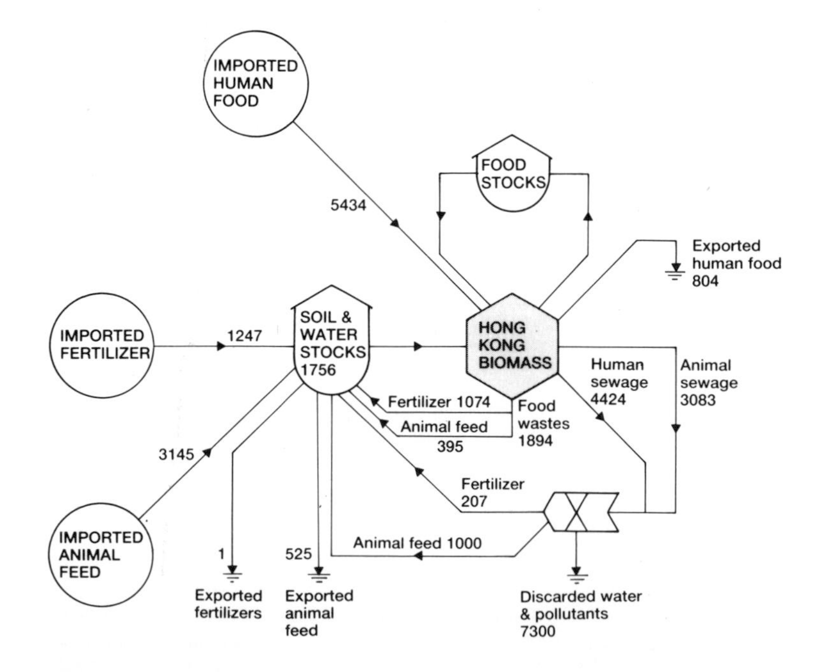

Figure 5.13 Parallel to the energy flow of fig. 5.12 are flows of materials, in this case phosphorus. This is an important plant nutrient and shows up in food supplies. In water, it is a contributor to eutrophication processes. Ideally, it should all be trapped and re-used. Units are kg/day. *Source*: I. Douglas, *The Urban Environment*.

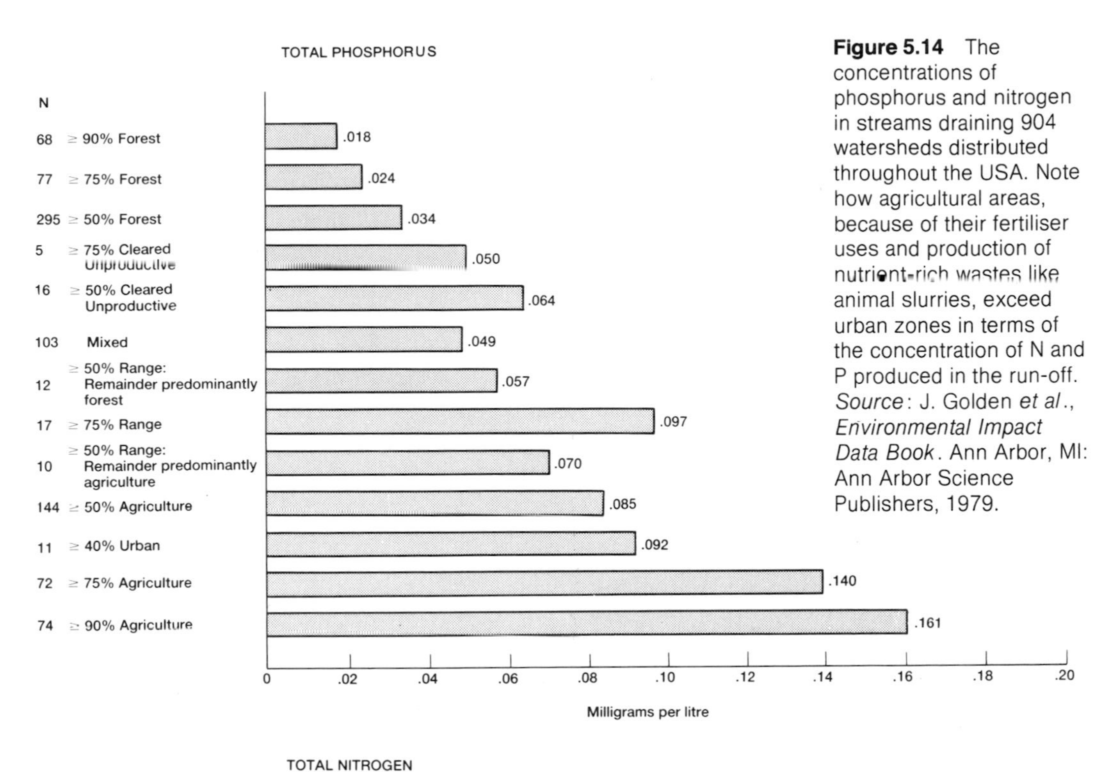

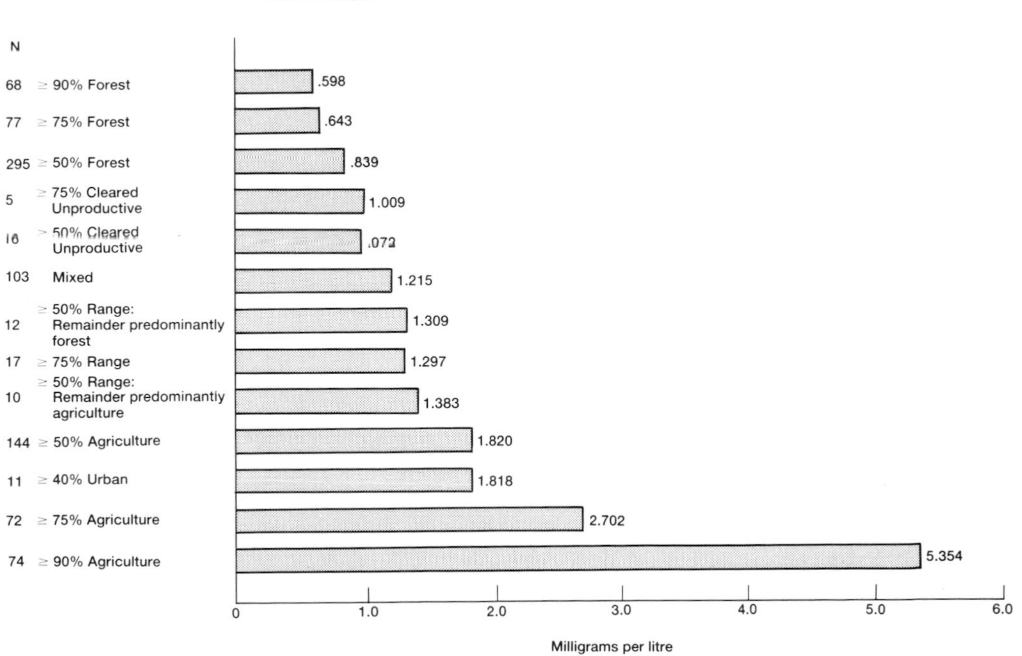

Figure 5.14 The concentrations of phosphorus and nitrogen in streams draining 904 watersheds distributed throughout the USA. Note how agricultural areas, because of their fertiliser uses and production of nutrient-rich wastes like animal slurries, exceed urban zones in terms of the concentration of N and P produced in the run-off. *Source*: J. Golden *et al*., *Environmental Impact Data Book*. Ann Arbor, MI: Ann Arbor Science Publishers, 1979.

the chief sources of nitrogen and phosphorus in the runoff. Figure 5.15 shows yields from different types of land use near Lake Dillon, Colorado, and it can be seen that even with sewage treatment the urban area yields high levels of phosphorus; a molybdenum mine yields high quantities of nitrogen from its workforce and possibly from ammonium nitrate used as an explosive. But all are very high compared with the natural and near-natural conditions. The disjecta of a modern city can also be gauged from the general materials budget summarized in table 5.15, and no doubt the refuse loadings would be higher were the mass of the population more affluent, since poorer people tend to re-use materials more often and for longer than the rich.

Rapidly urbanizing parts of LDCs are no less free than developed countries from the thrall of oil. A study of the city of Lae in Papua New Guinea shows that although it is set in the context of a country where fuelwood is the most important primary energy source, this

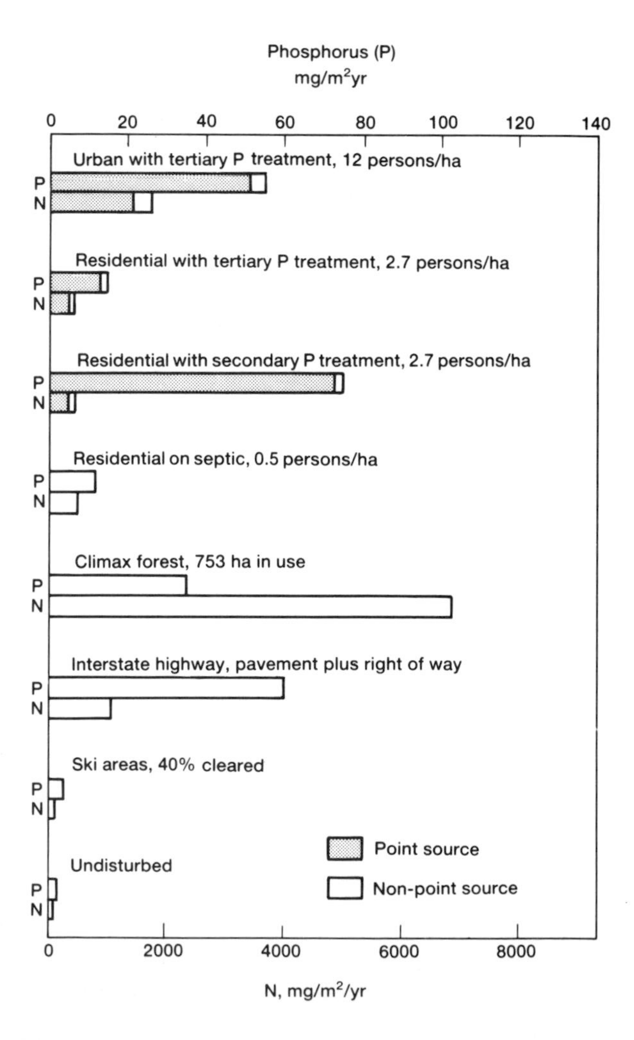

Figure 5.15 The yearly yield of phosphorus from different land uses in the Rock Mountains of the USA, assuming a runoff of 300 mm. The contrast of all used land with the undisturbed areas is striking but the high yields from the highways and the used forest are unexpected.
Source: W. M. Lewis *et al., Eutrophication and Land Use. Lake Dillon, Colorado*. New York: Springer-Verlag Ecological Studies 46, 1984.

Table 5.15 Generalized materials budget for Hong Kong, 1970s, t/day

	Imports	*Exports*	*Refuse*	*Atmospheric pollutants*
Liquid fuels	1100	612		639
Solid fuels	193	140		14
Glass	270	65	152	
Plastics	680	324	184	
Cement	0572	11		
Wood	1889	140	537	1
Iron and steel	1878	140	65	
Paper	1015	97	691	

Source: K. Newcombe et al., 'The metabolism of a city: the case of Hong Kong'. *Ambio* 7, 1978, 3–15.

urban area is dependent largely upon imported petroleum products. The flow chart for Lae (figure 5.16) indicates that these sources are supplemented by some HEP and firewood, but the city's energy metabolism rests on imported petroleum products which increased 6.6 times in price in the period 1956–76. If a country like Papua New Guinea intends therefore to support western-style cities then it imports economic, strategic and political dependence as part of the tankers' cargo. A shift to renewable energy sources would thus seem to be in the national interest and the potential for a mix of hydro-power, biomass resources (especially the sago and Nipa palms), forest residues, sawmill wastes, and direct solar capture seems considerable.

Overall environmental change is often most apparent at the urban fringe, when rapid growth of built structures may be intercalated with derelict land or land awaiting development. Recent work has suggested that intensive agriculture might be an excellent use of land here, especially in LDCs (less-developed countries), since it could provide food, recycle wastes, and provide employment and amenity.

The dependence of the modern city upon fossil fuels (together with electricity from nuclear power in some nations) has provoked examinations of the efficiency of the use of energy: given the immense impacts of city growth and the environmental costs of energy, are they a good bargain? A general trend is marked: as energy fell in price during the 1918–1973 period, cities became much more dispersed entities which are inefficient in energy terms since they lose heat quickly and absorb a lot of energy for the transport of goods and workers. Such cities may have poor public transport systems and thus whole groups are seriously disadvantaged: the young, the old, and inner-city dwellers who cannot easily get to suburban employment,

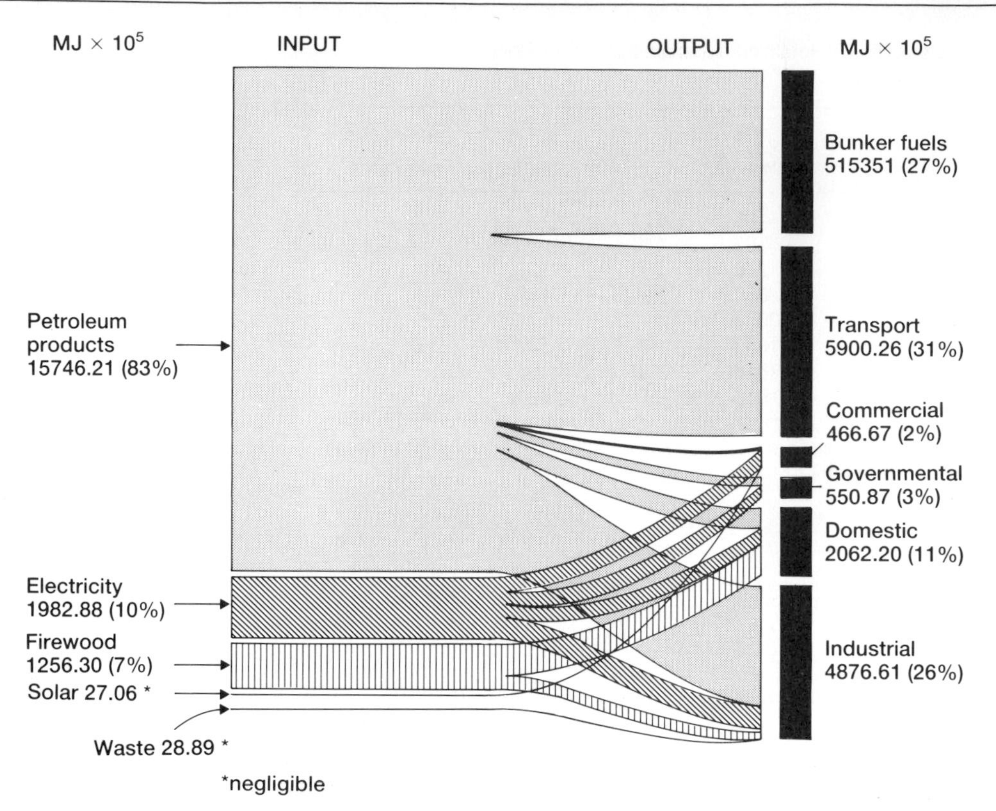

Figure 5.16 An energy flow chart for the city of Lae in Papua New Guinea for 1977 showing how a city in a developing country can nevertheless be dependent on imported fossil fuels. There is thus scope for looking at alternative sources of energy, especially in a tropical forest zone. *Source*: Unpublished original by Dr K. Newcombe of the World Bank, reproduced by kind permission.

for example. Social equity and energy conservation alike favour the older, denser pattern of cities, though to what extent the need for transport of people will be superseded by electronic communications is difficult to tell. These types of investigation lead to the question of whether energy considerations might indeed prove limiting to city growth and hence impose restrictions upon the types of environmental impact associated with urban expansion. Forecasts have predicted that in megalopolitan complexes, energy use may result in heat fluxes of 30–40 per cent of the incident solar energy by the year 2000, with consequences for water bodies, local atmospheric conditions (rises of mean annual temperature of the order of 2.5–7.0 °C) and the demand for air conditioning. Thermal energy releases, it appears, may also inhibit the removal of contaminants from city atmospheres. Thus the idea of an urban climate is given considerable strength (table 5.16).

Ideas for a lower-technology urban system for a highly populated island like Java at a time when population numbers become stable have been put forward by R. Meier. He suggests that the basic unit should be resource-conserving settlements with a power basis of electricity, centrally provided, which will power cooling fans, lights and water pumps, though perhaps not refrigerators for all. The community might focus round a water centre, giving baths and laundry facilities, collecting sewage, providing clean places for food preparation and perhaps even selling packaged water to take home.

Table 5.16 Climatic changes produced by cities

Element	Comparison with rural environs
Radiation	
Total horizontal	15–20% less
UV winter	30% less
UV summer	5% less
Cloudiness	
General cloud cover	5–15% more
Fog, winter	100% more
Fog, summer	30% more
Precipitation	
Totals	5–10% more
Heavy rain (>0.2 in/day), incl. thunderstorms	10% more
Temperature	
Annual mean	1–1.5°F more
Winter minimum	2–3°F more
Relative humidity	
Annual mean	4–6% less
Winter	2% less
Summer	8% less
Wind speed	
Annual mean	20–30% less
Extreme gusts	15–20% less
Calms	5–20% less
Contaminants	
Dust and particulates	10% more
SO_2	5 × or more
CO_2	10 ×
CO	25 ×

Source: R. T. Jaske, 'An evaluation of energy growth and use trends as a potential upper limit in metropolitan development'. *The Science of the Total Environment* 2, 1973, 45–60.

Thus scarce water can be controlled and sewage can be recycled. Energy is less critical in this type of urban future since people are not accustomed to access to large per capita usages. Energy subsidies are therefore likely to be low though a central role is envisaged for the petroleum-based plastic bag as the single most useful innovation in food processing and storage.

Cities are obviously major agents of environmental change, both on-site and outside their built-up limits where they reach out for resources and deposit their wastes. Furthermore, energy use is implicated in almost every aspect of form and function and no sensible changes in cities can be proposed without a thorough knowledge of its role in a given region. In terms of environmental change, we need to emphasize the fact that urban expansion consumes not only energy but land, and that one global forecast suggests that 24 million ha of cropland will be transformed to urban–industrial use by AD 2000. This is only 2 per cent of the world total but it is equivalent to the present-day food supply of some 84 million people.

Industry

This feature of human economic life is now so diverse that to chronicle all its possible environmental linkages is scarcely possible, although we can at least recognize an immense variation in the scale of alterations. At one end might be a hugh open-pit mineral working with an associated processing plant; at the other a small electrically powered factory employing a few people who walk to work. Energy supply and cost is critical in mineral extraction and refining, and we can note in passing that physical output in the US metal mining industries (in terms of product per kilocalorie of direct input) has declined 60 per cent since 1939 as the richness of ore grade has declined. Hence, the role of the mineral industry in energy use can be regionally dominant: the Bougainville Copper Company uses 21 per cent of the entire energy flow of Papua New Guinea to provide its electricity requirements.

An attempt to summarize the environmental consequences of mineral development is given in table 5.17. These stages will have been preceded by phases of development which are by no means impact-free, such as prospecting, exploration, and the intensive construction which accompanies the start of production; they will be followed in advanced countries by restoration work. The actual exploitation of a mineral resource starts with the extraction of the rock, and this produces an irreversible (though ameliorable) change in the environment unless a very deep shaft mine is used. Impacts are normally experienced on both surface and subsurface areas and are related to such factors as the chemical composition and spatial character of the deposit, the technology of extraction, and the location and mix of support industries and infrastructure (figure 5.17). As with oil and coal development almost every phase of the process can produce land use changes and environmental contamination: in open-pit mines, to give a single example, the overburden and spoil banks generally exceed the area of the mine by a factor of 3–5 times, and while they provide the raw materials for restoration, they may act

Table 5.17 Environmental impact of mineral industries: a summary

Qualitative aspects	*Spatial aspects*	*Temporal aspects*
A Impacts that modify pre-existing processes	Extent of impact depending on:	Character of impact, depending on:
1 Land alienation	1 Type of natural environment	1 Stage of development
2 Resource extraction	2 Magnitude of impact (energy/ unit area)	2 Source of impact
3 Contamination of air, soil, water	3 Location and extent deposit	3 Duration and type of impact e.g. mechanical, chemical, thermal, noise, ionizing, radiation
B New Linkages		
1 Reclamation and restoration	4 Character of extraction (surface or sub-surface)	
2 New settlements and industries	5 Distance from source of impact	

Source: Modified from T. B. Denisova, 'The environmental impact of mineral industries', *Soviet Geography* 19, 1977, 646–59.

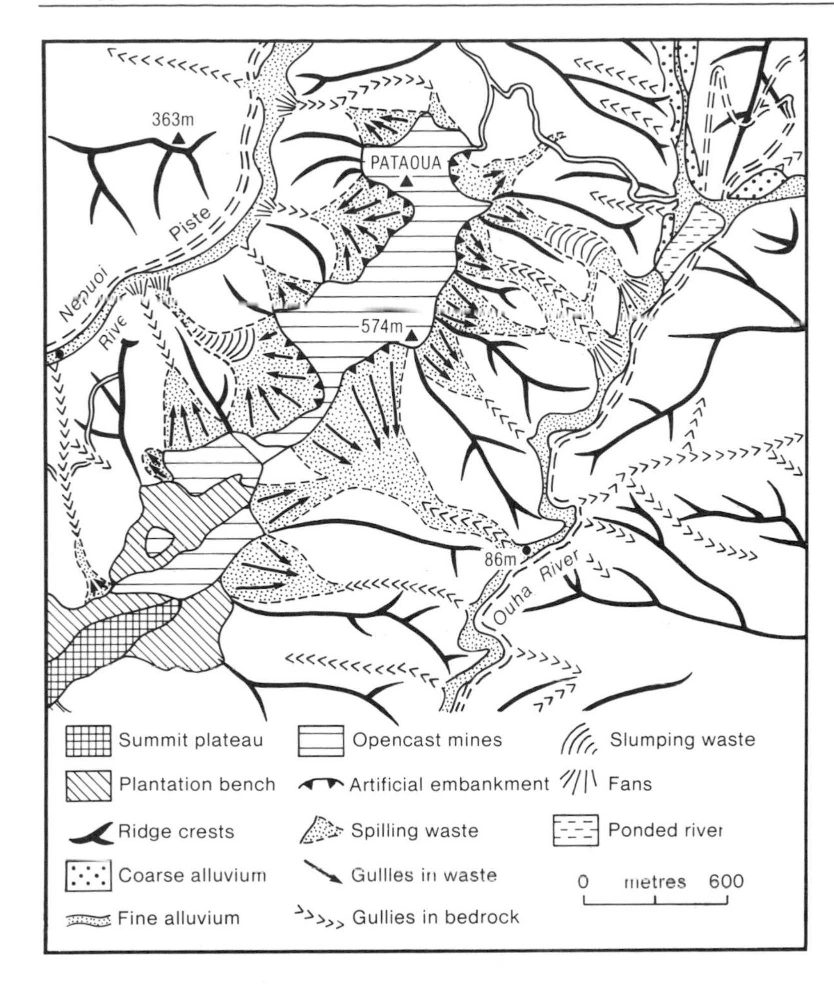

Figure 5.17 The effects of opencast mining on the environment of the Mt. Graunda area of New Caledonia. It can be seen that the mine itself has taken out most of a mountain ridge and thus its impact spills downwards and outwards to the drainage.
Source: E. C. F. Bird *et al., Impact of Open-cast mining on the Rivers and Coasts of New Caledonia*. Tokyo: United Nations University Publications no. 505, 1984.

as a source of contaminants (dust, silt, nutrients) during their lifetime. If the mine is large, then the labour may call forth urban development for its needs, perhaps creating a small town with the usual environmental impacts. In addition, the effect of developments will extend beyond the immediate vicinity of the mine: even a small-scale lead mine in the nineteenth century on the Mendip Hills of south-west England accelerated flood-plain sedimentation 2–3 km away by a factor of 3.5.

One example of a major enterprise with a high potential for environmental change is the chemical industry. Each year perhaps 1000 new chemicals enter commercial use (there are about 70 000 on the market) and wastes from production processes are very diverse in their characteristics. Most ought only to be dumped in specified places and some must be destroyed by high temperature combustion. Improved technology is yearly lowering the burden of toxic chemicals released to the environment but there are accidents, mistakes and the accumulated legacy of years of ignorance and neglect to provide releases of various substances, of which 12 000 are on the US government's toxic chemicals list, and 1800 of which are known carcinogens. Land and energy have to be devoted to separation of the toxins (by flotation, aeration or with activated sludge, for example), their neutralization where this is possible, sedimentation into a handlable form and perhaps storage under highly secure conditions. Even with less poisonous substances, the Biochemical Oxygen Demand of releases of chemical wastes into water is very high: glycerol and fatty acids 10 000 mg/l; phenols 6600 mg/l; herbicides like 2,-4-D, 15 000 mg/l and pharmaceuticals at 14 000 mg/l can all be compared with average domestic waste water at *c*. 200 mg/l.

To summarize the effects of industries upon the flow of materials and hence their likely and potential environmental linkages, table 5.13 has been presented. This covers most of the obvious effects, though we need to remember that the spatial outreach of industry is very wide. 'Downstream' in terms of contaminants and their penetration to all corners of the earth is in general obvious; 'upstream' less so but we may recall the manipulation needed to provide raw materials for industry, to ensure water supplies, and the necessary transport network, as well as the communications to keep the system functioning. A simple medium-scale example will demonstrate the scale of change: the installation of major chemical plants on Teesside in England in the 1960s led to the drowning of upland valleys in two distant locations (Upper Teesdale and Kielder, see figure 5.18) and the construction of an interbasin transfer system. This is a relatively medium-scale development: multiply it by a hundred times for some of the world's major industrial complexes.

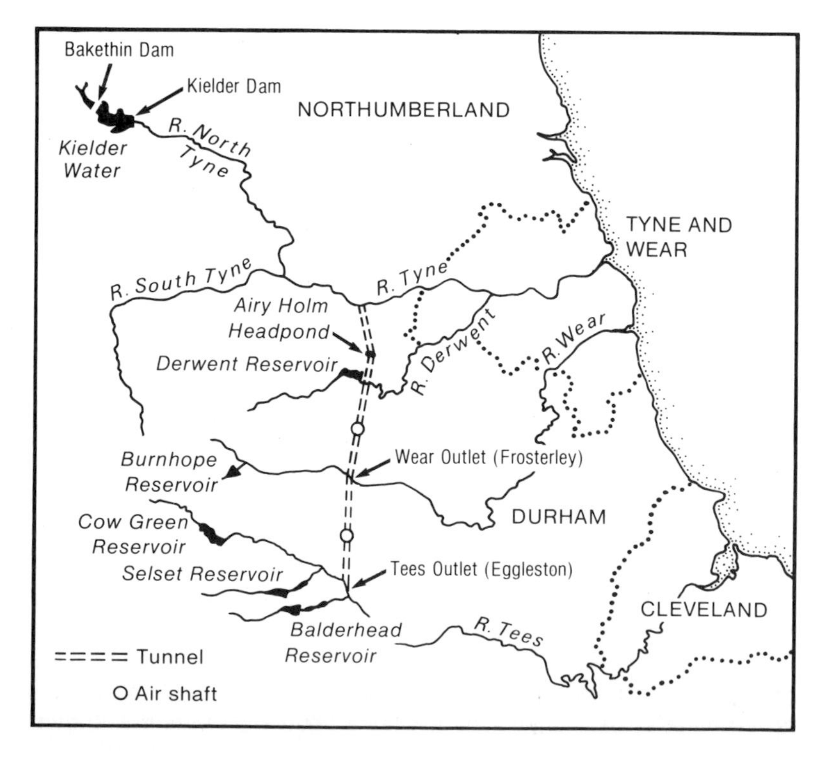

Figure 5.18 Large-scale water manipulation in north-east England: to a collection of 19th- and 20th-century medium-sized impoundments has been added Kielder Water, the biggest reservoir in Europe. Using inter-basin transfer, its function is to supply water for industrial development in Cleveland. Water management is thus responsible for environmental impact both 'upstream' and 'downstream' and temporarily in between during the construction of the tunnel.
Source: Northumbrian Water Authority, by kind permission.

Comparisons

In spite of the industrial growth of the last 150 years and its energy basis, the quantities directed by human societies are still small compared with the flows of nature. Table 5.4 on p. 216 set out global mean energy flows for a selection of processes and events; the cities come highest in the man-made category, with overall fossil fuel usage about 25 per cent of the city level. To accept this ranking at face value is, however, to ignore the variable resilience of the natural systems: it may be possible to perturb them by a relatively small input at a particularly sensitive point. In the case of the atmosphere we do not know where such places might be. There is also a history of ignorance of the characteristics of the various substances produced and transformations effected: the increase in concentration of CO_2 and other 'greenhouse gases' in the atmosphere (see p. 332) is not very high in absolute terms but its overall effects may well be greater than anything we have so far discussed.

The industrialization of related systems

Agriculture

Until about AD 1600 there was little difference in agricultural productivity between Europe, India and China. Then, even before industrialization, Europe began to break away as it began to eliminate fallow, introduce new roots, legumes and grains into crop rotations, and to change tenure systems, especially with the elimination of common fields. Yet thereafter the relatively slow penetration of industrialized methods of cultivation into even the technologically advanced regions of the world had a great deal to do with the unsuitability of steam for agricultural uses other than for pumping water. The infiltration of the threshing engine into an older, more solar-based system as described by Thomas Hardy in *Tess of the d'Urbervilles* (1891):

> He [the engineman] . . . was in the agricultural world but not of it. He served fire and smoke; these denizens of the field served vegetation, weather, frost and sun.

was destined to remain tangential; it was the fuller development of the internal combustion engine that brought about the revolution on the land, encapsulated by the data for the numbers of tractors that replaced horses in the USA and UK (figure 5.19). Although the first factory making tractors opened in 1907 in the USA, it was only in the interwar period that they replaced the horse in North America and Australia, and even later in Europe. Reaping machinery, the all-iron plough, chemical fertilizers and biocides all came in at various rates after the late nineteenth century, as did scientifically bred strains of plants and animals. But above all else the tractor has been the means of environmental change since it has pulled heavy ploughs that have

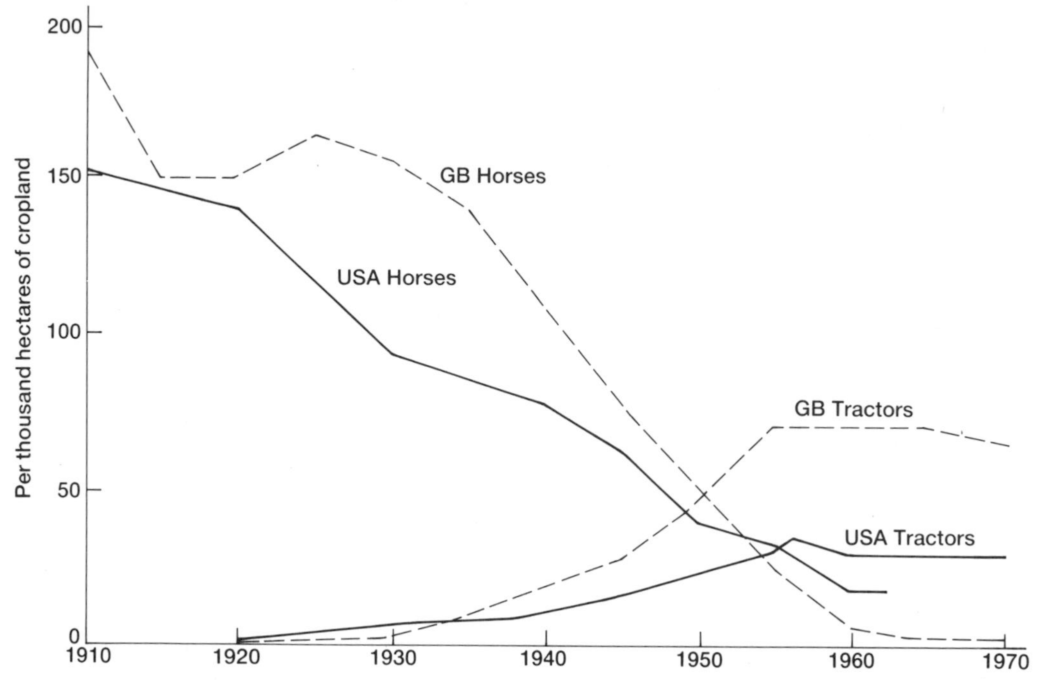

Figure 5.19 The industrialization of agriculture as indexed by the numbers of horses and tractors per thousand hectares of cropland. In Great Britain it is only in the 1930s that the change really takes off; the USA starts to lose horses earlier.
Source: D. B. Grigg, *The Dynamics of Agricultural Change*. London: Hutchinson, 1982.

changed the structure of soils as well as turned under the grassy or herbaceous vegetation above; it has, with chains attached, cleared shrubby vegetation and pulled out trees; it has allowed deep drains to be installed under wet land; it has pulled all kinds of machines including those which applied fertilizers and pesticides; and it is the source of power for machines which harvested crops, like potato lifters and combine harvesters. Apart from mechanical reapers, the general adoption of most of these techniques in developed countries dates from the 1930s. Indeed, none of the earlier innovations could stem some of the cataclysmic agricultural disasters of the nineteenth century, such as the potato blight of Ireland, Western Scotland, Norway and the Low Countries, and the phylloxera epidemic which devastated French vineyards so much that they had to be replanted with Californian rootstock.

The day-to-day application of energy in modern agriculture comes in a number of phases. One of these is obviously on the farm itself, where machinery, chemicals and knowledge are applied to the land where considerable environmental management and impact takes place (Plate 5.8). There are as well both 'upstream' and 'downstream' uses of energy in the total food-producing system (figure 5.20): the production of machinery and fertilizers, for example, is energy-intensive, as the processing and packaging of food and its delivery to the urban consumer. A numerical comparison between the 1820s and the 1970s shows the increase in upstream inputs upon farms. In figure 4.16 (p. 137) we saw the energy flows of an English farm in the 1820s where much of the power was human labour. Inputs from outside the farm itself were very small: T. Bayliss-Smith suggests that

Plate 5.8 Industrialized agriculture makes heavy use of machinery, as here in harvesting cereals in the UK. Energy has also been used, we may recall, in making the machines, and on inputs to the crop which may in turn be dried and/or stored at a constant temperature, even before it is processed for human food. (Countryside Commission)

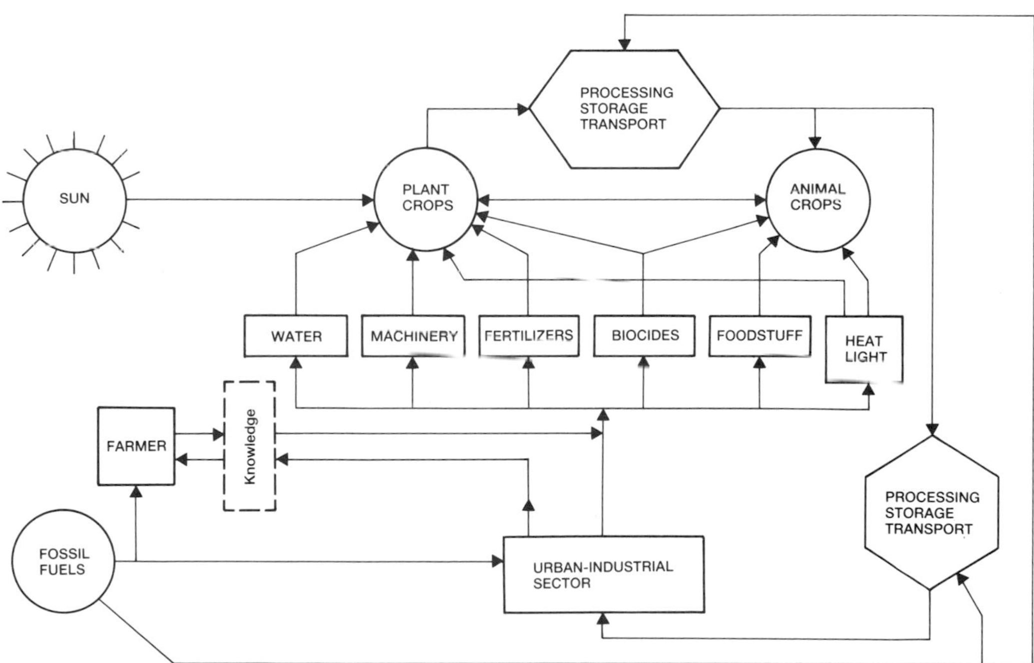

Figure 5.20 A simplified flow diagram of energy in modern agriculture. The addition of the fossil fuel means that agricultural workers can support about 32 times their own number in the urban-industrial sector.

about 2 per cent of the energy input may have come in the shape of purchased coal. But of the gross income of £3820, only £100 went on 'outside goods and services'. (Similarly, a Dutch study for 1800 suggests that the quantity of energy from outside the farm was negligible.) A farm of the same size in the same region of England in the 1970s is depicted in figure 5.21. Here a major difference is seen in

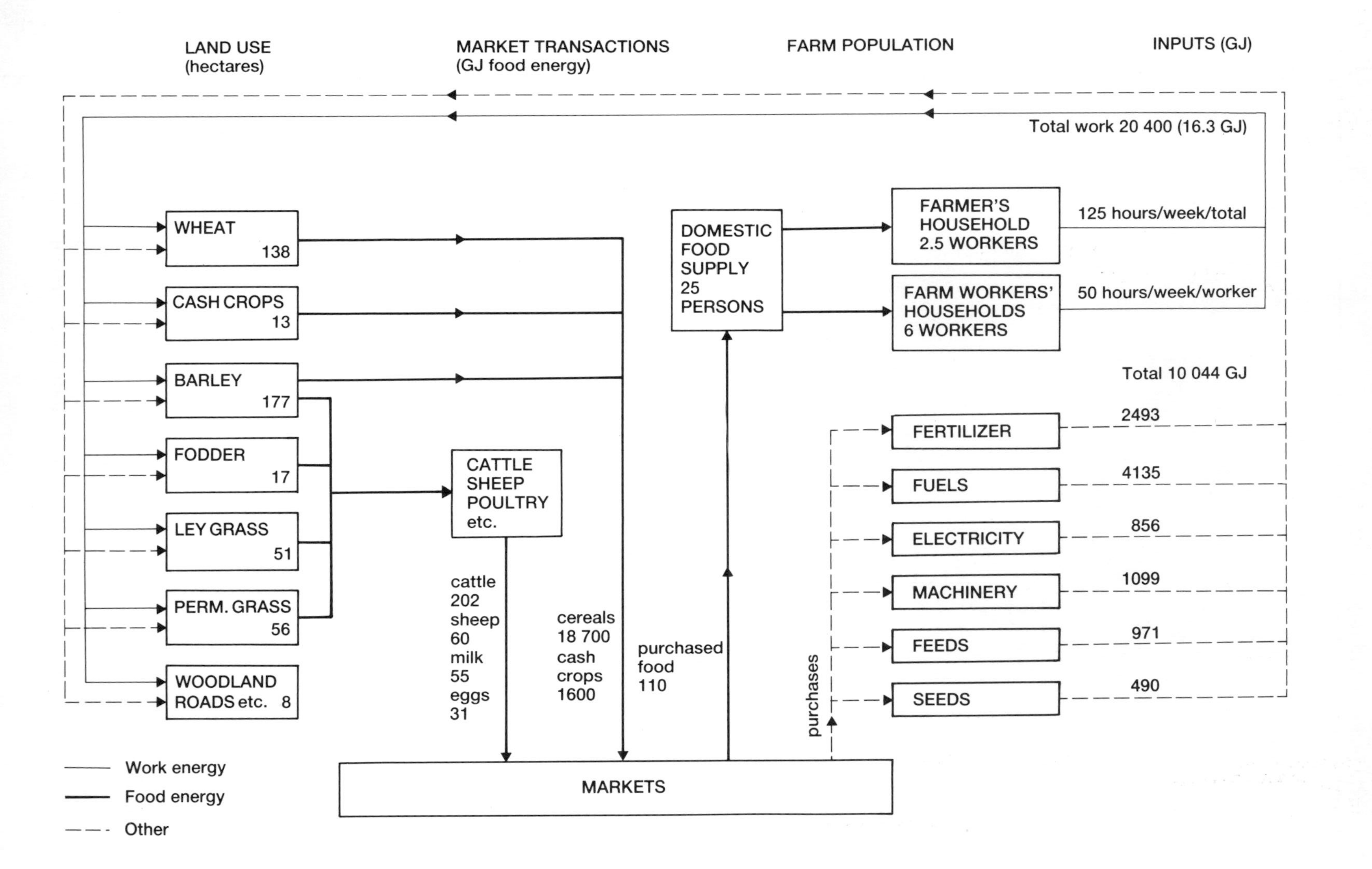

LAND USE
(hectares)
MARKET TRANSACTIONS
(GJ food energy)
FARM POPULATION
INPUTS (GJ)
Total work 20 400 (16.3 GJ)
WHEAT 138
CASH CROPS 13
BARLEY 177
FODDER 17
LEY GRASS 51
PERM. GRASS 56
WOODLAND ROADS etc. 8
CATTLE SHEEP POULTRY etc.
cattle 202 sheep 60 milk 55 eggs 31
cereals 18 700 cash crops 1600
purchased food 110
DOMESTIC FOOD SUPPLY 25 PERSONS
FARMER'S HOUSEHOLD 2.5 WORKERS
FARM WORKERS' HOUSEHOLDS 6 WORKERS
125 hours/week/total
50 hours/week/worker
Total 10 044 GJ
FERTILIZER
2493
FUELS
4135
ELECTRICITY
856
MACHINERY
1099
FEEDS
971
SEEDS
490
purchases
MARKETS
Work energy
Food energy
Other

the energy value of the purchases which at 10 044 GJ completely outstrip the human labour of 16.3 GJ, so 99.8 per cent of the energy is imported from beyond the confines of the farm. In the 1820s energy expended per ha of farmland was 140 MJ; in the 1970s it was 21 869 MJ, an increase of about 150-fold, much of which will have been aimed at the agroecosystem itself, though some is in the form of domestic electricity and, presumably, fuel for the farmer's car. In reality we should perhaps add some energy for the visits of the vet and perhaps for stored knowledge in the form of books or the *Farmers Weekly*. Hence in the context of environmental linkage, we can classify farms in terms of an 'energy density', i.e. the quantity of energy per unit area from all sources which is added to the solar flux, and table 5.18 gives some contemporary examples that show a predictable progression from the hunter–gatherer with no added energy to cattle feedlots which are practically nothing but added energy. (Our English example above at 21 GJ/ha rather extends the range of its mixed-farm category.)

Table 5.18 Energy density of some food producing systems

System	*Energy density* (GJ/ha)
Hunter–gatherer	zero
Andean village (Peru)	0.2
Hill sheep	0.6
Marginal farming	4.0
Open-range beef (NZ)	5.0
Mixed farm in developed country	12–15
Intensive crop production	15–20
Feedlot animal production	40

Source: M. Slesser, 'Energy requirements of agriculture'. In J. Lenihan and W. W. Fletcher (eds), *Food, Agriculture and the Environment*. Glasgow. Blackie, 1976, 1–20.

Three further points may be drawn out. The first is that the energy-rich matter also carries nutrients and that these may now be brought great distances from factories or quarries, and that they may equally be taken a great way in the form of crops before they end up as sewage. Thus the enclosed or 'tight' nutrient cycles of the pre-industrial farm are split wide open by industrialization and the farm's environmental linkages are spatially extended many-fold. The second is that the addition of fossil energy changes the ratios of energy input and output in farming systems. Crude statistics show that the ratio E_r (Energy out to energy in) is highest for pre-industrial systems since only human labour is used. In the tropics, E_r can be 12–40, with cassava yielding 60. This figure falls with the use of fertilizers and machinery (5–10 being typical) and becomes negative in fully industrial systems (0.6 for sugar beet in the UK; 0.47 all agriculture in the UK, 1968; 0.0023 for winter lettuce (Plate 5.9) in heated green-

Figure 5.21 The pattern of energy flows on a 460 ha arable farm in southern England, 1971–72. (Compare with fig. 4.16) Note here how the outside world contributes 10 044 GJ on energy and the input of human energy is 16.3 GJ. The eventual output of foodstuffs is 20 648 so the E_r is nearly 2.0 at the farm gate. *Source*: T. Bayliss-Smith, *The Ecology of Agricultural Systems*, CUP, 1982.

Plate 5.9 Even more energy-intensive is the practice of growing crops under glass in The Netherlands. This conserves solar heat but just as importantly keeps in hot air from fossil-fuel fired heaters so as to produce luxury crops during winter for rich economies: flowers and lettuce, for example. (Aerocamera-Bart Hofmeester)

houses). These data are at the 'farm gate' and appear even worse if the downstream energy inputs are considered. Wheat, for example, with an E_r of 3.5 at the farm gate, may turn up as sliced, wrapped bread with an E_r of 0.5. But E_r is not the whole picture: in the tropics high E_r figures conceal seasonal nutritional stresses because of lack of food storage, and the data may also be quoted for an individual worker rather than all the people he has to support. A useful parallel measure is the surplus energy which each farm worker produces. (Table 5.19) In pre-industrial systems this is of the order of 2.3 MJ/person/day but in fully industrial systems more like 4.0–19.0 MJ/person/day, with the large surpluses increasing the material possessions of the farmers and feeding a large urban population. Thirdly, the application of fossil fuels opened up the farms to market-orientation and often made them specialized. They have thus become vulnerable to outside economic processes and have had to adapt quickly to such changes, whereas before their general self-sufficiency in food for the household and labourers and a dependence upon local energy sources made them less responsive to events in the wider world.

Table 5.19 Energy relations of some agricultural systems

System	*Surplus energy* (MJ/cap/day)	E_r
New Guinea 1970s	2.3	14.2
England 1826	2.4 (labourer's household)	12.6
S. India	7.5 (high caste household)	13.0
Moscow collective 1970s	4.1	1.3
England 1971	18.8	2.1

Source: T. Bayliss-Smith, *The Ecology of Agricultural Systems*. Cambridge: CUP, 1982.

One of the consequences of the industrialization of agriculture has been to extend the cultivated area, by making possible for example the clearing of scrub and bushland, the ploughing of tough sods, and the carriage of water for irrigation. Table 5.20 shows the net extension of cultivated land from 485 to 982 million ha in the world in the period 1870–1970, bearing in mind that land has gone permanently out of cultivation in this period as well as having been taken in. Much of the increase has been in the mid-latitude temperate grasslands where some 218 million ha were cultivated for the first time ever, mostly for wheat. This great expansion came to an end in the 1930s except for the virgin and idle lands programme of 1954–56 in the USSR which put 36 million ha of grassland under spring wheat cultivation. The extension of the area of wet rice in south-east Asia was proportionately large (from 3 to 17 million ha from 1870 to 1970) but was achieved predominantly in a pre-industrial context. The desert may be the next frontier. In Saudi Arabia, a farm delivers 300,000 litres/day of dairy produce to retailers. These come from 14,000 cattle living in the desert under canopies, sprayed at intervals with cold water and fed on green fodder produced from irrigated land alongside the cattle 'sheds'.

Table 5.20 Changes in the arable area, 1870–1970

	1870 ($\times 10^6$ ha)	*1970* ($\times 10^6$ ha)	*% increase*
Europe	141	147	4.25
USSR	102	232	127
China	81	110	36
India and Pakistan	68	193	184
N. America	80	218	172
Argentina, Uruguay	0.4	32	800
Australia	0.4	19	475
Japan	3.2	5.5	72
SE Asia rice areas	3	17	466[1]
Total	485	902	102

[1] (Burma, Thailand, Vietnam)

Source: D. B. Grigg, 'The growth and distribution of the world's arable land 1870–1970', *Geography* 59, 1974, 104–10.

The great extension of agriculture into the temperate grasslands was made possible by industrial innovations of which John Deere's steel plough and the McCormick reaper of the 1830s were crucial, and the production of cheap wire for fencing (available after 1860, with the patent for barbed wire in 1873) extremely helpful. So the plough that broke the plains and prairies of the USA could do the same in Canada, Australia and Argentine–Uruguay, with eventually the same processes being applied to the Khazak SSR. in the southern USSR. Some further extensions were made possible by the introduction of crop varieties tolerant of, for example, cold (important in Canada) or of drought (necessary in Australia). Tougher varieties took agriculture into areas of high risk, so agriculture has in these areas suffered setbacks, some of which had environmental consequences. Such events have also converted some of the tillers of drought-prone areas to dry-farming and others to rotations involving mixed farming and introduced forage species.

Another way of extending the area of agricultural production is to irrigate land otherwise too dry for the desired crop. We see in table 5.21 that in certain parts of the world a very large proportion of the per capita water use is devoted to agriculture. Table 5.22 complements this for energy, showing that fertilizers and machines take the largest share of the total energy involved in agriculture, but in certain regions and in particular in the LDCs, the proportion of energy devoted to irrigation is high. Global historical land use figures do not permit an analysis of land extension made possible by it, but in arid areas like the Middle East (where 70 per cent of agricultural production comes from the 35 per cent of cultivated land which is artificially watered) and the Indian subcontinent, as well as wetter zones like south-east Asia, previously uncultivated land has soon been tilled when irrigation became available. Table 5.23 shows the increase in some of the items of new technology associated with irrigation, compared with the traditional water-lifting device, the bullock.

Table 5.21 Water use, 1965

	Yearly use (m^3/cap)	*Agricultural use* (%)
UK	200	3
Czechoslovakia	285	6
India	600	96
Japan	710	72
Mexico	930	91
USSR	1000	53
USA	2300	42

Source: UNO, *The Demand for Water.* Natural Resources, Water Series No. 3. NY: UN, 1976.

Table 5.22 Distribution of direct and indirect usages of commercial energy in agriculture

Region	*E in Ag.* (10^9J = GJ)	*Fertilizers* (%)	*Machinery* (%)	*Irrigation* (%)	*Pesticides* (%)
Africa	2	53	42	3	1.6
SE Asia	20	66	13	20	0.5
Latin America	11	48	46	4	1.6
China	15	71	9	16	4.3
LDCs	49	59	26	14	1.0
DCs	214	39	57	2	0.9
World	260	43	50	4	2.1

Source: J. Parikh, 'Energy and agriculture interactions'. In K. Parikh and F. Fabar, *Food for All in a Sustainable World: the IIASA Food and Agriculture Program.* Laxenburg, Austria: IIASA, 1981, 178–83.

Table 5.23 Inputs into irrigation in India

	1951	*1972*
Bullocks × 10^6	58.5	70.4
Electric motors × 10^3	26	1642
Diesel engines × 10^3	82	1000

Source: D. S. Pathak and D. Singh, 'Energy returns in agriculture, with special reference to developing countries'. *Energy* 3, 1978, 119–26.

Irrigation can demand large quantities of input energy, usually needing to be delivered at exactly the right time so that electric or diesel pumps have an added attractiveness. Countries like China and India seem to require between 900–6500 MJ/ha of energy input for irrigation schemes; A. Makhijani and A. Poole suggested a general LDC figure of *c*. 32.0 GJ per hectare of crop, which is consonant with data for the total of work input plus irrigation energy into agroecosystems in China and Mexico, where both rice and maize received *c*. 15–16 GJ/ha of irrigation energy and 10–18 GJ/ha of human and animal labour. In these examples, fertilizer input was much lower (2–10 GJ/ha) than the irrigation proportion shown in table 5.22. When energy is freely available and relatively cheap as in the USA, then more of it can be used in agriculture: in the 17 western states, natural gas and electricity contribute most of the 3.0 MJ/m^3 energy cost of irrigation water. It takes 14.0 GJ/ha to irrigate with energy-intensive technologies like the centre pivot sprinkler, where water has to be pumped up at least 60 m to provide enough head to keep the rotary arm moving at a constant speed. Increased application of irrigation energy has resulted from its lower prices: in 1945 the annual per hectare input of irrigation energy into corn (maize) production in the USA was 434 MJ but in 1975 this had risen to 3270 MJ. Animal rearing systems which rely upon irrigated land to produce foodstuffs for beasts like cattle are considerable consumers of energy: irrigated corn silage for beef production needed fossil fuel inputs of 318 GJ/ha, of which 178 GJ (56 per cent) was irrigation input. A maximum of 53 per cent of the total cultural input energy is recoverable as retail beef.

On a world scale, extension of the area of arable land is unlikely to increase a great deal. In most places it seems more profitable (in every sense) to intensify agriculture rather than to extend it. Intensification is the process of using existing agricultural land to produce more crop per unit area per unit time and may be achieved with highly selected strains, with heavy inputs of water, fertilizer or machinery, or by taking more than one crop per year. It will rely upon modern transport and road construction to facilitate the inputs and probably to market the crops. Irrigation may feature in agricultural intensification just as it did in extension of the cropped area. All of these processes are likely to be underpinned by energy derived from fossil fuels, especially oil. If as in figure 5.20 we regard the farm as a central point in the energy flows, then intensified agriculture requires a great deal of 'upstream' energy in the manufacture and transport of inputs like fertilizers and pesticides; knowledge about crop breeding is itself a transformation of fossil fuels. 'Downstream' from the point of production, more energy will also be required for transport, processing and possibly packaging. Thus the whole can be regarded as a 'food system' and its energetics studied in that light, but we shall emphasize the on-farm phase of the system here since the other parts are indistinguishable from the general run of industrial procedures. An example of intensification of cropping might include the transformation of traditional methods of growing rice by the adoption of high-yielding varieties which require regular and large-scale inputs of water, fertilizers and pesticides and access to urban markets to sell the crop. Another would be the transition from rearing beef cattle on open pastures with semi-natural vegetation to keeping them on paved 'feedlots' with little movement allowed to animals which are fed on maize and industrially produced concentrates; or battery-produced poultry and eggs where the environment is totally controlled and the feed an industrial concoction.

Clearly, it is in the developed nations where the most energy-consumptive agriculture is found: in the USA of the 1970s the food system used 17 per cent of all the energy used by the economy. Thus maize production, for example, consumed 0.67 per cent of the nation's aggregate use of energy which is about the same as that burned by cars while idling at traffic lights, in jams and outside the windows of drive-in banks. Twice as much energy was used for food preparation in the house as was consumed by on-farm production. For comparison, the New Zealand food system consumed 30 per cent of that nation's primary energy, with 15 per cent of the food system's energy used in the household and 28 per cent for on-farm production. In Australia, households used 42 per cent and production 24 per cent of the food system total. In the EEC, about 4 per cent of national energy budgets goes to agriculture as far as the farm gate. The flows onto the land in a modern economy like Denmark are shown in figure 5.22, and in the total food system of Switzerland in figure 5.23.

On the farms, most cropping systems are subject to energy subsidy in developed countries (DCs): table 5.24 shows the energy content of various crops in New Mexico and it can again be emphasized that irrigation is a considerable consumer of energy. This not only produces the difference between dryland and irrigated wheat but the surprising gap between range beef and feedlot beef where the former is fed upon pastures which have been irrigated. Apples consume energy via both irrigation and crop sprays, though we may reasonably doubt whether the resulting taste justifies the expenditures upon them. Table 5.25 compares extensive with intensive animal husbandry in the UK where it can be seen that the differences are not all that great in several instances; where buildings have to be provided it costs little extra to use them all day, and it may be much more energy-

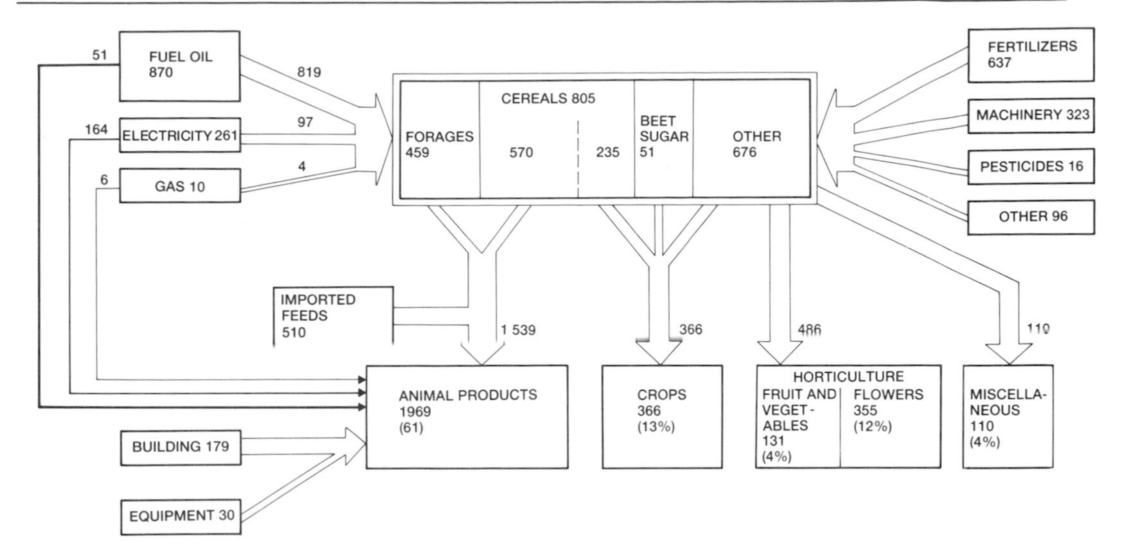

Figure 5.22 The flows of commercial (not solar) energy in the Danish agricultural system in 1974–75, in 10^3 tonnes oil equivalent. The developed-world status is shown by the high amounts of energy use (a) for animal products and (b) for fruit and flowers. *Source*: OECD, *The Energy problem and the Agro-food Sector*. Paris: OECD, 1982.

Table 5.24 Energy/acre or Energy/animal: New Mexico, 1970s

Crop	(*kcal* × 10^3)
Alfalfa	4252
Apples	122982
Cotton	6097
Corn	4355
Lettuce	5484
Dryland wheat	377
Irrigated wheat	2562
Sheep	7877
Range beef	14409
Feedlot beef	9910

Source: N. A. Patrick, 'Energy use patterns for agricultural production in New Mexico'. In W. Lockeretz (ed.), *Agriculture and Energy*. New York etc.. Academic Press, 1977, 31–40.

consuming to take foodstuffs to cattle in the fields than to deliver it to housed stock. Animals in fields need a lot of support energy in the form of food because they lose so much of it as heat in keeping a constant body temperature and because they have the curious and uneconomic habit of moving around. Thus DC agriculture is supported by a lot of energy subsidy, but at the farm level the quantities are not so great as is popularly supposed nor is the difference from more extensive (but energy-subsidized rather than 'traditional') systems so great. Nevertheless, at the level of a country like France, where agriculture uses 9.4 per cent of national energy consumption (figure 5.24), and where 68 per cent of the energy resource comes from imports, there is clearly some risk. Part of this seems unavoidable since nitrogen fertilizers are highly energy-intensive and not as

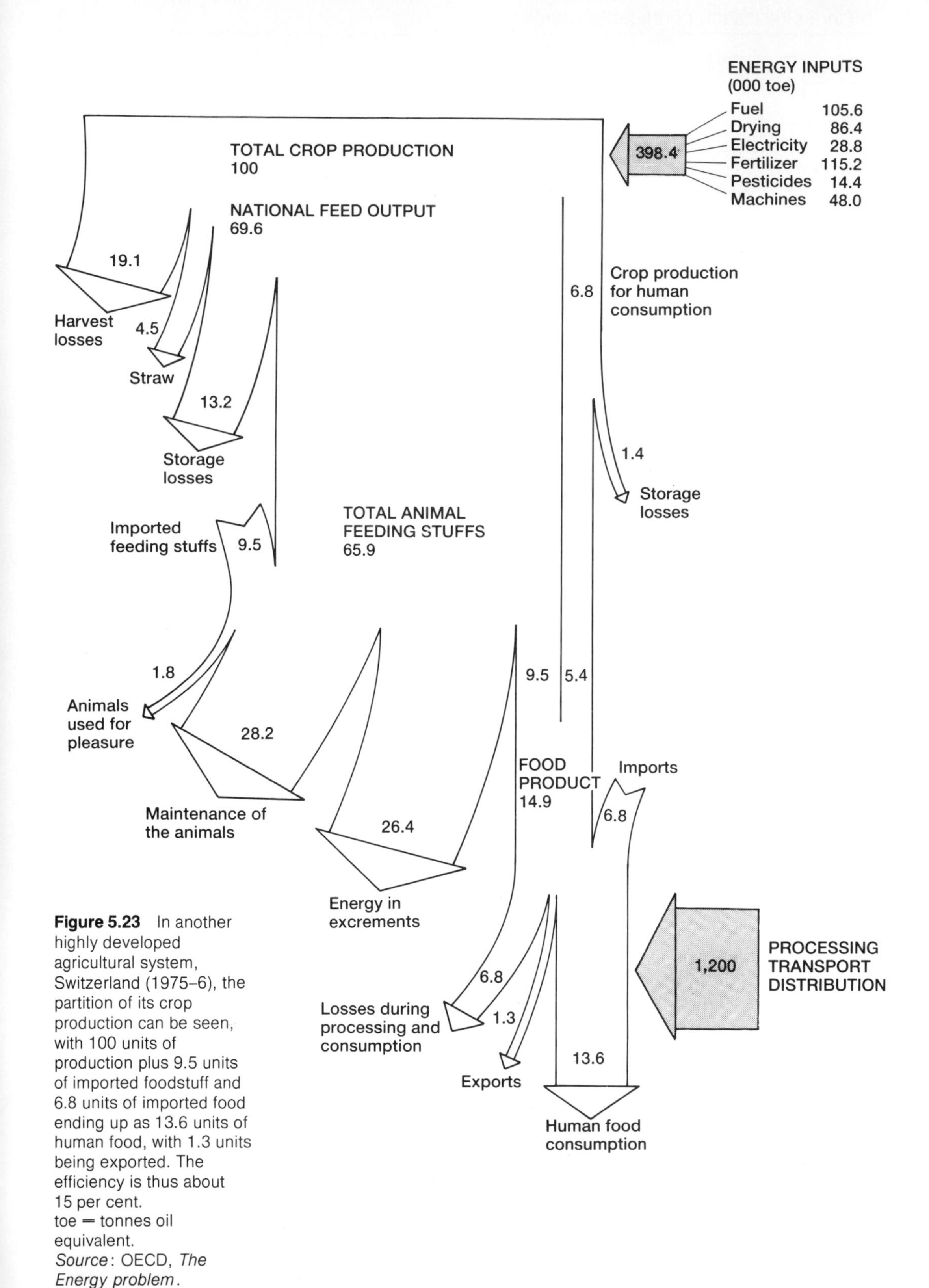

Figure 5.23 In another highly developed agricultural system, Switzerland (1975–6), the partition of its crop production can be seen, with 100 units of production plus 9.5 units of imported foodstuff and 6.8 units of imported food ending up as 13.6 units of human food, with 1.3 units being exported. The efficiency is thus about 15 per cent.
toe = tonnes oil equivalent.
Source: OECD, *The Energy problem*.

Table 5.25 Animal production in the UK: energy subsidy/kg of animal production

System	*MJ/kg*
Extensive systems	
Dairy	13.64
Beef	47.72
Breeding sows	40.11
Covered yard eggs	40.22
Intensive systems	
Dairy	9.12
Cereal beef	43.1
Breeding sows	51.7
Fattening pigs	32.42
Broilers	31.59
Battery hens	49.50

Source: C. R. W. Spedding et al., 'Energy and economics of intensive animal production'. *Agro-Ecosystems* 8, 1983, 169–81.

yet dispensable; on the other hand, over 50 per cent of cereal production goes to feed animals. (It may make little cultural sense, of course, to suggest that *filet de boeuf en croute* or *tripes à la mode de Caen* are dispensable in France.)

Even in rich nation-groups like the EEC, the consumption of so much oil-based energy in the agro-food system has evoked a close look at alternatives. There are possibilities for energy yield from biological wastes and crops grown especially for their calorific content. The environmental impacts resulting from large-scale adoption of these processes have not been explored in detail but would be considerable, e.g. the fermentation of animal wastes rather than their release to surface water, or the conversion of poor grassland to coppice. Where the land base is low but the population density high, as in Hong Kong, then the input emphasis (figure 5.13) is on fertilizers since they enable 6 to 8 crops a year of vegetables to be grown; application rates in the late 1960s were 35 times those of the USA, 5 times those of New Zealand and 107 times those of Australia. Relatively little energy is expended in packaging, processing and storage because rice and vegetables reach the never-distant markets in a fresh condition. But the system is dependent on imported energy which means that conservation measures, principally by reducing the quantity of nitrogenous fertilizer used (e.g. by returning to the traditional practice of using night soil) are indicated in the medium term. The food security of Hong Kong will probably be greatly enhanced by its accession to the People's Republic of China.

The energy flows in the agriculture of LDCs have been the subject of a great deal of study and concern. Following V. Smil, a classification can be elaborated in which the poorest countries use no 'modern' (i.e. fossil fuel derived) energy in the countryside: a solar-based economy (wood, dung, domestic animals and human labour) provides 2 meals a day and space heating for a few hours, at a level of

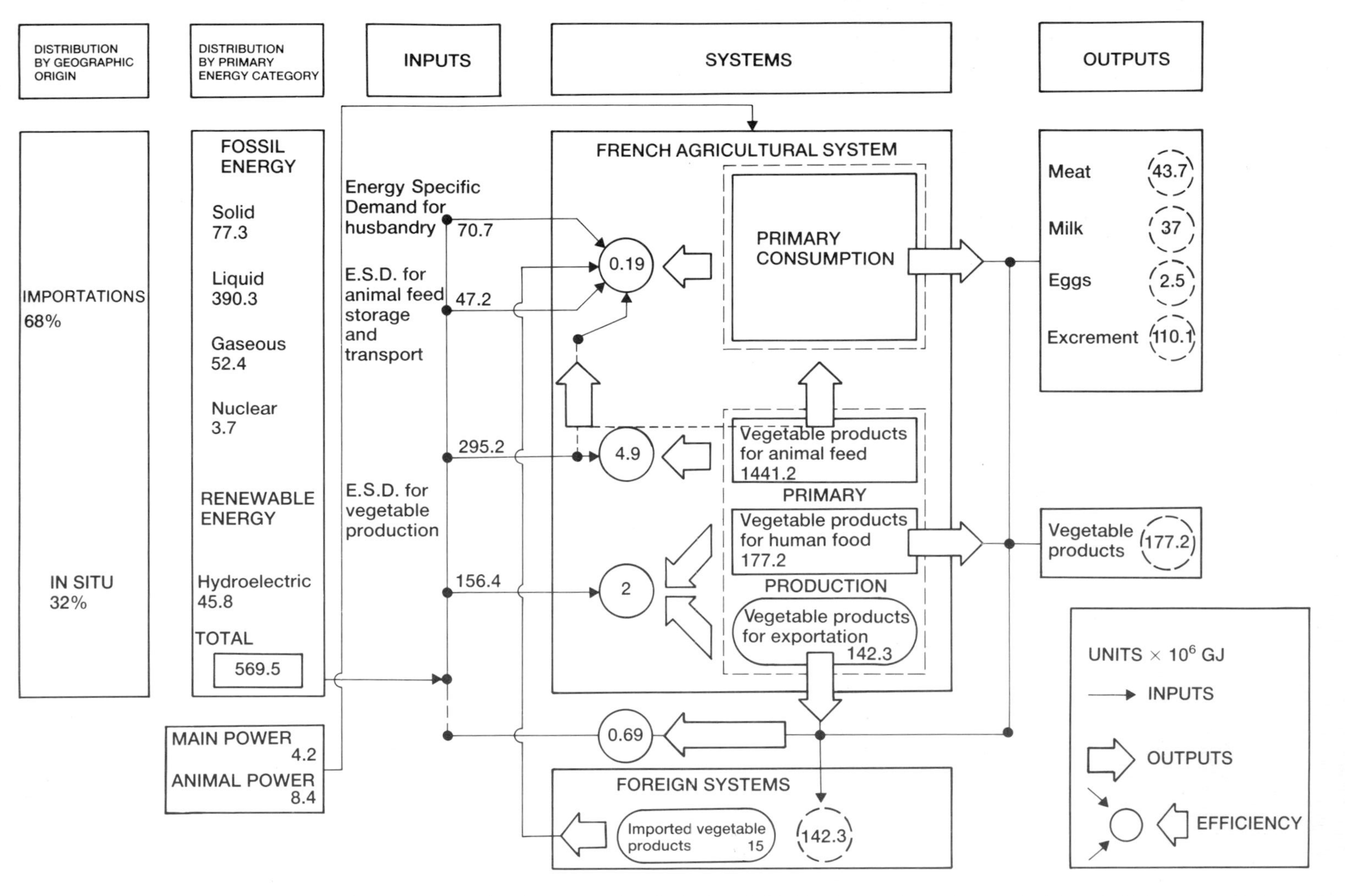

Figure 5.24 The energy flows through the French agricultural system in the 1970s, in 10^6 GJ. Here is a system, whatever its internal efficiencies and high-quality products, that is dependent on the outside world for 68% ('importations') of its commercial energy.
Source: J. P. Delage *et al*., 'Eco-energetics analysis of an agricultural system: the French case in 1970', *Agro-Ecosystems* 2, 1979, 345–65.

Plate 5.10 Lowlying and mainly agricultural economies such as that of Bangladesh are especially vulnerable to flooding at the time of the monsoon rains. As the slopes of the Himalaya have become disforested at human hands, these floods have increased in severity. (H. Mishra/WWF)

$12.5–27 \times 10^3$ KJ/cap/day, with 7.5×10^3 KJ representing a basic level of metabolic existence. Bangladesh, Ethiopia and the nations of the African Sahel are examples of this category (Plate 5.10) and even where political strife is absent, nutritional stress is common. A second category includes places like India, Pakistan and Thailand where modern energy is about as important as the traditional fuels (table 5.26) in rural areas, though plant materials and animal products are still the main providers of energy for some villagers. That the position is not static, however, can be glimpsed from Punjabi data which show that the consumption of commercial energy in agriculture (in the form of diesel oil, electricity, machinery, fertilizers and chemicals) increased by 387 per cent between 1965 and 1980, a rate of 27.6 per cent p.a. Thirdly, we may categorize

Table 5.26 Traditional and modern energies in some LDCs, 1977/78

Country	*Total traditional* (kcal/cap/day)	*Modern energies*	*% of total non-food consumption of energy filled by traditional energies*
Bangladesh	3200	500	86
Nigeria	7600	1700	82
Indonesia	6700	3400	66
India	3900	4200	48
Brazil	6900	12800	35
Turkey	5900	12100	33
China	3100	8300	27
Taiwan	300	29200	1

Source: V. Smil, 'Energy flows in the developing world'. *American Scientist* 67, 1979, 522–31.

nations like China and Brazil where the contribution of food and plant-based fuels is exceeded by modern energy sources. Rural populations may still be considerable users of solar products but those derived from petroleum are coming to the villages and fields in increasing quantities. However, if the energy is unevenly distributed, nutritional stress can still occur, as it does among the children of Brazilian sisal-workers where the money-earner has to take most of the food to have sufficient energy to feed machines with the raw sisal. Lastly, there are places like Taiwan which although not a totally industrialized nation has two-thirds of its population living in cities; in the outlying areas traditional fuels are still used but modern energies dominate the population. This progression seems to give good nutrition for most once the India–Pakistan level is reached and secure nutrition at the Brazil–China level, unless class differences distort the flow of energy as food.

Because nutrition seems better where more energy subsidy is involved, agricultural development in the last 40 years has often concentrated on replacing traditional fuels with their modern substitutes. Table 5.27 shows as an example the energy levels for producing *padi* rice by different methods. Possibly unexpectedly, this also shows that the highly technological Green Revolution type of development may not in fact be the most efficient way of intensifying the system, and that a transitional system might make it possible to achieve Green Revolution level yields with much lower inputs, thus achieving a distinctly higher efficiency. If a crop residue like rice straw were to be taken seriously as a rural energy source, then net energy yields by different methods need to be evaluated. On a world basis, the use of 30 per cent of the rice straw crop to yield energy would obviate the need for 60×10^6 t of oil, which is larger than the present amount of energy supplied from cattle dung. Indeed, if the energy content of the by-products of more machine-intensive farming in LDCs is calculated in the energy balance, it gives the system a much more favourable set of ratios than if only the food output is counted. Looked at in terms of comparative energy efficiency (table 5.27), to intensify to the Californian level would be a very expensive process and possibly an error; perhaps better would be the adoption of a south-east Asian model of 'vertical' intensification, i.e. producing more crops per year (4 to 12 a year are achieved in some parts of south-east Asia) with very little machinery. The dependence upon oil

Table 5.27 Commercial energy for rice production by various methods, 1970s

	Modern (USA)	*Transitional (Philippines)*	*Traditional (Philippines)*
Total input ($\times 10^6$J/ha)	64885	6386	173
Yield (kg/ha)	5800	2700	1250
Energy input/kg output (10^6J)	11.19	2.37	0.14

Source: FAO, *The State of Food and Agriculture 1976*. Rome: FAO, 1977, 93.

which many LDCs might do well to avoid is exemplified also in the flow of energy in China in the 1970s (figure 4.17). At that time the flow of fossil fuels was still 30 per cent less than that of plant-derived materials. To replace the plant fuels in the rural areas would require the use of about 100 million t/yr of oil, and one response has been to re-value biomass resources, small coal pits and mini-hydropower, and solar and wind energies, as well as to develop better technologies for biogas production and the use of wood in stoves.

Most LDC diets are dominantly vegetarian, since either pressure of population upon resources or the politically biased distribution of energy prevents the widespread consumption of meat. So relatively little attention is usually given to animals in discussions of agricultural intensification. Yet it may be more useful in the Third World to integrate more animals into the food-producing system rather than employ more machines. The general characteristics of successful systems might include features such as multiple cropping to maximize labour resources and to produce forage; land use patterns which cropped fertile land intensively but concentrated on forage production in less fertile areas; the use wherever possible of animals rather than machines for draught purposes and transport; and the employment of manure as a fertilizer rather than as a fuel. Some of these ideas have been tried as part of the task of improving shifting cultivation in the numerous places where it is breaking down. (Because of population growth, fallow periods are shorter so the system may produce poor nutrition and environmental degradation.) An intensification programme which incorporates domestic animals, is the theoretical answer but has not always been successful.

The great virtues of energy-rich intensification are the greater output per unit area and the way in which energy can often be substituted for land in places with very dense populations or where soil erosion threatens output. Consumer choice is usually enhanced so that DCs can offer their inhabitants meat, fresh vegetables and fruit as well as a choice of forms of carbohydrate. On the other hand, there are the problems of environmental impact which should be coupled with an awareness of the fact that energy accounting is simply a reductionist tool for describing some of the features of a system. It does not prescribe a healthy diet, nor quantify animal welfare, amenity, or landscape aesthetics, nor does it enhance the structure of soils, let alone societies, and in a wider political framework, intensification makes linkage to oil prices and supplies inevitable.

The environmental consequences of industrialized agriculture

The high productivity made possible by the addition of fossil fuel power has caused a number of shifts. Land has moved out of agriculture into urban growth and into recreational use, for example. Soil erosion and desertification may also be consequences of short-term high production levels, and wildlife populations as well as the genetic variety in cultivars have diminished as well. Looked at overall, the total terrestrial biomass seems to have been reduced by 7 per cent in the century to 1970.

Most natural terrestrial ecosystems lose soil through erosion but the advent of certain human activities increases the rate of loss by many times. Heavy grazing, deforestation, frequent fires, cultivation and the abandonment of soil conservation practices such as terracing or bunding can all have the same end result. Blowing, washing and gulleying have been the associates of human societies for at least ten millennia, with few if any zones immune. Soil erosion reduces NPP to very low levels and is only very slowly reversible, but has off-site effects as well. Suspended particles in watercourses exacerbate floods, silt up harbours and reservoirs, and change the components of aquatic ecosystems. Offshore features like coral reefs can get choked by heavy silt loads from incoming river mouths. These changes can be summarized by the estimate that under 'natural' conditions the world's rivers would carry to the sea some 9.3×10^9 t/yr of suspended sediment, but that in our own time the equivalent figure is 25×10^9 t/yr. This latter figure derives in part from the estimated 430×10^6 ha of soils destroyed by erosion. (For comparison the world arable area is 1500×10^6 ha.)

Modern technology has increased both the rates of soil erosion and the area subject to it. Deforestation, for example, can be carried out in areas hitherto inaccessible but now reached on roads constructed by bulldozers. Steep slopes formerly under forest may now be logged for their timber resource. Equally, they may be cleared for agricultural land under the pressure of population growth. Yet again, the opening up of the economy may mean the use of the flat land for cash crops, with the peasants being forced on to hill-lands which were formerly under forest, savanna or natural grassland. The Philippines provide an example of the first process, the southern flanks of the Himalaya of the second, and El Salvador the third. There is also the impact, mostly in the DCs, of modern agriculture in the shape of heavy machines and chemical fertilizers. The machines produce compression, shear, slip and vibration in soils at least to the depth of ploughing so the soils tend to become impermeable, leading to gullying since rain water cannot infiltrate at the point of impact; productivity is also reduced since root development is inhibited. The combined use of chemical fertilizers and modern tillage appears to allow the breakdown and removal of the organic matter which is so important in retaining soil moisture and in binding the particulates together. Its loss makes the soil more susceptible to wind and water erosion and in the long term the loss of carbon from the soils adds to the build-up of carbon dioxide in the atmosphere.

The most spectacular and well-publicized instances of soil erosion have been those which followed the breaking of mid-latitude grasslands in the High Plains of the USA like the 'Dust Bowl' of the 1920s and 1930s, and the similar events in Khazakstan in the 1950s during N. I. Kruschev's 'virgin and idle lands' campaign. Both these areas, and many like them, have undergone remedial treatment and are now

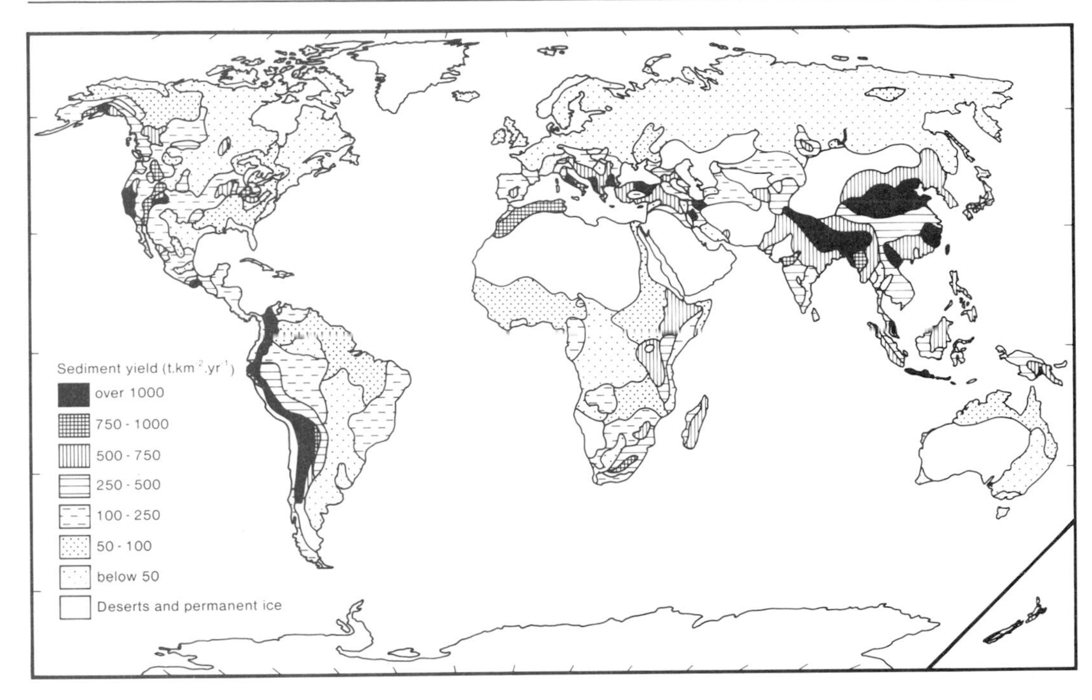

Figure 5.25 The differential rates of sediment loss in the world's river basins is depicted here. Obviously, being high is the first quality needed for rapid denudation, but land cover and subsoil type are also important: recent deforestation and loss are factors which elevate levels.
Source: D. E. Walling and B. W. Webb, 'Patterns of sediment yield', in K. J. Gregory (ed.) *Background to Palaeohydrology*. Chichester: Wiley, 1983.

cropped more rationally, but a world map of soil erosion (figure 5.25) shows that in spite of a good deal of scientific and technological knowledge the problems remain (Plate 5.11). Intensive agriculture in the USA, for example, produces soil losses of over 2 t/ha/yr on 84 per cent of the grain farms of the Midwest, Plains and Northwest regions; 4–8 t/ha/yr on corn farms in Iowa and Illinois, and up to 8 t/ha/yr on sloping areas growing cotton and millet. All these losses can at present be masked by increasing the quantities of fossil fuel derived

Plate 5.11 Even in rich economies, land which is climatically marginal for farming can become a virtual desert. Dust storms and the failure of rainfall have caused the abandonment of what was once an above-average farmstead near Elder, Colorado, USA. (US Department of Agriculture Soil Conservation Service)

energy applied to the land, especially in the form of fertilizers. But if this is the case in a nation where agronomic capability is perhaps the highest in the world, what likelihood is there of better results in nations without either the knowledge, the technology, or the political context in which to apply it? Rational farm practices such as the growth of windbreaks, retardation of erosion by planting and check dams, zero tillage, contour and strip ploughing, water and fire control are all techniques waiting to be used. The tragedy of soil erosion, perhaps, is that the knowledge of both prevention and cure is available but it gets employed rather too infrequently in North as well as South.

An equally well documented set of unwanted environmental impacts has been accumulated for irrigation projects in arid and semi-arid lands. The very act of irrigation requires a considerable amount of environmental change, as table 5.28 shows, and subsequent events in the life of many irrigation schemes have meant that such land is being abandoned at about the same rate as it is being created. Between 30 and 80 per cent of all irrigated lands are affected by major environmental difficulties, with countries like Syria, Jordan, Iraq, Haiti, Mexico, the USA and Afghanistan being the most seriously affected. That 50 per cent of all irrigated land in Iraq, 80 per cent in the Punjab, and 50 per cent in Syria suffers from salinity and waterlogging is a measure of the problems. The basis of these difficulties lies in the large-scale modification of soil and water conditions brought about by irrigation, which are summarized in table 5.29. Waterlogging, for example, occurs if drainage is inadequate so water-tables may rise rapidly, usually at a rate of several centimetres per year but sometimes at several metres per year. The waterlogging is usually accompanied by the deposition of salts from the saline groundwater or as a result of evapotranspiration. Rising fresh groundwater may mobilize solid phase salts and, added to highly saline groundwater, cause the secondary soil salinization known as alkalinization. In this condition the soil-absorbing complex becomes saturated with sodium, pH rises to 9–11, and compaction inevitably results.

Irrigation also exerts a powerful effect upon the associated aquatic ecosystems. Impoundments convert terrestrial ecosystems to acquatic ones, with flora and fauna, the turbidity of the water and its nutrient content all changing at various rates both in the reservoir and downstream from it. Shallow storage reservoirs in the tropics allow an explosive increase in the number and size of bottom-feeding fish (e.g. *Tilapia* and *Labeo*) during the early life of the lake. Deeper bodies of water may, however, have an anaerobic profundal zone that generates hydrogen sulphides which are released when water comes up from that depth. Downstream, the water is less silty so that photosynthesis may increase, giving rise to high levels of primary productivity. These may lead to higher fish stocks or to algal blooms which affect water quality: mammalian, avian and human deaths have been reported after the rapid growth and decay of blooms of *Anabaena*

Table 5.28 Physical and biological impacts of new irrigation schemes

Activity	*Physical effects*	*Biological effects*
Water storage	Changes in physical and chemical properties of water	Changes in biomass of aquatic biota
	Possible ground water recharge	
	Change in microclimate; increase in relative humidity	—
	Fish stocking	Increase in biomass, changes in fish species
	Water loss by seepage and evaporation	
	Water logging and increased salinity near canals	Poor plant growth, increased in disease vectors
On-farm water distribution	Effects as above where relevant	
Land levelling	Changes in physical and chemical properties of soil	Changes in soil fauna and flora
	Changes in fertility of soil	
Drainage system	Improved soil condition	
	Lowering of water table	
	Improved soil aeration	Better plant growth
	Removal of salts	
	Removal of chemical residues	
Initial leachings	Removal of salts	Increased biomass
	Changes in chemical and physical properties	
Land cultivation	Changes in fertility of soils	Increased or decrease in biomass
	Changes in other soil properties	
	Effect on low-lying land	
Introduction of animal husbandry	Improved soil properties	Increased animal products and waste
	Higher soil productivity	

Source: Adapted from Annex 5 of UNESCO/MAB Report No. 29, *Impact of Human Activities on the Dynamics of Arid and Semi-arid Zone Ecosystems with Particular Attention to the Effects of Irrigation.* Paris: UNESCO, 1976.

Table 5.29 Changes in soil and water regime caused by irrigation schemes

Modification of atmospheric moisture content
Increase of evapotranspiration
Higher atmospheric water vapour content
Changes in amount and pattern of precipitation
Modification of surface runoff
Increase in amount and intensity of catchment runoff so higher erosion potential and sediment transport
Control of river discharges by reservoirs and retention of sediments in reservoirs
Changes in river beds because of decrease of sediment in streams
Modification of groundwater regime
Rise of water table
Increase in accretion of salts, decrease in their loss in runoff
Horizontal groundwater flow towards non-irrigated areas, raising water table of latter and accumulating salts under non-irrigated areas
Modification of water quality
Increase of salts due to evaporation due to storage, conveyance and distribution of irrigation water
Changes in temperature and suspended load, concentrations of nutrients, biocides and salts

Source: modified from E. B. Worthington (ed.), *Arid Land Irrigation in Developing Countries. Environmental Problems and Effects.* Oxford: Pergamon Press, 1977.

and *Gymnodinium*. Less silt may also mean fewer flood-derived nutrients: the 1950 t/yr of silt deposited in Lake Nasser will have to be replaced by 13 000 t/yr of chemical fertilizer on the 1×10^6 hectares that it used to fecundate.

The public health balance of irrigation is delicate. On the one hand, there are gains from a more secure public water supply, and better nutritional standards; on the other, such schemes are very likely to encourage conditions conducive to the spread of malaria, filariasis, yellow fever, dengue, equine encephalitis, river blindness, sleeping sickness, schistosomiasis (bilharzia), liver fluke and guinea worm. Public health measures for prevention and cure of these conditions are often unavailable in developing countries; equally, the techniques for combatting poor agricultural productivity (salt-tolerant crops, deep drains and good drainage control, the addition of gypsum to soils) are known but not sufficiently applied. Given all the social and ecological problems of large irrigation schemes, it is not surprising that development agencies are re-evaluating the role of small-scale water management projects in Third World agriculture.

The penetration of industrialization into agriculture has been a force for standardization. The use of machines, on-farm and downstream, means that as homogeneous a product as possible is required so developed agriculture, including some attempts at modernizing tropical agriculture, relies upon a small number of varieties of each crop. Even though some genetic diversity may not have prevented large-scale crop failure in pre-industrial times, now there is even less genetic variety within each crop: the wheat production of the Canadian prairies comes mainly from four varieties and US soybean production derives ultimately from an Asiatic group of just six plants. Of 145 breeds of cattle in Europe and the Mediterranean, 115 are threatened with extinction. So habitat change and crop breeding combine to narrow the genetic material available in both wild biota and traditional domestic strains, although even within 'narrow-base' families of cultivars there is likely to be potentially useful genetic variation. Genetic uniformity means an undifferentiated reaction to climatic change or extremes and uniform response to pests and diseases. A common result is crop failure, instances of which have occurred in recent years with crops like tobacco, maize and wheat. Even in North America these are economically stressful although fought with the complete battery of modern agronomic techniques; in a developing country such events are likely to be disastrous. Counter-measures are being taken, in terms of collecting and preserving seeds and plasm of wild and cultivated plants. A new wave of plants is also on its way, which require fewer inputs than the high-yielding varieties of the Green Revolution years and which are mostly better varieties of traditional subsistence crops (e.g. sorghum and millets) rather than the crops of world trade. Genetic manipulation (to be discussed in chapter 6) may have a contribution to make here, although asexual cloning from cells will make the preservation of genetic variety even more imporant; further, 'new' organisms produced in laboratories may do poorly when placed in environments alongside 'evolved' organisms.

The success of industrial technology has permitted the production of large numbers of chemicals which can be applied to crops and the two major categories of these are fertilizers and biocides: the one group to make plants grow and the other to kill predators, competitors and diseases. Coupled with the production of large quantities of wastes by spatially confined animal husbandry, and the accumulation of metal ion concentrations on land fertilized with sewage sludge, they may contribute to a toxification of the biophysical environment. Chemical fertilizers deliver chiefly nitrogen, potassium and phosphorus to the crop and the plants rarely use it all: applications normally result in 40–50 per cent of the nitrogen entering the runoff. The residual nitrogen then entrained in the hydrological cycle finds its way into surface runoff or into groundwater. (The same is true of phosphorus but to a lesser extent; most of it found in fresh water comes from sewage and detergents.) Continual man-made additions of nutrients produce an over-enrichment of the surface water (eutrophication), which is characterized by algal blooms whose decay robs the water of oxygen as well as creating a smelly nuisance. The mean annual concentration of nitrates in rivers like the Tama in Japan rose from 2.87 ppm in 1965 to 3.31 ppm in 1975; of the Rhine in Germany 10.1 ppm to 13.8 ppm over the same period and the River Wear in England from 1.94 to 3.85 also in that decade. Agricultural nitrogen appears to be the main source of nitrates in groundwater and at concentrations above approximately 10–12 mg/l can cause health problems, especially to bottle-fed babies. All these problems are potentially capable of solution (slow-release fertilizers, biological nitrogen fixation, restoration of shallow lakes by removal of nutrient-rich sediments) but not without cost.

As little as 1 per cent of the chemicals applied to control insect pests, plant pathogens and weeds may hit the target organisms, and the rest reaches non-target species or surrounding ecosystems. In the case of pesticides the duration of the stress from a single exposure depends upon the toxicity of the chemical and its persistence. Most ecosystems start to recover when the stress disappears and with many modern formulations (especially with the highly persistent organochlorines) this takes one to five years. Given, though, that repeated applications are common, and that LDCs in particular still use large quantities of persistent substances such as DDT, then a series of environmental side-effects can be expected. These include biological amplification, in which an initially low concentration of a pesticide passes along a food chain and becomes concentrated in organisms that cannot excrete it quickly. Both lethal and sublethal effects have been noted: for example, predatory birds often suffer egg-shell thinning, with lower reproductive success. Natural selection ensures that resistant populations of the target organisms develop: 17 out of 25 insect pest species in California have developed tolerance of pesticides. Twenty-five species of arthropods that attack cotton in 36 countries now resist pesticides. Twenty-four species of malarial mosquito are now resistant to DDT, and one, *Anopheles albimanus*, is resistant to DDT, aldrin, dieldrin and malathion. By the year 2000, it has been suggested, there may be only pesticide-resistant pests: in 1965, FAO identified 182 resistant species of arthropods and this had become 428 in 1980. (Between 1970 and 1982 the number of arthropods resistant to DDT rose from 98 to 229.) Any extirpation of non-target organisms may mean that any natural controls on a pest population are eliminated, so that it continues to be a problem. In effect, the elimination of one pest species may mean simply that another taxon undergoes a rapid expansion in numbers and takes the place of the first. Thus ecosystem stability and resilience, ecosystem diversity and food web structure, energy flow and nutrient cycling can all be affected by biocides, especially where the systems lack resilience or where they regularly receive large quantities of pesticides. Many of these effects can be prevented by using more target-specific and less persistent chemicals, and above all by integrated pest management schemes for agroecosystems which use the evolved defence mechanisms of plants as well as modern chemicals. Biological control is usually an important part of such schemes, which are of course more expensive and more complicated than spraying a crop with whatever comes out of the cheapest drum. They do, though, have the added advantage of poisoning fewer farm workers.

In the developed nations, another aspect of modern agriculture has caused concern, especially in Britain. Here, the intimate network of small fields, hedgerows, hedgerow timber and winding lanes forms a characteristic and valued landscape. Energy-intensive agriculture, fuelled also by EEC-guaranteed prices, has promoted the creation of large fields, and the uprooting of hedgerows together with their trees. The landscape is thus changed from something created in the seventeenth and eighteenth centuries back towards a more medieval aspect. There are fewer wild animals: in the Netherlands 10 ha of hedgerowed agricultural land can support 1000 pairs of song birds and 30–50 per cent of some species of owls nest in pollarded willows. The scale of some of the changes is seen in table 5.30; cooperation of farmers and nature conservationists may keep some wildlife habitats but is unlikely to reverse the trend to large scale production, except perhaps on very sandy soils where wind-blow has become an increasing problem or where agricultural surpluses make it necessary for governments to pay farmers not to produce crops by intensive methods.

Table 5.30 Changes in landscape associated with intensification of farming in England

Enterprise	*Study area samples*						
Average field size 1945 (acres)	14	19	18	9	11	15	12
Average field size 1972 (acres)	32	45	21	13	16	19	15
Hedges removed, feet per acre	47	37	8	20	19	15	13
Hedgerow trees per 100 acres, 1947	39	59	17	50	49	51	69
Hedgerow trees per 100 acres, 1972	5	12	7	15	40	33	81

Source: R. Westmacott and T. Worthington: *New Agricultural Landscapes*. Cheltenham, UK: Countryside Commission, 1974.

We cannot ignore the duality of modern agriculture: it has coped reasonably well with recent population growth (some would point to it as the cause: C. O. Sauer maintained that the ploughing of the mid-latitude grasslands made the Industrial Revolution possible) and indeed provides surpluses in some parts of the world. On the other hand, it has alleviated little the problems of the very poor and has often spread disease (as with schistosomiasis in irrigation projects), caused birth defects (especially where dioxin, a contaminant of defoliants like 2,4,5-D is consumed) and even violent death, like that of 500 people in Iraq in 1971–72 who ate seed-corn rice dressed with a mercury fungicide.

One of the great challenges is to find ways in which sustainable and productive agroecosystems can be developed in the poorer countries, especially the tropics, where the nature of the environment has often been poorly understood by agronomists trained in temperate zones, where the role of domestic animals in nutrient cycles is usually underestimated, and where highly bred crops with an industrial-world background may not be the most appropriate. Equally, to 'hook' countries without fossil fuels to the world supply and price of oil would seem to be no great service and likely to increase their economic problems. At the most basic level, the LDCs would seem to need to increase their 1980 food output by five times to cope with the expected population in AD 2050. The environmental consequences are difficult to imagine but can scarcely preclude a manipulation of both wild and tame ecosystems on a scale achieved already by the well fed.

But for everybody, long-term stable, sustained agricultural productivity is an ineluctable requirement and considerable research effort is needed for all places into cropping and livestock systems which meet that requirement, coupled with the socio-political framework which will allow them to be introduced and to flourish. In a concise perspective, E. L. Jones summarizes the predicament:

> A grave imbalance in protein, raw materials, fertilizer and energy consumption marked the era of Europe's dominion over world economic and ecological history. Other politics have now begun to redress the balance, but the finite extent of ocean and grassland in the world implies that under any known technology they cannot replicate the means by which Europe rose.

Pastoralism

In today's world, pastoralists are unlikely to be isolated, the more so where agriculture can now impinge upon their lands, using for example pumped irrigation water. The most common process, however, is intensification: demand for meat from urban markets, local population growth, deeper and more permanent wells, and independence from colonial control have all caused animal numbers to rise (Plate 5.12). Yet in most Southern nations, the carrying capacity of the ranges has not been improved and so in dry climates the phenomenon of desertification has arisen (figures 5.26, 5.27), especially during years when the rainfall has been particularly low (table 5.31). Such areas comprise about 35 per cent of the land area

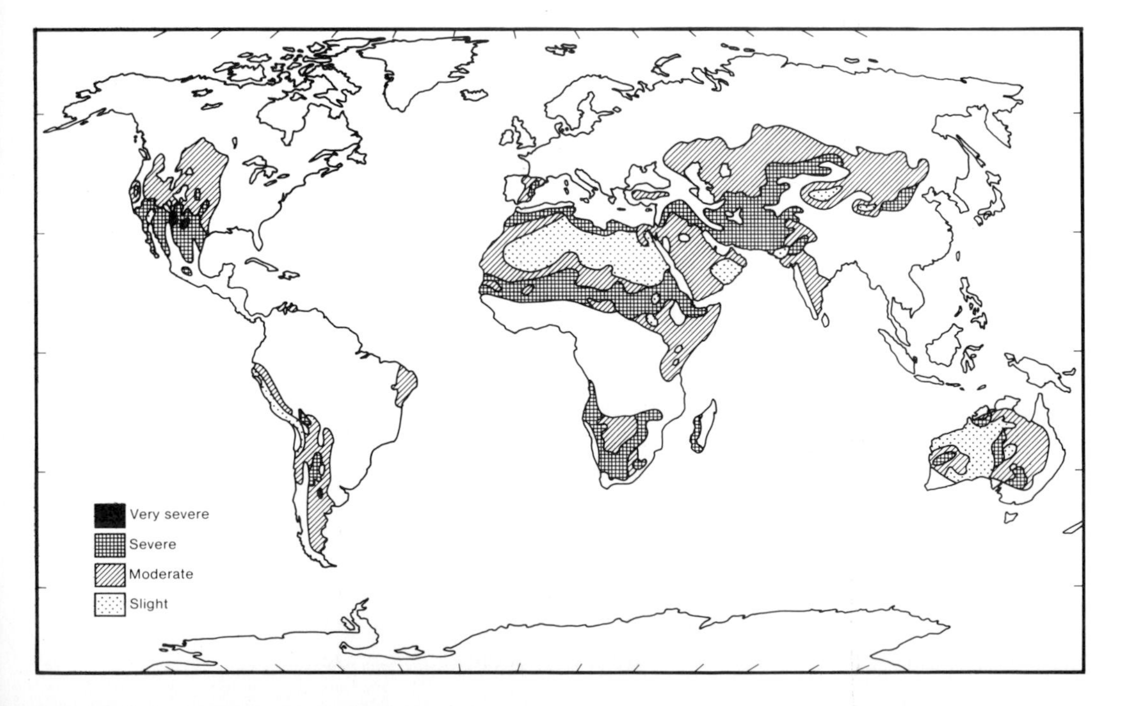

Figure 5.26 The distribution in the 1980s of desertification, showing especially that it is not a phenomenon merely of the poor nations nor confined to the Sahel zone of Africa.
Source: after an original provided by Dr A. Trilsbach and used by kind permission.

Plate 5.12 Kordofan province in the Sudan is grazed increasingly heavily by camels and other livestock. The lack of plant cover and the small mounds of sand suggest that incipient desertification is occurring. (Mark Edwards Still Pictures/Panos Pictures)

Table 5.31 Variation of plant production and theoretical carrying capacity (k) for cattle at Ferlo, N. Senegal

	Period of useful rainfall (days)	*Above-ground NPP* (kg/ha dry matter)	Theoretical k for cattle (No./1000 ha)
'Normal' good year	110	1300	187
'Normal' bad year	50	590	87
1972	0	0	0

Source: J. Swift, 'Sahelian pastoralists: under development, desertification and famine', *Ann. Rev. Anthropol.* 6, 1977, 457–78.

and in these places, the interaction of long periods of drought and heavy stocking rates has brought about animal deaths and human famine. The ecology of grazing in these circumstances produces a basal ground cover of only 0.1–5.0 per cent plant material. The loose sand thus created may be large enough in quantity to form moving dunes: 'deserts on the march'.

Such systems are often thought to be in need of modernization and one example of comparable energy flows is given for a scheme in Angola (figure 5.28). Both modern ranching and traditional systems seem to sell off the same quantities of hides and meat and support the same number of workers yet the one is underpinned by imported energy. Slaughter of the wild animals by the ranchers has meant the encroachment of bush which must be mechanically cleared. Other parts of the ranching world experience varying degrees of ecological impact as rotation grazing, fertilizer use, reseeding, partial irrigation, chemical and mechanical vegetation control and animal management from the all-terrain vehicle or the helicopter are tried in differnt mixtures (Plate 5.13). A modern cowboy has rather more technology than a lariat and a six-shooter.

Figure 5.27 A network schema of some of the factors in desert encroachment in northern Kenya, with an emphasis on the modern period, as evident from the material of the top two lines. *Source*: H. F. Lamprey, 'Pastoralism yesterday and today: the overgrazing problem', in J. Bourlière (ed.) *Tropical Savannas*. Ecosystems of the World vol. 13. Amsterdam: Elsevier, 1981, 643–66.

Ranching dates from about 1850 onwards. It was a response to the Industrial Revolution in the sense that it converted grasslands and steppe into meat and wool for Europe and eastern North America. Nowhere has its imprint been greater than in Australia. 'At the beginning of 1788 no hoof had ever been imprinted on Australian soil . . . By 1890 there were over a hundred million sheep and nearly eight million cattle . . .' writes G. Bolton, who chronicles the immense changes in vegetation that occurred consequently. The introduced rabbit was also an agent of vegetation change and was also the reason for the construction of a 'rabbit-proof fence' across Western Australia (figure 5.29 and plate 5.14).

Minor pastoralist economies are also likely to be subject to change. The Saami (better but less accurately known as the Lapps) have been afflicted with intrusions that have disrupted their herding of reindeer. Road construction, dam building, logging and the fallout from

Plate 5.13 One face of man-made Australia (cp Pl. 5.14). Southern Gippsland in the State of Victoria. Cattle herds in a humid environment where good grassland has been fashioned out of forest and scientific management of animals and their forage is practised. (Eric Wadsworth/Australian Overseas Information Service, London)

Plate 5.14 Another face (cp Pl. 5.13) of man-made Australia. The introduced and ill-controlled rabbit before myxomatosis. During periods of drought, as here, they converge in very large numbers on waterholes. (Australian Overseas Information Service, London)

Chernobyl have all resulted in smaller herds and so fewer Saami can be supported in the traditional ways. As in the South, they have migrated to the cities and also engaged in local tourism which in turn affects the suitability of the region for reindeer herding: cars and snowmobiles in particular frighten these animals (plate 5.15).

In legend, nomadic pastoralists have furnished many heroes, from Genghis Khan to Clint Eastwood. But those movie cowboys who fought to the death against the wire fences strung across the range were firing the first shots in a war which their spiritual kin are losing, often along with the land as well.

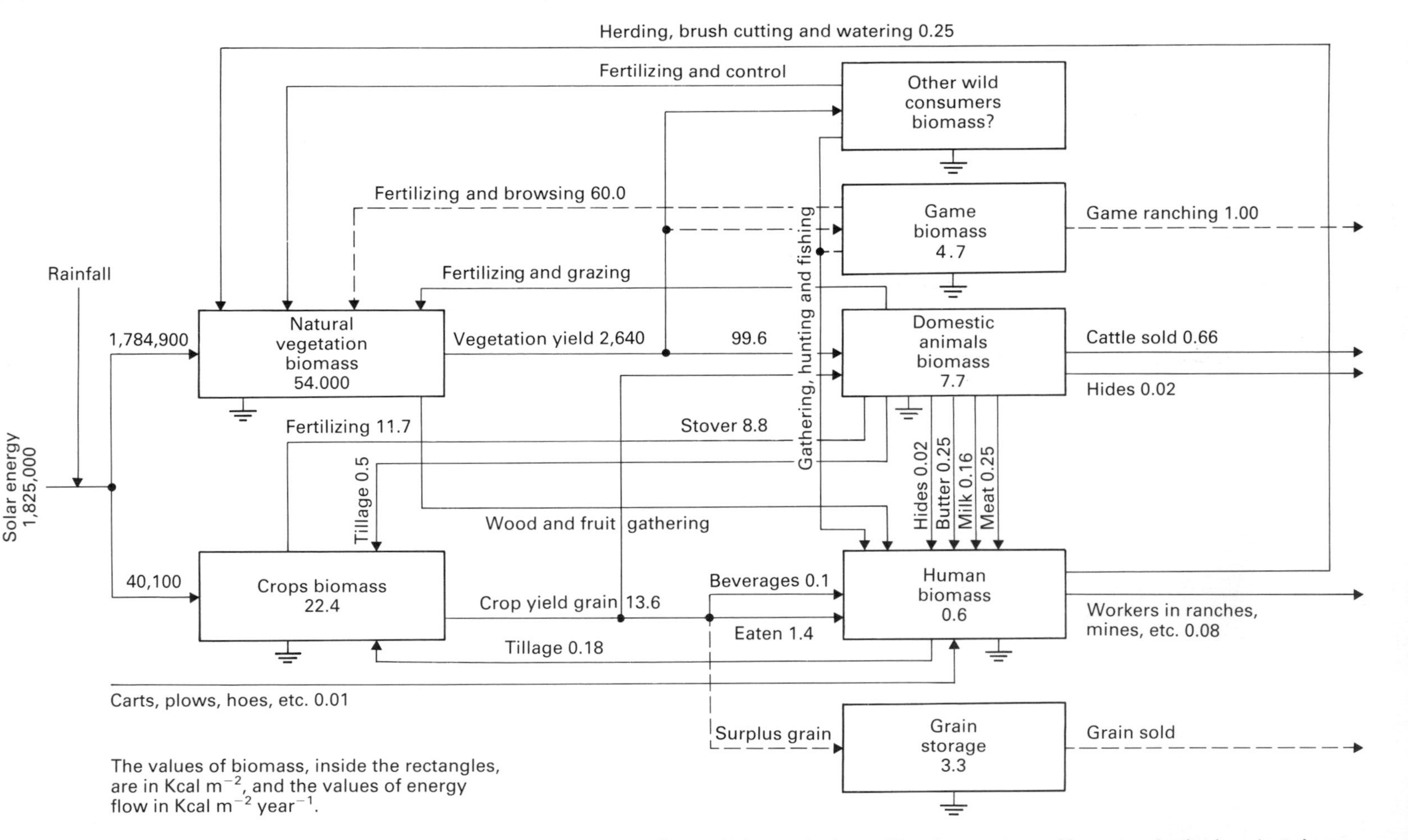

Figure 5.28: Pastoralism and ranching in Angola. The diagram shows energy flow and biomass in the traditional ecosystem, with meat and animal products from a variety of wild and domesticated sources and some grain. Surpluses were produced for trade. Also depicted is the modern ranching system which supports a lower number of humans but many more domesticated animals; the offtake of domesticated beasts appears not to be higher than the traditional system, however.
Source: E. Cruz de Carvalho and J. Viera da Silva, 'The Cuene region: ecological analysis of an African agroecosystem', in F-W. Heimer (ed.) *Social Change in Angola*. Munich: Weltforum Verlag, 1973.

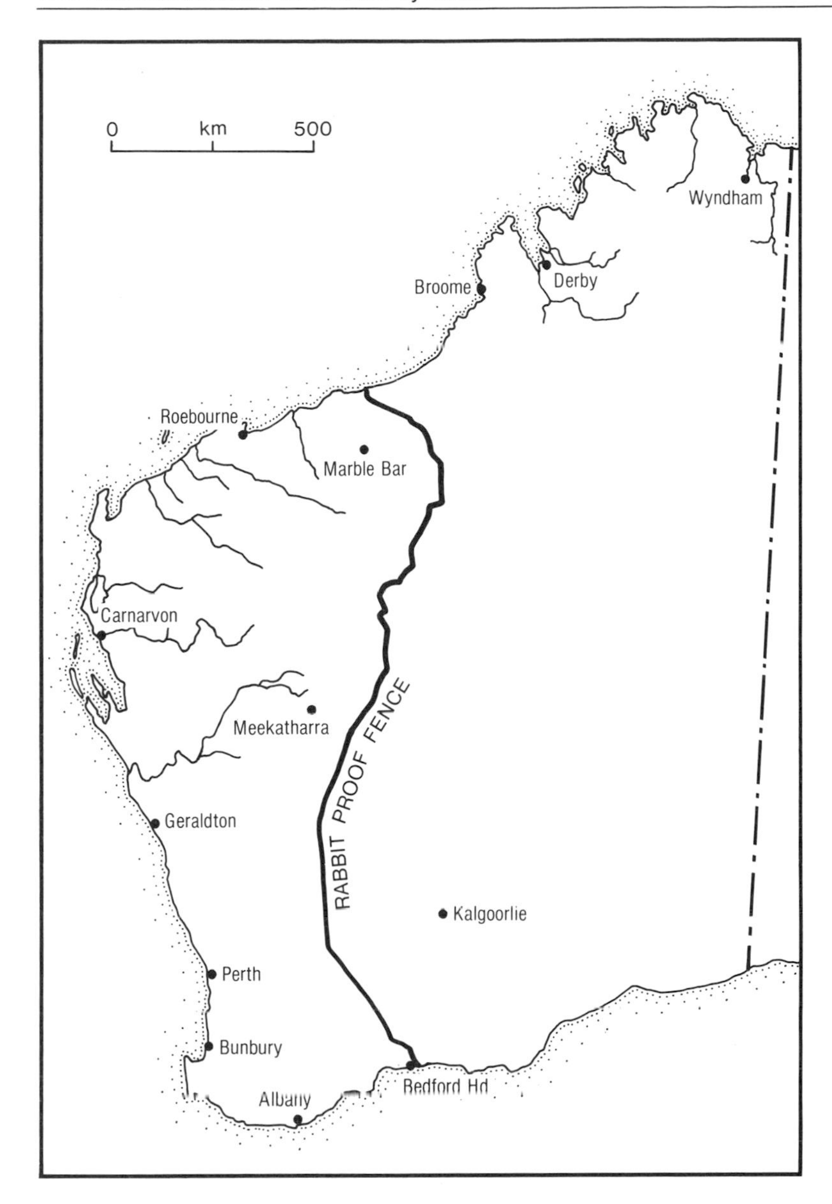

Figure 5.29 The line of the Western Australia rabbit-proof fence, analogous to the Great Wall of China in its attempts to keep out the unwanted.
Sources: G. Bolton, *Spoils and Spoilers: Australians Make their Environment, 1788–1980*. Sydney: Allen and Unwin, 1981.

Gardens

In industrialized places, gardens are the closest parts of many people's environment which are consciously non-industrial. Be they private plots or public parks, they symbolize a rural scene, with combinations of flowers, grass, trees, shrubs and water. Where they are formal, they hark back to the large estate-like gardens of the great villas of periods like the seventeenth and eighteenth centuries. Some nineteenth- and twentieth-century gardens try to be totally wild and thus go back beyond agriculture to hunting and gathering or, very likely, the Garden of Eden itself. (Some parts are like savannas, said

Plate 5.15 The environmental impact of the reindeer herding Saami in Arctic Fenno-Scandia has been changed (cp the Inuit in Pl. 3.3) by the advent of the snowmobile, as has the use of this environment by others such as recreationists. It did not however prevent the Saami suffering loss of livelihood when their animals' level of radioactivity rose dramatically after the Chernobyl incident. (Bryan and Cherry Alexander Photography)

by some to be the very environment of human evolution.) But in all these cases, various technological developments of the nineteenth and twentieth centuries are germane.

One example is the steamship. The faster passage afforded by this vehicle meant that plants spent less time in transit, and so the exotics of the Himalaya could be taken to England, Germany and the USA; vice versa, the upholder of the Empire in India could enjoy his *chota peg* overlooking gardens of English flowers: not for nothing do Paul Scott's *Raj Quartet* novels revolve round Rose Cottage. So the garden environments of Europe were diversified by *Fuchsia*, *Calceolaria* and *Rhododendron*, by alpines, aquilegias, narcissi and orchids. Some of these transfers depleted the supply end of the process: T. Whittle tells the story of a Prussian orchid-hunter who collected a few specimens of an Indian orchid and then destroyed the rest of the natural population so that he might have the monopoly on its introduction to Europe. Other innovations had their influence as well. The lawn-mower is an obvious example as are the huge conservatories and glasshouses that could be built with cast iron frames (cf. plate 4.7). At the Duke of Devonshire's house at Chatsworth in Derbyshire, England, a conservatory was built with an iron frame based on the leaf of the giant lily *Victoria amazonica* and it is perhaps not too fanciful to see the flowering of industrialism in the Great Exhibition of 1851 housed in the massive glasshouse of the Crystal Palace in London's Hyde Park. The exhibitionism mania which led many a European and North American mansion (whether at Home or in the colonies) to display the massed heads of animals slaughtered sportingly by the successive heads of the household was paralleled in the garden (more often a female preserve) by gardens which were collections rather than visual displays. In public parks, grass and bedding plants were usually thought a proper accompaniment to trees, a bandstand and a rather eutrophicated pond, but the class distinctions were maintained by the severe presence of large clumps of rhododendrons keeping order:

> Guarding and guarded by
> The Council's black
> Forbidding forbidden stones

as Ted Hughes's poem puts it.

The twentieth century sees the full weight of its agricultural and industrial technology applied to gardens, scaled down where necessary in the form of mini-tractors and small powered implements. Machinery, biocides, fertilizers and irrigation are all used and one study in the USA showed that a surburban lawn in California absorbed 578 kcal/m^2/yr of energy as fuel, carbon, water and fertilizers, which compared with 715 kcal/m^2/yr for maize. The primary productivity of the grass lawn was 1020 g/m^2/yr which compared with maize (1066 g/m^2/yr) and tall grass prairie (1000 g/m^2/yr). This intensity of use shows itself well in gardens which are used to produce fresh fruit and vegetables. An English study in the 1950s showed that suburban gardens had a food output better than any farmland: some allotment gardens produced 48 t/ha of vegetables which represented 60 GJ and 788 kg protein per hectare per year, reflecting the intensity of labour input and multiple cropping. The best way to increase food production was clearly to build houses on the land. So industrialization has seen the continuance of the garden as living space and producer, as playground, and as art, with less emphasis on the third as the availability of servants has declined.

The trends of the nineteenth century were largely consolidated in the twentieth, with exotic flora (produced by breeding rather than by import since Gregor Mendel's sweet peas of 1856–74 had by now yielded their most lasting harvest), the adaptation of old forms like Gertrude Jekyll's cottage gardens, and wild gardening where a pristine-looking environment was sought. To these functions has perhaps been added the role of nature reserve since private gardens, public parks and patches of undeveloped land in cities can all provide habitats for wild plants and animals. The full scope of twentieth-century technology can now be used to produce an illusion of its absence: roof gardens on large buildings are a simple example, landscaped gardens on a small scale another, as is the heavy use of powered equipment in all but the smallest patches of suburban land. Most people in DCs, after all, would be unhappy without the benefits of the industrial economy, but cherish the illusion that it is not all-pervasive by having rural second homes, vacations in poorer places, and above all by cultivating gardens.

Land reclamation

Like many other processes, the transformation of apparently 'useless' land to something more productive did not start with industrialization, and earlier examples are given in chapter 4. But modern technology confers the ability to drain, embank and infill on a much larger scale than ever before, and with more certainty about the permanence of the results and the continuation of the income. Hence, for example, mangroves, coastal salt marshes and estuarine mudflats have been converted to flat, dry land suitable for purposes such as agriculture, housing, industry, ports and airports. The efforts of the Dutch during the last 40 years in draining the IJsselmeer (Zuider Zee) and in damming off much of the southern half of the country behind the dikes of the Delta Plan (plate 5.16) are well known. San Francisco Bay has a plethora of reclaimed marshes supporting harbours, housing, industry, marinas and two large airports. The national shortage of flat land meant that the surge of industrial growth in Japan in the 1960s and 1970s was, where heavy industry was concerned, sited largely on reclaimed land on the Pacific coast and round the Inland Sea. In a similar process, the wish to exploit the oil resources of the North Sea led to reclamation projects in many of the estuaries of the east coast of England and Scotland.

The benefits have been undeniable, but there has been little attention paid to the long-term costs. In a narrowly economic sense, these are generally hidden and do not accrue to the reclaimers: for example estuaries are often vital habitats for the early stages of commercial fish populations, but their decline falls on another section of the community. The reclamation of coastal wetlands, too, may deprive a town of the chance to process wastes such as sewage for very little cost: in the tropics, mangrove-type vegetation will take up nitrogen and phosphorus as well as provide a substrate for the breaking down of organic solids. Mangroves, too, can act as a barrier against tropical storms, and their reclamation allows the full force of wind and waves to attack the rice fields or the suburbs built on the 'new' land; in this case the economic dislocation is more direct. In north-east China, the wetlands of the Sanjiang Plain have been the target of reclamation efforts during the last three decades and the resulting soils now exhibit increased wind erosion together with local desertification and soil salinization. Many wetlands, both salt- and fresh-water, are valuable for their wildlife and this is usually lost when reclamation takes place with wildfowl populations often the first to be threatened. In Europe, shallow areas like the Wadden Sea are proposed for enclosure and drainage and in Japan the desire to drain the Kushiro swamps in Hokkaido threatens the last remaining refuge of the *tsuru*, the sacred Japanese crane (*Grus japoniensis*). Wildlife can rarely be economically valued and so unless there is a strong tide of public opinion, its presence scarcely ever precludes reclamation projects.

Happily, one form of reclamation does meet with more or less universal approbation, and that is the transformation of derelict and

Plate 5.16 The sealing-off of the natural coastline of The Netherlands by the great sea wall of the Delta Plan would not have been possible without the access to the energy of fossil fuels mediated through heavy machinery. (Aerocamera-Bart Hofmeester)

degraded land, most of which derives from nineteenth- and twentieth-century industrial activity. Such land is dominated by 'holes and heaps', though flatter stretches may exist. It is likely to be inimical to life because of the physical nature of the material (a very fine ash, perhaps, or large lumps of slate, or a slag high in metals), the lack of nutrients, the toxicity of the substrate, fires, or simply the steep faces of a tip or quarry. In many cases, however, the power of industrial energy can be brought to bear upon such areas in order to reclaim them for the community. Levelling of heaps, spreading of topsoil, spraying of nutrients, construction of drainage channels, and the compilation and dissemination of knowledge about reclamation are all made possible by the availability of energy. According to A. Bradshaw and M. Chadwick, derelict land problems centre on such factors as texture and structure of surface layers, their stability, water supply, surface temperature and nutrient supply, and problems of acidity, toxic material, salinity and the availability of appropriate species to be planted. Solutions include large-scale treatments like levelling, landscaping and draining, the conservation or restoration of top soil and sometimes its replacement by pockets of high-nutrient materials. Seeding can vary in intensity from hydraulic methods where a mixture of seeds and substrate are sprayed over a reshaped tip, to the individual planting of young trees in pockets of soil or peat.

Once such areas are salvaged, their uses are many. Most land uses can take place on properly treated land: housing, industry, communications, recreation, agriculture and forestry are the main instances. Even water bodies such as gravel pits and areas of mining subsidence can be put to use as fish farms and recreation facilities. All over the world spoil from deep pits, overburden from strip mines, quarry faces and floors, urban refuse tips, and fuel ashes have been transformed. Such is the pace of transformation of industrial materials and energy, nevertheless, that even in the most reclamation-conscious countries, the creation of derelict land currently exceeds the rate of its reclamation.

The use of inland waters

If we imagine an early stage when *Homo sapiens* used water only for drinking, then each adult would require an average of 2.25 litres per day. If we contrast that metabolic need with our own time, when there is a much extended use of water for domestic purposes, agriculture and industry, then the need for environmental manipulation to ensure supplies needs no emphasis. Further, the quality of water is usually important. As appropriate for particular purposes, it may have to be free from bacteria, or silica, or nitrogen, and so the need to change its composition by using technological means may also increase intervention in the hydrological cycle.

The residential use of an individual is therefore no longer 2.25 l/day but is measured against a minimum need of 90 l/day for drinking, cooking, washing and sullage. This is the order of consumption in the households of Karachi, for instance. In London it is 263 l/day, in the USA 635 l/day. Modern agriculture, too, raises the demand for water. The adoption of irrigation increases the water usage by the equivalent of perhaps a rainfall of 500–600 mm/yr so that whereas 1.0 kg of rainfed wheat has needed about 500 litres of water for its production, 1.0 kg of irrigated rice has used up 1822 litres. The production of a litre of milk may absorb 4200 litres of water, since a lactating Jersey cow, for example, needs 27–48 kg of water to produce 2.25–14 kg of milk. In the food industry the unit consumption is correspondingly higher because of the multiplicity of processes which need water. A tonne of bread produced industrially consumes 2100–4200 litres of water; in the UK 1000 l of beer has needed 600–10 000 litres of water for its production, and in the USA the same quantity of beer has needed 15 200 litres of water: American beer is either cleaner or weaker than its British equivalent, we may suppose.

A tonne of steel may need 8000–12 000 litres of water at the steel mill, and a single automobile some 38 000 litres while at the factory, to say nothing of Sunday mornings thereafter. All these data are for the 1960s and the magnitudes are now probably lower because of improved efficiency, but the environmental manipulations they and their like produced are still with us and in view of population growth likely to remain so. In 1983 world annual average per capita runoff was 8.3×10^3 m^3, but one projection lowers this to 6.3 (−24%) by AD 2000. In this context schemes for increased re-use such as those which recycle 97 per cent of the water used in WCs take on an increased significance.

Increasing knowledge of scientific hygiene leads to raised requirements for clean water. In the DCs this usually means industrially based treatment, with filtration to remove sediment and organic matter, aeration, sterilization with ozone or chlorine to kill micro-organisms, and possibly further chemical treatment to adjust the pH. All this may well necessitate the use of energy, not only in constructing the facilities but in pumping the water between source, storage, treatment and use. In developing nations such water treatment is less common, with various diseases (especially typhoid, food poisoning and hepatitis) being a common result. Contamination of water is thus a factor in the deaths of perhaps five million people every year. To bring clean water to all users is a major task in many LDCs: in Tanzania it means getting it to 20 million rural inhabitants. These schemes will in some places compete with agriculture for water resources which are in limited supply.

The role of industrial technology since the nineteenth century has been to confer an enhanced ability upon human societies to intervene in most phases of the hydrological cycle in attempts to divert water for its various uses. The newest and most controversial has been that of cloud seeding, which attempts to intercept water in the atmospheric phase of the cycle. (Not for the first time: cannons were fired into clouds in eighteenth-century Europe in order to try and prevent hailstorms.) Since the 1960s attempts have been made to increase the number of condensation nuclei in clouds by scattering substances such as silver iodide from aircraft: it has been claimed that one gram of silver iodide can produce 250×10^3 m^3 of rain. The statistical problems of interpreting the results are considerable (would it have rained anyway?) and the legal difficulties are a paradise for lawyers who are called in to arbitrate (whose rain is it?). Although possessing potential (including that of a weapon of war), weather modification seems unlikely to be as important a resource as ground and surface water, and so environmental modification is the more productive in these latter modes. In many places in the world, water is still drawn by hand from a shallow well or stream and poured into a domestic container or irrigation channel, but industrial technology has added to this system such equipment as the motor-driven pump and cheap pipes (of steel, asbestos, cement or PVC) which increase the delivery rate many-fold. Such changes are adequate for small-scale domestic agricultural use but there often arises the need for a steadier supply at a higher volume so recourse is had to deeper ground water reserves or to surface impoundments. The latter (plate 5.17) are created by the construction of a dam, whose size may vary from the cosily domestic to that of the mega-project like the proposed Narmada Valley project (Madhya Pradesh, India) which envisages 30 major dams to irrigate 50×10^4 ha of land and generate 2700 MW of electricity. Even the World Bank had second thoughts in 1985 about this scale of change. One estimate suggests there are about 24 000 dams in the world,

Plate 5.17 Water impoundments may be made on a large scale: here is the largest in Europe, at Kielder in Northumberland, England, which replaced farm and forest land. It has not, though, led to many immediate land use changes as would have happened in an arid zone of the world, for instance. The water is primarily destined for urban-industrial use. (By kind permission of the Northumberland Water Authority)

with 41 man-made lakes above 1000 km^2 in area; Lake Nasser in Egypt, for instance, has an area of 5300 km^2, the Volta Dam in Ghana is 85 000 km^2 with a shoreline of 6400 km. In 1977, the USA had 5609 dams over 15 m high although 70 per cent of these had a water storage volume of less than 10×10^6 m^3.

The environmental consequences of large impoundment schemes have been studied in great detail and continue for some time after construction: the chemical changes in the vegetation and soils of a drowned area, for example, may affect water quality and fish species for many years. Figure 5.30 gives an outline of some of the local environmental effects of a large impoundment. While the immediate on-site effects are the strongest, other consequences are by no means negligible, like the changes in silt load both upstream and downstream and the impacts of this on the river channel. Attempts may be made to influence water quality and quantity by manipulating the vegetation of the catchment surface. Thus forests may be logged in strips on an east–west axis to catch and keep snow, or riverside vegetation dominated by plants which transpire large quantities of water may be removed. On the other hand, grassy or bare watersheds may yield a great deal of silt and so a type of riparian vegetation which filters this out may be encouraged; control of this vegetation can be used to affect such variables as silt load, temperature and nutrient loading of the stream.

Once captured, the value of water may be such that it is profitable to send it large distances and so pipelines are constructed. These are usually underground, with disturbance of the surface for variable periods of time where cut-and-cover is used. If inter-basin transfer is attempted, then the system is linked to the energy network of the area since pumping will be very likely needed. Such transfers may change river systems considerably if the conditions in the receiving system are appreciably different from the imported water: in temperature, silt load, nutrients, and effluent content, for example. Such consequences would be manifest in large-scale Chinese proposals to transfer water from south to north, e.g. from the Chang Jiang basin to the Huang-Huai-Hai plain and other areas of the north-west.

The coupling of the water use system to the energy supply system is also made plain when deep groundwater sources are to be tapped. Some of these have always been accessible through springs, but to get at deep aquifers, wells must be bored by machine and pumps are often needed. The actual extraction of the water, however, causes relatively little environmental change unless the aquifer is near to the surface, and has unconsolidated sediments on top of it. Then there wil be subsidence when the water is withdrawn. This has been noticed most in the case of cities such as Mexico City, Long Beach and Venice (table 5.32). The subsidence increases the gradient of inflowing streams, alters drainage patterns and stream orientations, and can be the cause of earth fissures and faults, all of which may have a severe impact upon built structures. Near the sea the risk of flooding becomes much greater: in the case of Venice, the sea-walls were breached in 1966 and expensive plans to close the lagoon entrance have had to be prepared, and wells in the city closed. Contamination of fresh water in aquifers by saline intrusion is

Table 5.32 Examples of land subsidence due to groundwater extraction

Locality	*Maximum subsidence* (m)	*Area affected* (km²)
San Joaquin Valley, CA	9	13500
Houston–Galveston, TX	2.75	12170
Eloy–Pichacho, AZ	3.6	8700
Tokyo	4.6	2400
Nobi plain, Japan	1.5	800
Po valley, Italy	3	780
Venice	0.14	400
London, UK	0.35	450
Mexico City	8.7	225

Source: D. R. Coates, 'Large-scale land subsidence'. In R. Gardner and H. Scoging (eds), *Mega-Geomorphology*. Oxford: Clarendon Press, 1983, 212–33.

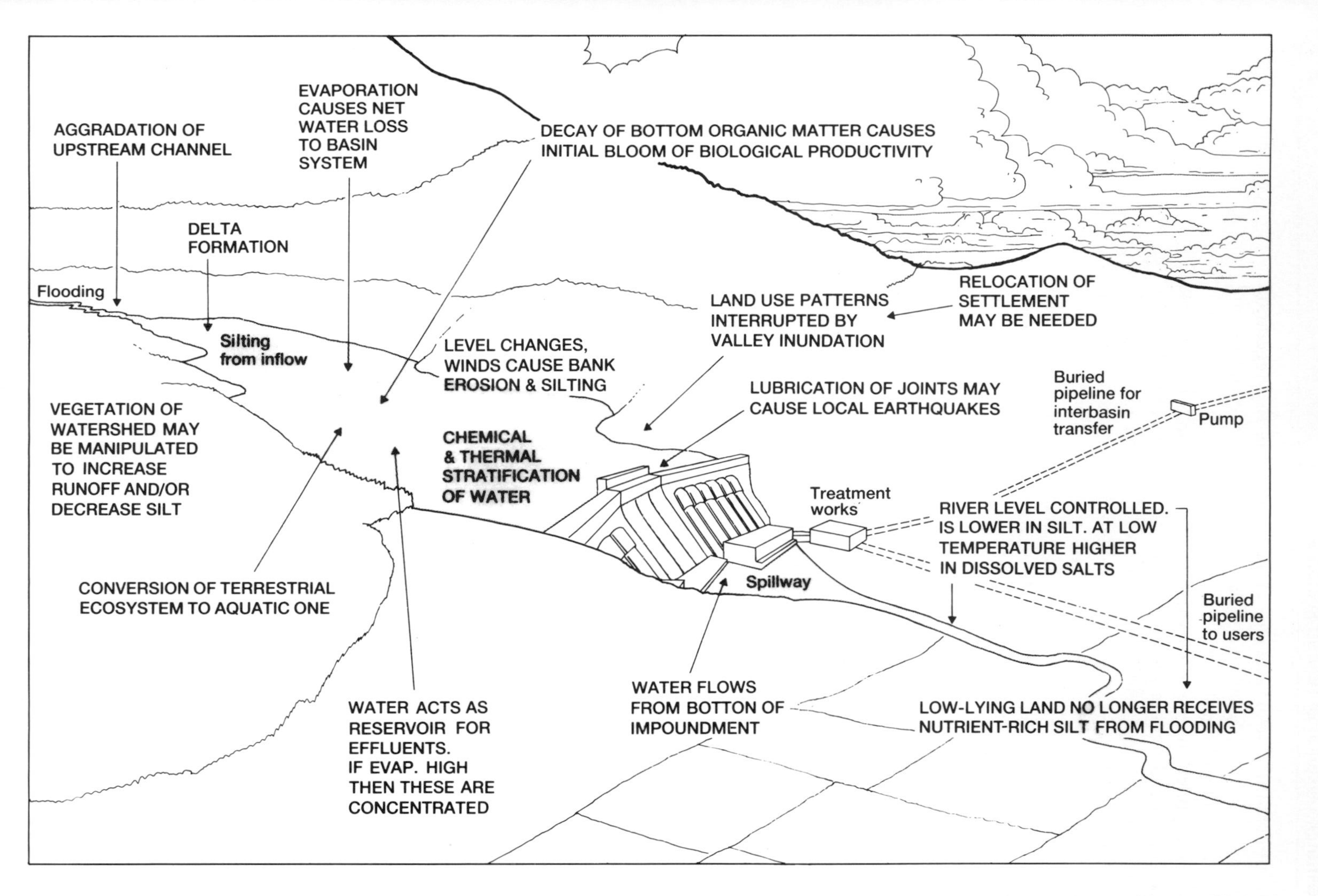

Figure 5.30 Some of the major effects of a large impoundment in its immediate vicinity. In addition, the control over water flows make possible many kinds of downstream environmental and land use changes.

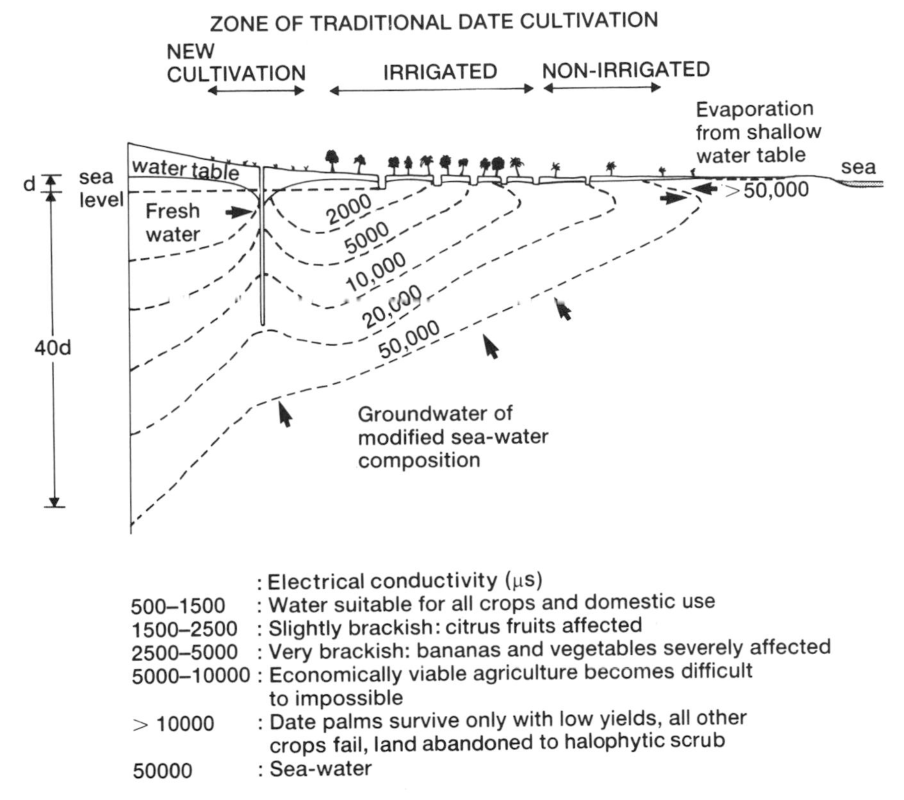

Figure 5.31 Cultivation of dates near the coast of Oman has changed the water table from fresh to saline, with progressive encroachment from the sea (to R of diagram) of water of high salinity, i.e. with high values for electrical conductivity. *Source*: G. Stanger, 'Coastal salination: a case history from Oman', *Agricultural Water Management* 9, 1984–85, 269–86.

common along coasts where wells have been bored even deeper since the availability of diesel pumps. The process is stopped when all the well water becomes saline; an example of this process is the coastal strip of Oman in the period 1950–80 (figure 5.31). Many waste disposal practices can thus affect the ground water system (figure 5.32).

Ground water may have accumulated over perhaps thousands of years but is extracted over tens of years. A key example is the Ogalala aquifer beneath the southern High Plains of the USA which were rapidly depleted in the 1950s and 1960s, leading to the use of sprinkler irrigation (plate 5.18) as a conservation technique and perhaps in the future back to dry farming. If heavily used, aquifers may be recharged with surplus or waste water on the assumption that the rocks will purify the water. This may be true in homogeneously porous strata but less so if transport within the rocks is by 'pipes', nor is it axiomatic that dissolved substances such as nitrates will be precipitated out or absorbed into the enclosing rocks. In westernized countries, ground water may contain, for example, high levels of nitrates (mostly from fertilizers) as well as industrial solvents (e.g. trichloroethylene and tetrachloroethylene which can be carcinogenic) at levels 10 or more times the WHO guidelines.

The effect of cities is plain since the construction and maintenance of the urban area changes water flow in many ways (pp. 229). The

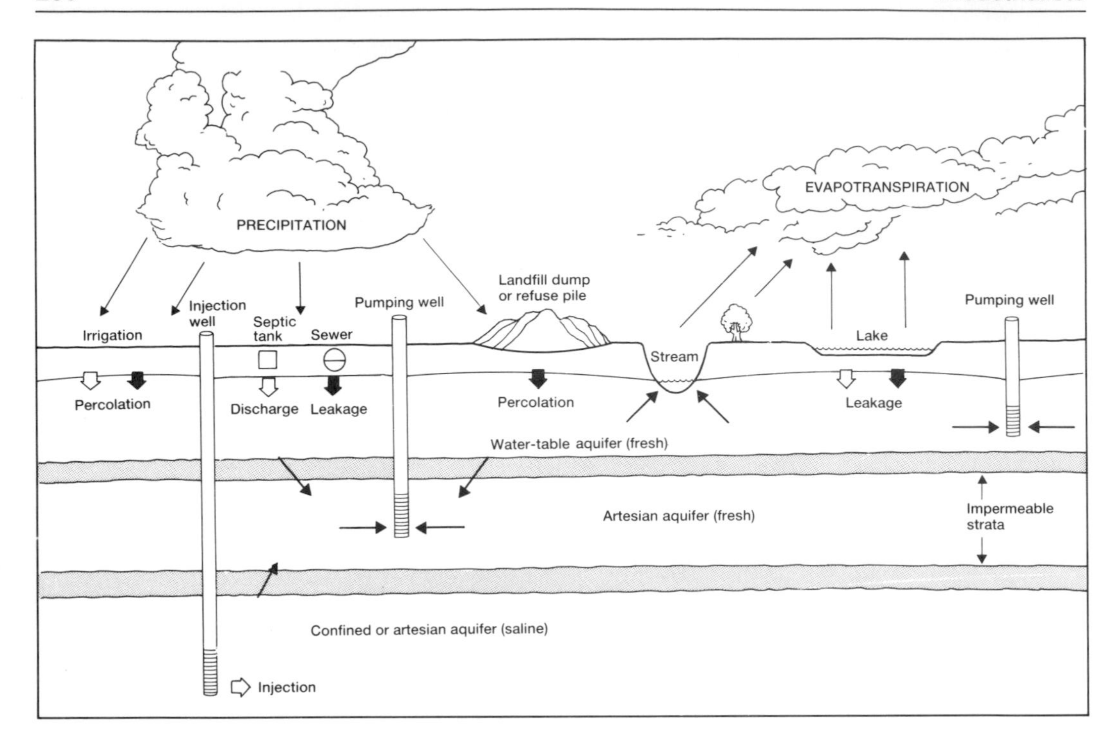

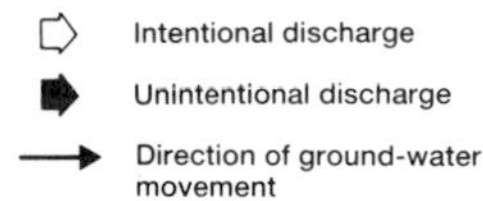

Figure 5.32 The 'leakage' from disposal of various wastes can easily affect the groundwater system at different levels. The nearer the surface, the greater the intensity, but wastes can be injected quite deep into the earth's crust.

Plate 5.18 A good head of water (in this case from a well 0.5 km away and 30 m deep) allows the use of self-propelled center-pivot irrigation sprinklers, with one arm serving both fields and being moved along the track in the centre of the picture. Each field now has an unwatered corner in contrast to traditional irrigation systems (cp Pls 4.3 and 4.4). (Russell V. Jongewaard/USDA Soil Conservation Service)

creeks which drain the built-up areas undergo considerable changes: catchment and riparian vegetation is removed, channels are regularized, and various structures are erected. In such altered habitats, exotic plants and fish may compete successfully with native species: a study of Brisbane, Australia, showed how the exotic para grass (*Brachiaria mutica*) has substantially reduced the area of open water in streams, reduced flow velocities and competed out most native plants. In the DCs there is an increased demand for water for hedonistic use: in Canberra, Australia, the use of water in gardens accounts for 70 per cent of domestic consumption and the runoff from such use can be seen in the hydrographs of local streams at times of overall low flow, partly because a lot of water from unattended sprinklers falls onto impermeable urban surfaces such as roads, footpaths and gutters. The concentration of technology in urban–industrial areas can also mean that they are the sites of developments in re-use: sewage treatment is likely to be highly efficient and return drinkable water to rivers; thermal power stations can put back to a river 95 per cent of the water they extract from it, albeit at a higher temperature than at the intake.

The oceans contain 97 per cent of all the water on the planet. For most human uses this water is too high in dissolved salts for agricultural, industrial or domestic purposes. One specialized response, now gaining popularity, is to breed crops that can grow in saline conditions; another, more general, is to desalinate the water. Here, some industrial processes use available commercial energy to distil the water and return a hot concentrated brine to the sea, and this (apart from the plant itself) is where direct environmental impact occurs. To the extent that desalination is commonest in such places as the Persian/Arab Gulf, where oil is cheap, then the temperature and salinity contrast is lower than it would be in say the Baltic. But so far, the economics of desalination have kept its spread to a level where only the most local impacts have occurred.

A last aspect of the environmental linkages of modern water management must be mentioned: the question of too much water. There is no doubt that many land use practices, past and present, may cause or exacerbate floods, as cities or deforested hills rapidly shed water, or intensive agriculture yields large quantities of silt. But modern technology has also drained a lot of land by machine-powered trenching and pipe-laying and has made possible various water diversion, storage and training schemes, all of which have diminished flood hazard. The channelization of rivers is a particular effect of the last 50 years and in England and Wales some 8504 km of river have been so treated, with maintenance of similar structures on another 35 500 km of river. This has produced a density of channelized river in England and Wales of 0.06 km/km^2, compared with 0.003 km/km^2 in the USA. The engineering works have changed flows, velocities, silt loads, channel morphology, substrates, benthic and fish populations, and have sometimes had effects upon the foundations of bridges. In a reverse direction, so to speak, contemporary engineering has provided the means of preventing storm and tidal surges getting far enough up estuaries and rivers to cause severe floods. The Thames now has a barrage downstream from central London, and one is planned for Venice.

Viewed overall, then, the application of large-scale technology to water use has produced an immense capacity to alter environments, whether it be bringing water to Los Angeles, taking it away from the Sudd via the Jonglei canal, or allowing shipping to avoid Cape Horn via the Panama Canal. Not only ecologies but whole ways of life can be altered at very short notice and in very radical ways. In a future perspective, the demands made upon water seem likely to enforce more thrifty use of it, especially in semi-arid and arid areas. This may lessen the technological impact on the hydrological cycle in some places (e.g. where cycling is practised) but increase it in others, if for example, desalination becomes more widespread.

Although not strictly relevant under this heading it is a convenient place to mention land subsidence caused by processes other than ground water withdrawal. The commonest causes are oil and gas extraction, geothermal heat production, mining, dewatering to allow deeper mining, and surface loading. None of these is as important as water extraction but locally may have severe impacts upon the land surface and human-built structures, e.g. oil and gas over 10 km^2 of the Goose Creek field in Texas and similarly the Long Beach area of California (now largely halted by salt water injection), and the Niigata City methane field in western Japan. Geothermal subsidence is reported from Wairakei in New Zealand over an area of 1.3 km^2 and from the Geysers area of California. An example of dewatering is the Far West Rand district near Johannesburg, RSA, which caused the formation of 8 sinkholes larger than 50 m in diameter, one of which was 125 m in diameter and 50 m in depth; in 1962 the sudden collapse of a similar sinkhole killed 29 people.

Aquaculture

This activity is the hydrological equivalent of agriculture: an attempt to manipulate the rates of growth, mortality and reproduction of aquatic organisms. To do this, transformation of the pre-existing environment may be necessary, less so in extensive 'open' systems like net systems and artificial reefs, and more so in intensive 'closed' systems like ponds and tanks. The environmental impacts of aquaculture centre upon the holding facilities: the ways in which the organisms (usually fish, crustacea, molluscs or algae) are confined, or in the case of sessile species given a substrate, may be negligible or temporary (suspended cages for example) or virtually total, as with concrete tanks or lined ponds. In environmental terms, ponds are the commonest causes of transformations, with concrete raceways, tanks and other enclosures coming a poor second. At a world scale, 82 per cent of the ponds are fed by fresh water and the rest by the tides. The habitats most likely to be changed are flat lands beside rivers (e.g. for

Plate 5.19 In Asia, the conversion of flat land is not all to rice-growing. Here at Tin Shui near Yuen Long is a stretch of fish farms. These are most common in estuaries and on low-lying coasts and may formerly have been salt-marsh or mangrove. (Reproduced by kind permission of the Hong Kong Government)

trout farms in developed countries), salt marshes and tidal flats along estuaries (e.g. for carp in China) and coastal mangroves (e.g. for milkfish in south-east Asia). Asia shows the greatest ecological effect of aquaculture (plate 5.19), though it is gaining in importance elsewhere, often for luxury production rather than that of an important protein source. 'Upstream' effects may include the diversion of surface flows to provide changes of fresh water for the holding facilities, and the purification of such waters to a particular quality; 'downstream', there are inevitably wastes, such as surplus food, eutrophicated water and dead but inedible plants and animals. A large tidal estuary is an apparently ideal place for disposing of all such by-products but elsewhere some form of processing may have to occur. In Hong Kong and Nepal, for instance, small ponds are used as fish–duck systems, with both animal and human excreta fertilizing the ponds which then yield heavy crops of animal protein; the aim is to create a more or less closed system, with only the saleable products leaving it.

Hunting

Even in the industrializing England of 1837–38, Charles Dickens (in *Oliver Twist*) could write that 'There is a passion for hunting something deeply implanted in the human breast', and we have seen in chapter 3 how a little of this has persisted in hunting and gathering groups of the present day. Even these remote people have not been immune from industrialization's outreach: in the case of the Indians of the Amazon Basin, for example, one scholar has contended that the modern shotgun is far more efficient than the traditional bow in procuring animals and hence in bringing about a reduction in populations of prey species. This observation is perhaps more important than it seems at first, for some anthropologists have concluded that for groups like the Yanomama, who both cultivate and hunt, access to animal protein is a limiting factor on the size, distribution and permanence of the human populations. More efficient hunting in which the studied groups sometimes join their neighbours the Ye'cuana, who possess battery-powered headlamps and motorized canoes, may make for rapid population increase. The position is further complicated by the traditional role of warfare as a homeostatic process, but other anthropologists go for less materialistic interpretations. The anthropologists seem confused; the Amazonians clearly get more food.

Less complication seems to attend the position in the high Arctic, where the Inuit have adopted technology such as the outboard motor, the high-velocity rifle and the snowmobile as hunting aids. With the boat motors, animals such as walrus and narwhal can be more successfully hunted than hitherto, indeed to the point of severely reducing narwhal populations. The use of rifles allows hunters to pick off seals at a great distance, though apparently the animals may learn to become more wary; snowmobiles transport the hunters over a wider daily range and if run in circles on the ice may 'drive' seals to a particular blowhole where they can be harpooned. Caribou can be hunted from the snowmobile over a wide area. So traditional nutritive sources are maintained and age-old skills to some extent taught to young people, although the centrality of the Hudson's Bay Company store is not now to be disputed. The work of Kemp on Baffin Island (figure 3.10) shows the energy value of modern technology in relation to traditional energy flows. Some biologists claim that native people have adopted twentieth-century killing technology but not scientific game management, and thus in Alaska populations of moose, caribou and walrus are all in danger of disappearance.

Hunting was very popular in LDCs when they were colonies of European powers. Accounts of life in Kenya, Tanganyika, India, the Belgian Congo, and the Dutch East Indies are often liberally scattered with photographs (especially perhaps in the period *c*. 1880–1935) of white hunters surrounded by their plentiful kill of game animals and, although accurate studies are scarce, it seems very likely that the populations of such groups as the rarer antelopes and the scarcer predators must have been permanently lowered. The importance of hunting seems to rise with the status of the hunter: British kings like Edward VII and George V (plate 5.20) seemed to glory in such slaughter and as late as the 1970s the President of France was said to

Plate 5.20 Big-game hunting in the Empire has always been popular with Europeans. Here, King George V of the United Kingdom inspects the four tigers killed in one morning in 1912. The modest number of deaths is inevitable given the nature of these animals. Back home, 600 pheasants might fall to the guns in a single day on a country estate. (*Illustrated London News*)

enjoy a special relationship with the 'Emperor' Boukassa of the Central African Republic because of the hunting afforded in his 'empire'. Perhaps the photographs were strongly symbolic of the relationship between Monarch and colonial peoples as well as wild animals. Today, most people in developed countries who hunt do so for recreation, and certain individuals are always willing to buy a lot of gear, travel long distances and generally divert considerable amounts of disposable income to it. In Europe, these people are usually less than 1 per cent of the population (though 3.95 in France), and in the USA, 8.5 per cent.

In some forms the chase may depend upon the survival of a pre-industrial landscape: fox hunting, whether in England or Virginia, depends upon the presence of hedges and wooden fences (neither wire fencing nor 'new prairie' landscapes are helpful) and small copses, all of which tend to disappear with modern agricultural intensification. On the other hand, deer hunting in the USA and some parts of Europe is favoured by the increasing amount of deciduous secondary woodland that results from the concentration of agricultural production upon the best soils. So explosive increases in the populations of small deer (e.g. roe deer in Europe and the whitetailed deer in North America) are sometimes found, and sport hunting can play a valuable role in preventing the deer from exceeding the winter carrying capacity of their habitat and then starving in large numbers in an especially bad season. Such is the interest in the sport in North America (and for an understanding of its symbolic role in the USA the reader can do no better than see the movie *The Deer Hunter*) that many state resource agencies manage the habitat in order to keep up high densities of deer: winter feeding may be practised, and in the Californian chaparral areas are deliberately burned from time to time because of the nutritive forage that ensues.

An example of thoroughgoing manipulation of habitat to ensure good sport is that of grouse-moor management in England and Scotland. The red grouse (*Lagopus scoticus*) is found mostly on the drier uplands of the eastern side of the UK where the vegetation is dominated by the ling or heather (*Calluna vulgaris*). In the early nineteenth century, grouse were shot from behind by muzzle-loading shotguns, after they had been put up by dogs. After about 1840 and the dissemination of the breech-loading shotgun, and the railway train that could quickly whisk the newly-rich industrialist to the moors for the weekend, it became common practice to burn the moors in rotation patches so that the vegetation became nearly 100 per cent heather and each pair of grouse could have a territory with different heights of ling within it to provide food for adults, food for chicks and nesting cover. Strict keepering kept people away and the gamekeepers shot any and all potential predators, reducing populations of falcons such as the merlin and peregrine. With careful management the territories could be more closely packed and so more game birds per moor were kept. These were then shot, in a season beginning on the 'Glorious Twelfth' of August (after the London Season and during the Parliamentary Recess), the birds being raised by 'beaters' and driven over a line of small open shelters called butts. Enormous bags often resulted: 1000 brace from one day's shooting of a moor was a good but not outstanding total. In the early years of the twentieth century, the densities of heather and birds were sometimes such that epidemics of heather bark beetle and grouse ringworm occurred, the latter evoking an official government inquiry. Later research has shown that considerable proportional losses of nutrients occur during and after the firing of an ecosystem not notably rich in nitrogen and phosphorus, and that silt load in the runoff is also high on grouse-moors. Nevertheless the system is still stable to the point of attracting large numbers of guns who may pay up to £2000 per day to be flown for a day's shooting from the Netherlands or Germany. On some large estates the profit from grouse subsidizes less viable sheep farming on unimproved moorland.

Sport hunting today is also a part of international tourism, and specialist agencies exist to provide a vacation hunting boar in central Europe, for example. In southern Europe, shooting of migratory birds, including the smallest, for prestige or the pot, is a continuing drain on their numbers and is pursued with a Mediterranean passion that might have surprised even Charles Dickens. We have to conclude that hunting is a symbolic act relating to pre-industrial times which gives the participant a feeling of high status in another, largely vanished, order of society: see the movie *The Shooting Party* for an eloquent display of this. It is, above all, for real men.

Species protection

In the course of time, humans have changed many ecosystems and introduced many species to parts of the globe to which they were not native. Like so many other processes, these transfers speeded up in the nineteenth century with the availability of steamship and railway transport. These displacements have produced many benefits but have had an adverse side as well since some immigrants have proved to be unwelcome citizens: few loyal Americans would now admit the European sparrow and starling if they had the choice of keeping them out, and Australians would cheerfully poison the last rabbit and burn the last blackberry bush.

Parallel to both of these processes has been a human role in the extirpation of species. Extinction is a normal part of evolutionary change and we can guess its rate (though this is a risky task given the incompleteness of the fossil record and the possibility that evolution proceeds not gradually but by a series of punctuated equilibria) to be in the order of 1 per cent of the given number of species every 2000–3000 years. The total number of biological species on this planet is at least 10 million, perhaps 30 million if insects are included. At least 10 per cent of all plant species are as yet unnamed and about 5 per cent of other groups are similarly undiscovered. New species continue to be discovered at a slow but steady rate including in the 1980s a 3 cm-long bat that had to be placed in a new family. Less hazy numbers can be put on extinctions (table 5.33). In total, 286 species and subspecies of mammals and birds have become extinct since AD 1600, at a rate of 1 species every 4 years in the period AD 1600–1900 and 1 each year thereafter; in 1987, one group of biologists asserted that the rate was several species per day and that by 2000 it will be several species per hour. 'Ball-park' estimates thus put the extinction rate of biota at human hands at 40–400 times the 'natural' rate. At present, the IUCN (The International Union for the Conservation of Nature and Natural Resources) estimates that 25 000 plant species plus 1000 vertebrate species and subspecies are in danger of extinction; if an estimate is made of threatened smaller animals then another 20 years could see 0.5 to 1.0 million species under threat.

The types of environmental impact which have produced these exterminations and endangered other species are well documented. There is firstly the overculling of a particular taxon which may be highly local in its distribution: flightless birds such as the great auk and the dodo (plate 5.21, plate 5.22) are examples; predatory birds whose eggs or feathers are valued another; rare orchids or alpines taken by collectors yet one more. Overkill is the sole mechanism by which marine mammals have been eliminated (4 species including a pelagic whale) and many more have been severely reduced in number or range. It has been the main cause of nearly all the modern extinctions of large mammals and about 15 per cent of bird extinctions. In the case of endangered predators like the tiger, concern over its toll of domesticated animals (and occasionally of people as well) has been a

Table 5.33 Species and subspecies of mammals and birds extinct since AD 1600

	Mammals	*Birds*
Continents		
Africa	11	0
Asia	11	6
Australia	22	0
Europe	7	0
North America	22	8
South America	0	2
Total	73	16
Pelagic	1	0
Islands		
Continental		
Africa	0	2
Asia	4	0
Australia	0	2
North America	4	3
Oceanic		
Pacific	4	109
Indian	4	18
Atlantic	23	20
Mediterranean	2	1
Total: All islands	41	155
Total: All locations	115	171

Source: J. M. Diamond, 'Historic extinctions: a Rosetta stone for understanding prehistoric extinctions'. In P. S. Martin and R. G. Klein (eds), *Quaternary Extinctions. A Prehistoric Revolution.* Tucson: University of Arizona Press, 1984, 824–62.

Plate 5.21 Some species are now so rare that their perpetuation is only possible if they are kept in zoos. Such places may also be the scene of attempts to breed enough of the species to release them back into the wild. These are European Bison in the Copenhagen Zoo. (Eric Parbst/WWF)

Plate 5.22 Though a somewhat hackneyed example, let the Dodo stand for all the species of plants and animals which have become extinct during the phases of agriculture and industrialization. The word comes from the Portuguese for 'stupid' which tells us more about ourselves than about the Dodo. (The Mansell Collection)

strong factor in its decline, though another derives from the second major category of impact, namely habitat change, like the replacement of forest by agricultural land.

Habitat destruction is now the most important cause of extinction. Within this category the outstanding processes are deforestation for timber, agriculture or stock grazing; introduced grazing and browsing animals, especially goats and rabbits; and drainage of wetlands. Agricultural changes, for example, together with shooting, have eliminated the Great Bustard from most of Europe, and wetland drainage threatens the small remnant population of the Japanese crane in Hokkaido. Presumably the ploughing of the Great Plains was an important factor in the demise of the passenger pigeon, and the El Segundo butterfly lives in a few hectares at the end of a runway at LA International at the mercy of extension of the tarmac; the 4000 remaining individuals of Abbott's booby live on 130 km^2 of Christmas Island which is also valuable for fertilizer, being dug at a rate of 2×10^6 t/yr: 25 years may see the bird's disappearance. A last powerful influence in extinction and endangerment is the intercontinental import of exotics, especially predators. Rats, goats, cats and pigs, for example, have reduced the diversity of many island faunas by direct predation and by eating out some of the vegetation, as in reducing forest to scrub; flightless birds on isolated islands have been particularly affected by rats, for instance. Lesser factors in extinction (though often important in the reduction of range or abundance) include the introduction of competitors which may drive out the native species: red deer in New Zealand were more efficient grazers of the same species of grass than the native takahe, a flightless bird which is now very rare. Australia stands out as a good example of a continental-sized area whose biota has been drastically changed by imported domestic and wild animals and plants: an expedition to the Murray–Darling junction region in 1856–57 recorded 31 species of native mammals (excluding bats), of which 22 species are now extinct. Introduced diseases have been implicated in the demise of chestnut and elm in North America, and the Norfolk Island parakeet declined due to a disease caught from domestic poultry.

Trophic cascades can also result in eliminations. When Barro Colorado Island was formed by the construction of the Panama Canal, the largest predators were lost and the resultant population explosion of smaller predators eliminated all the ground-nesting birds. Once the dodo had disappeared from Mauritius in about 1680 then a tree whose seeds needed to pass through the dodo's gizzard and gut in order to germinate began to decline in regenerative effectiveness.

The types of species which have died out or are at the margins of extinction mostly fall into certain categories. There are the rare species which are biological relicts, like the giant sequoia of California, already scarce long before human influence was important. Then there are those taxa which are scarce because they are at the end of long food chains, like large mammals and predatory birds. These often have a low reproductive rate, which ill-fits them for a man-disturbed world and which may mean that protected status will be too late for their population to recover from a low level: the California condor does not breed until at least 6 years of age, does not nest every year and when it does nest lays only one egg. (A quail, well adapted to agricultural land, breeds at 2 years and lays 15 eggs every year.) Equally at risk are the biota of islands: 90 per cent of bird extinctions in historic times have been from islands where they, their eggs and their young have been unable to escape the dangers of exotic animals and hungry sailors. In the sixteenth century, goats were introduced to St. Helena and many of its endemic species of plants were eaten out, with 33 of them recorded in 1810 but only 11 now. Wetlands are another vulnerable category of habitats, open to change from drainage, irrigation, impoundment and reclamation for more 'useful' purposes; coastal wetlands such as mud flats and tidal salt-marshes are especially liable to such manipulations. Apart from birds, groups like snails and amphibians are at risk but public concern is unlikely to extend to them as much as to tigers and falcons. Finally, the birds of tropical forests seem unable to adapt to secondary forest (unlike those of temperate regions) or even to smaller, lighter, patches of high forest. Thus out of some 400 endangered bird species nearly 300 are in the tropical moist forests of the world.

It is not one of the primary functions of this book to detail the reasons why the acceleration of extinctions should be retarded. In brief, there are, for example, sound commercial grounds for retaining the genetic variety contained in 'traditional' strains of crop plants: breeding programmes will often need such sources of variety. Many species may have undiscovered or undeveloped uses: the jojoba bean, whose oil (plate 5.23a & b) can replace sperm whale oil for most of

Plate 5.23 (a) Whale oil is used in cosmetics but its qualities can be mimicked by the use of oil from the jojoba plant; at a time of public concern for whales, such a substitution is essential for an environmentally-concerned chain of stores such as this. (b) The jojoba in its native habitat, the arid south-west of North America. It is hoped that cultivation of it will be a useful addition to the rural economy of indigenous peoples. (The Body Shop)

its uses, was only discovered to possess these properties in 1975. Species without any obvious value as resources may have recreational, aesthetic and scientific values, may act as stabilizers in ecosystems, demonstrate usefulness as environmental monitors, and are available for habitat reconstruction. So, more germane to our present theme, there has come about a movement for the protection of the diversity of species, both domesticated and wild. The conservation measures proposed are mostly of two types: the protection of an individual species, or the protection of an ecosystem either as the support system for a particular species or for an assemblage, or for its own holistic value. The World Conservation Strategy, for example, says comprehensively that we should, 'Preserve as many varieties as possible of crop plants, forage plants, timber trees, livestock, animals for agriculture, microbes and other domestic organisms and their wild relations.' It suggests that such protection can be either on-site, where the stock is kept in the ecosystem in which it occurs naturally, or in parts off-site as with seed and sperm banks. Whole organisms can be safeguarded off-site in plantations, zoological or botanical gardens or aquarium collections. Wild species should be protected according to certain priorities: those endangered throughout their whole range and those which are the sole member of their family or genus are at the top of the list. Like cultivars, both on-site and off-site preservation are possible but both are essential in this case, for it is unlikely that botanic gardens could cultivate the 25 000–40 000 species of plants which are currently estimated to be threatened. So allocation and management of land and water resources must be made for these purposes, and ecosystems must be managed for protection just as they are for production. In the case of wild species there is no choice, but for cultivars although parks and reserves sound attractive, they are so vulnerable to population growth and development pressures that 'gene banks' offer a much more sound alternative, albeit one in which no further natural evolution will take place.

The protection of ecosystems at a global scale is proceeding from IUCN's mapping of 193 biogeographical provinces and the pressing forward of the need to have national parks (plate 5.24) or biosphere reserves for representative samples of all of them. Some 35 of the provinces have no large protected areas at all; the 243 biosphere reserves designated by 1984 included 75 per cent of the 193 provinces, but there are gaps in subtropical rainforests, temperate needleleaf forests, cold deserts, tropical grasslands and savannas, temperate grasslands and freshwater systems. The programme also identifies areas especially at risk (islands, tropical moist forests, drylands, Mediterranean ecosystems, wetlands) and also areas of exceptional diversity (tropical moist forests, the tropical dry forests, Mediterranean-type ecosystems of southern Africa and western Australia, species-rich islands like New Caledonia and Hawaii, coral ecosystems and many freshwater ecosystems of most continents). The concept of a biosphere reserve goes beyond the traditional type of reserve devoted to species or habitat protection. Like the latter, the former will comprise representative examples of natural ecosystems or areas with unique features or species, but it may also contain examples of 'harmonious landscape' resulting from traditional patterns of land use or examples of modified or degraded ecosystems that are capable of being restored to near-natural conditions. Such biosphere reserves are often managed by zoning, with a central core

Plate 5.24 In industrialized nations, National Parks are often islands of less heavily manipulated ecosystems but they are rarely ecologically pristine. At the very least they will be subject to pressure from commercial forestry operations, sport hunters and recreational developers for e.g. winter sports. The Abruzzi N.P. in Italy is placed to be a focus of such pressures. (Photo: Fiorepress)

where human intervention is kept to a minimum and even research is confined to non-destructive observation and monitoring. Buffer zones, on the other hand, may contain man-modified areas or even degraded landscapes provided these are capable of restoration (figure 5.33). Complementary to these spatial plans, there are four international conventions on trade in threatened species (Washington 1973), wetlands (Ramsar 1971), migratory species (Bonn 1979) and the protection of the World Heritage (Paris 1972). But despite a number of efforts, no genuinely predictive theory of nature protection has been involved: even without the political dimension, the process would have largely been *ad hoc* and empirical. Recently some advocates of island biogeography theory have suggested its relevance for 'islands' of protected ecosystems, but it can scarcely be said that acceptance of the idea is universal among biologists, let alone land managers.

Within the general category of protected areas, a special category of 'World Heritage Sites' has been designated. Some such sites are purely cultural but a 1982 list comprises 219 areas which are labelled 'natural sites'. Some of these are totally wild like Ellesmere Island National Park in Canada, others like the Mesa Verde National Park in the USA, have a strong cultural content as well. Caves, as in the Mogao grottoes of Kansu province in China, may have equal significance as natural phenomena and cultural monuments.

It is one problem to place species in biological gardens and ecosystems in biosphere reserves, yet another to manage them for the kind of perpetuity that is desired. Captive propagation of animals, for

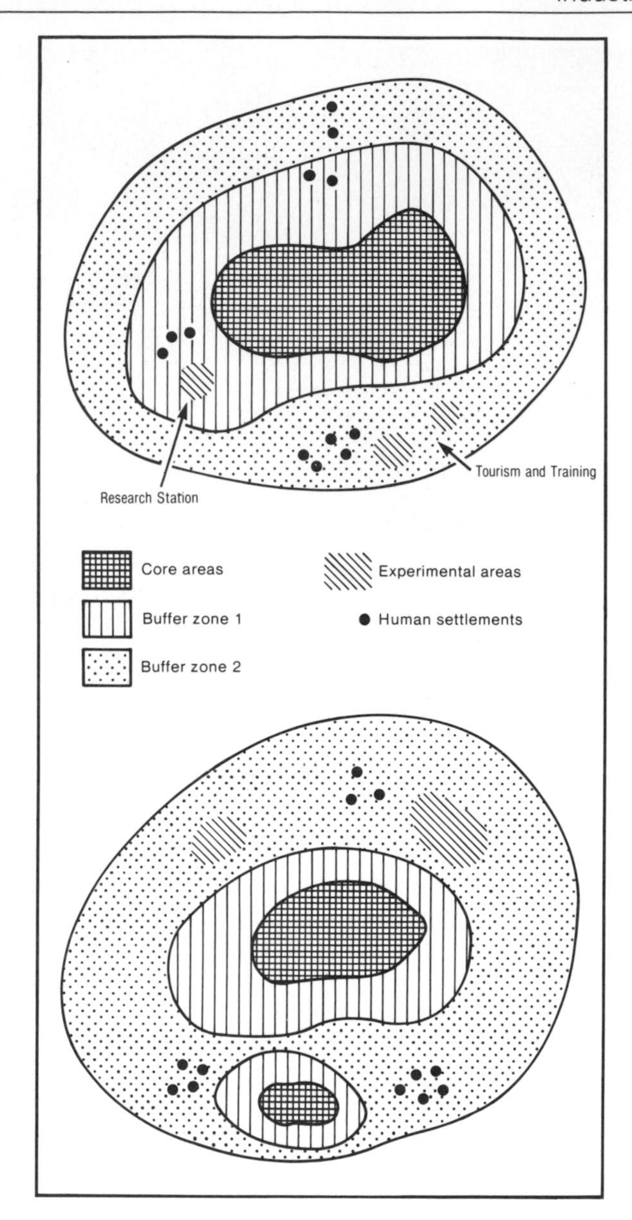

Figure 5.33 A generalized model of a biosphere reserve with concentric zoning to ensure that the core area is protected while allowing other compatible uses to be located in ecologically appropriate places.
Source: Anon, 'Looking at biosphere reserves – a 1983 perspective', *Nature and Resources* 19, 1983, 22–5.

example, may be made difficult by the shortfall of genetic variety in the zoo population, or by lack of appropriate social behaviour in the artificial environment. Successful reintroductions to the wild have been made (e.g. the European bison) but some species will clearly not survive at all unless in zoos: Père David's deer and Mongolian wild horse are of this latter character. An international system now watches the status of 2500 species of birds and mammals kept by 223 zoos in Europe and North America. As for the geopolitics of biosphere reserve designation and management, it must suffice here to say that such areas are vulnerable to a multiplicity of impacts: poachers, cultivators, loggers and developers all have their eye on

Plate 5.25 The Mediterranean littoral has been greatly changed down the centuries by both agriculture-based and industry-based processes. The monasteries of Mount Athos in Greece have, however, maintained a *cordon sanitaire* of little-manipulated vegetation around their establishments, as seen here from the monastery at Stavronikita. (Gerald Cubitt/Bruce Coleman Ltd)

some component of the ecosystem; if the reserve is surrounded by more intensively productive systems then it may get their spinoffs in the form of invasions of 'pest' species or biocide sprays (by drift or in the water) or fertilizer runoff. The reserve species may range outside the protected area and be killed for food or as pests. Alterations to the drainage of agricultural land surrounding a reserve, or industrial residues in air or water, may make equally difficult the perpetuation of an ecosystem and its constituents (plate 5.25).

Customarily, such reserves are regarded as more natural than other places but other circumstances are encountered. The first, relatively rare, is where a species has been bred in captivity but is then reintroduced into the wild, as with the bison species mentioned above, or transferred from an area of plentiful distribution to one of former extinction, as when caribou were reintroduced to Cape Breton National Park in eastern Canada. More common is an area which is valued for the landscape it presents but which is mostly a man-manipulated ecosystem although the component species are wild rather than domesticated. An example is the heathland areas of southern Britain, the Netherlands and northern Germany. Originally deciduous forest, these heaths (dominated by low shrub vegetation, mostly Ericaceae, and open scrub) were created in prehistoric time and subsequently have been maintained by fire and grazing. They have become a reserve of open land with some wild species of considerable scientific interest, and a recreational resource, and being highly valued as such are often placed under protective legislation.

That the basic philosophy and tenets of species protection are at odds with the types of land transformation and economic activity which are the main theme of chapters 4 and 5 of this book is no surprise. One replacement of this conflict views 'conservation' and 'development' not as antithetical but as necessary elements of a common process which will

(a) maintain essential ecological processes and life-support systems on which human survival and development depend, and
(b) preserve genetic diversity to protect and improve domesticated biota as well as provide a future resource, and
(c) ensure the sustainable utilization of species and ecosystems, notably fish, forests and grazing lands.

These are the main aims of the 1980 World Conservation Strategy which if widely adopted would change the course of environmental management in a number of countries and help with the kind of species protection discussed in this section.

Forests and forestry

As we saw in the equivalent section of chapter 4, human management of forests has a long history. Although the ecology of forests is dominated by trees, the other components of the ecosystem are closely linked: the soil parent material affects the supply of nutrients to the developing forest, as does the atmosphere, but in turn the tree species affects the nutrient return to the soil; small mammal populations may eat seeds but disperse some of them to places favourable for germination; large herbivores eat some saplings but may provide a seed bed for others by trampling out competing plants. Nearly all these processes are changed when human societies use the forest, and completely transformed when the forest is removed either temporarily or permanently.

The major economic change has been the growth of a relentless demand for forest products in the DCs, coupled with an equally strong need for fuelwood in the LDCs (table 5.34). The former has mainly been for construction timber, but is now ovewhelmingly for paper and paper products. In the first case, the timber goes into 'storage' for the lifetime of the building and is not usually then salvaged; in the second, paper is generally discarded and either burned or buried. So the use of wood products tends to be a flow

Table 5.34 World forest production

	1970 (× 10^6 m^3)	*1978* (× 10^6 m^3)
Roundwood total	2365	2601
Fuelwood + charcoal total	1690	1219
Fuelwood + charcoal:		
Developing market economies	760	895
Industrial roundwood:		
Developed market economies	713	740

Source: FAO *Yearbook of Forest Products*, Rome, 1980.

process rather than a cycle. The demand for wood products in DCs has been met by a few of them from their own resources but most import large quantities of wood or pulp from producer nations, especially those of the tropics and the boreal forest zone of Eurasia and North America. This has led to more intensive management of what had been remoter parts of the world, although Scandinavia, for example, has a very long history of exporting timber. Steam-powered communications also enhanced the rate at which species could be transferred intercontinentally with the aim of building up plantations in the receiving country. Good examples are the introduction of North American conifers such as Douglas Fir, Sitka Spruce and Lodgepole Pine to Europe where their growth rate outstrips native species. The Australian genus *Eucalyptus* has been taken almost everywhere where there is a climate with a dry season: imported into California to provide for a wood veneer industry, eucalypts are now naturalized and part of the characteristic scene; in China they are the centrepiece of reafforestation schemes in the south. Along with the transfer of species has gone the introduction of machinery to work the forests. Here, fossil fuel comes in many forms, such as transport of labour and extracted wood, machines which uproot trees whole and strip them of bark and branches *in situ*, ploughs which prepare a seed-bed or transplantation site as thoroughly as for wheat or rice, and aircraft which spray pesticides to attack outbreaks of predators or water to fight fire. As with modern agriculture, these are complemented by a considerable flow of energy into the processing and marketing of the product (figure 5.34). All these innovations alter the energy and nutrient flows characteristic of unmanaged or lightly managed woodlands.

The tension between forests and cultivated land is as strong today as it has been ever since the adoption of agriculture in prehistoric times, and has been discussed earlier in this chapter. At the same time, appreciation of other uses of forests has been growing, notably as regulators of water quality and quantity where they cover a catchment area and as aesthetic and recreational environments in DCs and in LDCs wishing to retain or develop a tourist industry. But if the encroachment of peasant agriculture upon forests is designated as one principal cause of the loss of forest cover, then another is the connection between similar populations and the use of wood for fuel with its environmental and other linkages (table 5.35 and plate 5.26). Most such groups of people have little choice but to use wood for cooking and heating; some may have access to some kerosene but this is increasingly expensive and is likely to be reserved for lighting and space heating. Calculations from countries like Tanzania, Gambia, Thailand and Iran suggest that consumption of fuelwood is of the order of 1–2 t/cap/yr, which multiplied up makes for a heavy and growing world demand (table 5.36). This can best be seen near cities where wood is still important: Bamato, the capital of Mali, is currently supplied by a zone of 100 km radius and by 1990 will need an area of 100 000 ha to supply its needs. Similarly, the areas of closed forest within a 100 km radius of India's cities fell sharply (an average of 35 per cent) in the 1972–82 period in spite of a 42 per cent rise in the real price of firewood. Villages demand less in absolute terms but in semi-arid scrub, some 4 per cent of all the 2.5 t/ha above-ground biomass may be taken for fuel; India lost 16 per cent of its forest cover in the period 1973–81. This is differentially cut in terms of the area nearest the settlement and in selection of species, some of which are uprooted rather than lopped. In such environments, grazing animals may exert a concomitant pressure upon woody vegetation, and in Iran a shift from camels to sheep and goats not only increased pressures upon the forage resource but sharply raised the demand for fuelwood because the processing of goat and sheep milk for yoghurt, buttermilk and butter requires heating. This type of problem is particularly worrying to Africans, because the use of petroleum products is highly restricted and the traditional hearths

Plate 5.26 The establishment of woodland is most crucial in semi-arid and arid areas of the LDCs where fuelwood is still the key energy source in rural, and some urban, areas. The current task is summed up in this picture of a lone load in a Niger landscape with little woody vegetation, though under controlled management more could be grown successfully. (Mark Edwards-Still Pictures/ Panos Pictures)

Table 5.35 Identifying socioeconomic concerns related to fuelwood use

Energy source	*Energy policy issue*	*Possible actions*	*Environmental concerns*	*Affected group*
Fuelwood	Should fuelwood be used as a major source of energy in Asia?	1 Continue present practices	Deforestation, leading to: loss of top-soil; erosion; silting of dams and channels; increased flooding; possible climate change	Farmers, irrigation agencies, electric utilities. Residents of flood plains, fuelwood suppliers
		2 Introduce rapidly growing 'exotic' tree plantations	Changes in rest of the ecosystem	Farmers, pest managers, hunters, water users
		3 Substitute with energy from animal and agricultural wastes	Possible loss of fertilizer from ecosystem. Damage to soil structure	Farmers, other residents, fertilizer manufacturers, kerosene and coal suppliers
		4 Substitute with other energy sources, e.g. kerosene, solar cookers	Construction of roads to supply petroleum and its products, or coal	Persons affected by clearing of land; road builders; manufacturers of solar cookers, coal and kerosene burners, etc. Taxpayers, if such fuels are subsidized (which is probably necessary)

Source: modified from W. H. Matthews and T. A. Siddiqi, 'Energy and environmental issues in the developing countries'. In M. Chatterji (ed.), *Energy and Environment in the Developing Countries*. Chichester: Wiley, 1981, 53–67.

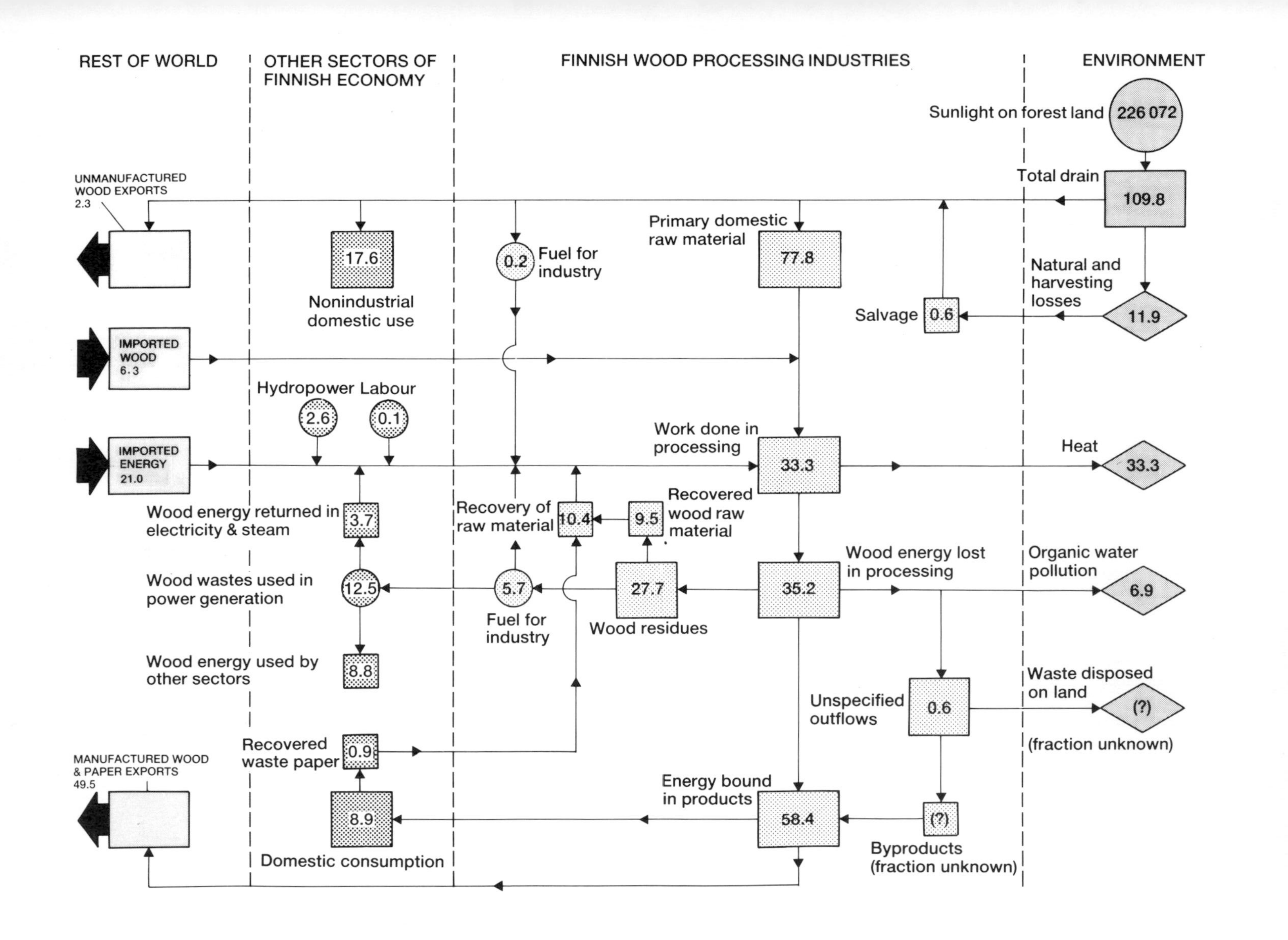

REST OF WORLD
OTHER SECTORS OF FINNISH ECONOMY
FINNISH WOOD PROCESSING INDUSTRIES
ENVIRONMENT
Sunlight on forest land
226 072
Total drain
109.8
Natural and harvesting losses
11.9
Salvage
0.6
UNMANUFACTURED WOOD EXPORTS 2.3
17.6
Nonindustrial domestic use
0.2
Fuel for industry
Primary domestic raw material
77.8
IMPORTED WOOD 6.3
Hydropower
Labour
2.6
0.1
IMPORTED ENERGY 21.0
Work done in processing
33.3
Heat
33.3
Wood energy returned in electricity & steam
3.7
Recovery of raw material
10.4
9.5
Recovered wood raw material
Wood wastes used in power generation
12.5
5.7
Fuel for industry
27.7
Wood residues
35.2
Wood energy lost in processing
Organic water pollution
6.9
Wood energy used by other sectors
8.8
Unspecified outflows
0.6
Waste disposed on land
(?)
(fraction unknown)
Recovered waste paper
0.9
MANUFACTURED WOOD & PAPER EXPORTS 49.5
Energy bound in products
58.4
8.9
Domestic consumption
(?)
Byproducts (fraction unknown)

Table 5.36 Production of fuelwood

	1975 (× 10^6 m^3)	*1978* (× 10^6 m^3)
Africa	285	298
N. America	17	18
Latin America	192	198
Asia	533	556
Europe + USSR	124	113
Oceania	7	6
Total	1158	1189

Source: UN *Statistical Papers*, J–22, 1979.

are especially inefficient; thus more forest is cleared in Africa for fuelwood than for agriculture. The types of stove used are a critical piece of technology in most LDCs which use fuelwood.

The demand for fuelwood and the long distances it has often to be carried have increased the demand for charcoal. Charcoal has nearly twice the calorific value of wood (4.7 × 10^6 kcal/t dry wood; 7.1 × 10^6 kcal/t charcoal), but since most conversion processes use 4 kg of wood to produce 1 kg of charcoal, its production may well exert an environmental impact at least twice as severe as that of wood collecting. Some governments have forbidden charcoal use in semi-arid scrub and woodland fringes, but efficient combustion technology together with planned use of plantations might make it a more viable option in some places.

The potential for fuelwood production ought to be reasonably good since woody vegetation grows in so many places. Problems are usually of a social rather than an ecological nature and attempts to create a reserve of growing wood as a community asset rather than under government pressure have had notable success in Gujerat and Bihar (India), South Korea and Nepal. Such programmes are often most successful as part of a rural development package which includes other energy-related elements (e.g. biogas from dairies) and income-generating occupations. Mistakes have been made: in India, for example, the choice of *Eucalyptus* species has brought the risk of the trees using so much water that streams used to water cattle will dry up. Experience of such a fast-growing species has also prompted some landlords to turn agricultural land over to *Eucalyptus* production for pulp.

Larger scale afforestation has also been characteristic of LDCs, especially those in the tropics (table 5.37), although in the tropics overall 10 trees are cut for every one planted and in Africa the ratio is 29:1. Colonial administrations sometimes undertook this practice (e.g. in Burma after 1855, Java after 1829) and provided an institutional framework within which it could develop. After 1900 there were often plantings of industrial tree crops where there was little indigenous forest, e.g. *Pinus*, *Eucalyptus* and *Araucaria* in South Africa. After 1945, huge afforestation programmes were started in

Figure 5.34 (*left*) Energy flow in the wood processing industries of Finland in 1970, in 10^6 GCal/yr. Although energy flows are useful in seeing where energy is expended in processing the timber, it must be remembered that energy output is not the purpose of the forest industries of Finland. *Source*: E. Jaatinen, 'Energy accounting in forestry – summary studies at the Finnish Forestry Research Institute', in C. O. Tamm (ed.) *Man and the Boreal Forest*. Stockholm: Ecological Bulletins 21, 1976, 87–94.

Table 5.37 Forest plantations in the Tropics, by region

Region	*1980* (× 10³ ha)	*1985 (estimated)* (× 10³ ha)
Africa	2595.2	3643.4
Asia	10322.7	15862.1
Oceania	261.5	383.9
Central America and Caribbean	509.5	758.5
South America	4210.6	6901.6

Source: J. D. Evans, *Plantation Forestry in the Tropics.* Oxford: Clarendon Press, 1982.

China, Kenya, Malawi and Papua New Guinea and in recent years *Pinus* and *Eucalyptus* have become important in Brazil and Fiji, for example. Nowadays, Eucalypts, pines and teak account for 85 per cent of all plantations in the tropics and 54 per cent of these world-wide are for industrial purposes, 18 per cent for fuelwood including charcoal, and 17 per cent for environmental protection against processes like erosion and desertification. There has also been a movement towards combining forestry with other productive land uses. Possibilities are agro-silviculture, concurrently producing agricultural and tree crops; sylvo-pastoralism which produces wood and domesticated animals, and combinations of these, all collectively known as agroforestry. Indeed 'three-dimensional forestry' (figure 5.35) has been successfully adopted in parts of Central America, Indonesia and the Philippines.

Large-scale projects inevitably involve environmental impacts, especially at the planting stage when the land is being prepared for the young trees. The previous habitat is destroyed, soils are exposed and then compacted by machinery; burning releases nitrogen in the smoke; watercourses receive large quantities of silt and nutrients. When growing, the plantations tend to release more nutrients than natural forests and accumulate more surface organic matter; at

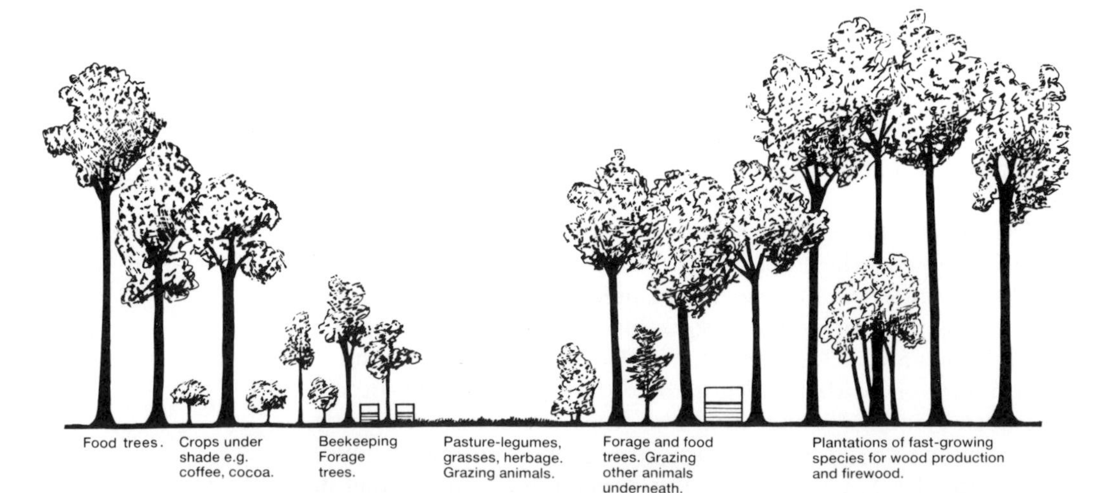

Figure 5.35 A cross-section through a tropical forest maintained not only for environmental protection but for many kinds of sustained yield as well: this is sometimes called 'three-dimensional forestry'.
Source: J. Evans, *Plantation Forestry in the Tropics*. Oxford: Clarendon Press, 1982.

harvest time significant nutrient losses from the site may be a consequence of the cropping. The plantations themselves may well support a less diverse fauna than natural forest unless the new forests replace grassland or sparse savanna. In Malawi, for instance, the leopard is found in plantations and in Venezuela, *Pinus* afforestation has led to an increase in deer populations and the return of the jaguar.

LDCs are not uniformly a picture of environmental degradation due to deforestation but the case of the lowland moist forests (sometimes called Tropical Moist Forests or TMFs) has received much public attention. These forests have the highest plant biomass (600–650 t/ha) and NPP (2200 $g/m^2/yr$) of any terrestrial biome, though the animal biomass (0.2 t/ha) is rather small. They are very variable in composition but the richest of all environments in diversity of species. They are being cleared by shifting cultivation, by commercial logging companies and by development projects for agriculture and cattle ranching. In Latin America, the conversion of TMF to pasture for beef cattle largely to supply the US hamburger market (34 ha of forest land converts to a million hamburgers) is a major aim of developers, including the Brazilian government.

Difficulties of definition and measurement make it difficult to state accurately the rate of deforestation: the forests themselves are estimated at an original area of between 750 and 1200 $\times 10^6$ ha, and the rate of loss for closed tropical forests at 7.4 $\times 10^6$ ha/yr in the 1980s but as high as 16–20 $\times 10^6$ ha/yr in the 1970s. Table 5.38 summarizes some of the absolute figures which are probably of the right order. On the other hand, there have been extensive plantings of some tropical forest areas with tropical pines, teak and eucalypts; and there are some 223 reserves in the TMFs totalling 23.1 $\times 10^6$ ha, of

Table 5.38 Regression of tropical moist forests

	Climax area ($\times 10^6$ ha)	*Actual area* ($\times 10^6$ ha)	*Actual/Potential* (%)
E. Africa	25	7	72
C. Africa	269	149	44.6
W. Africa	68	19	72
Total Africa	362	175	51.6
S. America	750	472	37.1
C. America	53	34	35.8
Total Latin America	803	506	37.0
Pacific region	48	36	25
SE Asia	302	187	38.1
S. Asia	85	31	63.5
Total Asia	435	254	41.6
Total World	1600	935	41.6

Source: A. Sommer, 'Attempts at an assessment of the world's tropical forests'. *Unasylva* 28, 1976, 112–13.

which some 12×10^6 ha are in the Neotropics (the tropical areas of the Americas). A major product is tropical hardwood: only 10–20 trees per hectare may be taken but forestry practices are often so bad that 30 to 60 per cent of the remaining forest is injured beyond recovery. In Kalimantan, a single fire in February–June of 1983 destroyed 3.5×10^6 ha (three times the area of Lebanon) of secondary forest which was regenerating after commercial logging and some illegal cutting. Where subistence cultivation is carried on (as it is by 250 million people in TMF zones), the rotation periods are now so short that the soils can only be used once.

Such pressures threaten the long-term resource value of the TMFs. The forests might be made to yield numerous minor forest products of considerable value (latexes, dyes and waxes, for instance) without the disruption of clearance. There is also potential for the sustainable development of edible oils, and other palm oils which might in future substitute for petroleum products as industrial lubricants. Four thousand TMF species are known to have been eaten; in Papua New Guinea alone, 256 species have edible fruit, and everywhere the tree leaves have a high protein content. The undomesticated relations of existing crops may be TMF species: in recent years, wild genetic material has been bred into commercial sugar cane, bananas, coffee and cocoa. The TMFs also present a reservoir of insects which might be of considerable value in the biological control of crop pests. Lastly, this biome contains more alkaloid-bearing plants than any other: current products include morphine, reserpine, cocaine and ephedrine. In north-west Amazonia, 1300 species are used indigenously as medicines and narcotics and it has been said that any prescription taken to a DC pharmacist stands a 25 per cent chance of having had its starting material originate in a TMF.

These forests also provide environmental services at a variety of scales: they act as a sponge for runoff (especially in south-east Asia) and if removed contribute heavy loads of silt to rivers and reservoirs; coastal forests and mangroves buffer the land against storm damage and tidal waves; clearance produces a change in albedo which some authors suggest might decrease rainfall in the equatorial zone; and TMFs are probably no longer a sink for carbon dioxide but a net source of carbon in the atmosphere. The disappearance of TMFs is often represented as irreversible but Quaternary ecologists point out that during the Pleistocene, TMFs colonized large areas which had formerly been savanna. Colonizing tree species able to capture nutrients from the atmosphere and to fix them in a shallow rooting layer were possibly the keys to this process. It is dangerous to suggest that conditions are now the same but history perhaps offers some hope that even if much of the TMF is converted to a form of grassland, then reversibility at some future time may be possible. The immediate remedies for TMF losses are, in essence, simple. They focus on the biome in the context of the development of renewable resources. The forests are unlikely to remain totally unmanaged but they might be highly productive mosaics of tree farms, wild forest and agriculture, and small scale tesselations would retain the essential characteristics of the TMF. Not only would environments remain stable but wealth and employment would be generated for LDC populations.

In the DCs the problems concerning forests are mostly different. At the utilitarian level, the trend is to make more intensive use of the forest and its wood products. In most places this means more wood per unit area per unit time is to be extracted from existing forests for pulping and by-product manufacture such as particle board. Short rotations are preferred and more machinery is to be used, and in essence forestry becomes much more like DC agriculture. This type of woodland management results in the increased loss of nutrients and carbon from the forest stand at harvest and often also through soil losses following large-scale cutting. In these circumstances, the practice of whole-tree harvesting means that for some time nutrient loss is unlikely to be compensated by input from the atmosphere or from soils (table 5.39). With nutrients, a number of questions have to be satisfactorily answered before intensive harvesting can be evaluated in terms of both its economic success and environmental demands. These involve knowing what proportion of the site's nutrient capital is removed in harvested material, how available to plants are the remaining nutrients, what the rates and sources of replenishment are, and whether they are affected by the harvesting mechanisms, what the nutrient requirements of the next crop are at various stages of its growth, and what is the proposed length of the crop rotation. Problems of short rotation are alleviated if there is a

Table 5.39 Nutrients removed by different forest harvesting methods in four US watersheds

	N	*P*	*K*	*Ca*
	(kg/ha)			
Mixed deciduous, 60-year-old				
Conventional harvest	135	8	90	307
Whole-tree harvest (WTH)	267	21	172	537
Annual atmospheric input	9	0.5	3	14
Douglas fir, 450-year-old				
Conventional	349	36	48	401
WTH	566	86	189	687
Annual atmospheric input	2	0.5	1	3
Northern hardwood, 60-year old				
Conventional	1325	11	71	193
WTH	367	33	154	402
Annual atmospheric input	6.5	0.05	1	2
Loblolly pine, 16-year-old				
Conventional	79	11	65	74
WTH	257	31	165	187
Annual atmospheric input	10	0.5	12	6

Source: R. I. Van Hook et al., 'Environmental effects of harvesting forests for energy'. *Forest Ecology and Management* 4, 1982, 79–84.

high rate of natural replacement of nutrients or if chemical fertilizers are added. The adoption of this last practice would indeed bring forestry much closer in agroecosystem terms to DC agriculture and further tie the system into the fossil fuel nexus; at present, forests are still heavily energy-positive even when linked to the whole industrial system.

A consequence of the move towards shorter rotations has been the replacement of broadleaved trees by conifers in the temperate deciduous forests of the DCs. A body of articulate opinion regrets the loss of the broadleaved forests as elements of landscape, as havens for wildlife and also as a source of high-value timber products. Given the taste for high quality which DCs are able to indulge, it should be possible to manage woodlands for the long-term production of timbers like oak, beech and elm and at the same time satisfy some non-material demands: traditional small-scale management promoted a high degree of variation in the structure of the vegetation, and this could be reproduced without significant economic loss. This last point emerges from both LDC and DC forest management studies: that small-scale operations produce the least environmental impact and the greatest sustainable yield of the resources.

The great and wide sea

Even before the days of marine steam and diesel engines, fishermen ranged the deep oceans in search of, for example, cod and whales. The coming of industrial technology enabled them not only to scour the remote corners of the high seas but to intensify their catches as well: in the 1970s the Faeroese topped the world league with a yearly catch of 5.4 t per inhabitant, most of which was then exported. The new technology of fisheries has centred on developments such as steam and diesel power, better aids to navigation and the location of fish, and synthetic fibres to give stronger and bigger nets which can be hauled up from greater depths. Refrigeration both on the vessels themselves and on land has supplemented these developments. The larger market thus created is underlain by a rapidly growing human population, some of whom like fresh fish while others have a taste for beef fed on fish meal. The first industrial fishery dates from the 1870s, when the menhaden was caught off the north-east seaboard of the USA by steam trawlers and then factory-processed for a fish oil used in tanning and as a lamp oil. The progress of mechanization of fisheries has been differential: in England and Wales it was completed by about 1900, some 50 years after the mechanization of the rest of the economy, with the building of 1573 steam trawlers between 1881 and 1902; in the Halland district of Sweden, mechanization was normal by 1914; in the Faeroes, the process did not end until the 1930s, and in Newfoundland it was the late 1940s. In the tropics it has been in the last 20 years that most change has taken place. In the Philippines, for example, some engines were introduced as early as 1900 but purse seining started only after 1945. (At about the same time, dynamiting also became popular and was not significantly reduced until after 1975.) In Japan, fishing in the Inland Sea was traditional in its techniques until 1925 when powered boats were introduced, and the conversion advanced further with the coming of nylon nets in the 1950s.

Given the world ratio of volume of water to size of animal, it is perhaps surprising that some marine populations have been easily affected by fishing: some stocks began to collapse in the late nineteenth century (figure 5.36), and others have followed at regular intervals as commercial trawlers have moved on from one fish population to another. But even in the open ocean, fish swim in schools and thus are relatively easy to locate by sonar, and then to capture in large nets. Bottom flatfish move little and so can be 'hoovered' by modern nets. If the catch exceeds the replacement rate of the fish stock, younger fish are caught so that increased fishing effort eventually fails to produce the same level of yield or indeed any catch at all unless progressively smaller mesh nets are used. Thus the fishery is both biologically depleted and economically unprofitable: abandonment may allow it to recover (as happened during World War II in the North Sea) but there are several instances of a fish species being unable to recover its former place in the food chains, having been ousted by a competitor. There is no guarantee in such circumstances that the newly abundant taxon will be commercially desirable, even if there is one: the Atlanto-Scandinavian herring's disappearance was not followed by an increase in any other species. Some marine resources are sessile, like molluscs, and anybody who has seen a rocky shore literally turned over at low tide by holidaymakers in France will appreciate the likely effect on mollusc and crustacean populations. Whales too have never been very difficult to

Figure 5.36 The onset of the signs of overfishing in the populations of the North Atlantic. It began as early as 1890 in the North Sea when increases in fishing effort for plaice no longer produced increased catches.
Source: P. Ehrlich *et al.*, *Ecoscience*. San Francisco: W. H. Freeman, 1983.

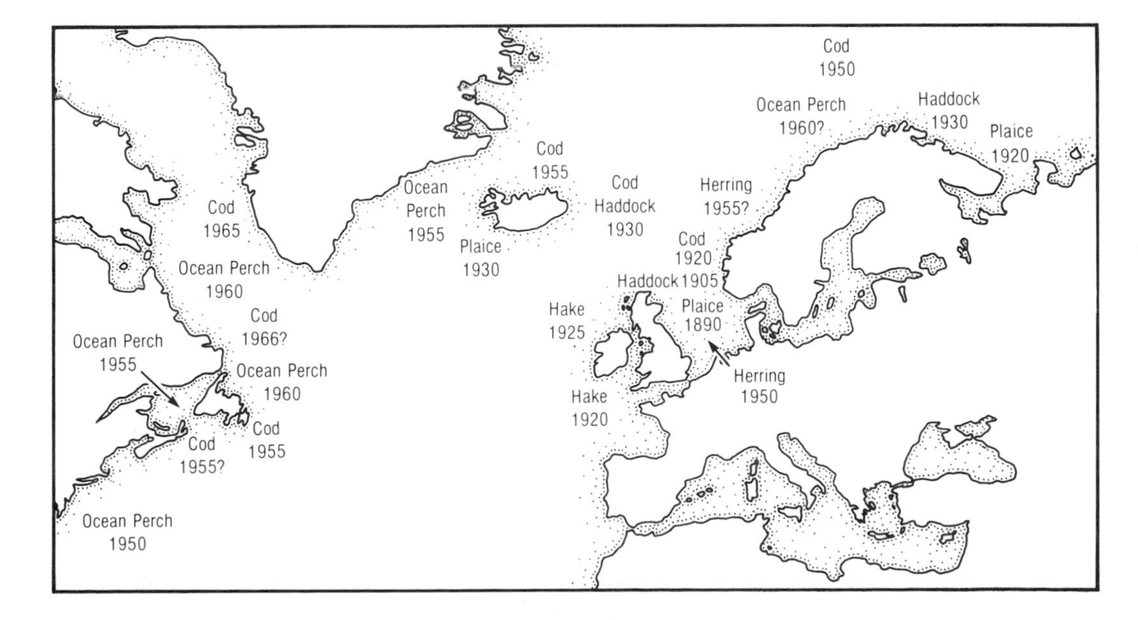

find and since they are mammals they have a slow reproductive rate which increases their vulnerability to overharvesting.

Indeed the history of whaling provides an example of an apparently irreversible impact. Whales have been hunted for centuries and as late as 1887 it could be said that the seas between 80 °N and 55 °S were thickly charted with whaling grounds. But the modern techniques of search and kill seem to have gone beyond replacement levels: in 1933 a total of 28 907 whales were caught, producing 2.6×10^6 barrels of whale oil. In 1966, 57 891 whales were killed, but the yield of 1.5×10^6 barrels of oil was only 60 per cent of the figure of 30 years earlier. This diminution of efficiency coincided with shifts from larger to smaller species (figure 5.37 and plate 5.27). Since some whale species (especially the blue whale but also the right and humpback whales) were hunted to the verge of extinction, regulation of the catch has been undertaken by an International Whaling Commission. This not altogether disinterested body has been put under severe pressure to reduce quotas and introduce moratoria, but perhaps more significant have been bans by national governments on the import and use of whale products, and the discovery that oil from the jojoba plant (a semi-desert shrub, plate 5.23) can replace whale oil in virtually all its remaining uses.

In essence, the story of whaling can be applied to many fish stocks. It is not difficult with modern technology to reduce a fish population to the point where it is pointless even to try to cull it, and this process

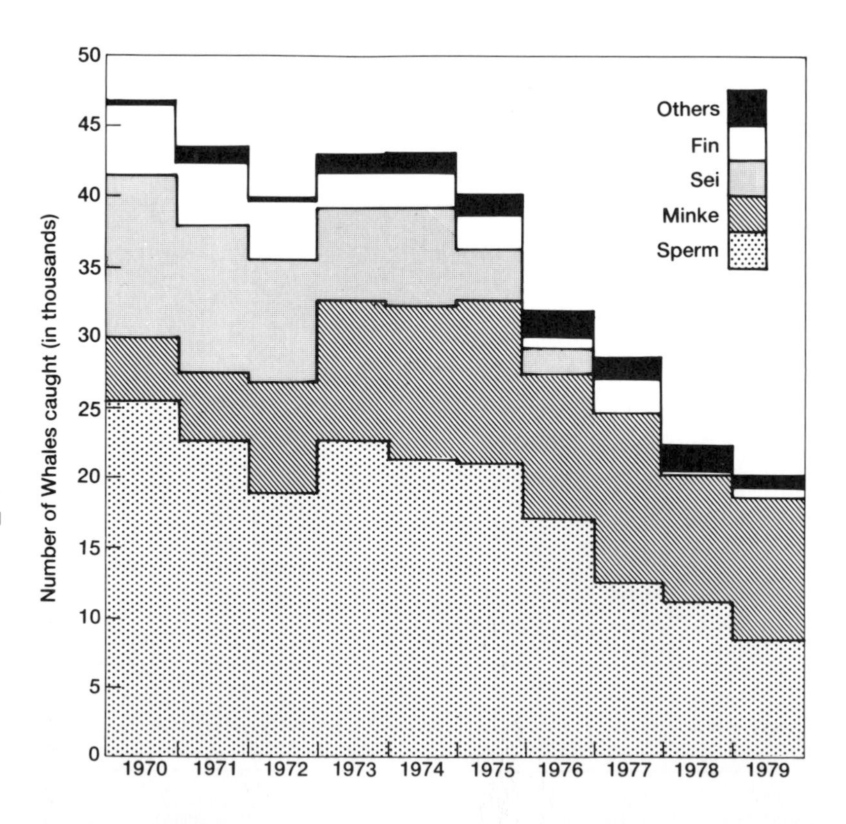

Figure 5.37 World catches of whales, 1970–79, in numbers of whales. These data start only after some groups had been much reduced by whaling. Most catches are declining in this period, but we can see the increase in the take of the minke group, the last major taxon to be exploited commercially. *Source*: M. Holdgate *et al., The World Environment 1972–82*. Dublin: Tycooly Press for UNEP, 1982.

Plate 5.27 A modern whaling ship hauling in a catch of minke whales. The minkes were the last group to come under commercial exploitation before only 'scientific' whaling was allowed in the 1980s. With stream power and modern navigational and detection aids, almost unimaginable inroads have been made on most whale species; other Cetaceans have suffered as well. (Philippa Scott/ International Centre for Conservation Education)

is made worse when the natural population is itself subject to considerable variations. These may have a detectable cause, as with the El Niño current off Peru which in some years prevents the upwelling of nutrient-rich water that is the basis for the anchoveta fishery, or may instead be due to unexplained cyclical fluctuations in fish populations. In this latter case, concepts of Maximum Sustained Yield (MSY) mean very little: the sustainable yield in fact fluctuates from year to year and few fishery organizations have the capability either to measure such a level or to enforce it. So maps of the fisheries of e.g. the North Atlantic (figure 5.36) show a series of dates when fish stocks have become overfished, and it is no coincidence that conficts between the UK and Iceland have been in the form of 'cod wars', and that fisheries policy has been one of the most contentious issues of all in the EEC. The North Atlantic is an obvious place for overfishing, being surrounded by DCs, but the much more isolated example of New Zealand can also be cited. Until that nation declared a 200-mile EEZ (Exclusive Economic Zone), unrestricted fishing was common in New Zealand waters, not only by locally based fishermen taking rock lobster and scallops for example but by deep-water trawlers from other nations as well (figure 5.38). Only the allocation of the resource to a nation by the declaration of an EEZ, and close supervision within that zone, has had any ameliorating effect.

It has proved possible, given the right kind of scientific knowledge, to manage fisheries to give some level of sustained yield: the North Pacific halibut fishery is a good example. Such cooperation and informed use might well be applied to new fisheries, such as Antarctic krill, and other fish of remote southern waters such as whitefish and southern blue whiting. The krill may provide an instructive example of the impact of contemporary technology. This small shrimp can now be caught at a rate of 500 t/day, with a total annual catch from Antarctic waters of about 0.5 million t/yr. If krill occupy a keystone position in the oceanic foodchains then a high exploitation rate (even if at a sustained-yield level, for which 100 million t/yr has been

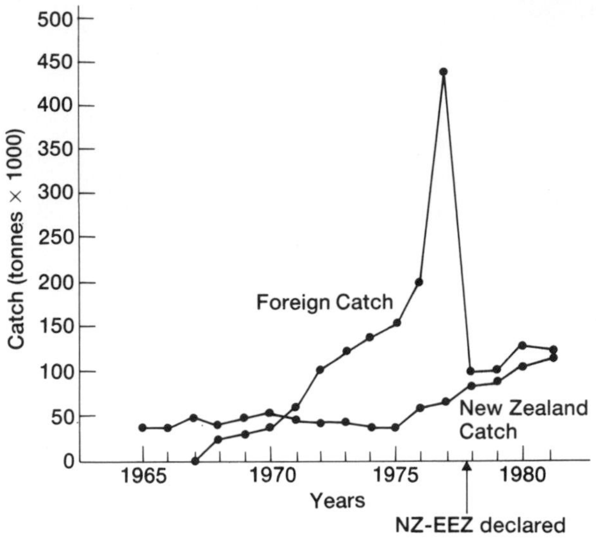

Figure 5.38 The effect of declaring an Exclusive Economic Zone (EEZ) in New Zealand waters upon domestic fish catch and on the use of the fishery by foreign trawlers, mainly from Japan, Korea and the USSR.
Source: G. Struik, 'Commercial fishing in New Zealand: an industry bent on extinction', *The Ecologist* 13, 1983, 323–47.

suggested) may have various unpredictable effects. The phytoplankton–krill–mammal chain might shift to a phytoplankton–copepod–fish chain; whale stocks might not recover even if whaling ceased; dependent populations of birds, seals, fish and squid might decline; and the krill population itself might not withstand the onslaught.

Overall, the world-wide fish catch seems to have stabilized at about 70×10^6 t/yr: the WCS argues that overfishing has prevented sustainable yields from rising to $85–90 \times 10^6$ t/yr but others reply that the southern oceans can yield another 10×10^6 t/yr without difficulty. It seems difficult though to envisage a standard fish catch of over 100×10^6 t/yr. The potential of other types of marine biota is difficult to assess: cephalopods and small crustacea (of which krill is an example) probably have the most potential: molluscs and mammals are very readily depleted if used commercially, witness the decline of abalone on the coast of California.

The living resources of the seas attract most attention but other materials are also extracted. Oil and natural gas may come from beneath the sea bed where the rigs and associated equipment may cause some environmental change: tidal scour may occur around rigs and equally some are said to attract shoals of fish. Flared-off methane attracts migratory birds, not usually to their advantage. Sea-water may be used as a source of fresh water and the result of distillation is a stream of hot brine which effectively douses life until the plume has diminished in concentration and temperature. Another widespread practice is marine mining, which is the discovery, extraction, refining and transport of ocean bed resources such as mineral ores, sand and gravel, and shell banks. Table 5.40 summarizes the effects of strip mining the sea floor: not all operations affect the ecosystem, but those which do have a very marked effect, especially upon the landforms of the sea bed and upon its fish populations. Where refining or sorting takes place at sea or immediately on shore, then another suite

Table 5.40 Detrimental environmental effects of mineral extraction of the sea bed

	Ecosystem components							
Effect	*Geomorphology*	*Water quality*	*Phytoplankton*	*Zooplankton*	*Benthos*	*Shellfish*	*Juvenile fish*	*Adult fish*
Changing the sea-bed landforms	0				0	0	0	+
Coastal erosion	0	0	0	+				
Translocation of biota					0	0	0	0
Change in particle size distribution		0	0	+	0	+	0	0
Oxygen depletion and release of sulphides		0	0	0	0	0	0	0

0 = measurable effect, direct
\+ = measurable effect, indirect

of effects has been observed, again with generally detrimental effects upon fish populations and plankton. It is possible, though, for enhanced nutrient levels to increase biological productivity provided that turbidity levels are not so high as to interfere with photosynthesis. Most such operations occur on continental shelves where biological productivity is high by oceanic standards, and thus contribute to the reduction of the renewable biological resources.

The reduction of the biological productivity of the seas is also likely to be the main outcome of the many uses of the ocean basins as a dump for unwanted materials and heat, collectively referred to as pollution. If some natural substance is present on the land then it will very likely end up in the seas, and this seems to be true of man-made materials as well. To list all the contaminants of the oceans is impossible as well as unreadable, and their spatial impact is highly variable: table 5.41 gives UNEP's assessment of the regional distribution of the main pollutants, and table 5.42 some details of the individual pollutants which present the greatest dangers to man and to other biota. These data are to some extent misleading, for although some of the substances are indeed globally distributed they are differentially concentrated: mercury for example. If we are forced into global assessments then the most that can be said is that certain groups of organisms are highly sensitive to the effects of toxins: monocellular algae, microzooplankton filter feeders and the developmental stages of benthic and nektonic organisms are the outstanding examples. The Russian scientist S. A. Patin suggests that large scale contamination of the oceans has caused a decline in the annual production of nekton (animals able to navigate at will rather than those carried on currents or tides) of about 20×10^6t. Compared with annual global fish landings of 60–70 $\times$ 10^6 t/yr, this is a

Table 5.41 Sources of pollution in various regions of the world ocean

Water discharge or other process of activity potentially causing contamination	*Baltic Sea*	*North Sea*	*Mediterranean*	*The Gulf*	*W. Africa*	*S. Africa*	*Indian Ocean*	*SE Asia*	*Japanese coasts*	*N. America*	*Caribbean*	*SW Atlantic*	*SE Pacific*	*Australia*	*NZ Coasts*
Sewage	X	X	X	X	X	X	X	X	X	X	X	X	X	X	X
Petroleum hydrocarbon (maritime transport)	X	X	X	X	X	X	X	X	X	X	X	X			
Petroleum hydrocarbon (exploration and exploitation)		X		X	X			X		X	X	X	X		
Petrochemical industry		X	X	X					X	X	X				
Mining			X					X		X			X	X	
Radioactive wastes	X	X	X				X		X	X		X			
Food and beverage processing	X	X	X			X				X	X	X	X	X	X
Metal industries		X	X		X				X	X		X			X
Chemical industries	X	X	X						X	X					
Pulp and paper manufacture	X				X					X			X	X	X
Agriculture runoff (pesticides and fertilizer)			X		X		X	X		X					X
Siltation from agriculture and coastal development						X	X	X			X				
Sea-salt extraction						X	X	X			X				
Thermal effluents						X	X		X	X	X	X			
Dumping of sewage sludge and dredge spoils		X							X	X					

Source: M. Holdgate et al., *The World Environment 1972–1982.* Dublin: Tycooly International, 1982.

significant amount. Most marine scientists agree that contamination of the oceans is far more significant at a local or regional level than globally and that solutions to various problems can be sought at those scales with greater chance of success than with major oceans like the South Pacific or the North Atlantic. The contamination of the Mediterranean has been such that all the fringing nations have concluded a Convention on clearing it up; as yet there is no agreement over the heavily used North Sea. Some allege that it is one of the most heavily polluted areas of the world, others that only 10 per cent of it can be so considered and that the interaction of environmental

Table 5.42 A ranked evaluation of major marine pollutants

Contaminant	*Hazard to biota*	*Hazard to human health*	*Hazard to amenity*	*Prevalence*
Sewage	1	1	1	Local
Pesticides	1	2	5	Global
Inorganic trade wastes	3	3 (exc. Hg)	5	Regional
Radioactive wastes	5	2	5	Global
Petroleum	2	2	1	Global
Organic chemicals	variable	2	variable	Regional
Military wastes	4	4	5	Local
Waste heat	1	5	5	Local
Detergents	4	5	1	Regional
Solid objects	3	5	2	Global
Inert solids and dredged spoil	2	5	2	Regional

1 require restrictive or preventative measures
2 merits evaluation for restriction or prevention
3 caution and further study required
4 apart from special situation, no restrictive action needed
5 no restrictive action needed
NB The fact that a distribution in col. 4 is 'Global' does not mean that there may not be local or regional concentrations of significance.

Source: R. I. Johnston (ed.), *Marine Pollution*. London: Academic Press, 1976. With additional material from S. A. Patin, *Pollution and the Biological Resources of the Oceans*. London: Butterworth Scientific, 1982.

pollution with fishing effort and natural variation in fish population make it impossible to say with accuracy whether the biological productivity has declined at all, let alone to segregate the input of wastes as the leading factor. As with many recent toxicants, of course, one problem is the unknown effects of long-lived substances which may accumulate in food chains or in sediments: to that extent there may be a number of 'time-bombs' ticking away in marine ecosystems.

A major problem of the oceanic environment has been that most of it belonged to no-one and therefore its use, whether for resource extraction or waste disposal, was in the hands of those technologically able to get at it, without much restriction. It was to some extent a global commons, with each unit of use benefitting the user but bringing harm to everybody and everything else, even though there were many informal sets of rules about fishing which conferred a type of 'ownership' on individuals or groups, together with a number of international agreements covering particular species. This period of an open frontier type of exploitation seems to be ending because so many operators recognize that it can have no future. Nevertheless, the conflict between the long-term perspectives and short-term profits is acute, and is shown particularly clearly by the long-term refusal of whaling nations to accept scientifically assessed catch limits and by the difficulties in the various attempts in the 1970s by the UN to bring up to date the Law of the Sea. Whatever perspective is adopted, nobody would now agree with the writer of the 'Harvest of the Seas' chapter of *Man's Role*, when he concluded that 'here at the beginning and the end is the great matrix that man can hardly sully and cannot appreciably despoil.'

Warfare and environment

As we noted in chapter 4, warfare is a deeply rooted habit of mankind, and our propensity to engage in it is scarcely getting smaller. There were, for instance, 133 armed conflicts between 1945 and 1981, involving 81 countries (mostly LDCs) and at least 12 of these caused significant environmental damage. Modern technology has greatly amplified the impact of warfare, though the effects so far are virtually insignificant compared with the potential of an exchange of nuclear weapons, discussed in the next chapter. The preparations for, and execution of, warfare have a tenacious hold on world economies: 5–6 per cent of world yearly oil consumption is military, as is 6.3 per cent of the consumption of aluminium and 11 per cent of copper. Some 20 per cent of the world's scientists and technologists are engaged in military research and development, and total global military expenditure has increased 30 times since 1900, meaning that current spending is about US$1 million per minute on arms.

The environmental impacts include the consequences of current and past wars, the preparation for war, and the hazards of possible future warfare must also be considered. Of past wars, there is usually very little environmental trace, with the exception of the potential of unexploded munitions. Except where deliberately preserved the land use pattern and landscape of Flanders do not reveal its condition during World War I as a morass of vast muddy pools which resulted from shelling, together with choked, overflowing streams turned into long stretches of mire (plate 5.28). The front line of battle was in fact a high-energy ecosystem, with the rapid transformation of many forms of energy into noise and heat, and a dense population of human beings, horses and lice, though not many other forms of life. World War II similarly left relatively few long-lasting traces in land use and landscape systems: at the time some 10 nations suffered losses of agricultural production up to 38 per cent, but it is now a fair bet they are among the overproducers of the EEC, and whereas Iwo Jima was reduced to a lunar landscape of volcanic dust in 1944, its landscape is now more or less normal again. Less visibly, seven metres of sediment accumulated in the ports of the Charente estuary of France as a result of the military operations of 1944 further upstream. However, one government, in Poland, which has already cleared some 14.5 million mines from its territory is still taking out 300–400,000 per year. Curiously enough, the technique which has produced one of the longest-lasting traces of hostility on the land is still used. The Great Wall of China and Hadrian's Wall in northern England were both designed to keep out barbarians; in recent times, Israel has put up bank and ditch defences along its northern flank (supplementing the traditional obstacles with modern electronics),

Plate 5.28 The concentration of fire-power of World War I (here at Ypres in 1917) produced immense environmental changes: 'A burnt space through ripe fields/A fair mouth's broken tooth', wrote Isaac Rosenberg (1890–1918). Much has been regraded and re-vegetated; the cemeteries are a permanent land use change. (Imperial War Museum)

and Morocco has done much the same in the Sahara to try and interdict Polisario guerrilla fighters.

It was perhaps the second Indo-China war of 1961–75 where non-nuclear weapon technology showed its greatest power. United States' General Westmoreland averred that 'technologically the Vietnam war has been a great success' and this meant *inter alia*, the delivery of 587 kg/ha of high explosive to South Vietnam, producing 350×10^6 craters, unevenly spread, displacing 3×10^9 m^3 of soil. (The USA expended 14.3×10^9 kg of munitions in this war, compared with 2.6×10^9 kg in Korea and 7.7×10^9 in World War II.) Some 325×10^3 ha (2 per cent) of the forest area of Vietnam was cleared mechanically, much of it either side of roads, destroying a great deal of forest. So much might have easily been predicted, though the 296 ha swathe of destruction from a single B-52 bomber mission was a new experience. What lingers in Indo-China is the effects of attempts to defoliate crops and forests by aerial spraying of herbicides. Some 72×10^6 litres of herbicides were dropped on an estimated 1.7×10^6 ha of South Vietnam, about 10 per cent of its area. The herbicides used were mixtures of defoliants: 2,4-D, 2,4,5-D, picloram and dimethyl arsenic acid. The first two contained the contaminant dioxin (TCDD) which is highly toxic to humans and a known teratogenic agent: between 170 and 500 kg of dioxin was actually dropped on Vietnam. The deforested areas have been invaded by tall grasses, especially *Imperata*, and the forest is very slow to recolonize since water tables are often high due to the lack of evapotranspiration. Mangroves suffered especially badly and recolonization is very sparse so that large areas of coast have little protection against storms. In all, some 14 per cent of the standing crop of timber of South Vietnam was destroyed and the rare kouprey (*Bos sauveti*) further endangered. Animals that increased in number were the malarial mosquitoes which were presented with immense new breeding grounds in water-filled craters, and tigers given to scavenging on corpses. Other technology employed in Vietnam included cloud-seeding (to try and flood the Ho Chi Minh trail) but the success or otherwise of the measure is disputed.

Vietnam remains the outstanding example of the deployment of a full arsenal of non-nuclear weapons. The boundary with nuclear warfare, it may be observed, is now getting more blurred in the sense that it is possible for a single aircraft carrying appropriate high explosive devices to deliver as much immediate devastation as a tactical nuclear weapon with a 1 kt warhead. Preparation for war involves, environmentally, the devotion of areas of land and sea to training and weapons testing. At the height of World War II, some 20 per cent of the land surface of the UK was devoted to that purpose, a figure which has now shrunk to 1.27 per cent. These uses are normally exclusive (though recreation may be allowed at weekends on some ranges) so may preserve biota and ecosystems which would otherwise be ploughed up or built upon. At Suffield, Alberta, the tank range preserves an area of virgin prairie grassland, and in Wiltshire, England, 2752 ha of limestone grassland and scrub have been neither ploughed, fertilized nor sprayed for decades and so are of considerable ecological interest (plate 5.29). Few studies have been made of the ecological impact of weapons testing and military training, but one exception is the careful analysis of energy flows at the White Sands missile range in New Mexico, USA. This work compares the natural primary and secondary production of the desert and mountain areas of the range with fossil fuel use for missile tests, construction and maintenance. It also charts the pathways of solid wastes, sewage, noise and air pollutants which may stress the natural biophysical systems and reduce energy flow through it. Figure 5.39 shows that gross primary production averages at 71.7×10^{11} kcal/year (30 PJ) for the range, secondary production is 5.3×10^{8} kcal/year (2.2 PJ) and that the total stresses are 0.747×10^{11} kcal/year (0.3 PJ), i.e. about 1.0 per cent of natural energy flow.

The study of energy flows at a much larger scale is naturally enough now a part of warfare itself: in World War II the German dependence upon coal and electricity was a vulnerability which could be exploited by trying to bomb 416 power plants in five main areas, and electricity was also at the heart of synthetic fuel plants which were concentrated in the Ruhr valley. If this interlinkage had been realized earlier, industrial production might have been retarded rather sooner, and at less cost. By contrast, Japan had a very decentralized energy network, with 78 per cent of the electricity coming from small hydropower plants, the largest of which generated

Plate 5.29 Some military activities preclude changes like ploughing and certainly reduce public access, thus creating a kind of nature reserve. In the UK there are few areas of Chalk grassland that have not been ploughed in modern times and these have a unique flora. Salisbury Plain Training Area, Wiltshire, England. (Cambridge University Collection of Air Photographs)

165 MW (figure 5.40). Thus no effort was made to bomb these plants nor the distribution system for electricity.

One last effect of war can be noted: there is nearly always a massive movement of people from the war zone. This migration may itself be environmentally significant, even if only temporarily, and the eventual area of settlement, having now a denser population, itself exerts more environmental impact. That population density may itself be the cause of war in the twentieth century is not normally thought to be true, though an attempt was made to depict it thus in the so-called 'soccer war' between El Salvador and Honduras in the 1970s. As W. Durham's study showed, however, its origins were more complex than simple Malthusian relationships. Causes apart, the effects were high in human suffering and our task here is to recognize that such traumas, there as elsewhere, may not be an end to the legacy of war but may involve the other components of the biosphere as well.

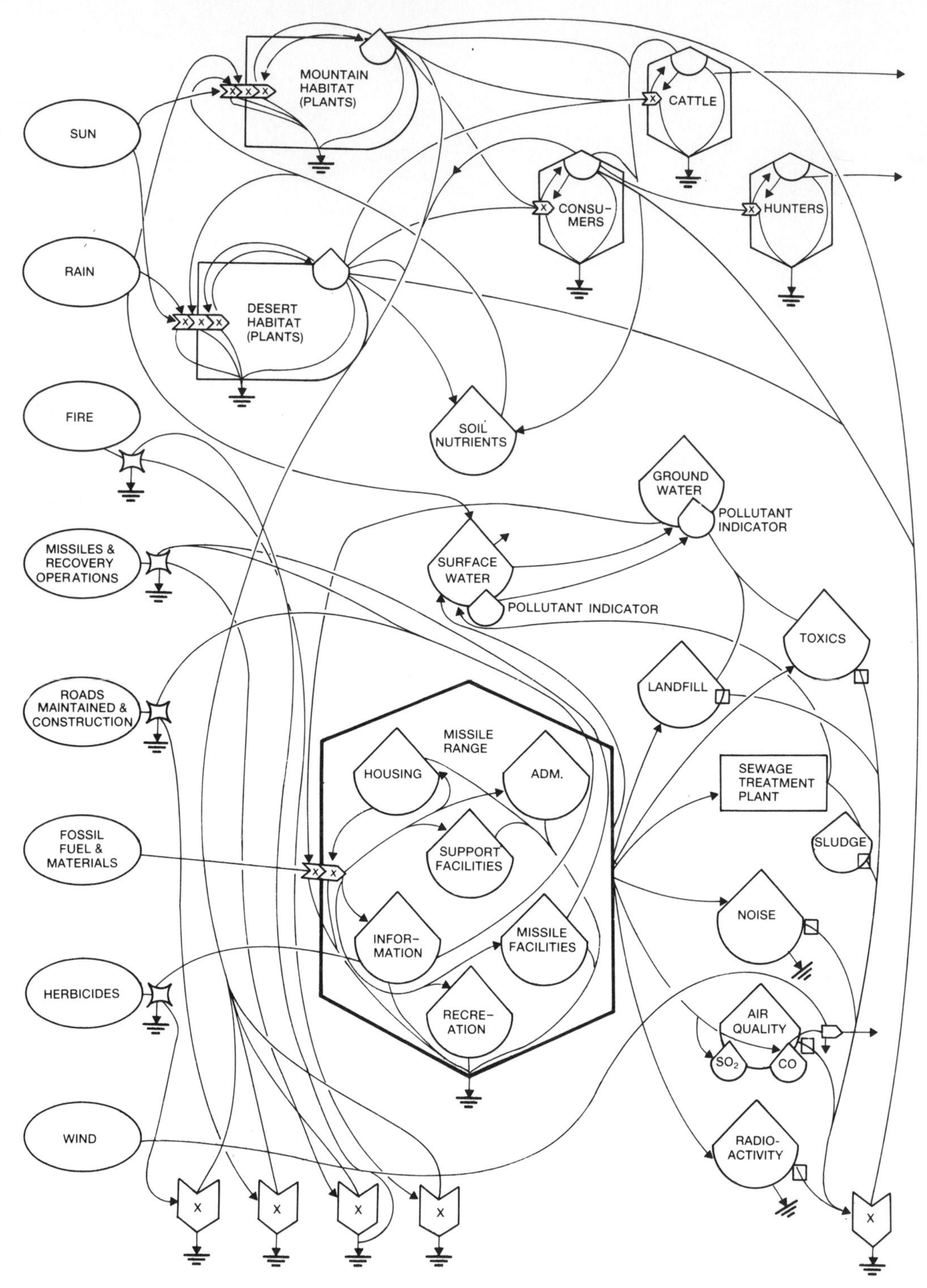

Figure 5.39 A systems diagram of that part of the New Mexico desert which houses the White Sands Missile range, showing the interactions present anyway and the impact of the weapons testing range and its activities.
Source: M. W. Gilliland and P. G. Risser, 'The use of systems diagrams for environmental impact assessment: procedures and an application', *Ecological Modelling* 3, 1977, 183–209.

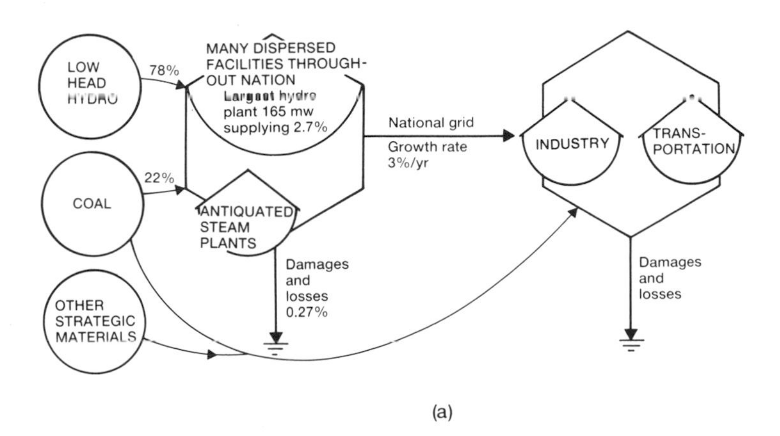

(a)

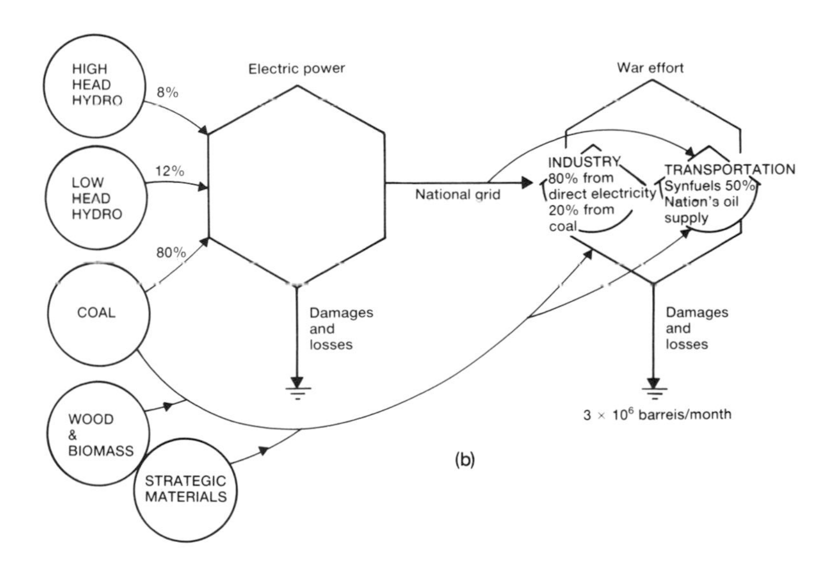

(b)

Figure 5.40 Simple systems diagrams of the energy generation and use systems of (a) Japan and (b) Germany during World War II. The dispersed system of Japan made it relatively invulnerable to attack at the generation stage (though industry and transport suffered heavy losses) whereas in Germany the larger plants made it possible for damage on a large scale to be inflicted on the generation plants as well as other parts of the system.
Source: A. D. Merriam and W. Clark, 'Energy and war – a survival strategy for both', in W. J. Mitsch *et al*. (eds) *Energy and Ecological Modelling*. Amsterdam: Elsevier Developments in Ecological Modelling 1, 1981, 797–825.

New environmental consequences of industrialism

Recreation and tourism

Travel, it is said, broadens the mind, and in *The Dry Salvages* no less a poet than T. S. Eliot admonishes us not to 'fare well,/But fare forward, voyagers'. Though not always perhaps in search of such lofty ideals, people have taken to leaving home for the day, for overnight trips and to go overseas as never before. An international tourist is defined as a temporary visitor to a country who stays more than 24 hours for pleasure or business, and the number of recorded arrivals grew from 140 million in 1967 to 286 million in 1980, with an expected growth rate of 4 per cent per annum. As one example, the number of tourists to Hawaii grew from 15 000 in 1964 to 3 million in 1974. Domestic tourism, crossing no national boundaries, runs at about four times the international rate, and day recreation exceeds that by many times. A very large share of the international tourism in the world takes place in Europe and North America; during the early 1980s the LDCs share was only 12.6 per cent. So the recreation phenomenon is largely associated with the rich and it is largely also an urban phenomenon in the sense that most travellers, even though they journey to remote places, seek urban levels of food, shelter and hygiene when they get there: much European sanitation is likely to fall short of the standards demanded by matrons from Eureka, SD.

Mass travel of pleasure-seekers as distinct from armies seems to date initially from the nineteenth-century advent of railway and steamship which brought down costs and made feasible conducted tours such as those arranged by Thomas Cook, and thereafter from the post-1945 period when increasing leisure time, disposable income and personal mobility created a wave of leisure activity. In its international manifestations it has been particularly aided by the cheap-to-operate jet aeroplane, where charter companies run package holidays using aircraft which are often superannuated from the front-line business carriers.

The environment, broadly interpreted, is an essential resource for tourism and in turn it receives many of the consequent impacts, planned and unplanned. At one extreme there is wilderness recreation where the natural environment, in as pristine a condition as possible, is the object of the vacation. To appreciate it properly, there should be little insulation between visitor and nature, and so foot travel with the minimum of equipment is prescribed, often with strict injunctions not to leave any wastes beyond the strictly physiological. Very fragile wildernesses may be visited only tangentially, as it were, as with tourists in Antarctica being housed in cruise ships and put ashore only at selected places where the influx of relatively few people is least likely to disturb the biota. At the other extreme there is the attraction of the totally man-made environment. This may be a cultural resource like one of the cities of the Italian renaissance, which has some kind of organic unity and historical continuity (Siena would be a good example), or an instant creation like Disneyland. Such complete conversions of environments are more common than we sometimes imagine; as well as imaginary worlds (the movie *Westworld* is an interesting commentary on western vacation-hedonism), we reconstruct historic worlds (like the Jorvik museum in York, England), and create desirable worlds from elsewhere as with the project to transform an ordinary island off Singapore into a proper 'desert island' from the South Seas. There is even a story that a US company tried to buy a small LDC in its entirety in order to run it as a vacation-land named Tarzania. In terms of preferences, however, no one environment has been more alluring than the 'sun, sand and sea' package (a fourth 's' is sometimes added), which has brought intensive tourism to established holiday areas like the Mediterranean (Plate 5.30), Madeira and Hawaii, and to formerly less accessible places like the Seychelles, the Indian Ocean coast of central Africa, Fiji, Samoa, Malaysia and The Gambia. In these and similar places the environmental relationships of tourism are highly visible because of their recent and rapid development.

In assessing the impact of tourism and recreation, or in planning for it, there are a number of forcing functions. There is first of all the necessary construction activity, whether of airports, hotels, roads or even the beaches themselves; secondly the influx of the visitors with their needs for food, water, something to do, and waste disposal; and thirdly the long-term maintenance of the system with its gradual spread of influence on adjacent land use and socio-economic systems. On-the-ground effects depend upon such factors as the intensity of the development, the resiliency of the local environment and the time perspective of the developer.

Plate 5.30 Many a littoral has been changed by the advent of mass tourism, building over the coastline and paving over the beach with people. This is part of the Costa Brava in Spain but could be many other parts of the world. (D & J McClurg/Bruce Coleman Ltd)

The idea of stress functions in producing environmental impact is the basis for the generalizations in table 5.43 which not only point to various categories of environmental impact but to some of the individual and collective responses of people to them. At a site level, figure 5.41 sets out the ways in which recreational or tourist pressure can affect particular ecosystems, both directly and indirectly. Taken together, these two bring some generalizations to the thousands of disparate examples of impact that have been chronicled. Vegetation and soils, for example, are often the first elements of an ecosystem to be affected. People may simply collect plants or by walking across a sward they gradually alter the vegetation in favour of those species that can resist trampling and that can grow in more compacted soils. Campgrounds in particular can become virtually devoid of plant life and have to be closed and given various treatments such as scarifying and mulching the topsoil. Trees are always vulnerable to root damage from soil compaction, to vandalism, and when in forests to fire caused by careless visitors. The more concentrated the people, the more intensive the changes, but even attempts to get away from mass recreation sites may cause equal damage: the off-road vehicle for example can cause significant damage in dunes and deserts since its tyres kill plants. The wind then produces blow-outs of the sandy areas. Winter usage of areas by people and vehicles (including the snowmobile) will compact snow and hence reduce its insulative properties so that shifts in plant species may result. Some animals flee the tourist and so become more rare; some become scarcer because of disruption of their behaviour patterns: just as wheeling vultures point to a carcass on the savannas of Africa, so wheeling minibuses point to a lion eating or sleeping. Rare birds may often be prevented from feeding because they are harried by bird-watchers, and ground-nesting species are vulnerable to tourists in places like the Galapagos Islands (Plate 5.31) and Antarctica. At a secondary level, animals may be illegally killed to provide souvenirs for the visitors, and others killed on the roads by heavy recreation traffic. Other species of animal may be attracted by visitors: those that feed on garbage for example. So insects, brown bears, jays, squirrels, yellowhammers and domesticated sheep all at various places scavenge through the tourists' detritus. Should these species carry pathogens (e.g. rabies) or be easily frightened (e.g. bears) then the stage is set for a less-than-happy holiday.

Not only species but whole habitats may be manipulated in the service of tourism. Great stretches of coastline have been built over completely; strings of hotels are often found in sand dune areas, utilizing the inter-dune slack as a garden but rarely controlling passage over the dunes to the sea. The the dunes break down and blown sand soon affects the paintwork as well as becoming a pervasive element in the bedrooms and the kitchens. If, in areas of seasonal or year-round aridity, water is pumped up for swimming pools and gardens then the freshwater table falls and is replaced by a saline wedge, often to the detriment of nearby small farms or

Table 5.43 A framework for the study of tourism and environmental stress

Stressor activities	*Stress*	*Primary environmental response*
1 Permanent environmental restructuring (a) Major construction activity urban expansion transport network tourist facilities marinas, ski-lifts, sea walls (b) Change in land use expansion of recreational lands	Restructuring of local environments expansion of built environments land taken out of primary production	Change in habitat Change in population of biological species Change in health and welfare of man Change in visual quality
2 Generation of waste residuals urbanization transportation	Pollution loadings emissions effluent discharges solid waste disposal noise (traffic, aircraft)	Change in quality of environmental media air water soil Health of biological organisms Health of humans
3 Tourist activities skiiing walking hunting trail-bike riding collecting	Trampling of vegetation and soils Destruction of species	Change in habitat Change in population of biological species
4 Effect on population dynamics	Population density (seasonal)	Congestion
Population growth		Demand for natural resources land and water energy

Source: D. Pearce, *Tourist Development*. London: Longman, 1981.

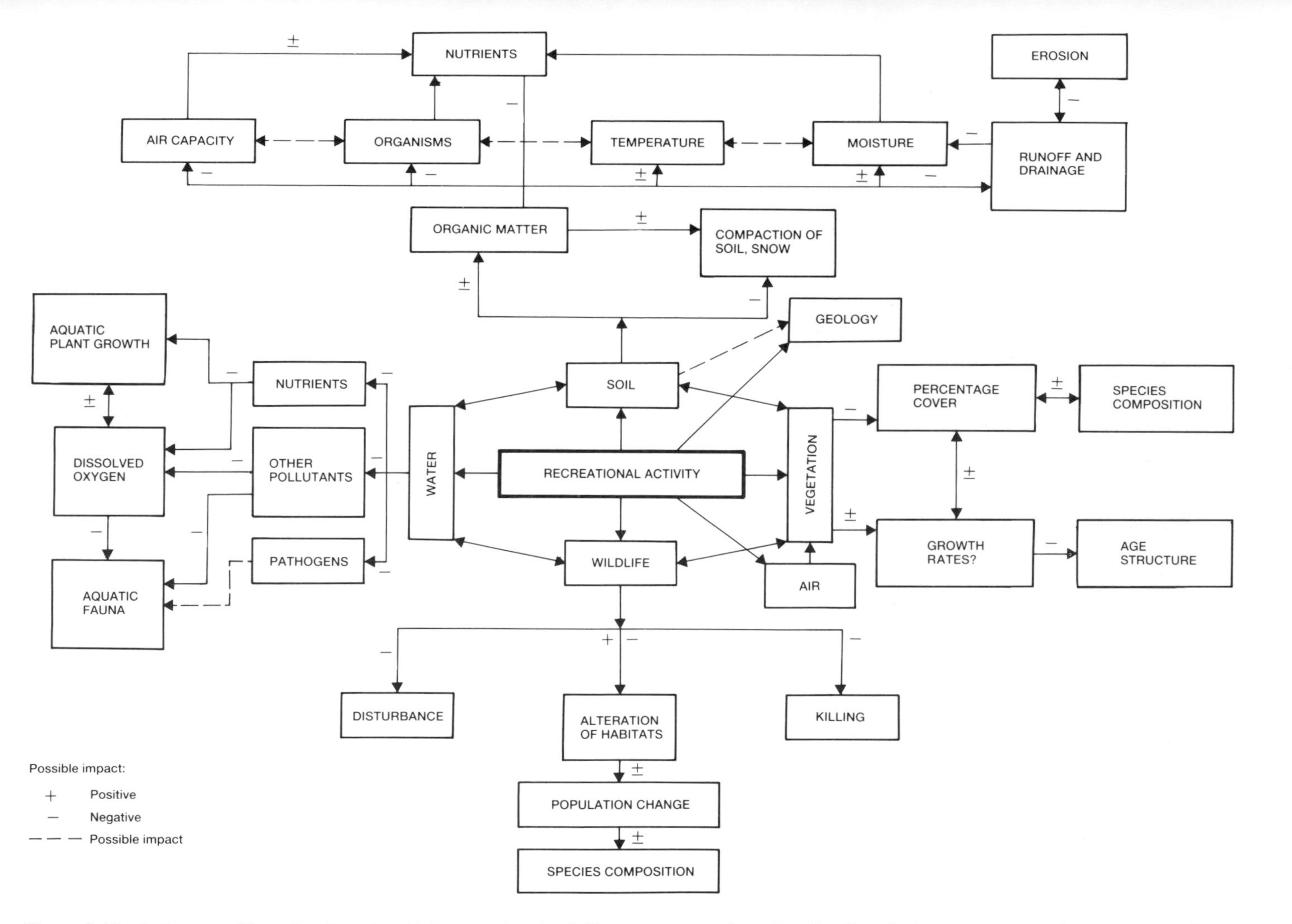

Figure 5.41 A diagram of four directions in which recreational activities can more or less directly affect the host systems: soil, water, vegetation and [animal] wildlife are treated. There are probably more linkages between the soil and the other systems than this type of diagram can portray. *Source*: A. Mathieson and G. Wall, *Tourism: Economic, Physical and Social Impacts*. London and New York: Longman, 1982.

Plate 5.31 Tourism reaches remote areas like Bartolone in the Galapagos where the ash landscape affords none of the pleasures of Pl. 5.30. Does whatever inspired Charles Darwin do the same for these latter-day beaglers? (Philip Steele: ICCE Photolibrary)

households unconnected to a mains supply. Any hints of erosion along cliffs are usually stamped upon by extensive concrete sea defences, irrespective of the wider geomorphological context, and it may in some places be thought essential to truck in sand to create a beach rather than leave the natural shingle or rock as it is, yet such creations do not usually last very long. Fragile habitats can suffer a great deal of damage quite quickly: coral reefs, for instance, can lose most of their species from trampling and by shell collecting (plate 5.32). If the latter is systematically carried out by merchants who sell mollusc shells to tourists (as in Papua New Guinea, for instance), then it is all the more severe. Mountain environments can be quick to suffer, too: the paraphernalia of downhill skiing can lead to considerable soil erosion in both constructional and operational phases, and in Nepal, trekkers have in places removed the last remaining vestiges of woodland in their search for fuel and cover.

The quality of air and water is seldom enhanced by mass recreation and tourism. Air contamination is exacerbated by airport development, for example, and attractions in places with atmospheric inversions can cause photochemical smog: when cars were permitted in Yosemite Valley in summer the smog levels were not that different from the Californian cities which were the source of so many of the visitors. Water is especially vulnerable to concentrations of people so that eutrophication from untreated sewage is common, and dry-season drawdowns of river to cater for additional water demand may exacerbate this problem. Sewage is perhaps the biggest problem of coastal recreation areas since rapid tourist development often saves costs by simply putting a pipe across the beach to low-water mark or, in more sophisticated developments, a few hundred metres out to sea. The aesthetics of water and beach contaminated by such materials if

Plate 5.32 Some environments are fragile enough to be damaged by even light pressures: this coral reef at Bali Barat in Indonesia has lost species and suffered fragmentation through the effects of tourists walking over it. (Sue Wells: ICCE Photolibrary)

washed towards the shore are not too good, but are insignificant beside the health hazards presented by pathogens from raw sewage, in spite of the purifying effects of salt water. Nevertheless, so long as industrialism is a viable economic mode, then the need to get away from it all, no matter what, is likely to perpetuate a strong demand to see faraway places with strange sounding names.

There have been attempts at management aimed at the amelioration of some of these problems. There may be general protective measures laid down by, for example, a national government, which might forbid building above say 2000 m above sea-level in mountain areas. There is the regulation of tourist development by a planning process which may often segregate tourist developments in a particular place. (This may be socially motivated, of course: some holiday brochures announce happily that 'this area is patrolled by security guards'.) Finally, there may be rationing of access to particularly scarce resources: museums and art galleries may close the doors when a certain number of people are inside; in the summer of 1987, the causeway into Venice was closed one weekend; and a permit is needed to enter some of the mountain wilderness areas of North America. The aim of sophisticated management is to recognize some kind of carrying capacity of the place for visitors and then to see that this is not exceeded. Carrying capacity is itself a difficult concept since it clearly has at least two components: that of the environment itself and the perception of it by the people. (A severely altered environment is not necessarily perceived as degraded by visitors.) Within a management plan, various measures are possible: people-control by zoning; the provision of walkways and surfaced paths, signboards and the like; and the provision of 'honeypots' to draw people away from vulnerable spots to ones which can absorb them with less damage. Indeed, it may be possible to provide alternative attractions which draw away people altogether: in both the UK and the USA visitor pressure on National Parks has been such that other types of recreational development have been nationally encouraged so as to try to leave the National Parks to those who appreciate their special qualities. (Not that there is much evidence that such an elitist approach has succeeded.) Most attempts at defining a carrying capacity are normally opposed by anybody making a profit from more visitors, no matter how destructive the extra influx appears to be.

Commentators generally reckon tourism to have brought more headaches than smiles, especially in an environmental context. But leisure is now as ingrained a habit as work in the West and could soon become a major use of time rather than a residual one. So sustainable systems of environmental management for tourism are not exactly a luxury either.

The environmental linkages of wastes

Human societies inevitably create wastes. In addition to those produced by the inhabitants' physiology, there are also the wastes which result from culture, whether this be of food collectors, food producers or industrialists. In truth, human societies have always produced wastes but these have become so pervasive since the nineteenth century that much more widespread notice has been taken of them. Certain effluents attracted unfavourable notice before the Industrial Revolution, but in three ways the position has radically changed in the last 200 years. The first of these concerns the sheer quantities. Not only has there been unprecedented population growth, but the range of materials acquired and the speed of their replacement by the wealthy have meant that resources are turned to garbage in greater amounts and at greater speeds than ever before, so that a DC city-dweller may now produce 1–2 kg/cap/day of solid wastes in the household, let alone those produced on his behalf in farm, factory, road and shop. (The equivalent in the Third World city is *c.* 0.5 kg/cap/day.) Secondly, the chemical nature of the wastes has changed. The success of science and technology has formulated many thousands of compounds and polymers which are unknown in the natural world. When these are led off into the biophysical systems of the planet, there may be no organisms which can break them down so their disappearance (plastics are a good example) is very slow. It may well not be possible or economically acceptable to determine in advance what effects such compounds may exert in their waste forms, as witness the length of the controversy over whether CFC aerosol propellants dissociate the ozone layer of the upper atmosphere and hence allow the transmission of more ultra-violet light. Thirdly, the concentrations of all kinds of wastes have increased, especially since industrial production became concentrated in factories which were located on or near energy sources like coalfields, and which in turn attracted large urban agglomerations. So while a river may cope with the new sewage of Farmer Giles and his family without much disturbance of the flora and fauna, it is unlikely to do so with the discharge of a city of 200 000 people, nor indeed with the effluent from a battery pig operation of 4000 porkers, if Farmer Giles chooses to intensify his operation. The tastes of the rich also have their impact upon the environment: table 5.44 compares the wastes produced in manufacturing pure white toilet paper compared with its unbleached equivalent.

Table 5.44 Wastes generated in the production of one ton of toilet paper

	Standard brightness (lb/ton)	*Unbleached* (lb/ton)
Gases		
Chlorine	1.2	0
Chlorine dioxide	0.6	0
Sulphium dioxide	5.6/20[a]	5.1/7.0[a]
Hydrogen sulphide	25.5	23.2
Particulates	57.5/1.0[a]	52.4/0.3[a]
Liquids		
Dissolved inorganic solids	263	22
Dissolved organic solids	244	41
Suspended organic solids	113	107
Suspended inorganic solids	4.5	4.1
BOD_5	147	41
Solids		
Inorganic solids	82	73.7
Organic solids	0	0

The output is an air-dry paper; the plant is fuelled by 1% sulphur fuel oil. In lines [a], the first figure is for production, the second for energy in plant use. It is assumed there are no constraints on discharges.

Source: A. V. Kneese and B. T. Bower, *Environmental Quality and Residuals Management*. Baltimore: John Hopkins Press, 1979.

The pre-industrial habit of regarding the environment as a convenient and largely bottomless sink for all kinds of wastes has persisted, often because it costs nothing in money terms to use (Plate 5.33). But the quantity, nature and concentration of wastes has in many places become so obtrusive that a cultural label is attached: the phenomenon of pollution. Informally, pollution can be regarded as something in the wrong place at the wrong time in the wrong quantity, or more formally, after M. Holdgate,

> The introduction by man into the environment of substances or energy liable to cause hazards to human health, harm to living resources and ecological systems, damage to structures or amenity, or interference with legitimate uses of the environment.

Pollution thus becomes a human increment to biogeochemical cycles, and follows various environmental pathways, with measurable effects upon people, other organisms, structures and ecosystems. The question of acceptability is also relevant (consider the tolerance of noise by different age-groups), so pollution has cultural as well as ecological dimensions.

Plate 5.33 Cleveland, Ohio, USA in the earlier years of this century, showing the 'smokestack' type of industry with its scant concern for the disposal of wastes in to the atmosphere. We like to think that there has been an improvement, but recent concerns like acid precipitation suggest that the environment is still regarded by many as a bottomless sink for wastes, free to users. (National Society for Clean Air)

Such is the complexity of technology that the compilation of complete lists of contaminants, their pathways and their effects cannot be attempted here: the specialist literature must be consulted. For simplicity it is customary either to classify the pollutants themselves (chemicals, particulates, metals, radioactivity, etc.), or the environments into which the wastes are led. Here, we shall adopt the latter at the cost of some repetition since some wastes are released into more than one compartment of the ecosphere. The core of the data is contained in table 5.45 which summarizes the main wastes affecting the atmosphere, the land, freshwater and the oceans, together with their biota. If there is a known effect on human health, this too is tabulated.

One feature not apparent from the table is the possible time-lag between the discharge of wastes and the detection of any environmental consequences. In the case of biocides, for example, there is often an immediate effect upon the target organisms, but the residual chemicals may not break down quickly and hence gradually accumulate in food chains. Perhaps a year or two later, these concentrations may induce a lethal effect in an organism far removed from the original target. Radioactivity, to give another instance, is subject to controlled release from nuclear power stations and reprocessing

Table 5.45 Major waste products and their receiving environments

Wastes	*Air*	*water*	*Fresh Oceans*	*Land*	*Clinical effects of residues on humans?*
Gases and associated particulate matter (e.g. SO_2, CO_2, CO, smoke, soot)	X	O	O	O	Yes
Photochemical compounds of exhaust gases	X	O	?	O	Yes
Urban–industrial solid wastes			X	X	No
Persistent inorganic residues, e.g. lead (Pb),	X	O	OX	X	Pb–probably
mercury (Hg),					Hg–definitely
cadmium (Cd)					Cd–definitely
Persistent organic compounds:					
Oil			X	O	No
Organochlorine residues	O	XO	X	X	Disputed
Pharmaceutical wastes		X	O		Unknown
Organic wastes:					
Sewage		X	X		Possible bacteria carrier
Fertilizer residues with N_2, P		O	O	X	Yes, specially N_2
Detergent with P		X	O		No
Radioactivity	X	O	X	O	Yes
Land dereliction				X	No
Heat	X	X	XO		No
Noise	X				Yes
Deliberate wasting, e.g. defoliation in CBW (chemical and biological warfare)		O	O	X	Yes

X – *Environments into which wastes are discharged*
O – *Environments into which wastes are transferred*

facilities but it may follow unusual pathways, e.g. from oceanic sediments or waters back onto the land via spray and thence from beaches through dust and sand into people's homes. Once there it is possible that cancers in humans may only become apparent after a latency period of many years. Nor does the table give any indication of the extent and seriousness of the problems in a global context as table 5.42 tried to do for the pollution of the oceans. 'Pollution on a global scale' is often seen as a book or chapter title but exactly what does it mean? Does it refer to substances which are found in most parts of the world but which are local in their observable effects, or to some truly global material, pervading every part of the ecosphere (figure 5.42) or at least every part of a major component of it?

A synoptic view of the world-wide emissions into water suggests that most of them are local or at most regional in their impacts. There are notable 'pollution hot spots' around the globe, such as the oil along the major tanker sea-lanes, the chemical alteration of shallow and enclosed seas (like the North Sea, Baltic and Mediterranean), the gross contamination of fresh waters like the Great Lakes and the Rhine, and possibly the quantity of radioactive effluent in the Irish Sea. But these are indeed regional and, moreover, have a high 'visibility'. At a wider scale, it is always possible that residual chlorinated hydrocarbons (mostly pesticides and polychlorinated biphenyls (PCBs)) in the oceans are causing global reductions in the NPP of phytoplankton but the detection of such a process at a planetary scale is not yet possible.

There seems no doubt that the atmosphere can be altered by human activities. Heat is released into the climatic system at power stations and points of end-use; particulate matter can act as condensation nuclei for cloud formation; and land use change can alter albedo. At regional and continental scales, the atmosphere is the vector of the substances, dominated by sulphur compounds, that fall to earth as 'acid deposition'. First recognized in England in 1661–2

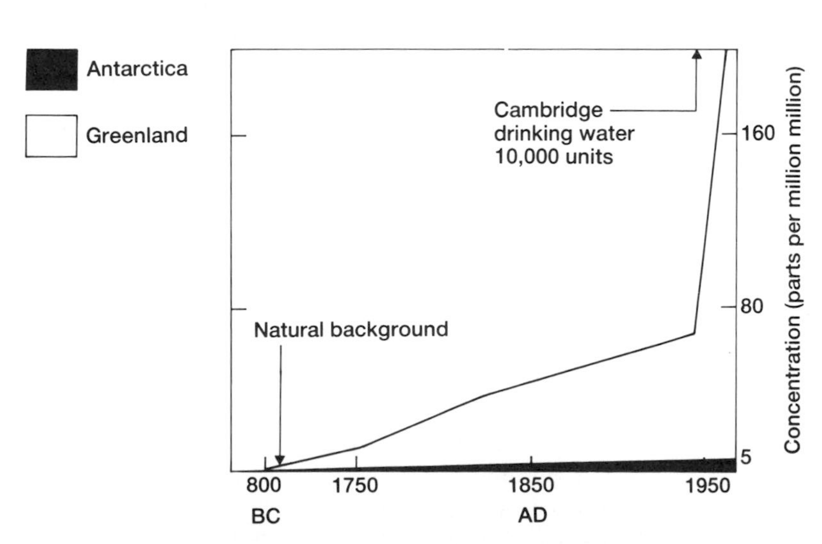

Figure 5.42 The growth of lead in the ice-sheets of both polar zones suggests that lead emitted into the atmosphere by human actions gets transported large distances in aerosol form but that more remains closer to the regions where the injections took place, i.e. the northern hemisphere. There appears to be little transfer between the two hemispheres.
Source: D. A. Peel, 'Is lead pollution of the atmosphere a global problem?' *Nature* 323, 1986, 200.

by John Evelyn and John Graunt, founder members of the Royal Society, this syndrome is held to be responsible for changes in the chemistry of surface water, decreased forest growth rates and high incidence of tree die-back, the leaching of toxic metals from soils into fresh water, and accelerated damage to materials. Cutting SO_2 emissions from power stations is a first step being taken in some nations to try and alleviate these problems. None of them, however, is apparently so important as the changes in the composition of the atmosphere brought about by the emission of substances such as chloroflurocarbons (CFCs), oxides of nitrogen, and carbon dioxide. CFCs are used as propellants in aerosol sprays, in refrigerants, as solvents in the electronics industry and as foaming agents in plastics; about 0.4 t/yr are emitted into the atmosphere and the rate of depletion by oxidation and uptake by the oceans is very slow, so that the residence times in the atmosphere are of the order of 50–100 years, and the concentration of CFCs is in the stratosphere is increasing at about 5 per cent per annum. In the stratosphere, UV radiation converts the CFCs to atomic chlorine and this is instrumental in breaking down the ozone (O_3) of this layer into oxygen. Depending on the models used and the level of future output, there could be a 3–12 per cent reduction in the ozone column during the next 70 years. Although detection of such changes is difficult, an ozone depletion of 2–3 per cent per decade was noticed in the 1970–80 period and more recently the existence of a spring-time ozone 'hole' over Antarctica (producing a 40 per cent reduction of O_3) has been confirmed. The effect of decreased ozone concentration is likely to have two major components. The first is to add to the global warming likely to be caused by the increased concentration of CO_2 and other 'greenhouse gases' (see below); the second is to allow the transmission to the surface of the earth of 2 per cent more UV-B radiation for every 1 per cent less ozone. Increased UV-B levels damage most living tissues; disruption of photosynthesis in phytoplankton and increases in human skin cancer are examples of likely effects and so in 1987 the first steps to international agreement to reduce CFC production were implemented. The significance of planktonic photosynthesis is further underlined by the hypothesis that the phytoplankton produce dimethylsulphide which is oxidized in the atmosphere to an aerosol. This aerosol is the source of cloud condensation nuclei over the oceans, so is critical in the albedo of clouds and hence in the earth's radiation budget. Biological regulation of climate is therefore possible via the physiology of phytoplankton, with all their sensitivities to human activity, especially waste products and increments in UV-B.

A distinctly global contaminant with consequences which would be genuinely world-wide is the rise in the concentration of CO_2 in the atmosphere. The global carbon cycle is affected by human activity in a number of ways. Combustion of fossil fuels adds CO_2 to the atmosphere, as does burning of forests and grasslands; some of this extra CO_2 is scavenged out by terrestrial vegetation and by phytoplankton but humans have also been responsible for clearing ecosystems with a high NPP such as forests. (Tropical deforestation may account for the release of $0.5–4.2 \times 10^{15}$ g/yr of carbon to the atmosphere, compared with 5.2×10^{15} g/yr from fossil fuels.) The net outcome seems to be that the atmospheric concentration of CO_2 has risen from 280 ppm in 1750 to a 1984 level of 345 ppm, which probably represents an increase of 1 ppm/yr in the last 20 years (figure 5.43). The significance of this trend lies in the property of carbon dioxide to enhance the transmission of incoming solar radiation and at the same time retard its radiation back to space: the so called 'greenhouse' effect. This effect can be exacerbated by other gases, notably the oxides of nitrogen coming from fertilizers and petrochemical combustion, and methane. Given various assumptions, CO_2 levels of 600 ppm might be reached by *c*. AD 2080. The problem at present becomes one of modelling whether and how much global warming is likely to result from the higher CO_2 concentrations. Rises in atmospheric temperatures of 1.5–3.0 °C consequent upon a doubling of the nineteenth-century level are suggested but there is no great agreement as yet on the climatic consequences of such an increase in global temperature. Polar regions would probably change most (a global rise of 3 °C might mean a rise of 7–10 °C in winter in the north polar region) with consequent effects upon global patterns of winds, precipitation and ocean currents. Added to this, the anti-transpirant effect of a doubling of CO_2 concentrations might increase streamflow by 40–60 per cent. Some models even suggest melting of part of the West Antarctica ice-sheet, that would give sea-level rises of 5–6 metres, on an uncertain time-scale. Though the exact values of the predictions are uncertain (and there is no necessary reason why any climatic response to increased CO_2 must be smooth since rapid changes from one quasi-stable state to another are in theory possible), it is clear that some areas of the world would benefit and others lose, although most would lose from rising sea-levels and all from a period of climatic instability; this latter would be harder to cope with than any gradual cooling or warming. Any action in the face of these possibilities (e.g. to limit CO_2 emissions, plant highly productive forests, encourage the development of nuclear power) is held back by the international agreements which would be necessary and by the uncertain nature of the actual outcome of the increases in CO_2 concentration. Additionally, the effects of carbon dioxide increases upon human societies can only be evaluated in the context of changes in population growth rates, energy and economic development, and technical metamorphosis.

Of all the difficulties produced for us by human alteration of the world, pollution has perhaps excited most attention and remedial action in the form of various control measures. Polluters have been made to pay for the costs which were formerly external to the economics of the processes, for example, and most developed countries have formidable batteries of anti-pollution legislation; there are too a growing number of international conventions on

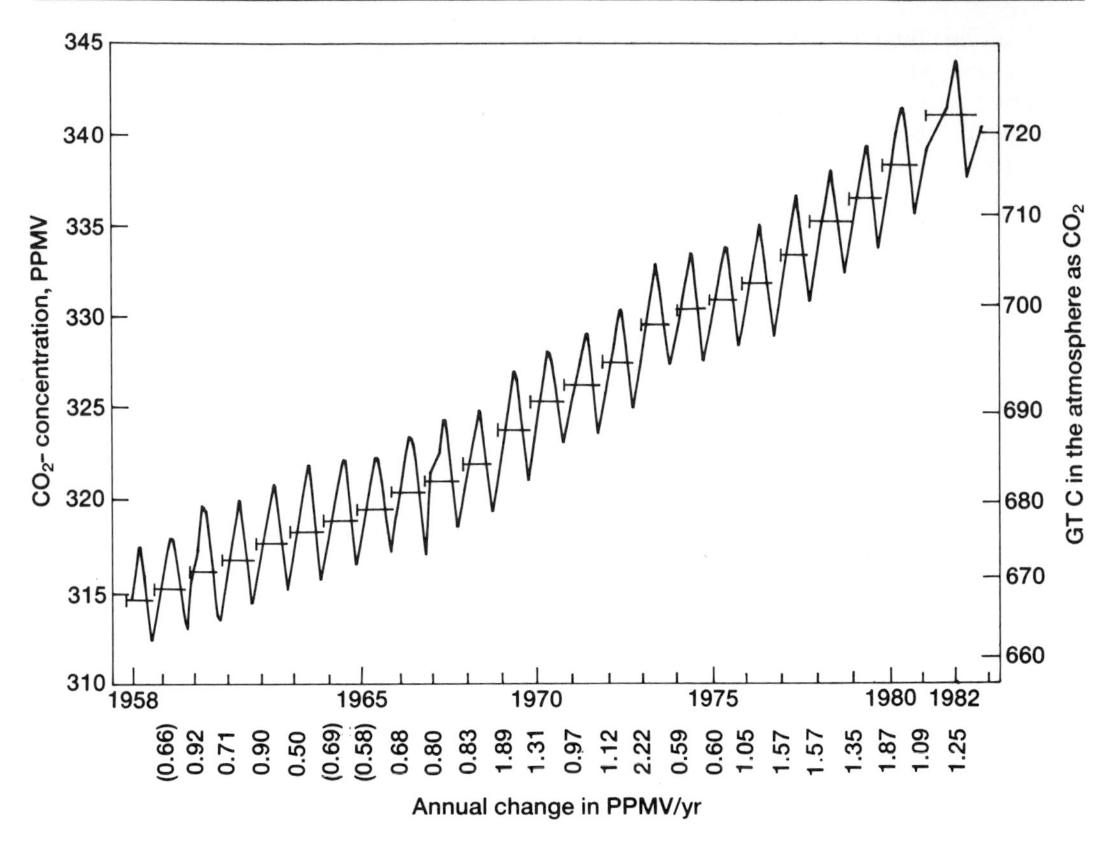

Figure 5.43 The trend in atmospheric carbon dioxide at Mauna Loa, Hawaii 1958–82. This curve is just part of an increasing concentration which is closely related to the quantities of fossil fuels burned and the removal of forest vegetation. Unlike the lead of fig. 5.42, this contaminant is truly global. *Source*: W. Bach, 'Carbon dioxide and climate: an update', *Progress in Physical Geography* 8, 1984, 82–93.

various aspects of pollution, e.g. for certain water bodies like the Mediterranean, and of certain substances such as radioactive isotopes. The technology for controlling contaminants is often available, but the economic and legal frameworks are rarely adequate and the political will is often distinctly feeble unless a short-term and highly visible effect upon people is concerned, as happened at Minamata. Rarer still is the consideration of the whole resource process of which waste production and emission are simply the last phase: if this process is unsustainable then mere clean-up measures resemble those in the story of the eighteenth-century proto-psychiatrist who had an infallible method of separating the mad from the sane. Those to be diagnosed were put in a locked room with a large number of water taps in it and a plentiful supply of mops and buckets. He then turned on all the taps and watched: the mad made for the mops and buckets, the sane turned off the taps.

Population, resources and environment in an industrialized world

The period since the middle of the nineteenth century can be seen as that of the great exponentials of growth. A central measure of these increases is that of the human population, rising from 1200 million in 1850 to 5000 million in 1987, with a growth rate in the latter year of 1.7 per cent p.a. which is equivalent to a doubling time of about 41 years. The take-off point of this surge from its previously much lower levels of increase seems to have been earlier, about AD 1650, but it is inconceivable that the Industrial Revolution and its consequences were not central factors thereafter, whether as cause, effect, or an iterative relationship of both. Most other measures that can be made of aggregate human activity also increased at high rates: GDP and GNP, for example, which quantify the sum of purchases of goods and services. Total energy use, which is a measure of the application of science and technology in a particular culture, has a similar upward trend. Not only have these indices risen as the population has surged upwards (table 5.46), in a linkage of a metabolic kind, but the per capita figures have risen as well, showing that globally the population has access to more materials and more energy than ever before.

Table 5.46 Growth in population and energy use, 1870–1970

Year	*World population* (millions)	*World use rate of industrial energy* (terawatts)	*Cumulative energy use since 1850* (terawatt-years)
1870	1300	0.2	3
1890	1500	0.5	10
1910	1700	1.1	25
1930	2000	2.0	55
1950	2500	2.9	100
1970	3600	7.1	200

1 terawatt (TW) = 10^{12} watts = 31.5×10^{18} J/yr = 14×10^{6} bbls/day of oil.

Source: J. P. Holdren, 'Energy and the global predicament: some elements of a sensible strategy'. In M. Levy and J. L. Robinson (eds), *Energy and Agriculture: their Interacting Futures. Policy Implications of Global Models.* Chur, Switzerland: Harwood Academic Press for UNU, 1984, 91–3.

As we have seen for agriculture, however, there was a divergence during the nineteenth century between those who were, and have stayed, high-level users of materials and energy (the DCs) and those who have become lower-level users of the new technologies and their driving forces (the LDCs). This gap persists, though there are some nations, and regions within nations, that have crossed the divide. As we have seen in this chapter, all this growth has taken a lot of natural resources to build and sustain it and consequently there has been an immense amount of manipulation of the environment in order to extract the resources and to accommodate the wastes that have been produced. (Plate 5.34) This manipulation has been most evident in the DCs and in the LDC areas under direct DC control (whether as

(a)

(b)

Plate 5.34 In temperate zones, industrialization has demonstrated how flexible is the land resource and its ecosystems. Here within a few kilometres of each other in England, are various environments differentiated out of the same pre-agricultural matrix of deciduous forest: (a) remnants of the original woodland on the lower slope; modern coniferous afforestation on the higher ground; (b) heathland, subject to some grazing by domesticated animals but easily reclaimed for agriculture or if unused rapidly reverting to woodland as in the background; (c) land similar to (b) under improved pasture and planted woodland; (d) arable crops are juxtaposed to a cokeworks and urban development. The cokeworks is now closed and the land being used for light industry. (Photos: Author)

(c)

(d)

colonies, or neo-colonies ruled by transnational companies), but is by no means confined to them.

Economic growth has been facilitated, above all, by movement up the scale of concentration of energy sources. The hunter–gatherer's 26 km^2 can be replaced by a barrel of oil and a few buckets of coal and, as we shall see, a few grammes of uranium so it becomes profitable in all senses to use energy to get energy: to invest some oil in opening up a huge open-cast coal mine, for example, or to use coal and energy to fabricate drilling rigs that open up vast quantities of oil. The ratios have always been positive, with several units of energy yield for every unit of energy input. Dependence upon cheap and easily convertible sources of energy (including perhaps food energy) is now so ingrained in the DCs that forecasts of shortfalls in the supply of oil, gas and coal are treated with considerable seriousness and alternatives sought (chapter 6) with an earnestness accorded to few other human problems.

The effects of all these processes have defied proper description since it is so difficult to give a synoptic picture of the whole. A few general points may none the less be made. We can note, for instance, the ubiquity of the effects of industrialization: the ocean deeps contain the remains of sunken iron ships which are only slowly biodegradable as the pictures of the remains of the *Titanic* have shown, together with barrels of radioactive waste which are intended to be scarcely biodegradable at all. Mount Everest and similar peaks are littered with the cast-off hardware of climbing expeditions, and the South Pole is host to a research station. The ice-caps of the northern hemisphere, if drilled and cored, show fall-out from industrialization in the form of particulate matter and aerosol lead from the combustion of hydrocarbons (figure 5.42), and many organisms remote from land contain traces of persistent biocides in their body fat. And, quietly, the gaseous composition of the global atmosphere is being changed by the addition of carbon dioxide and oxides of nitrogen. Ubiquity is therefore not in doubt, and almost equally universal today is the distribution of high intensities of manipulation. Such is the portability of energy, especially in the form of machines like diesel-engined bulldozers, that the massive changes to landform and land cover that can be made in order to build a road in the remoter parts of Zanskar are equal to those for a similar project in Bavaria or Alberta. Large-scale irrigation projects in India or Sudan are as manipulative of the pre-existing ecology as is intensive agribusiness in Belgium or in Iowa. Again, a large mineral working on Bougainville is environmentally as pervasive as one in Poland or Arizona. No corner of the world need be exempt from the potential impact of energy-using heavy technology, so that it becomes possible to talk of anthropogeomorphology, the artificial creation of landforms equal in magnitude to those made by natural processes. Probably the most widespread of these forms is land subsidence, which is reported alike from Venice, Osaka, Mexico, Lake Maracaibo, Pittsburgh and Long Beach. But subsistence is only one of a set of land management problems whose results are at a geomorphological scale rather than simply at the level of biota or soils: the clearing of forests, pastoralism, agriculture, mining, transport, river management, shoreline management and settlement can all produce effects on landforms. Some estimates for the quantities of soil, subsoil and rock materials mobilized (in million tonnes/year) include 800 from forest clearing, 50 000 from pastoralism, 156 800 from tillage, 5000 from mining and 10 000 from quarrying, 400 from roads and railways and 17 from urban–industrial construction. Globally, agriculture leads the field with 156 800 million t/yr, followed by mining at 15 000 million t/yr and urban–industrial building and construction at 417 milion t/yr. The total volume for the mid-1970s is thus 172×10^9 t/yr, which we may compare with the annual silt load of the Huang Ho of around 1.5×10^9 t/yr. Given that the impacts occur in different environments, it is not surprising that some countries are more at risk than others from a potentially catastrophic level of impact: D. Nir suggests that ten nations have received 'considerable harm' from man-made landform creation: Algeria, Morocco, Tunisia, Tanzania, Zambia, Iran, Israel, India, Nepal and Syria. At the other end of the scale, 'a small danger only of man's adverse geomorphological actions' exists in Korea, Thailand, Hungary, Poland, Yugoslavia and the USSR; it seems safest to be a mid-latitude socialist.

A further set of effects of the introduction of high-energy developments, especially into rural areas, is the de-localization of the economy. Populations which were dependent upon resources from nearby for their food and transport energy have been suddenly transformed by contact with high-energy systems such as mining, oil production, and hydropower projects, with their associated air and motor vehicle transport systems. These developments throw open the links of the local economy, force rising costs, encourage emigration by the young, change diets, introduce economic middlemen and facilitate tourism with its attendant cultural juxtapositions. So the whole local ecological system becomes linked to a much wider national network and perhaps indeed to the whole globe, especially if tourism is involved. It follows that any ecological perspective cannot be confined to the boundaries of the local ecosystems: the concept of the local carrying capacity no longer has much meaning even for a remote group of people if a mining company is paying some of its local labour in flour, sugar and free dental treatment. So assessment of local biophysical resources is only a part of their world: a study of the channels (social, economic, political and technical) through which the rest of the world can be brought to them is also essential.

It does no harm, though, to be reminded that the human use of energy (figure 5.44) is still on a small scale compared with that of nature. But our ability to appropriate and transform that part of it which becomes living matter is ever-growing. Table 5.47 summarizes the current intervention in the globe's NPP: this comes to 60.1 Pg

(1 Pg = 10^{15} g) of organic matter out of a total NPP of 224.5 Pg. But the impact of human activity falls most heavily on the land so that the appropriation is 58.1 Pg out of a total NPP of 149.6 Pg, i.e. 39 per cent. So the remaining fraction is left to sustain all other living species. To some, all such gradations present a humbling picture: they are a reminder that the human species, even with all its extra-

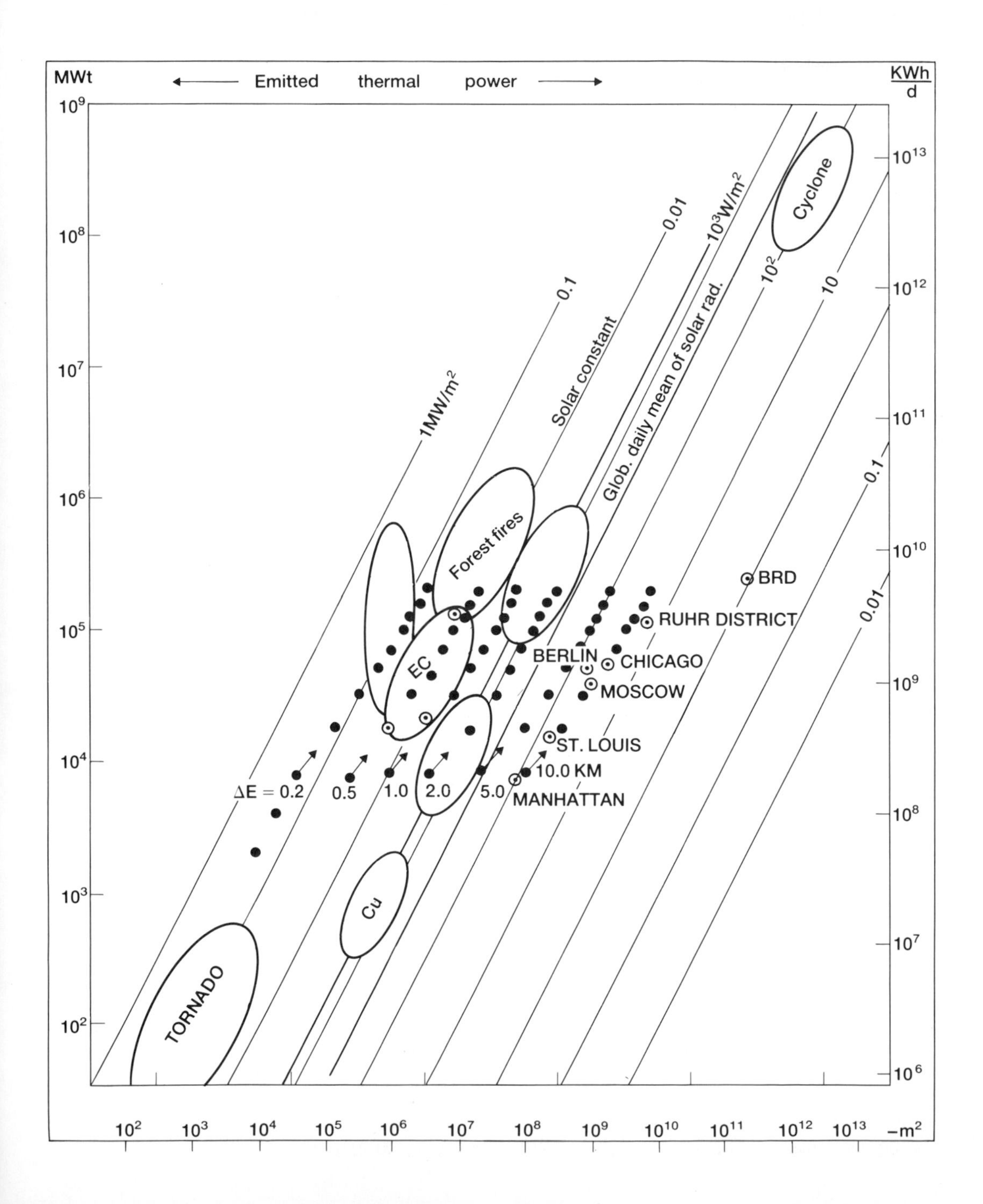

Table 5.47 Human appropriation of NPP, 1890s

	Organic matter in Pg (× 10^{15}g)
World NPP	
Terrestrial	132.1
Fresh water	0.8
Marine	91.6
Total	224.5
NPP Directly used by humans	
Plants eaten by humans	0.8
Plants eaten by domestic animals	2.2
Fish eaten by both	1.2
Wood used for timber or paper	1.2
Wood used for fire wood	1.00
Total	7.2
NPP Used or diverted by humans	
Cropland	15.0
Converted pastures	9.8
Others (cities, forest destruction, fish)	17.8
Total	42.6
NPP Used, diverted and reduced	
NPP Used or diverted	42.6
Reduced by conversion	17.5
Total	60.1

Source: J. M. Diamond, 'Human use of world resources', *Nature, London* 328, 1987, 479–80.

somatic technology, is still in some ways a minor feature of the planet in spite of all the changes of the kind summarized in table 5.48. To others, our subordinate position is a challenge to push forward to a point of greater control over natural processes, so that all of these biophysical entities can be remade in ways defined by human culture. Both points of view converge in a question about the industrial way of life: is it sustainable? After all, it is quite new compared with traditional agricultural systems which were the outcome of several thousand years of adaptations to natural and cultural circumstances. Is it then possible that industrialism is destined for a brief life, before there is an equally rapid reversion to a different economic mode like low-energy agriculture or hunting and gathering? Or is it more likely that the conversion of energy surpluses to knowledge will enable the human species to plan and adapt so that either industrialism becomes sustainable or an orderly transition to some other *genre de vie* becomes feasible? In these questions, the place of the supply, use and nature of energy supplies becomes a paramount question of public policy and interest. To one set of people, the fact that fossil fuels must by nature have a finite life, even if they are 'stretched' by technology and by the manipulation of demand, means that the obvious next stage is the large-scale development of civil nuclear energy, either by fission or fusion. To others, the societies that might be founded upon

Figure 5.44 The energy flux densities in the 1970s for some natural and some anthropogenetic phenomena, in terms of heat emitted. The vertical axis is a measure of the heat produced and over what period; the horizontal axis relates this to the spatial extent of the phenomenon. [Both axes are logarithmic.] Thus while cities are small compared with cyclones, some of them are achieving high levels of energy density. Compare with table 5.4.
Source: Simplified from H. G. Fortak, 'Entropy and climate', in W. Bach *et al*. (eds) *Man's Impact on Climate*. Amsterdam: Elsevier Developments in Atmospheric Science 10, 1979, 1–14 at p. 13.

Table 5.48 Major ecosystem characteristics as affected by industrialization: a summary

Characteristic	*Effects of industrialization*
Energy flow	Fundamental addition to the solar and gravitationally induced flows by the addition of fossil fuel (and later nuclear) based power via medium of metal-based technology. Steam power and electricity assume great importance, oil becomes something of a world currency. Energy sources much more concentrated than in previous stages, and so less E has to be expended to get E itself. Heat burden of atmosphere is increased regionally because of quantity and concentration of E flux from e.g. cities and large power stations – cities become 'heat islands'. E availability from fossil and other high-technology sources not distributed evenly within or between societies and appears to be a measure of disabling at lower and upper ends of provisions, but enabling in the middle. Industrial energy is characteristically substituted for human and animal labour whenever feasible.
Nutrient cycling	E availability means that rock fertilizers and atmospheric nitrogen can be added by industrial means to many agroecosystems, e.g. as fertilizers. Thus production of crops can be raised manyfold. But excessive applications plus long-distance transport of crops mean cycles are no longer 'tight' except in 'subsistence' societies. So nutrients frequently exported off-site and not returned as organic matter, hence the need for fertilizers. Some societies still consciously try to keep up organic cycle and some revival of this as 'organic farming' in DCs.
Biological productivity	Much enhanced by use of industrial energy, scientific knowledge, chemicals like fertilizers and pesticides where DC and some LDC agriculture is stable and well-tended. Where pollution and obliteration of ecosystems is found then may be diminished severely on a long- or short-term basis. Some ecosystems suffering particularly, e.g. coral reefs and estuaries from chemicals in water, corals from choking by silt, forests from air pollution by emissions from vehicles and thermal power stations.
Population Dynamics (Non-human populations)	Immense change by extirpation of many species and subspecies: rate of extinction higher than ever before. Rates of introduction around the world also at highest, both deliberate transfer and accidental changes, especially microorganisms. Crop species bred with new accuracy for appropriate conditions (e.g. HYVs of rice and wheat); chemical technology enables immense onslaught on weeds and pests but often with unforeseen ecological consequences. Genetic engineering promises new leaps forward in tailoring of crop plants, and presumably animals as well eventually.
Successional stages	Great increase in earlier stages on world basis: (a) much more bare area created by pollution, waste land, ploughers land, cities; (b) early successional stages of ecosystems mimicked by agricultural crops, including rotational grassland. Consequent diminution of mature ecosystems in all parts of terrestrial globe. Rise of movement to protect parts of these as National Parks and similar areas.
Biotic diversity	May be locally enhanced by mosaic of land uses and structures (e.g. higher diversity of birds in suburbs of city built in desert area) but world diversity being eroded as extinction occur at various scales. Islands particularly at risk, predators at end of long food chains and, the principal focus of concern, the tropical moist forests since their diversity is not yet even properly recorded.
Stability	Many more systems than ever before are kept stable by human intervention, often made possible by fossil fuel energy, as was probably the initial destabilization. Feedback loops thus part of cultural decision-making framework rather than natural homeostasis. For first time global instabilities possible via increased concentration of CO_2 in atmosphere and/or depleted ozone, 'Nuclear winter' scenario might bring about virtual extinction of agriculture in N. hemisphere and perhaps beyond.
Degree of modification	Very little of world is unmodified: some remote parts of land surface and oceans may look pristine but may well have chemical traces (e.g. aerosol lead on Greenland ice-sheet) of human activity. Litter on top of Everest, radioactive waste canisters on ocean floors, research base at South Pole. 'Wilderness' areas often scenes of battle between those who wish to protect them and those who wish to change them in some way even if only (at first) for limited tourism. At other end, some places e.g. mid-town Manhattan or some gardens represent total replacement of nature by the man-made except for atmosphere and occasional hazards like flood or earthquake.

'alternative' energies such as solar power, wind and tidal energy biomass and geothermal energy, are more attractive. The exploration of the environmental linkages of these two types of energy source, and other possible consequences form a major theme in the next chapter.

CHAPTER SIX

The Nuclear Age

We knew the world would not be the same.

J. Robert Oppenheimer, 16 July 1945, quoted in L. Giovanetti and F. Freed, *The Decision to Drop the Bomb* (London: Methuen, 1967, p. 197)

Introduction

One phase at least of human history can be accurately dated: it started at 2.30 p.m. on 2 December 1942, in Chicago, when Enrico Fermi achieved the first controlled fission of atomic nuclei in a chain reaction. This combination of science and technology has led to two developments: the use and spread of atomic fission as a source of civilian electrical energy, and the propagation of atomic fission and fusion in the form of thermonuclear weapons (plate 6.1). In terms of the organizing principle of this book that event marks access to a new source of energy and the potential for different relationships between human societies and environment. But to appreciate the full range of today's changing technological spectrum, two other scientific–technological advances will be considered. These are microelectronics, notably the advent of the large-capacity digital computer; and biotechnology, especially the possibilities inherent in the manipulation of recombinant DNA, or 'genetic engineering' as it is often

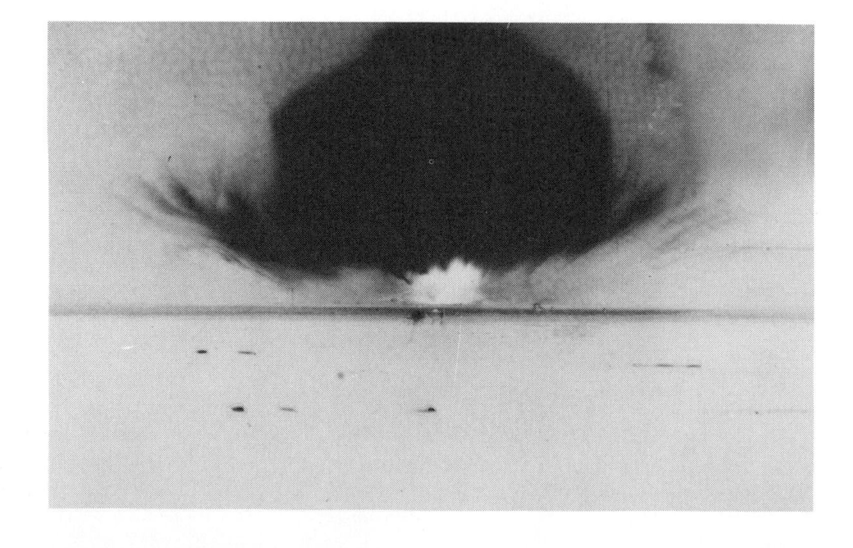

Plate 6.1 The environmental impact of this particular explosion was limited but it reminds us of the power that came into human hands on 16 July 1945. The Trinity explosion in the desert of New Mexico. (Los Alamos Photographic Laboratory: The American Institute of Physics)

called. Nuclear power is firmly entrenched but the other processes are so new that we have often to think in terms of potential and prospects rather than actuality, although some trends seem quite well established.

The new energy source

The discovery of radioactivity is credited to the French scientist Becquerel in 1896, and thereafter significant developments on the practical side included the isolation of polonium and radium from pitchblende by the Curies, and demonstrations of the fission of heavy nuclei by Rutherford, Bohr and others in the 1930s. In parallel, the theoretical work of Rutherford and Soddy on isotopes and Einstein on energy–mass relationships had confirmed the prevailing paradigm for nuclear physics. The technology of achieving chain fission reactions was greatly accelerated by the race to be the first to develop nuclear weapons during World War II; Enrico Fermi's success led to the successful test of an explosive device (plate 6.2) near Los Alamos in 1945, and soon after to the use of atomic bombs in August 1945 at Hiroshima and Nagasaki. The other way to gain access to nuclear energy is via the fusion of the nuclei of light elements and this too was developed for military use in the form of the 'hydrogen bomb', with the first successful explosions in 1953 and 1954. The civil use of atomic power has so far been entirely in the form of fission reactors, though the potential of nuclear fusion is such that a great deal of research and development is taking place in the leading centres of technology in the world.

The early development of civilian nuclear power as a source of electricity took place under the impetus of two concerns. There was the belief that it would be very cheap since the energy from 1 kg of uranium-235 was equivalent to 2500 t of coal, even though uranium-

Plate 6.2 Pioneers of atomic energy: (a) Enrico Fermi, who conducted the first controlled chain reaction in 1942, and (b) J. Robert Oppenheimer, director of the Los Alamos Laboratory during World War II and afterwards an opponent of fusion weapons. (Courtesy of AIP Niels Bohr Library)

235 (the fissionable isotope) forms only 0.7 per cent of natural uranium. Since most LDCs in the 1950s and 1960s were hoping to encourage manufacturing industry as the basis for development, they had hopes of the electricity being both cheap and ubiquitous. It was argued that the power stations would be quickly energy-positive, even though their construction and fuelling consumed a lot of energy and their lifetime was only about 25 years. Elsewhere, the feeling that fossil fuels were finite in supply for either technological, economic or political reasons, or at the very least likely to rise in price steeply, impelled many DCs towards the installation of nuclear generating capacity. The UK was first (in 1956) though the plant was a spinoff from the desire to produce plutonium for weapons use, and in that decade she was joined by the USA, USSR and France. The 1960s added West Germany, Canada, Italy, Sweden and Japan and by the end of 1985 there were 374 power stations in the world, generating 250 000 MW; another 157 plants were under construction bringing the generating capacity to 400 000 MW. In France, nearly 60 per cent of electricity generation comes from nuclear fission; about 30 per cent in West Germany and Japan, some 20 per cent in the UK and the OECD generally, and slightly less than that level in the USA and Canada. But the 1970s and 1980s have seen some slowing of the rate of ordering and installation so that forecasts from the early 1970s of a capacity of 2 million MW by AD 2000 now seem rather unrealistic. In fact many plants have performed poorly, the economics of them have looked increasingly precarious, and public acceptability has been dented by a number of accidents including a core melt-down at Chernobyl in the Ukraine in April 1986 which released a cloud of isotopes that raised radiation levels over much of central and northern Europe. This led, for example, to Iodine-131 levels in milk of 2000 Bq/l in Poland, 2900 Bq/l in Sweden, 1136 Bq/l in the UK and 6000 Bq/l in Italy; safety guidelines are mostly in the 1000–2000 Bq/l range.

All the commercial power has so far been generated by 'burner' reactors, whose fuel cycle consumes uranium; the availability and price of uranium ore are such that the next generation of reactors (currently at the development stage and proceeding relatively slowly) are increasingly likely to be 'breeder' reactors which produce more plutonium than they consume. Interest in fusion power is still strong although 25 years ago its commercial availability was forecast to be 30 years in the future, and the same lead time is still quoted today. There are many engineering problems to be tackled before commercial success can be contemplated, but if the energy turns out to be really inexpensive it holds out the promise of making anything from anything. Molecular bonds could cheaply be prised apart (granite in one end, fillet steak out the other, we may suppose), and if vanadium is used as a 'blanket' in the reactor, then radioactive wastes are much lower in volume and activity than with fission reactors. So fusion power is a great cornucopian hope of nuclear technologists.

Civil and miltiary uses share the process as far as the fabrication of either enriched uranium or plutonium (figure 6.1). In civil use the fissile uranium then goes through the rest of a 'fuel cycle'; in military use the end product is a rapid release of energy from plutonium fission which produces an explosion. The details of the civilian fuel cycle and its capacity for environmental interaction, and the actual and potential impacts of nuclear explosions, are given below in separate sections. Whereas we could debate whether nuclear electricity has merely reinforced the economic patterns set by fossil fuel usage or whether it may have introduced a different era, there can be no contradiction of the reality that nuclear weaponry can inflict damage and change environments with a force that outstrips its predecessors by many thousands of times; this material too is given a separate section for discussion of actuality and potential.

The civilian nuclear fuel cycle and its environmental relations

The flow of radioactive fuel and waste products is described in the first column of figure 6.1, where the starting point is the mining of uranium ore. The ore must then be milled to extract the uranium, of which there is *c*. 2 g per kilogram of ore. Enrichment (a highly energy-consumptive process) then follows in order to raise the concentration of the fissile isotope uranium-235. The enriched fuel is fabricated into fuel rods which are placed in the reactor and their carefully controlled fission produces heat which drives a turbine that generates electricity. The reactor's heat is conveyed to the turbine by a coolant, which may be water, heavy water, carbon dioxide or even liquid sodium metal. At the end of their useful life, the fuel rods are stored under water to cool and are then reprocessed so that any usable plutonium or uranium is recycled to an appropriate part of the flow. The fission of 1 tonne of fuel generates 5 m^3 of 'high-level' wastes. These are highly radioactive and also contain nitric acid, iron and organic solvents. So far no permanent resting place for this material has been found: it is stored in above-ground tanks. In the 1980s Hanford (Washington, USA) had 250 000 m^3 of this in store and Sellafield (Cumbria, England, plate 6.3) some 600 m^3, much of which was a by-product of the fabrication of weapons-grade plutonium rather than from civilian reprocessing. All these stages have to be tied together by transport: a nuclear power station requires 1–2 shipments per week of fuel, and 2-tonne loads of fuel are transported in 50 tonne casks. Eventually, the fission plant itself reaches the end of its life (after about 25 years) and has either to be dismantled or to be permanently entombed in concrete and sealed off from access virtually for ever, though no nuclear power station has yet come to this stage as part of a designed course of events.

The environmental links of each stage can now be considered. We need to think of two kinds of relationships: those which are planned, including the deliberate release of radioactivity, and those which are accidental, including possible incidents which might happen but which have so far been avoided. Although the nuclear industry claims

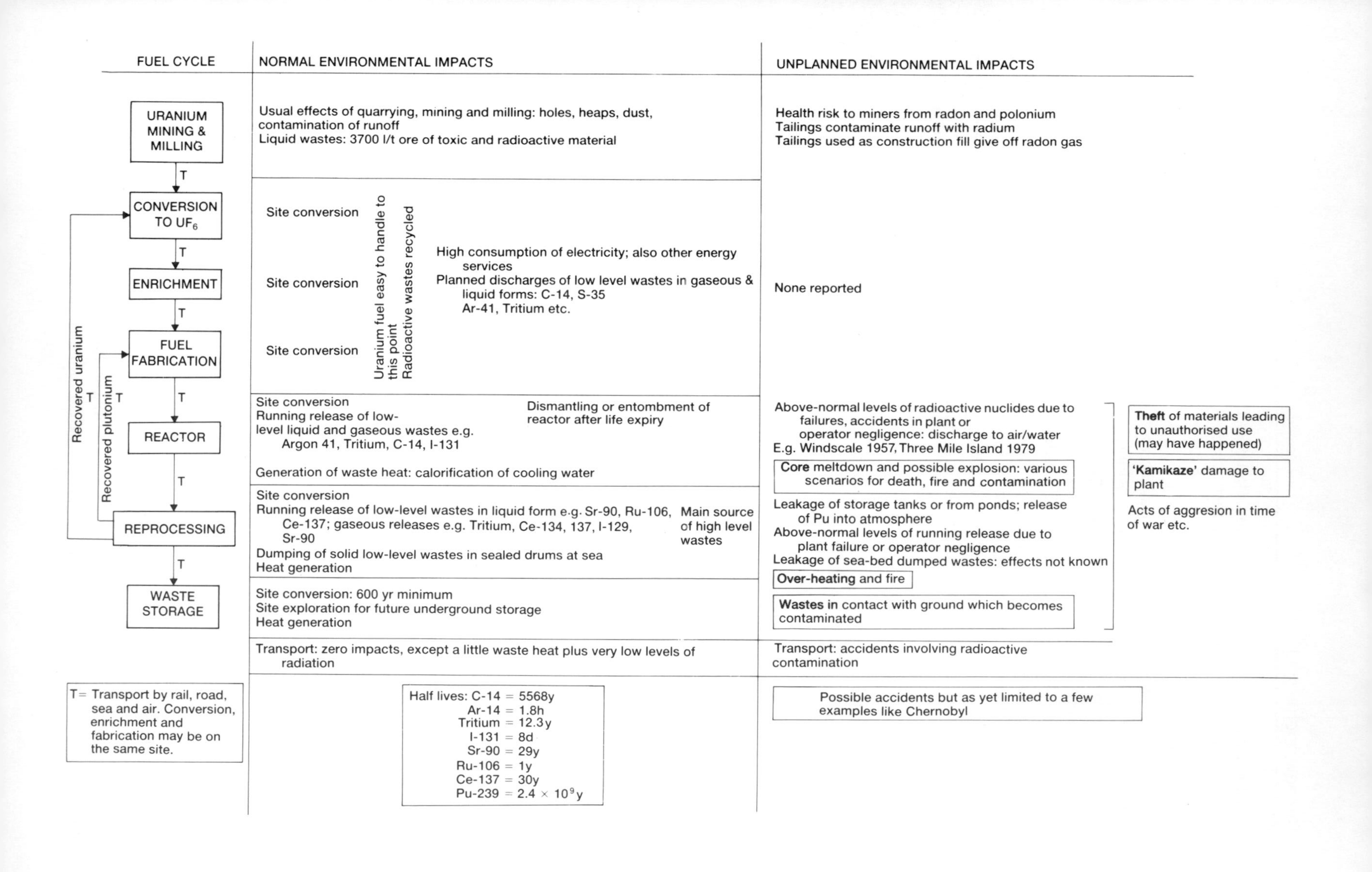
FUEL CYCLE
NORMAL ENVIRONMENTAL IMPACTS
UNPLANNED ENVIRONMENTAL IMPACTS
URANIUM MINING & MILLING
Usual effects of quarrying, mining and milling: holes, heaps, dust, contamination of runoff
Liquid wastes: 3700 l/t ore of toxic and radioactive material
Health risk to miners from radon and polonium
Tailings contaminate runoff with radium
Tailings used as construction fill give off radon gas
T
CONVERSION TO UF$_6$
Site conversion
T
ENRICHMENT
Site conversion
T
FUEL FABRICATION
Site conversion
Uranium fuel easy to handle to this point
Radioactive wastes recycled
High consumption of electricity; also other energy services
Planned discharges of low level wastes in gaseous & liquid forms: C-14, S-35 Ar-41, Tritium etc.
None reported
T
REACTOR
Site conversion
Running release of low-level liquid and gaseous wastes e.g. Argon 41, Tritium, C-14, I-131
Dismantling or entombment of reactor after life expiry
Generation of waste heat: calorification of cooling water
T
REPROCESSING
Site conversion
Running release of low-level wastes in liquid form e.g. Sr-90, Ru-106, Ce-137; gaseous releases e.g. Tritium, Ce-134, 137, I-129, Sr-90
Main source of high level wastes
Dumping of solid low-level wastes in sealed drums at sea
Heat generation
Recovered uranium
Recovered plutonium
T
T
T
WASTE STORAGE
Site conversion: 600 yr minimum
Site exploration for future underground storage
Heat generation
Above-normal levels of radioactive nuclides due to failures, accidents in plant or operator negligence: discharge to air/water
E.g. Windscale 1957, Three Mile Island 1979
Core meltdown and possible explosion: various scenarios for death, fire and contamination
Leakage of storage tanks or from ponds; release of Pu into atmosphere
Above-normal levels of running release due to plant failure or operator negligence
Leakage of sea-bed dumped wastes: effects not known
Over-heating and fire
Wastes in contact with ground which becomes contaminated
Theft of materials leading to unauthorised use (may have happened)
'Kamikaze' damage to plant
Acts of aggresion in time of war etc.
Transport: zero impacts, except a little waste heat plus very low levels of radiation
Transport: accidents involving radioactive contamination
T= Transport by rail, road, sea and air. Conversion, enrichment and fabrication may be on the same site.
Half lives: C-14 = 5568y
Ar-14 = 1.8h
Tritium = 12.3y
I-131 = 8d
Sr-90 = 29y
Ru-106 = 1y
Ce-137 = 30y
Pu-239 = 2.4 × 10^{9}y
Possible accidents but as yet limited to a few examples like Chernobyl

Plate 6.3 The nuclear complex at Sellafield in England, with both civilian and military uses and reprocessing of wastes. Controlled discharge into the Irish Sea of radionuclides is permitted. (Mike Hoggett: ICCE Photolibrary)

that the likelihood of large-scale accidents involving the release of radioactivity from reactors and reprocessing plants is very small, it is nevertheless governmental policy in most nations to site plants away from concentrations of human populations wherever possible. Inevitably this brings conflict with those who believe that these are the very places that should be kept free of industrialization: the more so since they are generally on rivers or coasts, where large volumes of cooling water can be cheaply obtained.

A summary of the environmental impacts of each stage is given in figure 6.1, and table 6.1 gives the units of measurement commonly used for radioactivity, and tables 6.2 to 6.5 give more detail for various parts of the fuel cycle. The major environmental concern in the nuclear energy industry is with wastes and these are of two kinds: non-radioactive and radioactive. The first category are not different in nature from other industries: tailing heaps which yield dust to the air on dry days and silt to the runoff when it rains are not unique to the nuclear industry. Similarly, heat generation at the power plant is

Table 6.1 Measures of radioactivity

Radioactivity	curie	Ci	3.7×10^{10} disintegrations per second
	bequerel[a]	Bq	1 disintegration per second
Absorbed dose	rad	rad	0.01 joule per kg
	gray[a]	Gy	1 joule per kg (= 100 rad)
Dose equivalent	rem	rem	rad × Q[b]
	sievert[a]	Sv	Gy × Q[b]

[a] SI unit
[b] Q = 'quality factor' of the radiation, i.e. its biological effect (Q = 1 for electrons, Q = 2 for fission neutrons and protons, Q = 20 for alpha particles and fission fragments)

Figure 6.1 (*left*) The civilian nuclear fuel cycle and its environmental linkages. These latter are divided into the planned impacts, including the permitted releases of radioactivity, and unplanned impacts, which include major accidents. Some of these have not yet been realized.

Table 6.2 Resource use per GWe/yr: pre-reactor fuel cycle

	Mining	*Milling*	*Conversion*	*Enrichment*	*Fabrication*
Land use					
Disturbed ha	34	3.8	0.12	0.15	0.023
Committed ha	1.1	3.2	9.3×10^3		
Water (m^3)		1.1×10^6	1.2×10^4	9.4×10^5	
Discharged to air					
Discharged to water			1.1×10^5	5.3×10^7	3.0×10^4
Discharged to ground	2.8×10^6				
Energy use (J)					
Coal				2.3×10^3 (840t)	
Gas		3.2×10^{14}	3.2×10^{13}		5.6×10^{12}
Oil	7.0×10^{13}			4.0×10^{11}	
Electricity	2.3×10^{13}	5.4×10^{13}	1.0×10^{13}	1.1×10^{15}	9.5×10^{12}

Source: J. M. Costello et al. 'A review of the environmental impact of mining and milling of radioactive ores, upgrading processes, and fabrication ofnuclear fuels'. In E. E. El-Hinnawi (ed.), *Nuclear Energy and the Environment*. Oxford: Pergamon Press, 1982, 15–51.

common to fission reactors and coal-burning power stations, and the land requirements are not very different: for an output of 1300 MW, a coal station might occupy 100 ha (excluding ash disposal area) and a nuclear plant some 60 ha. In bulk terms, 0.66 million t/yr of waste ash must be removed from a coal-fired plant, and about 20 loads/yr of fuels elements from a fission reactor; the corresponding inputs are 12 trains per day of coal and 1 lorry-load per fortnight of nuclear fuel.

The key difference is the production of radionuclides in the nuclear fuel cycle, with the various hazards they present to all living organisms, including humans. There are five main kinds of radioisotopes involved:

1 naturally occurring radionuclides rejected during the mining, milling and refining stages, including uranium, thorium, radium and radon gas;
2 heavy radionuclides formed in the fuel rods of the reactor from e.g. uranium. These are called actinides;
3 fission products: lighter radionuclides formed by the fission of uranium-235 , plutonium-239 or uranium-233 in the fuel rods;
4 solid radionuclides formed in the reactor structure by the absorption of neutrons, e.g. iron-59 and cobalt-60;
5 radionuclides formed by neutron absorption in and around the reactor: they can be liquid or gaseous, e.g. argon-41, carbon-14, and tritium (i.e. hydrogen-3).

Categories 1, 2, and 3 comprise the major hazards to humans and other animals, with 1 and 3 being potential public hazards rather than confined to those working in the plant. The nature of radioactivity is such that the particles undergo decay and so their 'impact' is measured in terms of the time it takes for 50 per cent of the radioactivity to dissipate: the half-life. That of argon-41, for example is 1.8 hr, of plutonium-239, 2.4×10^4 yr.

The main areas for environmental impact are at the beginning and the end of the fuel cycle. At the mining stage (the 'front end') of the cycle, the main hazard is in mining and milling wastes produced at a ratio of 1300 t of waste per tonne of uranium product. There is slow but constant leaching of radium-226 from this material into water (at the surface and below ground), and if milling wastes are used as landfill for construction, radon gas continues to be given off into whatever buildings are placed on the site. The next major area of concern is the reactor itself. Here, normal running produces low-level radioactive wastes that are emitted immediately, as are high-level wastes that must undergo reprocessing, and the plant itself becomes radioactive so that at the end of its life it must itself be treated as a radioactive waste and dismantled (either immediately or after an isolation period, table 6.5) or entombed in concrete. The total radioactive content of a dead AGR is expected to drop by a factor of about 10 in the first 10 years, by a factor of 4 in the following 20 years, but only by a further factor of 10 in the next 300 years. Steel contains 97 per cent of the radioactivity even though it is only half the mass of the plant (figure 6.2).

A hazard associated with reactors is that of accidental rather than planned releases of radioactivity: a number of accidents have occurred, of which Three Mile Island in Pennsylvania (USA) and Chernobyl (USSR) are the most notorious, but both the industry and

Figure 6.2 The decay time for the radioactive components of a nuclear power station. The total content drops by a factor of about 10 in the first 10 years, a factor of 4 in the following 20 years but only by a factor of 10 in the next 300 years. A graphite core has a half-life of 5700 years. The quantity of radiation will largely determine the timing and costs of the dismantling process and perhaps whether entombing in concrete is a better alternative.
Source: R. Milne, 'Breaking up is hard to do', *New Scientist* 112, 1986, 34–37.

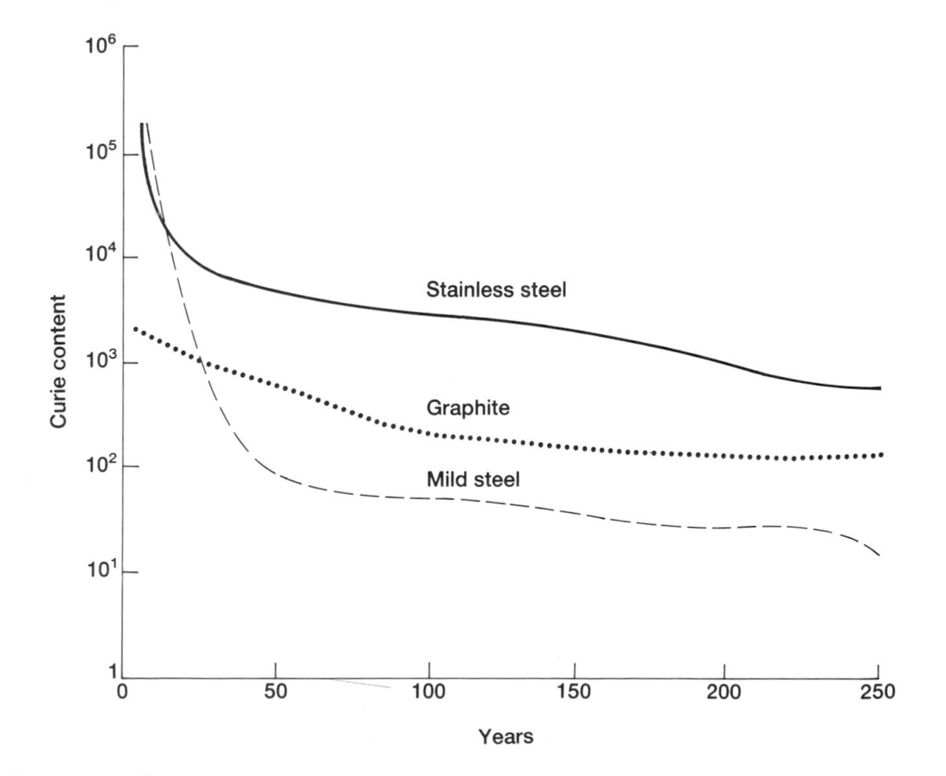

Table 6.3 Effluents discharged per GWe/yr: pre-reactor fuel cycle

	Mining	*Milling*	*Conversion*	*Enrichment*	*Fabrication*
Heat (J)	1.4×10^{14}	4.7×10^{14}	6.1×10^{13}	3.1×10^{15}	3.3×10^{13}
From coal equivalent of 66% of electricity[a] (tonnes)					
Sludge	210	500	100	1.0×10^4	84
SO_x to atmosphere	19	45	8.8	900	7.8
NO_x to atmosphere	15	35	7	730	6.0
CO to atmosphere	0.30	0.70	0.14	15	0.12
Particulates	0.93	2.1	0.43	42	0.38
Hydrocarbons	0.15	0.35	0.070	7.3	0.060
Plant effluent to atmosphere (tonnes)					
SO_x	21	0.14	8.5	12	–
NO_x	17	28	4.5	9.3	–
Particulates	2.5	1.4	0.14	0.55	8.8
CO	–	–	–	0.19	–
NH^3	–	–	0.12	–	–
Fluorides	–	–	0.070	0.035	7.8×10^3
Hydrocarbons	1.2	5.5	0.63	0.093	–
Radioactive effluent from plant (to atmosphere; TBq)					
Radon – 222	220	41	3.1×10^{-7}	–	–
Radium – 226	–	1.2×10^{-4}	9.3×10^{-8}	–	–
Uranium	–	4.9×10^{-3}	6.6×10^{-5}	3.6×10^{-5}	1.1×10^{-5}

[a] The source assumes that two thirds of the electrical requirements are derived from coal, the rest from nuclear power
Milling produces also 2.0×10^5 of tailings; there are 39 t of radioactive solids from the last 3 stages and small quantities of effluents to water, including uranium (C, E, F stages), thorium–230 (C), radium–226 (C)

Source: J. M. Costello et al., *op. cit.* (see table 6.2).

its opponents have calculated that there could be much more far-reaching effects from larger-scale breakdowns, including the MCA or Maximum Credible Accident. Resulting from a series of failures in the reactor, this usually assumes that a hydrogen fire induces the complete melting of the reactor core and so it sinks downwards (the so-called 'China syndrome') with the venting of a large plume of radioactive particles for about 30 minutes. The effects of such an accident would vary with meteorological conditions but there might be 1000 early fatalities, 1000 latent fatalities a year from cancer, and the need to decontaminate 26×10^3 km^2 of land from deposited caesium. A breeder reactor which experienced vaporization of 5 per cent of its core might cause 300 early deaths, and *c.* 4000 cancer deaths within 30 years. The nuclear industry calculates, however, that all such accidents are highly improbable: an accident causing more than 10 early fatalities is assigned a probability of 5×10^{-7} per reactor year, and with over 100 fatalities, 1×10^{-7} per reactor year. For comparison, the probability in the UK of death from all causes is 1.18×10^{-2} per year, from cancer 2.47×10^{-3}, road accidents 1.22×10^{-4} and fires 1.6×10^{-5}. These latter data are of course empirical, whereas the nuclear data are hypothetical.

Reprocessing plants extract re-usable uranium and plutonium and in the course of this, heat is given off and certain radionuclides are emitted to the air and to the sea or a large river. Accidents are not likely to be very large but might come about from loss of cooling, and if for instance 34×10^4 Ci of radiation were released into the air, evacuation of the local population and a ban on locally produced foodstuffs would be needed. Plant mismanagement may result in the planned release limits being exceeded, as happened at Sellafield in 1983, and for which British Nuclear Fuels were prosecuted in 1985. Planned release into the sea may result in the concentration of nuclides in sediments or food chains and constant vigilance is needed, since unpredictable concentrations may occur to the potential detriment of human and other forms of life.

Eventually (at the 'back end'), high-level wastes are formed from irradiated fuel and reprocessing wastes. These comprise contaminated metals and an acid solution containing fission products, actinides and unextracted plutonium. They are very hot and highly dangerous and so far they have been stored above ground. Active research and exploration is being undertaken for ways of ultimate (at least 600 yr and perhaps 244 000 yr, i.e. 10 half lives of plutonium-240, which for all practical purposes is for ever) disposal. The most favoured possibility seems to begin with sealing all the waste into an insoluble vitrified mass. Eventual disposal sites must be proof against mobilization of the wastes (especially by water), allow heat to escape (otherwise the radioactive containment will be ruptured), and be remote from earthquakes or volcanic activity. Homogeneous clays, crystalline rocks, salt strata and under the sea bed are currently favoured since the chief problem is the dispersal of radionuclides by

Table 6.4 Estimated discharge from a reprocessing plant

Type of discharge	*Quantity*
Atmospheric discharges	*Ci/yr*
krypton 85	1.4×10^7
tritium	1×10^4
carbon 14	600
iodine 129	0.015
ruthenium	0.4
strontium 90	0.1
alpha emitters (e.g. plutonium)	0.01
Liquid discharges	
caesium 134/137	1600
strontium 90	320
tritium	1×10^6
iodine	48
ruthenium	1500
zirconium/niobium 95	280
alpha emitters	360

Source: P. W. Mummery et al., 'The environmental impact of reprocessing'. In E. E. El-Hinnawi, *op. cit.* 1982 (see table 6.2), 139–67.

Table 6.5 Decommissioning data for a light water reactor

Contents		
Activated material	1190 m^3	
Contaminated material	16078 m^3	
Radioactive wastes	618 m^3	
Spent fuel	37 m^3	
Truck shipments	1363	
Rail shipments	28	
Process	*Immediate dismantlement*	*After 30 years*
Total cost (× 10^6 \$1978)	42	52
Time (years)	4	4.5
Occupational dose (man-rem)	1326	470
Public dose (man-rem)	22	3.5
Serious accidents, fatalities	0.097	0.107
Manpower (man-yr)	320	362

Source: E. E. El-Hinnawi, 'The environmental impacts of nuclear power plants'. In E. E. El-Hinnawi, *op. cit.* (see table 6.2), 52–72.

circulating water. Although there are considerable potential difficulties with this stage, it looks as if it may well be less hazardous to the environment than the inventories of fissile material at the surface, especially those of plutonium-239 where they are vulnerable to social threats from theft and terrorism.

Military nuclear power and the environment

The initial development of both nuclear fission and nuclear fusion was undertaken for military use, and the awesome amounts of energy that can be released in a few seconds from atomic and hydrogen bombs and shells have resulted in the acquisition of massive armouries of these weapons, principally under the control of the USA and USSR. Most of the discussion about their effects if used in anything but the most limited of exchanges has concentrated on the destruction of military bases, cities and civilian populations; here the emphasis is on the environment, but in no way is this intended to derogate the importance of people, their cities and their cultures.

The early examples of the use of nuclear weapons at Hiroshima (plate 6.4) and Nagasaki in August 1945 are not very instructive environmentally for today's conditions. The warheads were small (12.5 KT and 22 KT respectively), whereas today it is possible to make bombs of 58 MT or to deliver 350 KT as an 8-inch artillery shell. Apart from death and destruction, a meteorological effect was created in both cities in the form of heavy rain which resulted from moisture condensing around rising hot ash particles from the firestorms. The rain was black, oily and highly radioactive. At Hiroshima, there was also a 4-hour whirlwind at midday. There was little local fallout but leukemia has an incidence among survivors of 30 times the normal Japanese level and other neoplasms have higher

Plate 6.4 The impact of the first nuclear bomb: Hiroshima on 5 August, 1945. This was of course a small weapon by today's standards. (Imperial War Museum)

incidences. There is, however, an apparent absence of genetic damage. Various reasons have been put forward for this (and indeed the effects mey be yet to come), but all agree that the absence of observed genetic abnormalities does not preclude genetic damage. Fortunately many of the potential effects of large-scale nuclear warfare were not demonstrated since only two relatively low-yield bombs were exploded, in a social and environmental matrix unlikely to be characteristic of any future conflagration.

The arms race of the post-1945 period led to testing of nuclear devices on a considerable scale. Some of these tests were conducted underground but until the Partial Test Ban Treaty of 1963 the majority were above-ground. Numerous regional effects have come to light, though much information is still secret. In desert sites in the USA, for example, even early tests 'glazed' the land with melted and re-condensed sand. At a larger scale, the unexpectedly large yield of a device detonated at Bikini atoll in 1946 caused the middle part of the island to disappear entirely into a crater, thus exploding the word into the world's vocabulary, though this origin is not normally in the minds of either wearers or onlookers. Underground testing was carried out on Amchitka Island at the western end of the Aleutian chain in spite of worry, fortunately unfounded, that the circumpacific fault belt might undergo movement as a result. In many places, fall-out created regional problems because meteorological conditions were not accurately forecast, or because yield was underestimated. The Bikini islanders were evacuated temporarily to Rongerik and some returned in 1968, but tests in the 1970s show that land crabs and fresh water were still highly radioactive and they were moved out again in 1978. In 1954 a test contaminated a Japanese fishing vessel and its crew outside the designated test zone, and 236 Marshall

Islanders had to be hurriedly evacuated from Uterik, Rongerik and Rongelap atolls. The inhabitants of Rongelap were returned home in 1957 but the accumulation of radioactivity had by 1985 led the 258 islanders to request Greenpeace to re-evacuate them, to Ebadon at the northern end of Kwajalein atoll. Evidence of similar events from Nevada and Utah in the USA and from north Australia (scene of British testing) comes to light from time to time: a 1953 test of 43 KT at Yucca Flats, Nevada, deposited 35–70 Ci/km^2 of radioactivity around Troy and Albany, NY.

Radioactive debris from many of these tests was carried into the upper atmosphere and the circulation patterns ensured that much of it rained out at high latitudes, often into tundra. Here, the slow-growing lichens which accumulated radionuclides were the major food source for caribou and reindeer on which whole hunting populations in Scandinavia, Alaska, Canada and the USSR depended. Some isotopes were differentially concentrated in the animals' bone marrow, an item of food which was the prerogative of the paterfamilias. Some individuals thus built up body burdens of caesium-137 and strontium-90 which were 50–100 times higher than those of temperate zone inhabitants, although the total radiation doses fall within the range of variation of natural background activity. The fall-out rates have diminished considerably since the Treaty, although France and China continued atmospheric weapons testing after 1963. But the incidence of cancers amongst workers at military nuclear plants and nearby residents continues to cause concern. One provocative estimate avers that the 1985 rate of weapons production generated between 7000 and 15 000 victims yearly even without further nuclear testing.

This level of nuclear weapons testing, together with accidental releases of the kind experienced in the civilian sector, plus some major accidents, have demonstrated a number of concepts. There is firstly a reminder that some level of radioactivity is a normal feature of the planet's processes and that dilution of human-induced radioactivity may keep additional levels inside the 'natural' bounds. But such increments always add to, rather than replace, natural levels, and health physicists are uncertain about the meaning of threshold levels for the impact of radiation upon ecosystems and upon human health. Further, ecosystem processes can operate to concentrate apparently insignificant quantities of radioactive deposits within the compartments of an ecosystem. This is especially true of relatively simple systems with tightly conservative nutrient cycles, where the radioactive particles are channelled with the nutrient elements into a narrow array of species and a relatively small biomass. The Arctic provides a clear case, but in the oceans a scarce element like phosphorus will concentrate 2 million-fold over seawater in marine vertebrates, providing a pathway for a very high concentration of a shortlived radionuclide such as phosphorus-32, which locates in bone and is excreted very slowly.

In recent years, the continued build-up of stockpiles of nuclear weapons and the improvement of rock delivery vehicles has caused considerable speculation about the consequences of a large-scale exchange of nuclear weapons in the northern hemisphere. The energy released by a single warhead is now enough to destroy a very large city and to cause various effects far beyond the circle of immediate devastation (table 6.6). In the 1980s there has been available some 12 000 MT of yield carried on 17 000 warheads (the equivalent of 1 million Hiroshima bombs) plus some 30 000 low-yield tactical warheads. A full scale nuclear war might result in the detonation of 100–10 000 MT, depending upon the precise sequence of events, but a reasonable level for a scenario of consequences is 5000 MT, with 30 per cent of the total yield assigned to urban and industrial targets. There would be 10 400 explosions, most of which would be ground bursts on or near military targets.

Table 6.6 Immediate effects of nuclear weapons

	Ground-burst, within 24 hours of detonation: area of damage (ha)	
	20 kt *atomic bomb*	10 Mt *hydrogen bomb*
Cratering by blast wave	1	57
Vertebrates killed by blast wave	24	1540
All vegetation killed by radiation	43	121 000
Trees blown down by blast wave	362	52 500
Vertebrates killed by nuclear radiation	674	17 700
Dry vegetation ignited by thermal radiation	749	117 000
Vertebrates killed by thermal radiation	1000	150 000

(A circle with a diameter of 113m has an approx. area of 1 ha; a circle with a diameter of 1km has an approx. area of 78.5 ha. An area of damage of 150 000 ha if circular, would be 44km in diameter)

Source: J. P. Robinson, *The Effects of Weapons on Ecosystems*, Oxford: Pergamon, 1979, UNEP studies vol. 1.

The short-term consequences are not in doubt. Blast, fireballs, thermal radiation and radioactivity would simply destroy everything: blast alone will destroy all buildings within a radius of 7.5 km (1 MT) or 20 km (20 MT). A downwind plume of radioactive material will kill most living organisms outside this completely devastated area, some immediately and some later. Oil and gas wells would burn where extraction fields had been targeted, and nuclear power plants would contribute their quota of radioactive materials.

All this would combine to release huge quantities of smoke (perhaps 2.25×10^8 t from a 5000 MT exchange, of which 5 per cent would reach the stratosphere, which might then coalesce. Thus the sun might be blocked out and temperatures fall rapidly (figure 6.3). This is the 'nuclear winter' scenario and is of course a model. The immediate and longer-term biological and human consequences of such disastrous climatic changes are shown in table 6.7.

DAYS 5 - 10

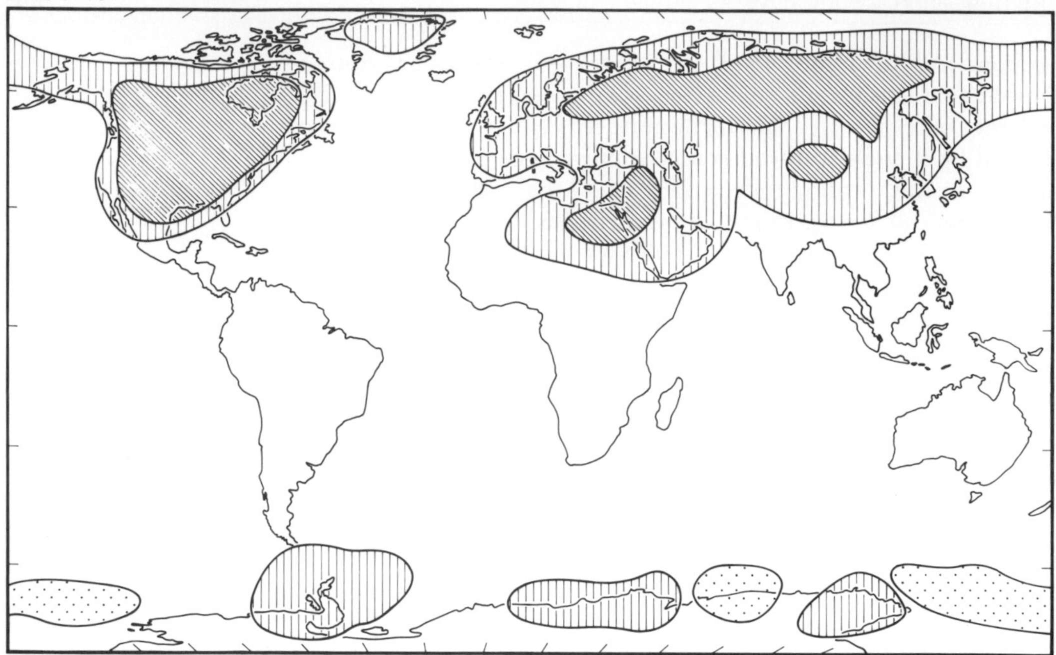

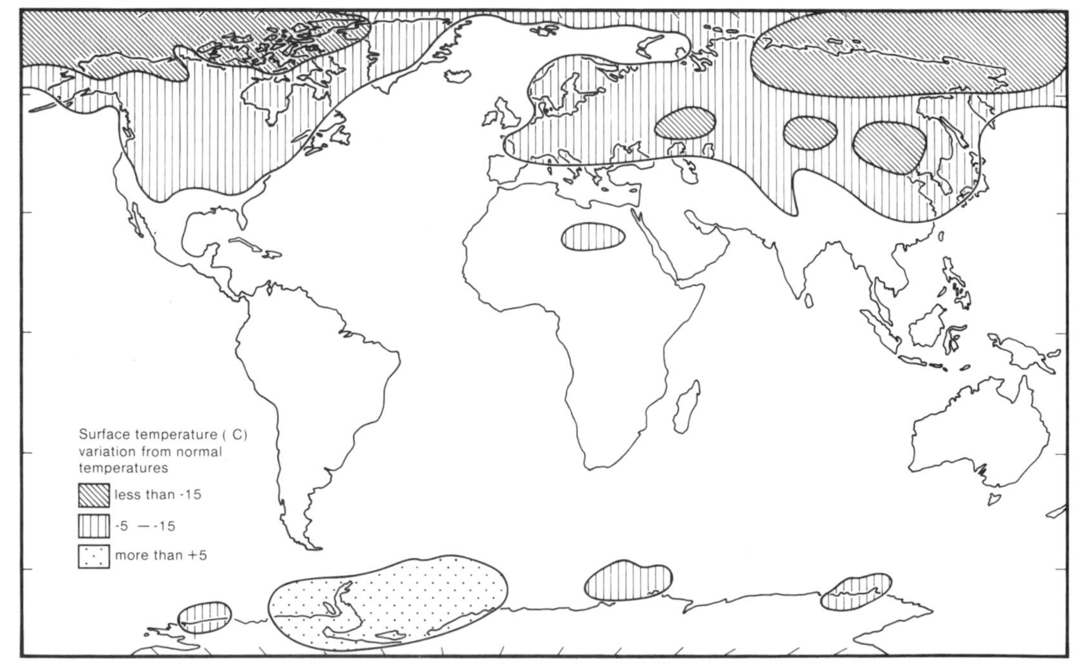

Figure 6.3 Simulations of the 'nuclear winter' produced by a 5000 MT exchange in the northern hemisphere during the summer, showing the drops in temperature over the continents caused by unscavenged clouds of soot in the stratosphere, after 5–10 days (top) and 35–40 days (bottom). Some models predict further coalesence of the clouds, beyond that implied in the lower map, keeping temperatures low; others that fallout of soot and dust will be rapid.
Source: L. Dotto: *Planet Earth in Jeopardy. Environmental Consequences of Nuclear War*. Chichester: Wiley, 1986.

Table 6.7 Potential Impacts of a nuclear war: effects of climatic changes

First few months	*End of first year*	*Next decade*
Extreme cold independent of season: severe damage to plants in Northern Hemisphere (NH) and tropics. Virtually no plant productivity. Animals stressed by darkness, cold and little free water. Storms at continental margins. These storms plus silt, toxic and radioactive loadings will eliminate support of many marine and freshwater animals.	*Terrestrial natural ecosystems* Plant productivity depressed, though hardy perennials and seeds would selectively survive. Animals would suffer competition for plant resources and many vertebrates would be extirpated. Tropical spp much at risk from low temperatures. As atmosphere clears, UV-B damages plants and impairs vision in many animals.	*Terrestrial natural ecosystems* Productivity recovers in context of irreversible damage to ecosystems. Homeostasis only slowly re-established: violent perturbations of populations more likely. Massive loss of spp, especially in tropics.
No net productivity of crops anywhere on earth; stored food destroyed, contaminated or quickly depleted. Survivors of immediate effects of explosions might be 50–75% of earth's population but would lose or suffer attenuation of all support systems such as food, energy, transport, medical care and communications, resulting in exposure, starvation, thirst, malnutrition and psychological stress, and very likely interpersonal and intergroup conflict.	*Aquatic natural ecosystems* Population collapses in many herbivore and carnivore species; benthic communities less disturbed. Many spp lost in freshwater systems because of period of freezing. Recovery in freshwaters quicker in temperate than tropical regions.	*Aquatic natural ecosystems* Recovery quicker than on land. Coastal marine ecosystems would begin to be harvestable though contamination by fallout still likely.
	Agroecosystems Crop productivity low: high UV-B rates depress NPP. Precipitation lower than prewar, soil loss high. No agricultural support systems and few surviving domestic (including draught) animals.	*Agroecosystems* Biotic potential restored: limiting factors for agriculture would be support systems of water, energy, fertilizers, pest control. 'Organic' agriculture only.
	Social systems With reduction of sunlight, eye damage from UV-B occurs. Psychological stress, radiation exposure and synegistic stress. Food likely to be a problem and so health effects widespread: epidemics and pandemics likely.	*Social systems* Human populations likely to remain severely depressed for a very long time; *Homo* might become extinct.

Source: Adapted and summarized from P. Ehrlich et al., 'Long-term biological consequences of nuclear war', *Science* 222, 1983, 1293–1300.

The result for the human population would be totally catastrophic in the northern hemisphere and possibly in the southern as well: it depends which version of the scenario is used. Even if these models are wrong by a factor of a hundred, they serve as a warning that nuclear war is a potential agent of environmental change of a highly undesirable kind.

Hi-sci, hi-tech

There are many known environmental consequences of the development of science and technology, but it seems appropriate to mention the potential for environmental change of a few rapidly growing developments. To continue where we have just left off, the military use of the discoveries of physics may not stop with nuclear weapons. Apart from the plans to develop space-based lasers and other 'Star Wars' weapons, there are possibilities for using geophysical hazards as weapons of war. During the war in Vietnam, there were largely unsuccessful attempts to flood the Ho Chi Minh trail by cloud-seeding, and the reverse of this could always be used to try and deprive a downwind enemy of precipitation. Various monomolecular films designed to decrease evaporation from bodies of stored water might perhaps be used at sea to steer hurricanes towards unfriendly shores. Seismic prediction and management, if more precisely achieved, could possibly be employed to produce earthquakes in foreign territories. Intense laser beams might burn precisely defined holes in the ozone layer so that ideological opponents could be subjected to intense UV radiation.

These are just possibilities and we may hope, though not perhaps with all that much optimism, that they never come to pass. In other fields we may look at some current developments in advanced science which are becoming usable. The whole world of microelectronics is doubtless replete with considerable environmental consequences, few of which are yet apparent with the important exception of the field of remote sensing. This capability is now mostly satellite-based and a great number of the earth's features can be systematically and synoptically plotted, from rocks and land use to cloud cover. Radar techniques are making it possible to sense the land surface (and for example its moisture content) through cloud. It is already the case that more data about the earth have been collected than anybody knows what to do with, even aided by very powerful computers. The analysis of these and other data on a very large scale will be aided by the so-called 'fifth generation' computers which will have sufficient power that they will appear to be able to 'think'. Since the complexity of the systems involved is one of the perplexities of environmental management, these computers may make possible rational environmental change on a scale hitherto unknown, especially if the predictability of the outcomes of various uses of resources and disposal of wastes can be greatly improved. It is perhaps a little unlikely that we shall see whole areas of landscape whose basic parameters are controlled by microprocessors, but not at all impossible to see say a river wholly managed in terms of rate of flow, dissolved salts, microbial activity and silt content, by an interactive computer linking a weather satellite with dams, extraction stations, groundwater wells, sewage works and the like. There may well be secondary environmental effects as the 'microelectronics revolution' brings about changes in settlement pattern and transport mode which result from different distributions of wealth, work and leisure. Some commentators forecast changes from an 'information revolution' as far-reaching as those which flowed from the Industrial Revolution of the nineteenth-century.

The rapidly growing area of biotechnology must not be ignored. This is not an entirely new development, for techniques such as brewing, the production of antibodies, sewage treatment and enzyme catalysis are well established, and their further elaboration will doubtless continue since there is great scope for their increased application in for example, the conservation of foods and fertilizers, in the scavenging of toxic substances from effluents and in the conversion of cellulose and other organic wastes to fuel feedstocks like ethanol, methane and hydrogen. But it is the recombination of the genetic material of organisms (so far mostly done with micro-organisms) which has excited the greatest interest in commercial, scientific and indeed ethical fields. In broad terms, foreign DNA can be introduced into another organism ('genetic engineering') in which it expresses itself and is capable of continued propagation. These organisms with new properties can be tailored and indeed some of them have been patented, so that a whole biological taxon can now be the property of an individual or corporation. Some of the most readily foreseen applications of these developments are set out in table 6.8. Some doubts about this type of work have been expressed since it is feared that engineered micro-organisms might 'get loose' and cause harm in the environment and to humans; in 1984 a US

Table 6.8 Some features of advanced biotechnology

Process	*Basis*	*Examples of applications*	*Possible environmental significance*
Gene splicing	Recombinant DNA used to insert a foreign gene into an organism, so that it is expressed more abundantly than native genes. Used on bacteria, yeast	Production of vaccines, interferon (A). Control mechanisms on DNA (P)	Unknown
Monoclonal antibodies	Long-lived cloned cell lines producing antibodies	Antibodies against viruses bacteria, fungi and parasites. Diagnostic aids (A). Selective cellular delivery vehicle for anti-cancer toxins (P)	Unknown
Protein engineering	Genes give instruction to organism to synthesize proteins	Production of enzymes as catalysts or in high-volume use in industry (P)	Acceleration of natural rates of evolution
Agricultural research	(a) Tissue culture (A) (b) Intergeneric hybridization (A) (c) Recombinant DNA	(a) Superior forest trees, 'tailored' plants (A) (b) Crops tolerant to poor soils, saline conditions (A) (c) N_2 fixation in plants. Improvement of photosynthesis, pest and pathogen resistance (P)	Considerable in forest and agricultural areas
Microbiological engineering	Use of micro-organisms to produce commercial products, inc. uses of gene splicing, fermentation	Production of antibiotics, alcohol, single-cell protein (A)	Secondary effects on resource processes

A = actual, P = potential

judge granted an injunction preventing a DNA researcher from releasing engineered micro-organisms onto a crop whose productivity they were designed to enhance, and some genetic engineering firms are likely to concentrate experimental work outdoors in countries with the least legislation. Doubtless, too, research into military applications is well under way.

The potential for environmental change can be seen in such possibilities as the introduction into higher plants of the capability of fixing atmospheric nitrogen. If this could be done widely then many plants could be grown without chemical or organic fertilizers, with considerable consequences for energy consumption in some places and night soil disposal in others, and with subsequent effects upon N_2 levels in rivers and ground water. Another example might be cloning forest trees to produce rapidly growing varieties. Some of today's monocultural forests would then seem positive havens of diversity by comparison. So many global and regional ecosystem biogeochemical processes are mediated by living matter (especially micro-organisms) that there must be a potential for impact on e.g. the nitrogen and carbon cycles. Large-scale application of biodegradation, fermentation and biomass energy technologies could conceivably increase the flux of carbon dioxide into the atmosphere, to add to the surplus already present from fossil fuel consumption, or by contrast organisms could perhaps be engineered to sequester atmospheric carbon dioxide in solid form.

Although presenting considerable horizons for circumventing the currently normal processes of both organic and man-made evolution, genetic engineering does not, it would appear, pre-empt the need to conserve historically engendered variety. The need to conserve genetic resources is still important since the bigger the base, the more chances of being able to find precisely the quality which it is desired to introduce into another variety of organism. Nevertheless, the potential for something of a revolution in the process of evolution, not to say an acceleration of a type not seen since the early days of domestication, is there and is unlikely to disappear.

Softer paths

In a humanist perspective on much of what is represented in this and the preceding chapter, the Welsh poet R. S. Thomas talks of the modernized farmer riding his tractor. From him (whom we might compare with Hardy's engineman on p. 239) is now

> Gone the old look that yoked him to the soil;
> He's a new man now, part of the machine,
> His nerves of metal and his blood oil.

But not everybody subscribes to the world-view that assumes a completely industrialized, energy-intensive economy with the concomitant environmental changes. In some places, notably the richest and most recently industrialized parts of the world like California and Australia, there has developed a set of ideas for an 'alternative' world-view. The synergy of various disparate strands into a coherent 'environmentalist' school of thought probably owes its inception to meetings and discussions in the Bay Area of northern California in 1964–65 and from there the viewpoint has spread to all parts of the world, although it is not very strong in many LDCs, and has flourished in a somewhat *sub rosa* form in the centrally planned economies of eastern Europe. Nor is it in fact dominant in the ideology of other countries, but it constantly presents a different set of pathways which is offered by its advocates as another opinion. The reactions which form the core of alternative thinking spring from processes such as rapid landscape change and urbanization, especially in industrialized lands occupied by European culture only in the nineteenth century, the incidence of contamination of air and water by toxic and unpleasant wastes, worry over the safety of nuclear reactors and the connections between civilian and military nuclear industries, concern at the effects upon nutrition and health of the upsurge in population growth rates in many Third World countries, and a re-evaluation of the role of women, especially in developing nations.

The main direction of these alternative paths has been to offer ways of garnering resources which are less transformational than those currently used. Some of them are said to follow natural processes rather than to replace them, use income rather than capital (especially in the case of energy), and re-use materials wherever possible. The key for many advocates in the 1960s and 1970s was a strongly neo-Malthusian attitude to population growth which saw its retardation as a prerequisite to all other environmental policies (an attitude which has largely been abandoned in the later 1970s and thereafter) and which was in practical terms reflected by the grass-roots work of many agencies during those years, with a UN Conference on Population in 1974 at Bucharest.

Recognizing also the binding nature of the energy resource, and impelled by the oil price rises of the 1970s, advocates of 'soft paths' have suggested many alternatives to oil, gas, coal and nuclear power which could be developed to the point of sustaining an industrial base of some kind in DCs and of permitting development in LDCs without hooking them firmly to coal supplies (which most do not have), or the world price of oil (which most cannot afford to pay: before 1973, Costa Rica exported 28 kg of bananas to pay for a barrel of oil, but in 1983 the same amount of oil required the export of 420 kg), or nuclear developments which are expensive and rarely come without economic or political strings attached. A summary of 'alternative' energy sources is presented in table 6.9: what they have in common is a relatively low environmental impact compared with the customary post-nineteenth-century sources of energy, and that they are renewable: they do not depend upon an exhaustible mineral like uranium or on fossil photosynthesis like oil, coal and natural gas. But none of them is totally free of environmental impact (plate 6.5) or other liabilities (table 6.9), with the possible exception of small-scale installations such as solar panels on the roofs of buildings or very

Table 6.9 Some environmental liabilities of alternative energies (see also table 6.8)

Energy source	*Liability*
Hydropower	Destruction of river systems and fertile valleys Dam failure
Geothermal energy	Water pollution by dissolved salts and toxic elements Air pollution by H_2S–SO_2 system
Solar heating	Fires from collector overheating Leaks of collector fluid
Wind power	Accidents from towers, vanes Aesthetic intrusion
Photovoltaics	Toxic substances released in manufacture, breakage or disposal of equipment
OTEC	Marine ecosystem disruption by altered temperatures and currents Worker hazards on collecting plants
Energy conservation	Indoor air pollution by chemicals from insulation and from reduced ventilation Carcinogens from diesel exhausts (where diesel fuel used in preference to more refined fuels) Decreased crash safety in small vehicles

Source: adapted from J. P. Holdren, 'Energy and the global predicament: some elements of a sensible strategy'. In M. Levy and J. L. Robinson (eds), *Energy and Agriculture: their Interacting Futures. Policy Implications of Global Models*. Chur, Switzerland: Harwood Academic Press for UNU, 1984, 41–93.

small turbines inserted into streams and providing electricity for perhaps one household. One wind pump, too, is unlikely to produce a significant impact, and the gathering of biological wastes for fuel may, if there is no alternative use for them, be more or less free of environmental significance. But once it is desired to tap any of these energy sources for what one protagonist calls 'bulk utility-interconnected applications' then like all other ways of harnessing energy,

Plate 6.5 Renewable energy: the Altamont wind farm in California. Such installations are not free of environmental impact, particularly of a visual kind; further electricity has to be led off to its users. (© Geoff Barnard/Panos Pictures)

the linkages with other systems become important. These are summarized in table 6.9 and will not be repeated here, except to emphasize that nearly all will create noise and dust during installation, and that for tidal power using estuarine barrages the list of impacts is very high (figure 6.4). Desk studies for a projected barrage across the estuary of the Severn in England (tidal range 7.3 m) have resulted in a large list of potential impacts upon general regional development patterns, ports, freshwater regions, landscape and amenity, recreation, the ecology of coastal habitats, freshwater fisheries, nature conservation, pollution, agriculture and flooding. Even geothermal power is not immune: the released gases are usually high in hydrogen sulphide, the waste water stream is highly toxic and corrosive, and if reinjected may increase the likelihood of seismic movements. The more seasoned of campaigners for 'alternative' energies point out that none of them can be totally free from environmental and health-related effects, and the sceptical remind us too that direct solar energy collectors, for example, have not yet been built that have produced as much energy as they took to build.

Energy from biomass may seem relatively benign and indeed it may be so if it uses plant and animal wastes which have no other potential (figure 6.5). 'No other potential' is a wide-ranging phrase, however, for even straw or twigs may be necessary to the nutrient cycling of a site and their diversion into energy production may then rob the soil of some of its fertility. For some years this has been true of animal dung in places like rural India, Bangladesh and Peru: if the dung was burned for cooking and space heating because there were no other fuels, then the fields were deprived of an important source of nutrients. The dichotomy of use becomes more acute if landowners decide to divert the use of land from food crops to growing biomass for energy. There is no particular reason why a crop cannot be found which yields both, but if an energy crop commands higher prices then economics demand that it will dominate, as it has with many other cash crops. At all events, energy plantations are currently landscape features in places such as South Korea, China, Indonesia, the Philippines and Bangladesh, among others. A rather bald summary of the biomass contribution would suggest that it is easily possible to overrate it and that the conservation of genetic variety in the earth's biomass ought to have priority over necessarily harvesting all the products of photosynthesis. But the planting of trees, especially drought-tolerant nitrogen fixers, would be an enormous gain in energy-deficient rural areas, especially if the scale was local and the species diverse. In dry places, drought-tolerant plants might be useful energy sources: Russian thistle (*Salsola kali*), creosote bush (*Larrea tridentata*) and milkweed (*Asclepias* spp.) can all produce logs, pellets and liquid fuels or resins. Water plants contribute very little at present but may assume a locally greater value: water hyacinth, water lettuce, pennywort and *Salvinia* are floating species that might fuel biogas plants, as might emergent species like cattail (reed-mace), soft rush and bulrush. Both groups might remove dissolved eutrophicants

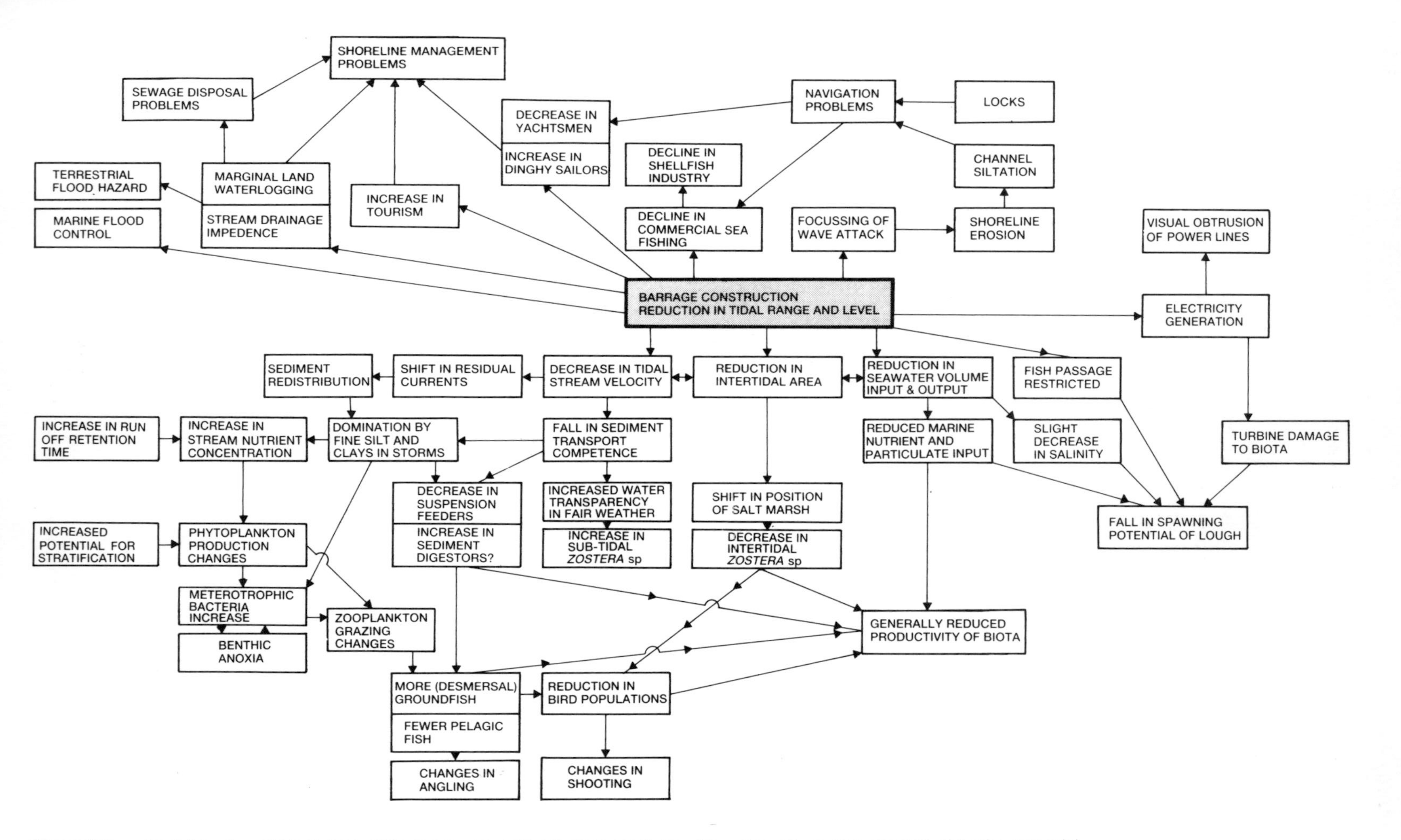

Figure 6.4 Lest it be thought that renewable energy sources are necessarily low-impact in nature, this diagram plots the potential environmental changes that could result from the operation of a tidal barrage across Strangford Lough in Northern Ireland. An estuary which held a major port would undergo many more commercial effects as well.
Source: R. Carter and P. J. Newbould, 'Environmental impact assessment of the Strangford Lough tidal power barrage scheme in Northern Ireland', *Water Science and Technology* 18, 1984, 455–62.

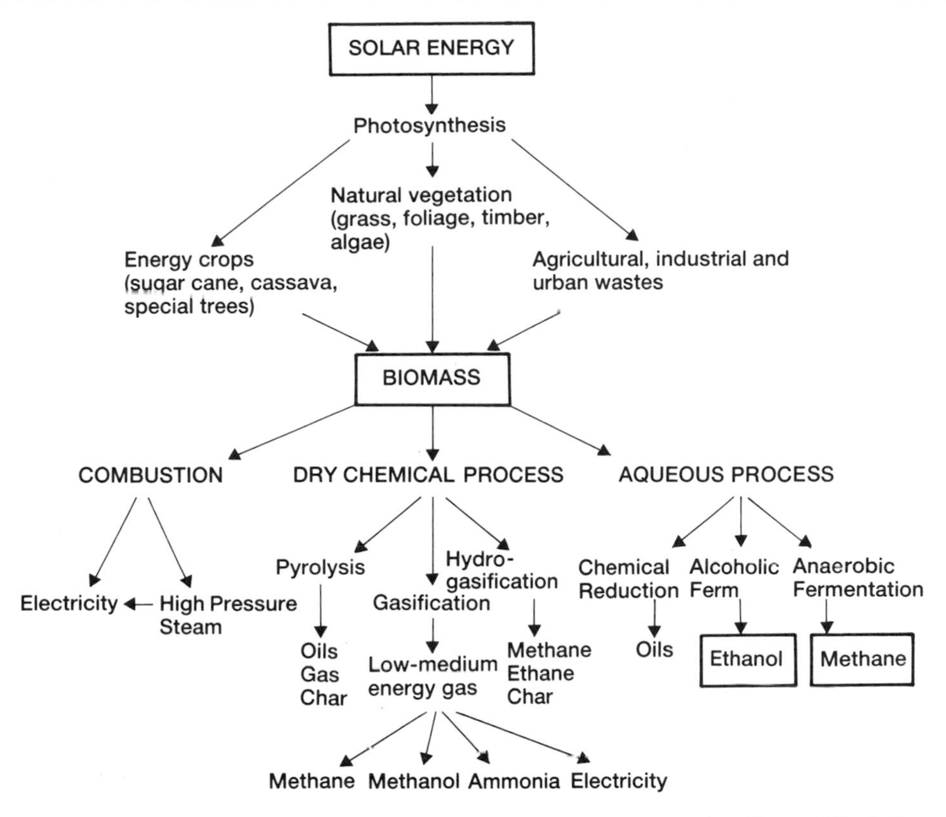

Figure 6.5 Pathways in which biomass can be converted to energy for human use. Photosynthesis (cf. fig. 1.1) is central to all the processes even if fermentation is used at a later stage.
Source: J. E. Smith, *Biotechnology*. London: Edward Arnold Studies in Biology 136, 1981.

such as nitrogen and phosphorus from fresh waters. Algae have for long been a favoured group but are not significant in environmental change except where there is large-scale production, e.g. of *Spirulina* on a 48 ha site covered with ponds in California, producing 20 g/m^2/day and recycling all its water. The prospects for making methane out of animal and plant wastes are relatively good, and the solid residues are still good fertilizers after anaerobic fermentation (plate 6.6), but the chemistry needs careful management: bacteria (like Professors of Geography) are susceptible to chambers that are too cold, too undernourished, and in too acid or too basic an environment.

In a global context, therefore, biomass is perhaps best employed in fulfilling two irreplaceable functions: acting as a reservoir of genetic variety and regulating the flow of carbon through its biogeochemical cycle. But the special needs of the poor for energy, especially but not totally in rural areas, cannot be pushed aside. One calculation suggests that basic human needs can be satisfied in the range 27 000–37 000 kcal/cap/day (113–155 MJ), a figure which can be compared with a world average consumption of commercial energy in 1975 of 42 500 (178 MJ), an LDC average of 18 000 (75 MJ), with nations like Bangladesh at 2300 (9.6 MJ) and India at 11 000 (46 MJ), and the USA at 243 000 (1018 MJ). In rural areas without access to commercial energy, 'alternatives' especially if based on solar and biomass power might provide a great deal of the difference between outright poverty and greater dignity. One study suggests that a

Plate 6.6 Part of a biogas plant in Yucatan, Mexico. The controlled fermentation of excreta not only produces methane which can be burned for heat but also concentrated organic fertiliser. (Liba Taylor/Panos Pictures)

careful mix of windmills, solar pumps and biogas-powered pumps might allow Indian irrigation systems to be run on renewable sources of energy. Certainly, it does not seem inevitable that bringing electricity to the rural people of developing countries necessarily means hooking them to national grids. Some similar considerations might be applied to poor urban areas, too: we know that 'slums' in LDC cities use about the same total energy as the surrounding rural areas but that they are more efficient in its use, achieving more services with the same amount of energy. The potential for upgrading their access to energy relatively easily seems therefore to be very high, and development now in progress is, we may hope, towards ways which do not produce the type of environmental impact which itself contributes to the poverty of rural societies as is the case with fuelwood deficiencies in semi-arid zones.

The term 'alternative agriculture' brings to mind the drop-outs of the 1960s engaging in more or less (usually less) successful 'self-sufficiency', but the term now has much wider connotations. It embraces all those agricultural techniques which are not dependent upon very high inputs of fossil fuel energy, especially chemical fertilizers and biocides. It does not generally eschew machinery altogether but advocates methods such as zero tillage which reduce the need for power-driven implements and which generate additional energy for various purposes from crop residues. The label 'alternative' is actually an umbrella term for several different kinds of agriculture but is often equated with 'organic' methods which rest on such practices as varied crop rotation, the encouragement of birds and other predators of crop pests, pesticides of vegetable origin, and the use of 'organic' fertilizers from farm animals or sewage sludge. Studies in the USA and the Netherlands, for example, suggest that crop yields are equal to or only slightly lower than current levels. In one US study, organic wheat and maize production was 29–70 per cent more energy efficient than conventional production but potato and apple production were 7–93 per cent more energy efficient in the conventional mode. For all farm crops, labour input per unit of yield was higher in the organic systems. The organic farms are, however, to some extent dependent upon conventional farms for their surplus manure and upon seeds and breeding strains developed in the 'normal' sector. In the case of the USA, home food demand could be met, but export potential would decrease.

The environmental impacts of such farming systems are more precisely controlled than with conventional agriculture and are likely to introduce fewer potentially toxic chemicals into local ecosystems. On the other hand, nitrogen applications may still result in eutrophication of runoff and groundwater and use of sewage sludge and pig farm residues may lead to high concentrations of copper and heavy metals in the soils. Overall, not enough evidence exists for a proper evaluation of 'alternative agriculture' in the DCs since it exists in a matrix of other types. The present consensus seems to be that it can feed large DC populations only if certain dietary shifts take place: since these would be in the direction of a high-fibre, low-fat nutrition, then there may be considerable scope for change once people realize that such foods are good for everybody and not just for cranks. In the LDCs 'organic agriculture' is still normal for many regimes; however, the term is usually applied in the sense of a developing food system which gives higher yields without yoking local agriculture to the need for, and the price of, oil. Not all LDC food production systems are amenable to the kind of intensification that keeps pace with growing populations unless energy is imported into the system, and some analysts hold the view that it is the limited use of commercial energy in agriculture (about 25 per cent of that used in DCs) which keeps down the productivity of land and labour. The application of such energy, mostly through fertilizers and machinery, but also through irrigation and pesticides, is deemed essential to strategies such as those of FAO, but appeals little to those who prefer to see more local self-sufficiency, with small groups taking the decisions that so strongly affect their lives. Nevertheless, one FAO-based author concludes that 'land requirements for renewable sources of energy need to be taken into account when planning horizontal expansion of agricultural production'. An example of the kind of thinking now possible comes from a very poor area of Rajasthan, India, where inventories showed that indigenous renewable energies were available to meet the needs of the population provided they were properly managed: a transition from firewood to biogas and simple solar devices relieved pressures on the ecosystem and provided employment as well.

Given the attraction of the organic farming idea, it is not surprising that advocates of 'alternative cities' have put forward the idea that urban areas should be less parasitic upon other places for their food and energy supplies. Thus while we can point to no actual example of a self-sufficient city, there are nevertheless movements to increase the quantities of food being grown by individuals or small groups on patches of unused land, in gardens and allotments, and on window-sills and roofs, for instance. A cautious movement towards local energy supplies is made by cities which incinerate solid wastes and power steam generators with the heat thus released, but more radical moves are in most places unlikely because demand levels for electricity cannot be met by these techniques. Biogas from sewage might help, but again generation of the levels to which we are currently accustomed seems impossible. If such alternative cities were possible then their environmental linkages would be considerably different from those at present, but actual forecasts are scarcely possible because of unknown variables like the density of vehicles and their power source, and the proximity of incinerators to the actual built area.

In this general field falls the whole set of ideas and techniques which are generally called 'alternative' or 'appropriate' technology (AT). This means machinery which is powered by human energy or locally renewable energy sources and has a low environmental impact as well as being labour-intensive. Such technology is seen as very important in reducing the migration in LDCs of people to cities and may well prove more attractive in DCs if the number of people employed in manufacturing falls radically. In LDCs, examples of AT include solar and biogas generators, brick-making machines, water pumps and irrigation ditch liners, to name but a few. The claims for lower environmental impact than their diesel or kerosene powered equivalents are not generally very great (it is the local control which is important), if we except energy for cooking and space heating where any technique which reduces the demand for fuelwood or dung is likely to be significant in its outward connections.

The example of LDCs with regard to material use has been one of the inspirations of an 'alternative' attitude in the DCs. It has often been reported that in poorer cultures, any article will find a second or even third use in its lifetime: food tins become cooking utensils and vehicle tyres are made into sandals, for example. This is contrasted with the 'throwaway' society of the West, with each of us producing a few kilogrammes per day of solid wastes (plate 6.7). One outgrowth of this attitude is that of seeking solutions to energy supply problems, or the forecasts of them, by using energy more efficiently. More use of public transport, better insulation of buildings, and more efficient machines are part of this syndrome. It extends to the re-use of

(a)

(b)

Plate 6.7 Western world-views of rubbish. (a) Business as usual: the environment as a dump for all kinds of materials; (b) a different view in which a citizen takes responsibility for the recycling of glass. The latter would be much more energy-efficient if the bottle were re-used instead of having to be melted down and re-cast. (Photos: D. Cooke)

materials where energy costs are a large proportion of manufacturing costs: glass bottles are a common example and even countries like the UK, where the public consciousness of the potential for recycling is very low, usually have 'bottle banks' in conspicuous places. Other countries are now encouraging drink can recycling (possible if the containers are all aluminium) but any visit to a landfill site gives an idea of what affluent societies are prepared to see buried. (Though not always forgotten: such tips create dust, are nesting places for rats, attract thousands of gulls which are now more common as inland scavengers than as coastal birds, sometimes leak toxic wastes to the runoff, and occasionally catch fire. This last is due to methane gas production which is now sometimes tapped as an energy source.) So recycling of materials and thermodynamic thrift, like organic agriculture, are most often identified with the beards–jeans–nuclear-disarmament syndrome, but clearly have a considerable future if energy prices rise. DC governments seem to divide into two on energy and material conservation: those who encourage it by all possible means, and those who leave it to the market where at present the dictates of economics are such that neither materials recycling nor energy conservation is immediately attractive.

The whole nexus of softer paths has its ideological base in what is usually termed the environmentalist movement, though other appellations have been heard on the lips of government officials and 'realistic' businessmen. The underpinnings of environmentalism are discussed in the concluding chapter and here we have only to ask questions about what is to replace manufacturing industry as a creator of wealth and a provider of employment. Indeed, can industry be rendered environmentally benign except by some technological *tour de force* of decoupling it entirely from the ecosphere (perspex domes, ice islands and deep burial all surface occasionally in the appropriate literature) rather than simply adopt the DC strategy of shipping off a polluting process to an LDC whose population is less fussy so long as they have a job? We have to conclude that at present the softer paths offer considerable potential but that adoption rates are slow: possibly the social scientists will tell us why.

Soft energy: more detail

The quest for renewable sources of energy which are also more benign environmentally than fossil and nuclear fuels has led in two main directions at national and regional level. The first of these is an attempt by certain governmental units (usually relatively small and often on islands) to achieve energy self-sufficiency on a renewable basis. The second, usually national in scale and therefore not fully adopted anywhere, is based on an awareness of the potential of energy conservation and efficiency combined with shifts in the mix of fuels to increase self-sufficiency and decrease vulnerability to outside suppliers. The primary motivation for these views are usually economic and political rather than environmental, but there is no doubt that they have been accompanied by a concern for the biophysical linkages of energy production and use. An example of the first type is that of an island group like Hawaii which is at present almost totally dependent on imported oil. It differs from many other places in not requiring much energy for space heating, though some air conditioning is required, and in consuming large quantities of petroleum products to support its tourist industry. After imported oil, the second largest source of energy is bagasse, the fibrous residue from sugar cane. Burning of this fibrous material provides more than 7 per cent of total energy consumption. In recent developments scientists and engineers have begun to look carefully at the commercial potential of the use of direct solar radiation to provide electricity via photovoltaic systems, at the use of energy embedded in the sea, and at various biomass sources other than bagasse, including *Eucalyptus* and *Leucaena* trees, wind energy and geothermal resources. Wind, geothermal power and solar energy are already cost-competitive with oil, and given continued market penetration of the new technologies, one authority indicates that 50 per cent self-sufficiency in electricity can be achieved by 1990 and 50 per cent total energy self-sufficiency (i.e. excluding the transportation of fuels) by 2000. It is worth reflecting that in terms of installations and processing, such a change would very likely increase the environmental impact of energy production and use in Hawaii, for at present only the residuals are significant, nearly all the extraction and refining of petroleum being carried out elsewhere. Bringing home the energy production would trade off lower smog levels in Honolulu, for example, for more windmills, photovoltaic installations and geothermal plants.

Not many parts of the world are so well furnished with warm seas, abundant sunshine, steady winds and predictable tectonics as Hawaii, and so the thrust of those advocating alternatives to nuclear and fossil energies have usually concentrated on lowering demand for energy and shifting the mix so that renewables, although unlikely to be dominant, are much more important than they are today. In 1979, for example, the International Institute for Environment and Development published a low energy strategy for the UK. This holds that there could be 50 years of economic growth but less primary energy need be used than in the 1970s, so nuclear power becomes peripheral and can be abandoned although it is suggested that it is kept at a 'tick-over' level as a precautionary measure. The twin pillars of the case are enormous savings in energy made by technological innovation, rational use, and insulation, together with a reliance on coal, a greater proportion of which will be used to make synthetic oil from 1995 onwards. ('Alternatives' play little part in the IIED scenarios.) Environmentally, the IIED proposals do not seem to mean great differences in the UK, with the possible exception of shifts in the composition of atmospheric contaminants.

Since coal is so important, presumably desulphurization at power stations and conversion plants cannot be ignored as it largely is at present. A study of renewables by J. Mustoe produces an energy mix which contrasts strongly with the present situation (figure 6.6). The total energy demand (in AD 2025) is reduced to 63 per cent of its present level, and the fossil fuels account for 72 per cent of total primary energy, rather than the current 98 per cent. Instead of being negligible, the renewables supply 28 per cent of primary energy. In terms of environment, the renewables are not without effect and in this scenario coal is relatively more important among the fossil fuels, though its absolute contribution declines. The same author has produced similar comparisons for the mainland USA, where renewables might contribute 18 per cent of total supply which is a more realistic estimate, perhaps, than some of the boosterist interpretations of biomass and wind energy contributions.

The application of ecological thinking to energy flow in a city was particularly close in the case of a project on the energetics of Lae, a

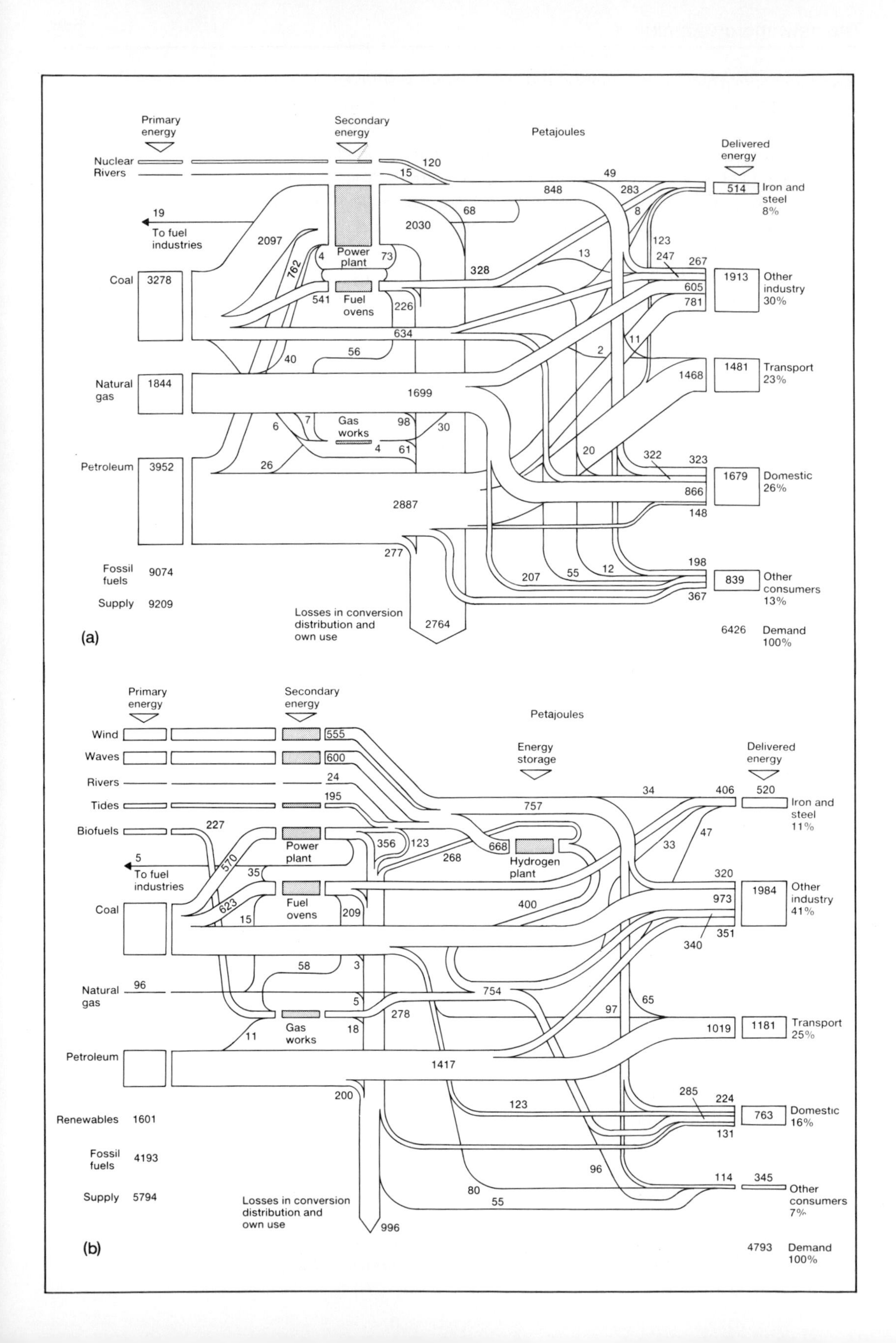

Primary energy
Secondary energy
Petajoules
Delivered energy
Nuclear
Rivers
120
15
49
848
283
514
Iron and steel 8%
19
To fuel industries
2097
2030
68
8
762
4
Power plant
73
123
13
247
267
328
Coal
3278
541
Fuel ovens
226
1913
Other industry 30%
605
781
634
11
56
2
40
Natural gas
1844
1699
1468
1481
Transport 23%
7
Gas works
98
6
30
4
61
Petroleum
3952
26
20
322
323
1679
Domestic 26%
866
2887
148
277
198
207
55
12
839
Other consumers 13%
367
Fossil fuels 9074
Supply 9209
Losses in conversion distribution and own use
2764
6426 Demand 100%
(a)
Primary energy
Secondary energy
Petajoules
Wind
555
Waves
600
Energy storage
Delivered energy
Rivers
24
Tides
195
34
406
520
757
Iron and steel 11%
Biofuels
227
356
123
668
Hydrogen plant
33
47
5
To fuel industries
570
Power plant
268
35
320
623
Fuel ovens
209
400
973
1984
Other industry 41%
Coal
15
340
351
58
3
Natural gas 96
754
5
97
65
278
18
Gas works
11
1019
1181
Transport 25%
Petroleum
1417
200
285
224
123
763
Domestic 16%
131
Renewables 1601
Fossil fuels 4193
96
114
345
Other consumers 7%
Supply 5794
80
55
Losses in conversion distribution and own use
996
(b)
4793 Demand 100%

city of 62 000 people in Papua New Guinea. In the 1960s and 1970s the heavily forested hills around the city were under pressure for cultivation and the deforested terrain shed water rapidly, washing away suburban allotments, roads and bridges. Eventually, the cultivated land became scarce, the refuse disposal system was close to breakdown, and raw sewage was dumped onto the ocean foreshore. The energy scene was dominated by imported oil and kerosene and the burning off of wood wastes. A programme of research was aimed at the identification of alternative energy flows and land use patterns, particularly those in which more rational land and energy use might benefit the lower income groups. The basic experiments were in fuelwood production, conservation forestry, intensive gardening for food, subsistence level agroforestry systems, composting of city wastes, and the planting of fruit trees in the urban area. Charcoal production, fuelwood production and industrial biogas production all began after two years of the project, and were joined by solar hot-water heating and alcohol fuel production. The whole project is still developing with active association from the local and national governments and is being extended to Port Moresby; the only resistance, apparently, has been from professionals in engineering and agriculture with a westernized outlook. The general lessons seem clear. Energy conservation cannot be but helpful (it has useful social spinoffs as well in terms of employment and health) and the soft energies have considerable potential, not the least of which is likely to be lower levels of atmospheric contamination. Spinoffs for DCs seem to be higher levels of decentralization of services (fear of which may be inhibiting governments from encouraging these 'alternative' sources) and less political vulnerability in terms of foreign suppliers. It is therefore perhaps not too far-fetched to say that greater efficiency of energy use is a vital ingredient of both economic progress (in both DCs and LDCs) and environmental protection.

Figure 6.6 Two studies of energy flow (PetaJoules = 10^{15}J) through the UK economy. (a) The actual flows in 1979, and (b) a scenario for 2025 which keeps an industrial economy but boosts the contribution of renewable sources and above all reduces domestic and similar consumption by a massive investment in conservation measures and increases in efficiency of use. Nuclear power becomes irrelevant in this picture.
Source: J. Mustoe, *An Atlas of Renewable Energy Sources in the United Kingdom and North America*. Chichester: Wiley, 1984.

Population, resources and environment in the 'post-modern' years

Like most of this chapter, it is necessary to speculate here: little of what can be said is accessible to objective verification. The outcome of new technologies is by no means certain in either environmental or social terms; for example, we can make forecasts about the role of computers or the effects of genetic engineering but can say little about the synergisms that will result from the interaction of the two. Looking at the last few decades, there has perhaps been one demonstrable effect of the meshing of technology and the kind of resource–environmental knowledge central to this book. This is in the field of the reduction of rates of population growth. Granted that environmental concerns are rarely uppermost in the minds of those taking their pill at bedtime or lying on the vasectomist's couch, the consequences for pressure upon resources and upon environmental systems must nevertheless eventually be very great indeed. In 1970 the world annual rate of increase of population was 1.9 per cent, but by 1975 it had come down to 1.64 and has remained below 1.7 thereafter. Within that aggregate, several countries have had periods during which they reduced their birth rates by 2 points per year, for instance China brought it down from 32 per 1000 in 1970 to 19 per 1000 in 1975, and it is currently stable at about 22 with further steady reductions being vigorously sought. Other rapid reducers include Taiwan, Tunisia, Barbados, Hong Kong, Singapore, Costa Rica and Egypt. The greatest reductions of birth rate have thus been in East Asia, North America and Western Europe, and in the last-mentioned some six countries, comprising about 4 per cent of world population, are at or near population stability. (These are the two Germanys, Luxembourg, Austria, the UK and Belgium.) Several other small European nations, the USA and USSR, Japan, Australia, New Zealand and Canada could easily join that group within the next few years, so that 25 per cent of global population would live at a stabilized level.

To add to this, the evidence of recessions, of lowered demand forecasts for energy and for some materials, gives hints that perhaps the end of the great period of exponential growth of everything (but especially population, energy demand, resource use and environmental impact) may be starting to happen, that an upper asymptote of such phenomena is likely to come about. Such a statement is even more speculative than most, because the signs are straws in the wind rather than flashing neon, and because the feedbacks to their perception could go either way: some will welcome them and try to make sure they are self-fulfilling prophecies, whereas others will feel that their vested interests are threatened and do their best to propel the curves onto an ever-upward track once again.

If we add these recent changes to all that has gone before, and especially the developments of the last 150 years, does it seem likely that the immediate pathways for environment–humanity relations will be more destructive than at present, or more sustainable? At a global level, the accumulation of CO_2 poses an apparent threat and possibly there is also toxification by persistent chemicals such as biocides and PCBs. The potential for local and regional scales of destruction by toxification and land degradation of all kinds has been generally enhanced in the last 150 years, and the most recent changes seem to have done little to reverse such trends. What the last few years have notably contributed is a capacity to render the earth virtually uninhabitable, and perhaps uninhabited, by our species via the agency of nuclear war. Hopes of sustainability are provided by the great resourcefulness and adaptability of technology once it is applied for something other than the making of fast money for a few individuals or multinational corporations. The depth of knowledge and foresight contained in documents like the World Conservation Strategy and the Bruntland Report, and the slow but apparently steady rise of political groups whose ideology centres upon ecospheric stability are signs: the Greens entered the European Parliament for the first time in 1984. The only acceptable judgement on all these tendencies is that it is too soon to tell: with the exception of global nuclear war, any one trend is never likely to oust another completely so we have to live with the sort of tension in which there are simultaneously signs of hope and of tribulation.

Conclusions

Of all man's miseries, the bitterest is this, to know so much and to be impotent to act.

Herodotus, *The Histories* (fifth century BC)

Let your reward be in the actions themselves; never in their fruits . . . casting off attachment, perform your actions, indifferent alike to gain or loss.

Bhagavad-Gita (second century BC)

An overview

We have established beyond doubt that human societies have altered their biophysical environments since very early times, even if we cannot assign a precise beginning to the process. From the start, the metamorphoses were of two kinds: those which were deliberate and those which were accidental. Sometimes, no doubt, the two have been virtually indistinguishable: a fire lit to flush out game from one habitat may have incidentally run into an area of vegetation valued for its fruits and seeds, to quote an imaginary example. We have used the terms 'environmental management' for desired changes, and 'environmental impact' for incidental transformations but these are recent coinages and so difficult to apply in historical contexts.

These niceties apart, it seems possible to recognize distinct phase shifts in the intensity of the sum total of environmental transmutation brought about by human societies. The first of these was the ability to manage fire, and includes the capability to kindle it rather than merely to take it from natural sources. (It is said that as late as the eighteenth century AD the Tasmanian aborigines did not possess fire-making skills: they carried glowing coals with them and had to get fire from neighbours if their own went out.) This accomplishment had many effects, but the main one was the permanent alteration of vegetation communities (and their dependent animal populations) which repeated fire produced, both as a management tool and by way of unplanned impacts. At some stage in human history, fire seems to have affected every ecosystem type (even, perhaps, the lowland moist forests of the tropical zones), so the spatial extent of the influence of burning has been very wide indeed.

A second key transition was the domestication of plants and animals. Though certain areas of the world, notably south-west Asia, the Far East, and Meso-America, seem to have been foci of the earliest domestications, many breeding ventures took place in other locations and over other periods of time, so that old models of circumscribed hearths of domestication now need amendment. But the spread of economies based upon varieties of plants and animals in which human choice had replaced natural selection was rapid and near-ubiquitous, omitting only those areas where only wild plants and animals could flourish, as in the oceans, the tundra and very high mountains. As with fire, there were both deliberate effects (of which terracing is the most obvious manifestation) and accidental changes, of which the gradual transformation of plant communities by herds of grazing animals can be an epitome.

Although the fossil fuels were known before the nineteenth century, it was in those years that the industrial way of life, based first on coal and then on natural gas and oil, was nurtured in Europe; it spread initially to those nations now regarded as developed and is more slowly reaching the less-developed countries. Once fuels can be transported cheaply, there is nowhere on earth where they cannot be used, so that we have military and mineral installations in the high Arctic, observatories on high mountain peaks, and cities deep in tropical forests, as well as the more obvious representations of fossil fuel use in urban–industrial areas. The patterns thus established have been intensified in places where electricity is important by the addition of civil nuclear power to the suite of energy sources. The great changes in biophysical systems made possible after the Industrial Revolution need little re-emphasis here, except to reiterate that in distribution and intensity they have far exceeded anything that went before. Both management and impact have sometimes led to the same phenomena: radioactive nuclides are deliberately released to air and water as part of the management of nuclear wastes and they get there incidentally when there is an accident at a reactor or reprocessing plant.

It is possible, in broad terms, to put figures to these various stages of cultural evolution. The initial stage is the somatic energy needed to keep a single human being alive, which is about 10 MJ per day. At each phase-change and sometimes in between, technological progress has enabled a human society to add extra-somatic energy, as with fire, domestic biota, wind and water, fossil fuels and nuclear power. Given that the per capita commercial energy consumption in the USA is now getting towards 1000 MJ per day, there has been a rise of two orders of magnitude between the Palaeolithic and the present. The total quantity of energy now being converted is unprecedented, and the rate of growth in recent decades has been twice that of the human population, although some slowing up of both rates has been characteristic of the last few years.

So the nineteenth century can be seen as the great discontinuity of the past: before then, energy use per capita was about ten times that of somatic energy and this allowed quite high levels of environmental manipulation, especially when it was cumulative over a long period of time. Access to energy trebled in the early years of industrialization and in some places is now a hundred times that of somatic energy, with the concomitant ability to change biophysical systems, and often to transform them very quickly. It is the moment, then, to ask what future pathways of these relationships are possible, for this kind of history must inevitably point forwards.

The regional distribution of environmental alteration

The biomes altered

It is by now clear that those maps of natural regions or major biomes which are found in atlases and textbooks are misleading since they present pictures of what would have been present had not the transformations caused by human societies taken place. Here we will very briefly review some of the main alterations that have occurred, recognizing that only accounts of much smaller regions can approach the actuality of much of the change, since it is often so diverse and multi-layered.

The section on forests in chapter 5 will have made it clear that the *tropical forests* of the world are rapidly shrinking in area and that they are being replaced with secondary woodland, savanna, and grassland, together with smaller patches of quasi-permanent cultivation. In all of them, logging is at an unprecedented level, often without proper re-seeding: this has been especially noticeable in the south-east Asian part of the biome. African forests are now being converted to agricultural land under the pressures of population growth, and Brazilian attempts to develop Amazonia for commercial pastoralism as well as peasant settlement are well known. The margins of this biome have of course been attractive for development since the nineteenth century when the differential extraction of timbers such as teak became common, and when large areas of forest were converted to plantations of rubber and oil palm. Population growth since the nineteenth century has also accelerated the rate of creation of secondary forest since forest fallow periods have become shorter.

The representation of the *savanna* biome is as accurate as any in the sense that it still comprises a great deal of grassland with interspersed trees at varying densities. It has come under greater pressure for change in recent years from population growth, especially in Africa, and more and more of the wild vegetation is converted to agriculture. In some parts of Africa this process clashes with the designation of National Parks which earn a lot of income from tourism. At the dry margins of the savanna, a combination of dry years and heavy pastoralism has often brought about desertification, a phenomenon also seen where aridity has caused the failure of traditional cropping economies. Attempts at wholescale transformation of savanna environments, for example, to commercial cropping, have usually failed since the soils become lateritic quite quickly if permanently cleared of grass and bush vegetation.

The areas of *temperate grasslands* have undergone a great deal of change. Most of them have been subject to fire by hunter–gatherers and then were converted either to commercial pastoralism or to crop agriculture in the nineteenth century. Both of these have carried the risk of a form of desertification in periods of dry years so abandonment of land has occurred in Australia, for example, and North America, though the industrial-age context of both of these has enabled recovery of the economies to take place. Much conversion to commercial agriculture has happened where water could be made plentiful from irrigation schemes. Some of these have 'mined' the water so may eventually have to fall back to dry farming or to a form of pastoralism. Today's commercial pastoralism may manipulate the vegetation a great deal, using machines and chemicals along with range ecology to ensure the maximum availability and use of forage for domestic stock.

Deserts are among the least changed from their natural condition. They have often expanded at the edges but in their interiors changes have been relatively minor and connected with, for example, oil extraction sites and their subsequent transportation networks. If water can be easily provided then large-scale irrigation can be attempted as in Libya today; in the south-western USA water availability makes desert colonizable for settlement in the form of cities and their suburbs or by vacation and second homes. Recreation is popular in industrialized areas and may lead to dune formation and shifting if the vegetation breaks up.

One of the most altered of all the major biomes are the *temperate deciduous forests*, for they have been the site of the early industrialized parts of the world that are now advanced industrial nations, as in Europe, North America and Japan. Not only that, but they were the scene of alteration by hunter–gatherers who used fire to manage animal populations, and of agriculturalists, both shifting and permanent. Thus many of these forests have disappeared and where they remain are usually heavily managed. The more natural-seeming of them are often nature reserves of some kind, such as the Białowicza Forest in eastern Poland. These areas, then, have seen the greatest development of the age-old tension between the forest as a store of resources and as a land bank to be exploited for agriculture once the trees have gone. Quite a large area remains under tree cover but of a different kind, as where faster-growing species have replaced native taxa; sometimes trees from another biome such as the Boreal Forest have proved highly successful, as with the introduction of exotic conifers into western Europe. In the southern hemisphere, large areas of this biome have been transformed, for example in New Zealand, to commercial forests with conifers and *Eucalyptus* as the dominant types of tree.

Sclerophyll vegetation of the 'Mediterranean' type (found also in places like Chile and California) is also often altered for hunting, agriculture and industrial-age purposes. Easily fired, it has been managed for centuries to make sure that there is browse available to mammal herbivores such as deer and to encourage species such as oak which yield edible crops. Equally, the vegetation can be converted to arboriculture (vines, olives) and cleared for agriculture, often accompanied by terraces. It is an attractive recreational environment, too, and can contain for instance a high density of vacation homes or be the backdrop to coastal hotel and resort development. These latter uses increase the fire risk considerably and in climates that have concentrated periods of heavy rainfall, this may lead to mudslides that engulf settlement and communications.

A biome with a large but shrinking presence of near-natural vegetation and animal communities is the *Boreal Forest*. Impact upon it from hunter–gatherers seems to have been quite low; and indeed the large-scale and very damaging fires that now occur are probably features of protection policies that allow organic matter to accumulate on the forest floor and provide an enhanced source of fuel for human- or lightning-caused conflagrations. Use of the forests for wood products (especially for pulp production) exerts its own pressures for change in the direction of monospecific stands (quite common anyway in this biome) and the provision of transport links. The southern edge of the biome (which is largely confined to the northern hemisphere) is the more affected by these various alterations but is also popular for recreation, especially the siting of vacation homes in zones where small lakes are frequent: parts of Ontario and Quebec in Canada, and southern Sweden and Finland are good examples.

Transformation of the *tundra* may seem unlikely, but the presence of increased numbers of people in the post-1945 era, when air transport became easy, has led to much more fire on such dry parts as exposed lichen-clad ridges. Defence installations, oil rigs and other exploration sites and the well-known Alaska pipeline from the North Slope to Valdez have all provided environmental impact, even though great care has usually been taken to minimize the effect of the installations especially on the permafrost and on the animal populations. Imported animals like the reindeer herds provided for Inuit people in Alaska have sometimes led to a form of overgrazing of lichens and a kind of 'polar desertification' but only in limited areas. Much more widespread has been the accumulation of radionuclides in the slow-growing lichen and moss plants as a result of fall-out from atmospheric nuclear testing; some of this has found its way in to the body burdens of Saami and Inuit peoples. The Chernobyl accident has intensified this problem in Lapland and the Saami have not been allowed to market their reindeer.

Aquatic environments of many types have undergone ecological metamorphosis. Rivers and lakes have had all manner of their parameters changed by damming, channel straightening, toxification, bridge-building, eutrophication, water extraction and myriad other processes. Coastal wetlands have been altered by reclamation since early agricultural times, with an acceleration of these changes since the coming of industrial economies; salt marshes have become airports, factory sites, shipyards, and marinas, for instance. The whole syndrome has spilled over into offshore waters, where coral reefs are particularly vulnerable to shell collecting, silting and toxic runoff. Such waters generally are likely to be affected by runoff from the adjacent land and rivers, and also to be subject to the impact of heavy fishing which may result in the economic disappearance of favoured species, especially those which are sessile or at any rate not very mobile, like molluscs and crustacea.

Two continents deserve a brief but special mention. The first of these is Africa, which according to many commentators now presents its inhabitants with the worst set of environmental constraints and problems of any of the major land masses of the world. The age of its soils, stochastic aridity, rapid deforestation and high rates of population growth all combine to make it an area of growing poverty in both the English derivations of *oikos*. The second is Europe. Altered since Upper Palaeolithic and Mesolithic times, subject to the immense changes wrought by the removal of the forests and by the coming of industrialization, it nevertheless has until relatively recently been a stable environment in which very dense and now largely urbanized populations lived at a high material standard and had an attractive set of surroundings to boot. This was partly made possible, of course, by 'ghost acreages' offshore and in overseas empires, circumstances which continue today in the shape of economic hegemony. This historic harmony is threatened by transnational pollution of the atmosphere at present, though the scale of the reaction may confirm J. Lovelock's hypothesis that industrially based pollution in the northern hemisphere is the least likely to inflict anything other than transient damage to the ecosphere since (a) it will be noticed, and (b) the nations are rich enough to do something about it.

Trajectories

Types of future

Now we enter the realms of forecasting, with its uncertainties. There is no intention to present, therefore, a series of 'future options', each tightly characterized: any prospects are unlikely to be like that. But a number of different kinds of population–resource–environment relationships, in each of which energy resources play an important role, have been set out by numerous writers. If for exposition purposes we filter out shades of grey, the main types seem to cluster as follows:

1 A high-technology pathway in which current trends are continued. Economic growth as at present understood is a major goal and increased levels of access to technology and extra-somatic energy are envisaged, especially in LDCs where these are seen as the keys to an enhanced quality of life.
2 An environmentalist future. In this model, stasis is the main aim: of population levels, of throughput of materials and of energy consumption. Development in LDCs should be based on

local self-sufficiency and not hooked to world fossil fuel supplies and prices; DCs would benefit from more small-scale enterprises and more local levels of control.

3 A decoupled future. High technology is used to decouple the ecosphere and the 'econosphere'. This involves the isolation of human- and machine-dominated systems from those characterized by lower levels of manipulation. At its extreme, this might perhaps involve putting urban–industrial areas under domes and recycling all wastes within that space. The outside world would be used partly for food production but largely allowed to remain in a little-manipulated state. A less thoroughgoing version might involve floating cities and underground power stations, for example.

All the possibilities outlined above appear to have drawbacks as well as opportunities. The first promises greater material wealth for all and such a future relies upon an abundance of cheap energy, ubiquitously distributed. Where is this to come from? Atomic fusion always seems 30 years into the future, and industrial-scale solar power is not yet available nor is it likely to confer the adaptability of oil and its products. If enhanced per capita consumption of materials and energy is a fundamental plank of this future, what are the environmental consequences likely to be? Can biophysical systems absorb the extra impact of all the extraction, the manufacture, the use, and the eventual disposal of wastes? Can in fact technology be used to render all these processes environmentally benign, as in the 'decoupled' future?

The position is to some extent reversed in the environmentalist future. Here the planet's other systems are accorded greater importance than human systems: their sustainability is paramount, it is argued, and therefore we must adapt to them. It is, in effect, a deterministic scenario. Foremost, this means limiting the human population to a level that can be supported on the available 'income' of energy and nutrients, rather than on mined 'capital'. It means also the limiting of material possessions by those already rich in those terms and the deliberate channelling of wealth towards those currently poor. The purpose is Utopian and usually very attractive, but how feasible is this set of ideas? Can population growth rates be slowed down without considerable increases in authoritarianism in LDCs? How could DC populations be persuaded to give to the world's poor the money spent on weapons or pet food or whatever our individual anathemas are? Would the environmental consequences of a set of self-sufficient worlds necessarily be more stable and sustainable than one in which there was interdependence: one with every place doing what it did best, adapted to the natural conditions of the region?

The decoupled future (plate 7.1) may seem the province of the professional futurist and the science fiction author. We see at present adumbrations of it in plans for floating cities in Japan's waters which will take some of the pressure off its limited space for housing,

Plate 7.1 These oil rigs off the coast of Mexico contain a hint of a possible future in which an attempt is made (more thoroughgoing than at present) to decouple economic production from the biosphere. (Liba Taylor/Panos Pictures)

industry and agriculture. In a minor way, buildings like the Houston (Texas) astrodome, which isolates sports (and rock concerts: what a good idea) from the open air, are also tendencies in that direction. Suggested but not executed are plans to isolate nuclear power stations by putting them offshore (like an oil rig) or burying them under solid rock. A whole world of this type is imaginatively constructed by N. Calder in his book *The Environment Game* (1967); here the population lives under plastic domes, eats algae converted into any kind of food by cell culture, and people gain their status from their roles as protectors of the natural environment: reserve wardens have the sort of standing currently reserved for property magnates. We can only boggle at the political complexities of getting from here to there.

None of these alternative futures is concerned specifically with the LDCs, yet it seems proper that they should receive very special consideration. These are the nations and regions which have been down different tracks from the DCs since the nineteenth century and now are regarded as backward both by themselves and the DCs. Their conditions are often unenviable, both in terms of basic human dignity and of pressures upon the environment. Whereas the major environmental issues of the DCs centre on the relatively short list of the problems of the disposal of toxic and nuclear wastes, of acid rain, and of occupational health, the LDC list is rather longer. It encompasses food and water supply, fuelwood availability, deforestation, desertification, the extinction of species, together with droughts and floods, as well as the dispersal of the pathogens of human waste (plate 7.2). Yet some overall plans to help the LDC nations, such as the Brandt Report (1980), virtually ignore environmental and energetic considerations, emphasizing fiscal processes instead. There is probably no agreement at either academic or political level as to whether the higher per capita energy levels which would enable the quality of life in LDCs to be raised would best be supplied by nuclear-generated electricity, by fossil fuels (especially oil), or by locally based renewables such as solar power or biomass energy. And indeed, given the contemporary states of technology and politics, how feasible are enhanced levels of access? The search for a more desirable future

Plate 7.2 Just one of the problems experienced by many poorer countries: rapid soil erosion. We now know that institutional factors are also a critical element in this process: it is not simply an 'environmental problem'. (Mark Edwards Still Pictures/Panos Pictures)

might well look also at the social consequences of higher energy levels from fossil fuels, for these may be profound and do nothing to help either social or environmental problems. Socially as well as environmentally, local energy self-sufficiency seems to have a lot to offer if the absolute level per capita can be high enough.

Does it matter?

The implicit question behind the preceding paragraphs is 'do the amount and nature of environmental change matter?', which in more detail amounts to saying 'Are there in some ways natural limits to the amount of environmental change that the planet's systems can withstand without undergoing unpredictable fluctuations in the amplitudes of their parameters, and if so how can human societies adjust to such limits?' The boundaries, if they exist, are to be found within the two extremes of (a) an earth affected only slightly by human activity (either as a Golden Age to which it is impossible to return or as part of a technological decoupling), and (b) a planet which as in a scenario postulated in 1964 by J. Fremlin saw the result of 875 years' population growth at 2 per cent p.a. (60×10^{15} people) housed in a 2000-storey building covering the entire surface of the planet and limited in size by the availability of materials strong enough to bear loads at the temperature which the roof would reach in radiating out all the heat produced by the inhabitants and their machinery. Both these extremes look rather absurd in practical terms, and indeed Fremlin's idea was produced to show the absurdity of the idea of maintenance of a 2 per cent p.a. population growth rate for much longer.

Within these two unlikely extremes, it may be useful to employ the concepts of inner and outer limits. Though these have been variously defined, let us take them to mean boundaries to environmental alteration by human agency on a local or regional scale and on the global scale, respectively. Furthermore, we acknowledge that though such limits are both a creation and a conception of human societies, they also have a biophysical reality which will often manifest itself in the form of ecosystem breakdown and the drastic lowering of primary productivity. Thus it comes about that there is a very low quantity of energy in the biomass and a very low level of complexity of organization of the ecosystems. Examples of such transgression of inner limits are not difficult to find. Sometimes they appear to be temporary and can be reversed if certain causative processes are themselves deflected: a toxic waste poured into a lake may render the water body and its sediments virtually abiotic but if the effluent is then stored in tanks on land, the lake may well recover gradually. Others appear to be virtually irreversible: toxic gases from a metal smelter may kill all living organisms downwind from the chimney to the point where revegetation takes decades, for example, or over-intensive cropping of a soil will cause it to erode and be washed away, leaving an infertile subsoil or even bare rock. As an historical example, Hughes and Thirgood have argued that deforestation in ancient Greece and Rome was a cause of their collapse: not necessarily the only begetter but a major element in the economic and political demise of the classical world. Perhaps the point to be made is that in many of these cases the breakdown is reversible but the cost may well be enormous (and the technology not always available in the right place) unless the degradation process is halted before it is too advanced. Soil erosion is a good example: if arrested by soil conservation techniques then it can be more or less stopped, but if it is allowed to run its full course then reclamation for tillage becomes virtually impossible. Along with soil erosion, the processes of toxification of water bodies and desertification provide probably the most significant examples of the exceeding of 'inner' limits.

The transgression of inner limits as described above is nothing new: in each chapter of this book, some example has been shown to have happened, yet the world has gone on supporting more and more people. What is now feared is that the developments of the nineteenth and twentieth centuries have produced the capability to infringe the 'outer' limits: those of the planet as a whole may suffer so much perturbation that unpredictable and unwanted changes in the whole human habitat occur. Such developments might happen as a result of the transgression of apparently 'inner' limits on a very wide and persistent scale, or as the outcome of some restricted but intensive process which hits the planetary systems in some sensitive place. The kinds of limits foreseen by most authors include such phenomena as the shortage of vital resources like food, energy and minerals; the creation of climatic instability by the incremental addition of carbon dioxide and perhaps dust particles as well; the depletion of the ozone layer of the stratosphere by gases such as oxides of nitrogen and CFCs, allowing harmful intensities of UV radiation to reach the earth's surface; the perturbation of the great biogeochemical cycles of the planet by flooding parts of them with nitrogen or phosphorus; the poisoning of large areas of shallow seas which are the main sites of marine photosynthesis and hence affecting the albedo of the oceans via the rate of production of cloud-condensing nuclei; and the unleashing of even a medium-scale nuclear war.

Many of these are matters of controversy because not all the scientifically determinable facts about past and present processes are known, and even if they were somebody would dispute them for political or economic reasons as happened in the late 1980s over sulphur dioxide and CFC production in Europe. For each of these possible oversteps of global limits there are unknowns and alternative possibilities: food resources would be adequate if sociopolitical conditions were different; CFCs may not affect global ozone distribution on a measurable scale; the CO_2-produced 'greenhouse effect' might only produce minor and foreseeable climatic change; new methods of getting nitrogen to crop plants will soon eliminate the need for massive injections of N_2 into its global cycle at certain points; poisons will be amenable to degradation by bacteria that will be engineered to render them harmless; the studies on the effects of nuclear war are the result of unilateralists seeking a new form of propaganda. Each factor can be viewed from a different angle, yet it is apparently beyond any doubt that some infringements, if they work in predicted fashion, could produce effects which were very deleterious indeed for this globe as a human habitat. It seems at least prudent to watch all the potentially sensitive areas for possible outcomes and to try and forestall global difficulties, as is being done now with carbon dioxide, biogeochemical cycles, and the 'nuclear winter' debate.

Part of the problem is that of the unpredictability of global systems. In ecology at virtually any scale, prediction is probabilistic rather than deterministic and this is especially so where the global interaction of man and biophysical systems is concerned. The natural systems themselves are difficult enough to comprehend scientifically, but add the varied activities of human groups and the results tend to be stochastic: nobody is sure how the world's ecology hangs together in its totality.

One holistic model is that put forward by J. E. Lovelock and L. Margulis and called by them the Gaia hypothesis, after the Greek name for the earth goddess (plate 7.3). This view starts from the idea that the climate and chemistry of the earth have for aeons been optimal for life. It seems unlikely that this could have happened by chance, and the usual explanation is that organic evolution adapted to the physical chemistry of a cooling planet. Going further, the Gaia hypothesis suggests that the chemical compositions of the atmosphere and oceans are the product of life, not that life is adapted to them. Living systems, therefore, produce a condition of homeostasis which exemplifies the optimum conditions for life. Physical and chemical entities while not themselves living become extensions of vital systems designed to maintain a good environment for life, in the manner of a cat's fur or a wasp's nest. More formally, Gaia can be defined as 'a complex entity involving the Earth's biosphere, atmosphere, oceans and soil; the totality constituting a feedback or cybernetic system which seeks an optimal physical and chemical environment for life on this planet.' Lovelock believes that any threats to outer limits will not come from regional industrial pollution in the northern hemisphere. Firstly, this is where the decision-makers live and where knowledge is best so correction can be most readily applied; secondly, the evidence of Pleistocene glaciations shows that perhaps 30 per cent of the earth's surface, outside 45 °N and 45 °S, can become sterile, with virtually nothing but ice, snow and bare rock. Yet the basic stabilities of the planet allowed the return of living systems once the ice had become water once again. Thus, runs the argument, industrially produced toxicities or low biological productivities are unlikely to equal the scale of Pleistocene perturbations and so, when coupled with the rapid feedback possible in highly organized nations, do not constitute a threat to global integrity. The core area of planetary coherence is the 45–50 degree belt and it is here

Plate 7.3 Pictures of the Earth from space (Apollo 17 in this case) have had a powerful effect in promoting the idea of the planet as a single system, as in the Gaia hypothesis. (NASA/Bruce Coleman Ltd)

that interference on a large scale with biogeochemical cycles might disrupt global unity. The areas that need watching, therefore, are the land surfaces and the near-shore waters of the tropics. Also of importance are the continental shelves of temperate and even sub-arctic zones. In these places it is possible to reduce global productivity and delete key species in the life-support system, and to exacerbate the situation by releasing into the air or the sea smaller quantities of certain compounds like dimeythl sulphide. Hence the ways in which the outer limits may be first breached come from a combination of the clearance of the tropical moist forests (and other vegetation types of the tropics which fix a lot of carbon), and the emission of carbon dioxide in the industrialized nations, coupled with any processes which diminish primary productivity on the continental shelves of the world, again especially in the tropics. Pesticide runoff, nutrient loadings which lead to stifling blooms of small animals, and silt from unhampered land clearance for any number of purposes, may play these roles. The Gaia model is not only plausible historically, but is predictive at a variety of spatial scales and can point to processes which ought to be the concern of the global human community, with the added complication that, as Lovelock puts it, 'what we do to our planet may depend greatly on where we do it'.

Last thoughts

Energy, entropy, culture

In his biologically based account of human history, P. Colinvaux asserts that 'Happiness, liberty and war depend upon changes in our own numbers alongside changes in the resources we can release'. So strategies of human and environmental development are likely to be very important. The whole question of energy could well be central. It is, as we have seen, instrumental in differentiating between an economic–technical pathway and the environmentalist alternative: both the types of energy used and the intensities with which it is used are envisaged in different ways. In particular, the concept of entropy as applied by N. Georgescu-Roegen and by E. Jantsch needs attention. The latter author draws our attention to the fact that societies are self-organizing 'dissipative structures', that is to say their complexity and their coherence are bought at the cost of using high-grade or low-entropy energy, largely in either solar or stored solar form. To erect a complex economic unit is to set in motion a higher rate of entropy formation than a simple society and there is no apparent escape from that process. There are nevertheless ways in which the creation of entropy by human societies can be slowed down and in particular there are paths which avoid the waste of negative entropy (i.e. high-grade energy which can do work) in goods and services which add little to either our dignity, our happiness or our potential for survival: conspicuous consumption in the form of energy-intensive possessions like sophisticated weapons and very large motor cars are obvious examples. Georgescu-Roegen goes so far as to suggest a set of activities which will in their wake inevitably bring about different rates of environmental manipulation. These he calls a 'bioeconomic' programme and it attempts to slow down the rates of entropy formation and the accompanying dissipation of matter into an unrecoverable state. The first step is not only to outlaw war but to prohibit the manufacture of instruments of war since apart from anything else, modern war represents a high level of energy dissipation without any useful result. In the direct field of population and environment, he suggests the quite rapid development of the LDCs, together with the attempt to lower the world's population to that which can be fed by 'organic' agriculture, i.e. food-producing systems not underpinned by constant inputs of fossil fuel. Until solar energy or atomic fusion energy becomes available to human societies in virtually limitless quantities, he proposes that energy wastage is something of a luxury which should be stopped. One part of the programme is a revised attitude to material possessions in which we cure ourselves of a morbid craving for extravagant gadgetry, acknowledge that 'fashion' is a disease of the human mind and above all insist that goods should be durable rather than ephemeral. Lastly, with many modern social thinkers, we need to acknowledge that work is not the only end of life (and there are signs that its importance is being re-valued since Georgescu-Roegen set out his programme in the 1970s, and that the spending of substantial leisure in an intelligent manner is a requisite of a good life.

In the long term, it seems as if the widespread use of a plentiful source of energy (like solar power or nuclear fusion) which renders the entropy problem virtually irrelevant, might hold out a hope for a good life for all of earth's citizens. At the moment societies with very low extra-somatic energy intensities are in general poor in every material sense, with low levels of nutrition and health. On the other hand, those nations at the opposite end of the spectrum exhibit a whole set of pathologies of affluence. It is simplistic to assert that these problems are caused by energy availability rather than called forth by some set of cultural traits, but none the less they are made possible by access to large stores of power, largely derived from fossil fuel. So at these extremes, energy availability appears as a disabling feature of the structure of society, whereas in the middle ranks, it is an enabling feature which allows a society to achieve adequate nutrition and health standards and to give individuals some meaningful work as well as adequate leisure. It might thus be a lesson to take away from energy studies that in all kinds of ways, and not just those of environmental impact, a sane society will decide to limit the per capita quantities of energy used. Too much ice-cream causes stomach ache, no matter how varied the flavours.

If in a more immediate sense we are looking for process criteria to regulate the actual year-to-year decisions that have to be made about man–environment relations then two come obviously to hand. The first is sustainability, which is in the long term linked with the process of entropy creation discussed above. In fact, it is usually employed as a middle-range term since planning as we know it rarely goes beyond that time-frame. The idea simply elicits the question, 'Is the process we are about to create one which can be maintained into the foreseeable future without dependence upon non-renewable resources?' If for example the process gives rise to a biological product, as in farming or forestry, then we also want to know whether significant inputs of fossil fuel are required as well. If a city is to be built, then are its inhabitants to be fed, watered, warmed, cooled and drained by the use of renewable resources or are they essentially to be consumers of an exhaustible set of stocks? Of course it is not quite so simple since we must ask 'sustainable at what level?' This may involve a cultural judgement: the population sustainable on a vegetarian diet may well be much larger than one nourished on hamburgers, french fries and greenhouse lettuce. The second concept is even more complex: that of reversibility. It is relatively easy to say whether a proposed or existing process leads to irreversible environmental changes, but much more difficult to answer the question 'does that matter?'. It is quite often argued that the latter is not a significant question, provided that we can happily shelter under the umbrella of the judgement that we ought not, by our actions, to foreclose options for future generations. This is a version of the duty to posterity argument, and naturally enough there are counterproposals to the effect that the best situation we can create for our successors is to do as much as we can with everything now so as to create a pool of possibilities. But irreversibility involves irreplaceable loss: most frequently of genetic material due to biological extirpation (at regional as well as global scales), but also of inorganic materials which are so dissipated as to make their recovery impossible in the sort of time-scales which we can contemplate, like the quantity of phosphorus put into the ocean basins. Thus as advocates of home motor-cycle maintenance might put it, if you're going to tinker with the engine, make sure you keep all the bits since Planet Earth has no spare-parts shops in the next neighbourhood. Appeals to reversibility as a criterion for environmental management need, then, to differentiate between an absolute irreversibility, in which it is completely impossible now or in the near future to reverse a process (to recreate genetic material for example, or to sieve out the residual DDT from the oceans), and an economic irreversibility, in which nobody at present thinks it worth the cost of doing, like clearing up the sites of abandoned factories and restoring them to food production or trying to re-establish forest on eroded slopes. So while sustainability and reversibility have obvious appeal in the short term, their meanings are not absolute.

Transformations

In the search for new beginnings and new ends for the mutual involvements of man and nature the role of metaphor has become important, and in this field of study the figures of St. Francis of Assisi and St. Benedict of Nursia have come to represent two different types of creative thinking in the West (and of the West: the contribution of modern scholars from elsewhere has not been very great) which try to move beyond the conventional western world-view prevalent since the nineteenth century. In recent years the icon of St. Francis has been held before both scholarly and lay communities by the historian Lynn White. In his much discussed and quoted paper of 1967, he proposed that there should be a rivival of the outlook and *mores* of St. Francis of Assisi (1181–1226), whom he proposed as the patron saint of ecologists. White's original argument has been attacked but the powerful image remains of a way of life in which the natural is given an equal or perhaps superior place alongside the human and the man-made, and in which simplicity in the material sense is exalted as a sustainable and hence superior life-style. Against this outlook, the medical scientist and humanist René Dubos put forward the claims of St. Benedict of Nursia (480–550), best known for the Benedictine Rule and for the numerous religious communities which have put it into practice. Dubos has claimed for St. Benedict that his followers

(a)

(b)

Plate 7.4 Symbols of two attitudes to the environment: (a) Benedict of Nursia, associated with constructive 'improvement' of nature; (b) Francis of Assisi, associated with a more quietist approach which accords an intrinsic value to the non-human. (Alinari/The Mansell Collection)

have engaged not in a romantic and atavistic flight to the 'natural', but have in a practical way brought about creative transformations of nature: turning waste places into productive agriculture, for instance. In this tradition he sees the great landscape gardens of eighteenth-century England, and the mixed farmland, woodland and small-scale settlement of Wisconsin or Lancaster County, Pennsylvania. He claims that the Benedictine tradition is one of stewardship, care and reverence for the qualities of the land and that given such a positive and beneficial type of change, nobody ought to want to expend energy on preserving the wild for its own sake. Just as White's interpretations of the influence of religious history have been challenged, so have Dubos's views: they are, it can be argued, too much based on the attractive visual outcome of processes in which social injustice was often manifest. Cistercian abbeys, for example, accumulated great wealth and were not above dispossessing peasants of their land and removing their settlements. Likewise, the great landscape gardens of William Kent and 'Capability' Brown (p. 160) may have occupied land which formerly supported yeoman farmers or smallholders.

If we are to incorporate these ideas but still to transcend the apparent dichotomy they present then perhaps we have to eschew the notion of a religious calendar in which there is only one saint. Thus at both intellectual and practical levels what we need is an environmental Book of Hours with both luminaries in their due seasons and in their appropriate places. After all, transformation of the earth's surface seems to be one of the inevitable concomitants of human societies down the millennia if we believe the story told here in chapters 3–6, and we should therefore be concerned that such metamorphoses ought to be creative rather than destructive. Yet there seems a persuasive case that no alterations ought to be total. On practical grounds, we need the unchanged, the wild, the natural because they are an evolving store of new materials (especially where living organisms are involved), may well be governors of biogeochemical cycles, and because the behaviour of natural systems may afford us clues as to the 'best' way of using ones like them for our own purposes. Beyond that, most industrial cultures now place a considerable value upon the wild, as 'being something other'. People may wish to visit such places for recreation, or to watch TV programmes about them, or scarcely consciously just to know that they are there. Such a deeply held set of views cannot simply be dismissed as 'emotional' by advocates of change.

This still leaves us with the problem of distinguishing creative transformation from its destructive opposite. Since no magic formula can be produced here, we can only resort to hints and guesses. The

Plate 7.5 Beyond duality: the garden of the Ryoan-ji in Kyoto. (Japanese Information Centre)

most important is, perhaps, to be open to innovations in terms not only of new technology, but also of new ways of applying it and new social structures. As in organic evolution, many of these will wither, but others may flourish and form the nurseries of the long-term viability which we seek. Being open must mean being ready to receive ideas from sources other than the orthodox repositories of scientific and technological information. T. Roszak argues that if the Gaia hypothesis is a reality then the planet must be able in some way to communicate with her inhabitants and tell them how to behave. Such knowledge is as likely to come intuitively as rationally and even those who would find Roszak and his followers too mystical (or downright fanciful) might accept that we place too much emphasis on the development of one brain hemisphere at the expense of the other. If we were to be 'greened' as individuals, with the more formal societal structures following only as a consequence, it is then possible to glimpse, albeit dimly, the kind of new self-organizing, co-evolutionary relationship in which, Jantsch wrote, 'Learning . . . would be a creative game played with reality . . . creative processes would be permitted to unfold and form new structures.' Happily, there is even some extravagance allowed, since evolution can be interpreted as being not merely functional. Just what all this might look like (in terms, say, of the kind of systems diagrams used frequently in this book) is not possible to predict, and we can console ourselves that most forecasts made just before a major phase change have turned out to be widely wrong, as in the classic case of the prognosis of major cities disappearing early in the twentieth century beneath great piles of horse dung. For each present moment, if we study the systems of the planet with all the rigour and care that science dictates and learn both its potentials and limits, and in a more meditative posture attend to its emergent properties (to which we may perhaps attach the term its way or Tao), and come in that way to explore our own horizons, then these, together with the prudent practical measures of the kind cited earlier in this chapter, may form one pathway:

For most of us, this is the aim
Never here to be realized
Who are only undefeated
Because we have gone on trying

Further Reading

Chapter 1 General Introduction

Ideas about the relations of man and nature

The basic work for the West from classical antiquity to the eighteenth century is C. Glacken, *Traces on the Rhodian Shore*. Berkeley and Los Angeles: University of California Press, 1967. Later ideas of interest are in R. Pryce, *Approaches to the Study of Man and Environment*. Open University Course D204, units 1–3, 1977, 47–90. Other works of relevance are S. R. Eyre and G. R. J. Jones (eds), *Geography in Human Ecology*, London: Edward Arnold; F. Braudel, *The Structures of Everyday Life: Civilization and Capitalism 15th–18th centuries*, London: Collins, 1981, vol. I; G. P. Marsh, *Man and Nature, or Physical Geography as Modified by Human Action*, New York: Scribner, 1864 (new edition, ed. D. Lowenthal, Cambridge Mass.: Belknap Press, 1965); and the monumental commemorative volume for Marsh, W. L. Thomas (ed.), *Man's Role in Changing the Face of the Earth*, Chicago: Chicago UP, 1956. See also J. Iversen, 'Landnam i Danmarks stenalder', *Denmarks Geologiske Undersøgelse* 2, no. 66, 1941; Yi Fu Tuan, 'Our treatment of the environment in ideal and actuality', *Amer Sci* 58, 1970, 244–49; D. J. Herlihy, 'Attitudes towards the environment in Medieval Society', in L. J. Bitsky (ed.), *Historical Ecology: Essays on Environmental and Social Change*, London and Port Washington, NY: Kennikat Press, 1980, 100–16; V. Westhoff, 'Man's attitude towards vegetation', in W. Holzner, M. J. A. Werger and I. Ikusima (eds), *Man's Impact on Vegetation*, The Hague, Boston and London: Junk, 1983, 7–24. A socio-political history of the world between ecological covers is A. Toynbee, *Mankind and Mother Earth*, Oxford University Press, 1976. Two popular accounts are W. M. S. Russell, *Man, Nature and History*, London: Aldus Books, 1967; and B. Campbell, *Human Ecology: The story of our Place in Nature from Prehistory to the Present*, London: Heinemann, 1983. More scholarly interpretations of nature and history include S. Boyden, *Western Civilization in Biological Perspective: Patterns in Biohistory*, Oxford: Clarendon Press, 1987; L. D. Levine (ed.), *Man in Nature: Historical Perspectives on Man in his Environment*, Toronto: Royal Ontario Museum, 1976; and also A. Maczak and W. N. Parker (eds), *Natural Resources in European History*, Washington DC: RFF Research Paper R-13,

1978. The best account, both scholarly and readable, is E. L. Jones, *The European Miracle: Environments, Economies and Geopolitics in the History of Europe and Asia*, Cambridge: CUP, 1981. The quotation from J. M. Roberts is at the end of his *The Triumph of the West*, London: BBC publications, 1985.

The interaction of energy and society is explored in a number of works. Basic sources used here are H. T. and E. C. Odum, *Energy Basis for Man and Nature*, New York: McGraw-Hill, 1976; K. Newcombe, *A Brief History of Concepts of Energy and the Use of Energy by Humankind*, Canberra: ANU Centre for Resource and Environment Studies, publication HEG 1–76, 1976; F. Cottrell, *Energy and Society*, Westport, Conn.: Greenwood Press 1955; R. N. Adams, *Energy and Structure: A Theory of Social Power*, Austin and London: University of Texas Press, 1975; N. Georgescu-Roegen, 'The entropy law and the economic problem', in *idem* (ed.), *Energy and Economic Myths*, Oxford: Pergamon, 1976, 53–60; T. C. Edens, 'Cassandra and the horn of plenty: ecological and thermodynamic constraints and economic goals', *Urban Ecology* 2, 1976, 15–31; H. T. Odum, 'Macroscopic minimodels of man and nature', *Systems Analysis and Simulation in Ecology* 4, 1976, 249–79; H. T. Odum, 'Enmergy (*sic*) in ecosystems', in N. Polunin (ed.), *Ecosystem Theory and Application*, Chichester: Wiley, 1986, 337–69. J. M. Alier and J. M. Naredo, 'A Marxist precursor of energy economics: Podolinsky', *J. Peasant Studies* 9, 1982, 208–24; B. M. Hannon, 'An energy standard of value', *Ann Amer Acad Pol & Soc Sci* 405, 1973, 139–53; G. M. Woodwell, 'Success, succession and Adam Smith', *BioScience* 24, 1974, 81–7; F. H. Bormann, 'An inseparable linkage: conservation of natural ecosystems and the conservation of fossil energy', *BioScience* 26, 1976, 754–60; R. N. Adams, *Paradoxical Harvest: Energy and Explanation in British History 1870–1914*, Cambridge: CUP, 1982; D. M. Gates, *Energy and Ecology*, Sunderland, Mass.: Sinauer, 1985; L. A. White, 'Energy and the evolution of culture', *Amer Anthropol* 45, 1943, 335–56; *idem*, *The Evolution of Culture*, New York: McGraw-Hill, 1959. The special case of agriculture is examined in technical detail in G. Stanhill (ed.), *Energy and Agriculture*, Berlin: Springer-Verlag, 1984, Advanced Series in Agriculture 14. The quotation from Sir Frederick Soddy FRS is in his *Wealth, Virtual Wealth and Debt: The Solution of the Economic Paradox*, London: Allen and Unwin, 1933, 2nd edn. The regional studies mentioned are: J. Browden, C. Littlejohn and D. Young, *The South Florida Study*, Tallahassee: University of Florida Center for Wetland and State Bureau of Comprehensive Planning, 1976; C. S. Holling (ed.), *Adaptive Environmental Assessment and Management*, New York and Chichester: Wiley, 1978 (the Obergurgl study); J. W. Shupe, 'Energy self-sufficiency for Hawaii', *Science* 216, 1982, 1193–9; J. Zuchetto and A-M. Jansson, *Resources and Society: a Systems Ecology Study of the Island of Gotland, Sweden*, New York: Springer-Verlag, Ecological Studies 56, 1985; see also A.-M. Jansson and B.-O. Jansson, 'Energy analysis approach to ecosystem redevelopment in the Baltic Sea and Great Lakes', *Ambio* 17, 1988, 131–36; the Hong Kong studies are referenced with chapter 5.

Organizing concepts

Modern ecology is dominated in its functional approach by the work of E. P. Odum, *Basic Ecology*, Philadelphia: Saunders, 1983, which follows his *Fundamentals of Ecology*, Philadelphia: Saunders, 3rd edn, 1971, and the shorter version, *Ecology*, New York: Holt, Rinehart and Winston, 2nd edn, 1975. See also H. T. Odum, *Systems Ecology*, Chichester: Wiley, 1983. A wider view of ecology can be seen in P. Colinvaux, *The Fates of Nations: A Biological Theory of History*, NY: Simon & Schuster, 1980; Harmondsworth: Pelican Books, 1980.

Biogeochemical cycles are dealt with in G. E. Likens (ed.), *Some Perspectives of the Major Biogeochemical Cycles*, Chichester: Wiley, 1981, SCOPE vol. 17; B. Bolin and R. B. Cook (eds), *The Major Biogeochemical Cycles and their Interactions*, Chichester: Wiley, 1983, SCOPE vol. 21. Applications of ecological theory are explored in G. H. Orians (ed.), *Ecological Knowledge and Environmental Problem-Solving: Concepts and Case Studies*, Washington DC: National Academy Press, 1986; N. Polunin (ed.), *Ecosystem Theory and Application*, Chichester: Wiley, 1986.

The adoption of systems-based ecological ideas into anthropology is summed up in J. W. Bennett, *The Ecological Transition: Cultural Anthropology and Human Adaptation*, Oxford: Pergamon, 1976; but a number of shorter works are in places illuminating, e.g., J. Dow, 'Systems models of cultural ecology', *Soc Sci Inform* 15, 1976, 953–76; G. L. Young, 'Human ecology as an interdisciplinary concept: a critical enquiry', *Adv Ecol Res* 8, 1974, 1–105; S. Cook, 'Production ecology and economic anthropology: notes towards an integrated frame of reference', *Soc Sci Inform* 12, 1973, 25–52; B. S. Orlove, 'Ecological anthropology', *Ann Rev Anthropol* 9, 1980, 235–73; P. J. Richerson, 'Ecology and human ecology: a comparison of theories in the biological and social sciences', *Amer Ethnol* 4, 1977, 1–27. Vol. 109, 1986, of *Int Soc Sci J* is entitled 'Environmental Awareness': see the papers by F. H. Buttel, 'Sociology and the environment: the winding road toward human ecology', 337–56, and S. M. Macgill, 'Environmental questions and human geography', 357–76. Another compilation of considerable interest is E. F. Moran (ed.), *The Ecosystem Concept in Anthropology*, Boulder, Co.: Westview Press, 1984, AAAS Selected Symposia 92. There is a useful chapter on the history of ideas about human/habitat interactions, also by E. F. Moran, in his *Human Adaptability: An Introduction to Ecological Anthropology*, North Scituate, Mass.: Duxbury Press, 1979.

The ecosystem concept as used here derives from E. P. Odum, *Basic Ecology*, the founding document being R. I. Lindeman, 'The trophic–dynamic aspect of ecology', *Ecology* 23, 339–418. The work of the IBP on production ecology is summed up in H. Lieth and R. H. Whittaker (eds), *Primary Productivity of the Biosphere*, Berlin, Heidelberg and New York: Springer-Verlag, 1975, Ecological Studies 14. The stability question is illuminated by, *inter alia*, R. Margalef, *Perspectives in Ecological Theory*, Chicago: Chicago UP, 1968; C. S. Holling, 'Resilience and stability of ecosystems', in E. Jantsch and C. H. Waddington (eds), *Evolution and Consciousness: Human Systems in Transition*, Reading, Mass.: Addison-Wesley, 1976, 73–92; S. L. Pimm, 'The complexity and stability of ecosystems', *Nature, Lond.* 307, 1984, 321–6; A. R. Hill, 'Ecosystem stability: some recent perspectives', *Progr Phys Geog* 11, 1987, 315–33. See also G. H. Orians, 'Diversity, stability and maturity in natural ecosystems', in W. H. van Dobben and R. H. Lowe-McConnell (eds), *Unifying Concepts in Ecology*, The Hague: Junk, 1975, 139–50. A recent exploration of ecological theory of relevance is R. V. O'Neill et al., *A Hierarchical Concept of Ecosystems*, Princeton and Guildford: Princeton UP, 1986, Monographs in Population Biology 23. For more applied work see W. A. Reiners, 'Disturbance and basic properties of ecosystem energetics', in H. A Mooney and M. Godron (eds), *Disturbance and*

Ecosystems: Components of Response, Berlin: Springer-Verlag, 1983, Ecological Studies no 44, 83–93; G. H. Orians (ed.), *Ecological Knowledge and Environmental Problem-Solving*, Washington DC: National Academy Press, 1986.

General works of interest include A. S. Goudie, *The Human Impact on the Natural Environment*, Oxford: Basil Blackwell, 2nd edn, 1986; C. F. Bennett, *Man and Earth's Ecosystems*, Chichester: Wiley, 1975; P. Danserau, 'Ecology and the escalation of human impact', *Int J Soc Sci* 22, 1970, 628–47. See also W. Holzner et al., *Man's Impact on Vegetation*, and the interesting check-list of rural impacts for the UK 'Chart of human impacts on the countryside', in *The Countryside in 1970*, London: HMSO, 1962, 113–51. The population history of mankind is treated in C. Cipolla, *An Economic History of World Population*, Harmondsworth, Pelican Books, 1970, 5th edn; and in C. McEvedy and R. Jones (eds), *Atlas of Population History*, Harmondsworth: Pelican Books, 1978.

Climate and history is documented in narrative form by E. Le Roy Ladurie, *Times of Feast, Times of Famine: a History of Climate since the Year 1000*, London: Allen and Unwin, 1972. More analytical approaches can be found in M. J. Ingram, G. Farmer and T. M. L. Wigley, 'Past climates and their impact upon man: a review', in T. M. L. Wigley, M. J. Ingram and G. Farmer (eds), *Climate and History: Studies in Past Climates and their Impact on Man*, Cambridge: CUP, 1981, 3–50; M. J. Bowden et al., 'The effect of climate fluctuations on human populations: two hypotheses', in T. M. L. Wigley et al., *op. cit.*, 479–513; J. de Vries, 'Measuring the impact of climate on history: the search for appropriate methodologies', *J. Interdisciplinary History* 10, 1980, 599–630. The account of disasters is condensed from E. L. Jones, *The European Miracle: Environments, Economics and Geopolitics in the History of Europe and Asia*, Cambridge: CUP, 1981, ch 2.

Chapter 2 Primitive Man and his Surroundings

General

A basic introduction is given in the *Cambridge Encyclopedia of Archaeology*, Cambridge: CUP, 1980, 51–184. Accessible accounts of prehistory in Africa and the earliest men in Europe and Asia can be found in J. G. D. Clark, *World Prehistory in New Perspective*, Cambridge: CUP, 1977; and at rather greater length in R. J. Wenke, *Patterns in Prehistory*, New York: OUP, 1980, where alternative *Homo* lineages then current are given. See also J. Desmond Clark (ed.), *The Cambridge History of Africa. Vol I: From the Earliest Times to c. 500 BC*, Cambridge: CUP, 1982. J. Desmond Clark gives a long temporal perspective in his 'Early human occupation of African savanna environments', in D. R. Harris (ed.), *Human Ecology in Savanna Environments*, London: Academic Press, 1979, 41–71. For the general environmental background see A. S. Goudie, *Environmental Change*, Oxford: Clarendon Press, 2nd edn, 1983, where there are some very useful palaeoenvironmental and chronological charts. A basically biological approach to the emergence of man is taken in J. S. Weiner, *Man's Natural History*, London: Weidenfeld & Nicolson, 1977. Some nice photographs of Palaeolithic tools are found in D. Birdsall and C. M. Cipolla, *The Technology of Man*, London: Wildwood House, 1980, 21–32. An excellent popular but up-to-date account of the Palaeolithic and the early Neolithic civilizations is

J. A. J. Gowlett, *Ascent to Civilization: The Archaeology of Early Man*, London: Collins, 1984.

Mankind evolving and spreading

The standard work of synthesis on Africa is J. Desmond Clark, *The Prehistory of Africa*, London: Thames and Hudson, 1970: see also H. M. McHenry, 'Fossils and the mosaic nature of human evolution', *Science* 190, 1975, 425–31. A lineage for mankind with wide, though not universal, scholarly acceptance is described in P. Andrews, 'The descent of man', *New Scientist* 102 (No. 1405), 1984, 24–5; an equally lucid though more technical account is J. E. Cronin et al., 'Tempo and mode in hominid evolution', *Nature, Lond.* 292, 1981, 113–22; then see E. Delson, 'Human phylogeny revised again', *Nature, Lond.* 322, 1986, 496–97, but the story keeps changing. Very early evidence of burned bones associated with tools is seen in J. E. Kalb et al., 'Fossil mammals and artifacts from the Middle Awash Valley, Ethiopia', *Nature, Lond.* 298, 1982, 25–9, and very early Acheulian tools in C. A. Repenning and O. Fejfar, 'Evidence for earlier date of Ubeidiya, Israel, hominid site', *Nature, Lond.* 299, 1982, 344–7. The dating of a *Homo erectus* skeleton to 1.6 Myr is in F. Brown et al., 'Early *Homo erectus* skeleton from West Lake Turkana, Kenya', *Nature, Lond.* 316, 1985, 788–92. Wider approaches are in G. E. Kennedy, 'The emergence of modern man', *Nature, Lond.* 284, 1980, 11–12; and K. Luchterhand, 'Late Cenozoic climate, mammalian evolutionary patterns and Middle Pleistocene human adaptation in eastern Asia', in L. G. Freeman (ed.), *Views of the Past*, The Hague: Mouton, 1978, 363–421. Environmental settings are described in several pieces by G. Ll. Isaac, notably 'The activities of early African hominids', in G. Ll. Isaac and E. R. McCown (eds), *Human Origins*, Menlo Park, Calif.: Staples Press, 1976, 483–514; and 'The diet of early man: aspects of archaeological evidence from lower and middle Pleistocene sites in Africa', *World Archaeology* 2, 1971, 278–98. See also the relevant chapters of *The Cambridge Encyclopedia of Archaeology*, Cambridge: CUP, 1980; and the text by J. J. Wymer, *The Palaeolithic Age*, London: Croom-Helm, 1982, which is strong on environmental matters. The technology of early *Homo* species is dealt with by J. G. D. Clark, *World Prehistory*, 1977 and G. Ll. Isaac, 'Early stone tools – an adaptive threshold', originally published in *Scientific American* in April 1978, and reproduced in C. C. Lamberg-Karlovsky (ed.), *Hunters, Farmers and Civilizations*, San Francisco: Freeman, 1979, 22–35; also R. G. Klein, 'The ecology of early man in southern Africa', *Science* 197, 1977, 115–26. The case of early man further east is the special topic of F. Ikawa-Smith (ed.), *Early Palaeolithic in South and East Asia*, The Hague: Mouton, 1978; see also Luchterhand, 'Late Cenozoic climate', op. cit., 1978. The closest paper in spirit to this chapter is K. W. Butzer, 'Environment, culture and human evolution', *American Scientist* 65, 1977, 572–84.

Early man, technology and environment

These sections are compiled from a variety of sources whose purpose was mostly something else. G. Ll. Isaac, op. cit., 1971 and 1979 were also used, as was his *Olorgesailie: Archaeological Studies of a Middle Pleistocene Lake Basin in Kenya* Chicago and London: Chicago UP, 1977. A broad review by Isaac is 'Casting the new wide: a review of archaeological evidence for early hominid land-use and ecological relations', in L.-K. Konigsson (ed.), *Current Argument on Early Man*, Oxford: Pergamon Press for the Royal Swedish Academy of Sciences, 1980, 226–51. See also C. E. Read-Martin and D. W. Read, 'Australopithecine scavenging and human evolution: an approach from faunal analysis', *Current Anthropology* 16, 1975, 359–68; a 'scavenging' bone assemblage is reported by H. Bunn et al., 'FxJj50: an early Pleistocene site in northern Kenya', *World Archaeology* 12, 1980, 109–36; see also J. B. Birdsell, 'Some predictions for the Pleistocene based on equilibrium systems among recent hunter–gatherers', in R. B. Lee and I. DeVore (eds), *Man the Hunter*, Chicago: Aldine Press, 1968, 229–40; J. D. Speth and D. D. Davis, 'Seasonal variability in early hominid predation', *Science*, 192, 1976, 441–5; R. G. Klein, 'The ecology of early man in southern Africa', *Science* 197, 1977, 115–26; T. Hatley and J. Kappelman, 'Bears, pigs and Plio-Pleistocene hominids: a case for the exploitation of belowground food resources', *Human Ecology* 8, 1980, 371–87. The population estimate is from C. McEvedy and R. Jones, *Atlas of World Population History*, Harmondsworth: Penguin, 1978. Butzer's statements are from the second edition of his *Environment and Archaeology*, London: Methuen, 1972. The overkill debate in Africa starts with P. S. Martin, 'Africa and Pleistocene overkill', *Nature, Lond.* 212, 1966, 339–42; reply by L. S. B. Leakey, *ibid*. 212, 1615–16 and a further rally in *ibid*. 215, 212–13 and *ibid*. 215, 213. Some relevant remarks on *Homo* as a predator are seen in S. L. Washburn and C. S. Lancaster, 'The evolution of hunting', in Lee and DeVore, *Man the Hunter*, op. cit., 293–303.

Homo erectus sites in Europe and Asia are neatly summarized in Wencke 1980; he had not kept up with M. J. Bishop, 'Earliest record of man's presence in Britain', *Nature, Lond.* 253, 1975, 95–7, for Westbury-sub-Mendip; but for Europe generally at this period a good, though dated, source is F. C. Howell, 'Observations on the earlier phases of the European Lower Palaeolithic', *Amer Anthropol* 66, 1966, 88–201. An accessible account of Terra Amata is by H. de Lumley in *Scientific American* 220 (5), 1969, 92–9, and reproduced in Lamberg-Karlovsky, *Hunters, Farmers and Civilizations*, S. F. Freeman, 1979, 57–65. The basic information on Zhoukoudian (Choukoutien) is in H. L. Movius, 'The Lower Palaeolithic culture of Southern and Eastern Asia', *Trans Amer Philos Soc* 38, 1948, 329–420; see also P. Tobias, 'New developments in hominid palaeontology in South and East Asia', *Ann Rev Anthropol* 2, 1973, 311–34. The fire-related episodes in the Hoxnian (Holstein) interglacial are described by R. G. West and C. M. B. McBurney, 'The Quaternary deposits at Hoxne, Suffolk, and their archaeology', *Proc Prehist Soc* 20, 1954, 131–54; and by C. Turner, 'The Middle Pleistocene deposits at Marks Tey, Essex', *Phil Trans Roy Soc* B 257, 1970, 373–437. They are put in a wider context (but with the same cautious conclusions) by R. G. West, 'Pleistocene forest history in East Anglia', *New Phytol* 85, 1980, 571–622, and by C. Turner, 'The correlation and duration of Middle Pleistocene periods in Northwest Europe', in K. W. Butzer and G. Ll. Isaac (eds), *After the Australopithecines*, The Hague and Paris: Mouton, 1975, 259–308. Interesting ideas on early hunter–gatherers can be found in A. C. Hamilton, *Environmental History of East Africa: A Study of the Quaternary*, London: Academic Press, esp. ch. 11.

A number of specialized symposia on the environment and evolution of early man are generally detailed reports of data, useful only to those interested in the fine detail. Examples are Y. Coppens et al. (eds), *Earliest Man and Environments in the Lake Rudolf Basin*, Chicago: Chicago UP,

1976; F. C. Howell and F. Bourlière (eds), *African Ecology and Human Evolution*, Chicago: Aldine, 1963; W. W. Bishop and J. D. Clark (eds), *Background to Evolution in Africa*, Chicago: Chicago UP, 1967; volumes describing the various finds at Olduvai Gorge by one or more of the Leakeys and their co-workers appear at regular intervals. A detailed account of Torralba and Ambrona is given by F. Clark Howell, 'Observations on the earlier phases of the European Lower Palaeolithic', *Amer Anthropol* 68 (1966) 88–201; and summaries in Butzer, *Environment and Archaeology*, London: Methuen, 2nd edn, 1972; and 'Environment, culture and human evolution', *Amer Sci* 65, 1977, 572–84. The later reinterpretation by L. R. Binford is, 'Were there elephant hunters at Torralba?' in M. H. and D. V. Nitecki (eds), *The Evolution of Human Hunting*, New York and London: Plenum Press, 1987, 47–105. The Clacton yew-wood spear mentioned in passing is illustrated in Birdsall and Cipolla, op. cit., 1980, 24–5.

The history and ecology of fire

Most of the evidence for early use comes from the appropriate sites, see above, especially J. A. J. Gowlett et al., 'Early archaeological sites, hominid remains and traces of fire from Chesowanja, Kenya', *Nature, Lond.* 294, 125–9; M. Barbetti et al., 'Palaeomagnetism and the search for very ancient fireplaces in Africa', *Anthropologie*, 18, 1980, 299–304. A review paper on fire in the early Pleistocene of Africa is J. D. Clark and J. W. K. Harris, 'Fire and its roles in early hominid lifeways', *The African Archaeological Review* 3, 1985, 3–27. There is a general account by K. P. Oakley, 'The earliest fire-makers', *Antiquity* 30, 1956, 102–7, which introduces the collector/producer distinction. The ecology of fire is dealt with in most ecology and biogeography texts but see also R. Daubenmire, 'Ecology of fire in grasslands', *Adv Ecol Res* 5, 1968, 209–66; T. T. Kozlowski and C. E. Ahlgren (eds), *Fire and Ecosystems*, London and New York: Academic Press, 1974; and the annual publications (1962 onwards) of the Tall Timbers Fire Research Station, Tallahassee, Florida, USA, edited by E. Komarek. One overview for a lay audience is by C. Perlès, 'L'homme préhistorique et le feu', *La Récherche* 60, 1975, 829–39; another is P. D. Moore, 'Fire: catastrophic or creative force?', *Impact of Science on Society* 32(1), 1982, 5–14. The ideas (though not the precise wording) of the last sentence come from N. Georgescu-Roegen, 'Feasible recipes versus viable technologies', *Atlantic Economic J.* 12, 1984, 21–31.

Chapter 3 Advanced Hunters

Evolution and culture in the late Pleistocene

A general volume which refers to both near-recent and Pleistocene hunters is R. B. Lee and I. DeVore (eds), *Man and the Hunter*, Chicago: Aldine Press, 1968. The Pleistocene environment and cultural sequence is treated succinctly but excellently by K. W. Butzer, *Environment and Archaeology*, London: Methuen, 1971, 2nd edn; and the Upper Palaeolithic developments are put into an overall Pleistocene perspective by Butzer in 'Environment, culture and human evolution', *Amer Sci* 65, 1977, 572–84. The spread of man into east and south-east Asia is chronicled by J. S. Aigner, 'Pleistocene archaeological remains for South China', *Asian Perspectives* 16, 1972, 16–38. Various aspects of the life of the neanderthals are found in E. Trinkhaus and W. W. Howells, 'The Neanderthals', *Sci Amer* 24(6), 1979, 94–105;

A. M. ApSimon, 'The last neanderthal in France?', *Nature, Lond.* 287, 271–2; P. Mellars, 'A new chronology for the French Mousterian period', *Nature, Lond.* 322, 1986, 410–11; and H. Watanabe, 'Periglacial ecology and the emergence of *Homo sapiens*', in F. Bordes (ed.), *The Origin of Homo sapiens*, Paris: UNESCO, 1972, 271–85. The general chronology and technological development of the Upper Pleistocene is dealt with in J. G. D. Clarke, *World Prehistory*, Cambridge: CUP, 1977, and R. Wenke, *Patterns in Prehistory*, Oxford: OUP, 1980 and some indication of the progress of technology is in M. Daumas (ed.), *A History of Technology and Invention*, London: John Murray, 1969, vol. I, although the hominid terminology is rather badly out of date. The history of malaria is outlined by L. J. Bruce-Chwatt, 'Paleogenesis and paleo-epidemiology of Primate malaria', *Bull World Health Org* 32, 1965, 363–87. An interesting account of the experimental study of tools and artefacts and their use in Palaeolithic times is S. A. Semenov, *Prehistoric Technology* (trans. M. W. Thompson), London: Cory, Adams and Mackay, 1964. There are some relevant illustrations in Birdsall and Cipolla, op. cit., 1980. Other aspects of Mousterian and later technology are found in F. Bordes, *A Tale of Two Caves*, New York: Harper and Row, 1972; S. R. and L. R. Binford, 'Stone tools and human behaviour', in Lamberg-Karlovsky, *Hunters, Farmers and Civilizations*, op. cit., 1969, 12–21; P. Mellars, 'The character of the Middle-Upper Palaeolithic transition in south-west France', in C. Renfrew (ed.), *The Explanation of Culture Change: Models in Prehistory*, London: Duckworth, 1973, 255–76; J. J. Wymer and R. Singer, 'Middle Stone Age occupational settlements on the Tzitzikama coast, east Cape Province, South Africa', in G. W. Dimbleby, P. Ucko and R. Tringham (eds), *Man, Settlement and Urbanism*, London: Duckworth, 1972, 207–10; N. David, 'On Upper Palaeolithic society, ecology and technological changes: the Noaillian case', in C. Renfrew (ed.), op. cit., 276–303; E. Shimkin, 'The upper palaeolithic in North-Central Asia: evidence and problems', in L. G. Freeman (ed.), *Views of the Past*, The Hague: Mouton, 1975, 193–315. A concise general review for the Palaeolithic can be found in chs 2 and 3 of C. Gamble, *The Palaeolithic Settlement of Europe*, Cambridge: CUP, 1986. Recent work in Brazil is described in N. Guidon and G. Delibrias, 'Carbon-14 dates points to man in the Americas 32,000 years ago', *Nature, Lond.* 321, 1986, 769–71. The most southerly men of the last glaciation are reported in K. Kiernan, R. Jones and D. Ranson, 'New evidence from Fraser Cave for glacial age man in south-west Tasmania', *Nature, Lond.* 301, 1983, 28–32. This paper is put into a global context by G. Bailey, 'Late Pleistocene life in Tasmania', *Nature, Lond.* 301, 1983, 13. For a discussion of Pleistocene climatic change in both Europe and the tropics see A. S. Goudie, *Environmental Change*, op. cit. Holocene fire in Amazonian forests is in P. A. Colinvaux, 'Amazon diversity in light of the palaeoecological record', *Quat Sci Rev* 6, 1987, 93–114.

The rise and fall of hunter-gatherers

The world-wide spread of hunters is chronicled in maps in the *Cambridge Encyclopedia of Archaeology*, Cambridge: CUP, 1980. Accounts of the transition from Palaeolithic to Mesolithic and the development of Mesolithic economies are given in *inter alia*, J. G. D. Clark, *World Prehistory*, Cambridge: CUP, 3rd edn, 1977; *idem*, *Prehistoric Europe: The Economic Basis*, Cambridge: CUP, 1952, reprinted 1971; *idem*, *The Earlier Stone Age Settlement of Scandinavia*, Cambridge: CUP, 1975; and *idem*, *Mesolithic Prelude*, Edinburgh UP, 1980.

An inclusive volume on the nature of the hunter–gatherer way of life past and present and on the status of recent hunting groups is R. B. Lee and I. DeVore (eds), *Man the Hunter*, Chicago: Aldine Publishing Co., 1968; particularly relevant at this point is G. P. Murdock's contribution 'The current status of the world's hunting and gathering peoples', ibid., 13–20. More detailed accounts of several contemporary and near-recent groups are given in M. C. Bicchieri (ed.), *Hunters and Gatherers Today*, New York: Holt, Rinehart and Winston, 1972. A more general introductory overview of their anthropology is given by E. R. Service, *The Hunters*, Englewood Cliffs, NJ: Prentice Hall Foundations of Modern Anthropology series, 1966; and a standard anthropological work is C. S. Coon, *The Hunting Peoples*, Harmondsworth: Penguin, 1976. A concise account of the economic relations of modern hunters is B. Hayden, 'Subsistence and ecological adaptations of modern hunter/gatherers', in R. S. O. Harding and G. Teleki (eds), *Omnivorous Primates*, NY: Columbia UP, 1981, 344–421. The 'footnote' on reviewed interest in traditional hunting ways is derived from M. Nowak, 'Subsistence trends in a modern Eskimo community', *Arctic* 28, 1975, 21–34; T. C. Meredith and L. Müller-Wille, *Man and Caribou: the Economics of Nakaspi hunting in Northeastern Quebec*, Montreal: McGill Subarctic Research Paper No 36, 1982; and P. K. Latz and G. F. Griffin, 'Changes in aboriginal land management in relation to fire and to food plants in central Australia', in B. S. Hetzel and H. J. Frith (eds), *The Nutrition of Aborigines in Relation to the Ecosystems of Central Australia*, Melbourne: CSIRO, 1978, 77–85.

Environmental relations of hunter–gatherers

The general context of the Upper Palaeolithic is given in J. G. D. Clark, *World History*, and more detail in J. M. Coles and E. S. Higgs, *The Archaeology of Early Man*, Harmondsworth: Penguin Books, 1975. The sites in the southern USSR are treated in R. S. Klein, *Ice-Age Hunters of the Ukraine*, Chicago and London: Chicago UP, 1973. The data for Upper Palaeolithic economies in France and Spain come mostly from P. Mellars, op. cit., 1973; A. Sieveking, I. H. Longworth and K. E. Wilson (eds), *Problems in Economic and Social Archaeology*, London: Duckworth, 1976, 583–603; and L. G. Freeman, 'The significance of mammalian fauna from Palaeolithic occupations in Cantabrian Spain', *Amer Antiq* 38, 1973, 3–44. An outline discussion of the human use of flora in one region is J. A. Tyldesley and P. G. Bahn, 'Use of plants in the European Palaeolithic: a review of the evidence', *Quat Sci Rev* 2, 1983, 53–81. The comparisons from the MSA and LSA in the RSA are given by R. Klein, 'Stone age predation on small African bovids', *South African Archaeological Bulletin* 36, 1981, 55–65; *idem*, 'Environmental and ecological implications of large mammals from Upper Pleistocene and Holocene sites in southern Africa', *Ann. S. African Museums* 81(7), 1980, 223–83. An interesting palaeoecological study is A. E. Speiss, *Reindeer and Caribou Hunters: An Archaeological Study*, London: Academic Press, 1979. The very different environment of the Nile in the late Pleistocene, at a time when the gathering of wild grasses was added to the subsistence repertoire, is the subject of D. R. Connor and A. E. Marks, 'The terminal Pleistocene on the Nile: the final Nilotic adjustment', in L. G. Straus (ed.), *The End of the Palaeolithic in the Old World*, Oxford: BAR International Series 284, 1986, 171–99. The assertion of a population avalanche in Upper Palaeolithic times is made by J. Lawrence Angel,

'Palaeoecology, palaeodemography and health', in S. Polgar (ed.), *Population, Ecology and Social Evolution*, Paris and The Hague: Mouton, 1975, 167–90.

An overview of the penetration of hunters into the Americas is given by J. Taylor, 'The earliest hunters, gatherers and farmers of North America', in J. V. S. Megaw (ed.), *Hunters, Gatherers and First Farmers beyond Europe*, Leicester: University Press, 1977, 199–224. The general economic archaeology of the High Plains is summarized in F. Wendorf and J. J. Hester, 'Early man's utilization of the Great Plains environment', *Amer Antiq* 28, 1962, 159–71, though it lacks some of the later evidence (see below). A massive collection of material about the bison is to be found in F. G. Roe, *The North American Buffalo: A Critical Study of the Species in its Wild State*, Newton Abbot: David and Charles, 2nd edn, 1972. Early hunting on the plains is exemplified in G. C. Frison, D. N. Walker, S. D. Webb and G. M. Ziemens, 'Palaeo-Indian procurement of *Camelops* on the north western Plains', *Quat Res* 10, 1978, 385–400; D. C. Fisher, 'Mastodon butchery by North American Paleo-Indians', *Nature, Lond.* 308, 1984, 271–2. Buffalo kill site descriptions in order of publication are, G. C. Frison, 'The Glenrock buffalo jump 48CD304: Late Prehistoric period buffalo procurement and butchering on the Northwestern Plains', *Plains Anthropologist Memoir no* 7, 1970, 1066; *idem*, 'The bison pound in North-Western Plains prehistory: site 48CA302, Wyoming', *Amer Antiq* 36, 1971, 77–91; Joe Ben Wheat, 'The Olsen–Chubbuck site: A paleo-Indian bison kill', *Amer Antiq* 37, 1972, 1–180. An article based on this paper appeared in *Scientific American* in January 1967 (216, 1, 44–52) and is reprinted in R. S. MacNeish (ed.), *Early Man in America*, San Francisco: Freeman, 1972, 80–9; also G. C. Frison, 'The Wardell buffalo trap 48SU301: communal procurement in the upper Green River basin, Wyoming', *University of Michigan Museum of Anthropology: Anthropological Papers no. 48*, Ann Arbor, Mich., 1973; *idem* (ed.), *The Casper Site: A Hell Gap Bison Kill on the High Plains*, New York and London: Academic Press, 1974; G. C. Frison, M. Wilson and D. J. Wilson, 'Fossil bison and artifacts from an early altithermal period arroyo trap in Wyoming', *Amer Antiq* 41, 1976, 28–57. Much of this is summed up in G. C. Frison, *Prehistoric Hunters of the High Plains*, NY etc.: Academic Press, 1978. Taphonomic analysis of spring kills of buffalo in New Mexico is found in J. D. Speth, *Bison Kills and Bone Counts: Decision Making by Ancient Hunters*, Chicago: Chicago UP, 1983. The prehistoric hunting patterns of the *puña* are discussed in J. W. Pires-Ferreira, E. Pires-Ferreira and P. Kaulicke, 'Preceramic animal utilization in the central Peruvian Andes', *Science* 194, 1976, 483–90.

Ethnographic accounts are summarized in the works cited above by Lee and DeVore, Bicchieri, Coon and Service, and a technological perspective is given by Darryl Forde, 'Foraging, hunting and fishing', in C. Singer et al. (eds), *A History of Technology, vol. I: From Early Times to Fall of Ancient Empires*, Oxford: Clarendon Press, 1954, 154–86; *idem*, *Habitat, Economy and Society*, London: Methuen, 1945 and subsequent editions, Part I. The calculation of the relative importance of meat and plant foods is by R. B. Lee, 'What hunters do for a living, or, how to make out on scarce resources', in R. B. Lee and I. DeVore (eds), *Man the Hunter*, Chicago: Aldine Press, 1968, 30–48.

The detailed accounts of the !Kung Bushmen are by R. B. Lee and his co-workers and are brought together in his book *The !Kung San: Men, Women and Work in a Foraging Society*, Cambridge: CUP, 1979. The somewhat similar G/wi are treated at shorter length by G. B. Silberhauer, 'The G/wi Bushmen', in M. G. Bicchieri (ed.), *Hunters and Gatherers Today*, New York: Holt, Rinehart and Winston, 1972, 271–325; and there is a wider-casting volume edited by R. B. Lee and I. DeVore, *Kalahari Hunter-Gatherers: Studies of the !Kung San and their Neighbours*, Cambridge, Mass.: Harvard UP, 1976.

The relevant work of Colin Turnbull is found in three popular books and a specialized monograph: *The Forest People*, London: Jonathan Cape, 1961; *Wayward Servants: The Two Worlds of the African Pygmies*, London: Eyre & Spottiswoode, 1965; *The Mbuti Pygmies: Adaptation and Change*, London and Philadelphia: Holt-Saunders, 1983; 'The Mbuti pygmies: an ethnographic survey', *Anthrop Papers Amer Mus Nat Hist* 50(3), 1965, 139–282. See also R. Harako, 'The Mbuti as hunters', *Kyoto Univ Afr Stud* 10, 1976, 37–99; T. Tanno, 'The Mbuti net-hunters in the Ituri forest, Eastern Zaïre – their hunting activities and band composition', ibid., 101–35. The scholarly controversy over the Yanomamo is scarcely relevant here, but a concise account can be read in L. E. Sponsel, 'Yanomamo Warfare, protein capture, and cultural ecology: a critical analysis of the arguments of the opponents', *Interciencia* 8, 1983, 204–10. For the Boreal forests see H. A. Feit, 1973, *infra*, and R. R. Gadacz, 'Montagnais hunting dynamics in historicoecological perspective', *Anthropologica* 17, 1975, 149–67; and J. G. E. Smith, 'Economic uncertainty in an "original affluent society": caribou eater Chipewyan adaptive strategies', *Arctic Anthropology*, 15, 1978, 68–88.

Energy flow

An overview of the entrance of energy use in hunters (and comparison with other economies) is given in M. D. Sahlins, *Stone Age Economics*, London: Tavistock Publications, 1974 (first published in the USA in 1972); data for the !Kung are from R. B. Lee, *The !Kung San*, 1979. A general model is specifically tested by D. R. Harris, 'Resource distribution and foraging effort in hunter–gatherer subsistence', in G. A. Harrison (ed.), *Energy and Effort*, London: Taylor and Francis, 1982, 189–208. See as well G. B. Silberhauer, 'The G/wi bushmen', in M. G. Bicchieri (ed.), *Hunters and Gatherers Today*, NY: Holt, Rinehart and Winston, 1972, 271–326; J. F. Eder, 'The caloric returns to food collecting: disruption and change among the Batak of the Philippine tropical forest', *Human Ecology* 6, 1978, 55–69; W. H. Townsend, 'Stone and steel tool use in a New Guinea society', *Ethnology* 8, 1969, 199–205; P. B. Dwyer, 'The price of protein: five hundred hours of hunting in the New Guinea highlands', *Oceania* 44, 1974, 278–93; R. F. Ellen, 'Non-domesticated resources in Nuaulu ecological relations', *Soc Sci Inform* 14, 1975, 127–50. The Inuit example is from W. B. Kemp, 'The flow of energy in a hunting society', *Sci Amer* 224(3), 1971, 104–15. Sub-arctic Cree Indians are described by A. Tanner, *Bringing Home Animals: Indigenous Ideology and Mode of Production of the Mistassini Cree Hunters*, New York: St. Martin's Press, 1979; and J. S. Savishinsky, *The Trail of the Hare: Life and Stress in an Arctic Community*, New York and London: Gordon and Breach, 1974.

The Waswapini Cree are described by H. A. Feit, 'The ethno-ecology of the Waswapini Cree; or how hunters can manage their resources', in B. Cox (ed.), *Cultural Ecology: Readings on the Canadian Indians and Eskimos*, Toronto: McClelland and Stewart, 1973, 115–25; see also R. Paine, 'Animals as capital: comparisons among northern nomadic herders and hunters', in Cox, op. cit., 301–14. The account of the Koyukon Indians referred to is R. K. Nelson, *Make Prayers to the Raven: A Koyukon View of the Northern Forest*, Chicago and London: Chicago UP, 1983. Reindeer hunters past and present are discussed by E. S. Burch, 'The caribou/wild reindeer as a human resource', *Amer Antiq* 37(3), 1972, 339–68; T. Ingold, *Hunters, Pastoralists and Ranchers*, Cambridge: CUP, 1980; O. Blehr, 'Traditional reindeer hunting and social change in the local communities surrounding Hardangervidda', *Norwegian Archaeol Rev* 6, 1977, 102–12; J. H. Steward, 'The Great Basin Shoshonean Indians: an example of a family level of sociocultural integration', in Y. A. Cohen (ed.), *Man in Adaptation: The Cultural Present*, Chicago: Aldine, 1968, 68–81.

Other North Americans are discussed by D. Wishart, *The Fur Trade of the American West 1807–40: A Geographical Synthesis*, London: Croom-Helm, 1979; A. J. Ray, 'Some conservation schemes of the Hudson's Bay Company, 1821–50: an examination of the problems of resource management in the fur trade', *J Hist Geog* 1, 1975, 49–68; and C. A. Bishop, *The Northern Ojibwa and the Fur Trade: an Historical and Ecological Study*, Toronto: Holt, Rinehart and Winston, 1974; C. Martin, *Keepers of the Game: Indian–Animal Relationships in the Fur Trade*, Berkeley and Los Angeles: University of California Press, 1978; S. Krech (ed.), *Indians, Animals and the Fur Trade*, Athens, Ga: University of Georgia Press, 1981. A detailed paper on one region is J. Kay, 'Native Americans in the fur trade and wildlife depletion', *Environmental Review* 9, 1985, 118–30. See also, for considerable detail, J. D. Hughes, *American Indian Ecology*, El Paso: Texas Western Press, 1983; and R. White, 'Native Americans and the environment', in W. R. Swagerty (ed.), *Scholars and the Indian Experience*, Critical Reviews of Recent Writing in the Social Sciences, Bloomington: Indiana University Press, 1984, 179–204. See also J. A. Hart, 'From subsistence to market: a case study of the Mbuti net hunters', *Human Ecology* 6, 1978, 55–69; B. Neitschmann, *Between Land and Water: the Subsistence Ecology of the Miskito Indians*, New York: Seminar Press, 1973.

Examples of ways in which impact has occurred

Star Carr is comprehensively discussed by J. G. D. Clark, *Star Carr: an essay in bioarchaeology*, Reading, Mass.: Addison-Wesley Modules in Anthropology, 1972, but there have been several later and conflicting interpretations of the economic nature of the site. The Alaskan example is C. A. Simenstadt, J. A. Estes and K. W. Kenyon, 'Aleuts, sea otters and alternate stable-state communities', *Science* 200, 1978, 403–11. 'Alternate' here means 'alternative', not 'alternating'.

The Waswapini Cree are discussed by H. A. Feit, 'The ethno-ecology of the Waswapini Cree', *op. cit.* For the musk-ox in the European Palaeolithic, see P. F. Wilkinson, 'The relevance of musk-ox exploitation to the study of prehistoric animal economies', in E. S. Higgs (ed.), *Palaeoeconomy*, London: CUP, 1975, 9–53. The 'Pleistocene overkill' hypothesis may be found in P. S. Martin, 'Prehistoric overkill', in P. S. Martin and H. E. Wright (eds), *Pleistocene Extinctions: the Search for a Cause*, New Haven: Yale UP, 1967, 75–120; and in J. E. Mosimann and P. S. Martin, 'Simulating overkill by Palaeoindians', *Amer Sci* 63, 1975, 304–13; P. S. Martin, 'The discovery of America', *Science* 179, 969–74. The overkill hypothesis for southern Africa is supported by R. G. Klein, 'Environmental and ecological implications of large mammals from Upper Pleistocene and Holocene sites in Southern Africa', *Ann. S. African Museums* 81(7), 1980, 223–83. For Australian 'overkill' see D. Merrilees, 'Man the destroyer: late Quaternary changes in the Australian marsupial fauna', *J Roy Soc Western Australia* 51, 1968, 1–24; and J. H. Calaby, 'Man, fauna and climate in aboriginal Australia', in D. J. Mulraney and J. Golson, *Aboriginal Man and Environment in Australia*, ANU Press, 1971, 81–93; R. Gillespie et al., 'Lancefield swamp and the extinction of the Australian megafauna', *Science* 200, 1978, 1044–8 gives an alternative view. 'Secondary overkill' comes from G. S. Krantz, 'Human activities and megafaunal extinctions', *Amer Sci* 58, 1970, 167–70. An expression of general dissent from the hypothesis is D. Webster, 'Late Pleistocene extinction and human predation: a critical overview', in R. S. O. Harding and G. Teleki, *Omnivorous Primates*, Columbia UP, 1981, 556–94. The whole idea is reviewed and discussed at considerable length (892 pages) in P. S. Martin and R. G. Klein (eds), *Quarternary Extinctions: A Prehistoric Revolution*, Tucson: University of Arizona Press, 1984, especially P. S. Martin, 'Prehistoric overkill: the global model', 354–403, and J. M. Diamond, 'Historic extinctions: a Rosetta stone for understanding prehistoric extinctions', 824–62. The Chipewyans and their use of aircraft come in L. Müller-Wille, 'Caribou never die! Modern caribou hunting economy of the Dene (Chipewyan) of Fond du Lac, Saskatchewan and N.W.T.', *Musk Ox*, 14, 1974, 7–19. Ivy pollen in the European Mesolithic is discussed by I. G. Simmons and G. W. Dimbleby, 'The possible role of ivy (*Hedera helix* L.) in the Mesolithic economy of Western Europe', *J Archaeol Sci* 1, 1974, 291–6. Crib-biting is the hypothesis of P. G. Bahn, 'Crib-biting: tethered horses in the Palaeolithic?', *World Archaeology* 12, 1980, 212–17.

Man as the successor to the sabre-tooth tiger is postulated by S. L. Washburn and C. S. Lancaster, 'The evolution of hunting', in R. B. Lee and I. DeVore (eds), *Man the Hunter*, op. cit., 293–303. On fire see P. Mellars, 'Fire ecology, animal populations and man: a study of some ecological relationships in prehistory', *Proc Prehist Soc* 42, 1976, 15–45; and for a detailed study of the near-recent, H. T. Lewis, 'Maskuta: the ecology of Indian fires in Northern Alberta', *W Canad J Anthropol* 7, 1977, 15–52. A detailed piece of Antipodean palaeoecology is G. Singh et al., 'Quaternary vegetation and fire history in Australia', in A. M. Gill, R. H. Groves and R. Noble (eds), *Fire and the Australian Biota*, Canberra: Australian Academy of Science, 1981, 23–54. See also H. T. Lewis, 'Fire technology and resource management in aboriginal North America and Australia', in N. M. Williams and E. S. Hunn (eds), *Resource Managers: North American and Australian Hunter–gatherers*, Boulder, Colo.: Westview Press, AAAS Selected Symposia no 67, 1982, 45–67. See also M. J. Rowlands, 'Aborigines and environment in holocene Australia: changing paradigms', *Aust Aboriginal Studs* 1983(2), 62–77, and the review in the same issue (89–90) by D. R. Horton of J. P. White and J. F. O'Connell, *A Prehistory of Australia*, Sydney: Academic Press, 1982. The Cape York area of Australia is described by D. R. Harris, 'Subsistence strategies across Torres Strait', in J. Allen, J. Golson and R. Jones (eds), *Sunda and Sahul: Prehistoric Studies in Southeast Asia, Melanesia and Australia*, London: Academic Press, 421–63; and a very full paper on the Antipodes as a whole is R. Jones, 'The Neolithic, Palaeolithic and the hunting gardeners: man and land in the Antipodes', in R. P. Suggate and M. M. Cresswell (eds), *Quarternary Studies*, Wellington: RSNZ, 1975, 21–34. Plant use in the early prehistory of Europe is discussed in J. A. Tyldesley and P. G. Bahn, 'Use of plants in the European palaeolithic: a review of the evidence', *Quat Sci Rev* 2, 1983, 53–81. Water management in Australia derives from H. Lourandos, 'Change or stability?

Hydraulics, hunter-gatherers and population in temperate Australia', *World Archaeology* 11, 1980, 245–64. Evidence of elementary dam construction in order to keep a lagoon full with water, and of cultivating grubs in old logs exposed to salt water is sketched in A. H. Campbell, 'Elementary food production by the Australian aborigines', *Mankind* 6, 1965, 206–7. Wide-ranging papers on early man and environment in Australia, including discussion of 'overkill', are N. B. Tindale, 'Ecology of primitive aboriginal man in Australia', in A. Keast, R. L. Crocker and C. S. Christian (eds), *Biogeography and Ecology in Australia*, The Hague: Junk, 1959, vol. I of Monographae Biologicae VIII, 36–5; R. Jones, 'The geographical background to the arrival of man in Australia and Tasmania', *Archaeology and Physical Anthropology in Oceania* 3, 1968, 186–215; *idem*, 'The Neolithic, Palaeolithic and the hunting gardens: man and land in the Antipodes', in R. P. Suggate and M. M. Creswell (eds), *Quaternary Studies*, Wellington: Royal Society of New Zealand, 1975, 21–34; J. H. Calaby, 'Man, fauna and climate in aboriginal Australia', in D. J. Mulvaney and J. Golson (eds), *Aboriginal Man and Environment in Australia*, Canberra: ANU Press, 1971, 81–93; D. Merrilees, 'Man the destroyer: late Quaternary changes in the Australian marsupial fauna', *J Roy Soc Western Australia* 51, 1968, 1–24; N. B. Tindale, 'Prehistory of the aborigines: some interesting considerations', in A. Keast et al. (eds), *Ecological Biogeography of Australia*, The Hague, Boston and London: Junk (vol. 3 of *Monographiae Biologicae*, vol. VIII, 1763–96). A relatively up to date review is J. P. White and J. F. O'Connell, 'Australian prehistory: new aspects of antiquity', *Science* 203, 1979, 21–8. The last hunter–gatherer culture in England is discussed in I. G. Simmons, G. W. Dimbleby and C. Grigson, 'The Mesolithic', in I. G. Simmons and M. J. Tooley (eds), *The Environment in British Prehistory*, London: Duckworth, 1981, 82–124.

Population, resources and environment in hunter–gatherers

The idea of carrying capacity is treated in M. A. Glassow, 'The concept of carrying capacity in the study of culture process', *Adv. in Archaeol. Theory and Method* 1, 1978, 31–48; B. Hayden, 'The carrying capacity dilemma: an alternate approach', *Amer Antiq* 40, 1975, 11–21. Population changes in the late Pleistocene are discussed in J. L. Angel, 'Palaeoecology, palaeodemography and health', in S. Polgar (ed.), *Population, Ecology and Social Evolution*, Paris and The Hague: Mouton, 1975; in F. A. Hassan, 'Demographic archaeology', *Adv. in Archaeol. Theory and Method* 1, 49–103; A. J. Ammermann, 'Late Pleistocene population dynamics: an alternative view', *Human Ecology* 3, 1975, 219–33; with an overview of the whole process of population by D. E. Dumond, 'The limitation of human population: a natural history', *Science* 187, 1975, 713–21. Population limitation and 'pressure' are dealt with respectively by W. T. Divale, 'Systematic population control in the Middle and Upper Palaeolithic: inferences based on contemporary hunter–gatherers', *World Archaeology* 4, 1972, 222–43; B. Hayden, 'Population control among hunter/gatherers', *World Archaeology* 4, 1972, 205–21. P. E. L. Smith, 'Changes in population pressure in archaeological explanation', *World Archaeology* 4, 1972, 5–18; see also J. B. Birdsell, 'Some predictions for the Pleistocene based on equilibrium systems among recent hunter–gatherers', in R. B. Lee and I. De Vore (eds), *Man the Hunter*, op. cit., 229–40; F. A. Hassan, 'Determination of the size, density, and growth rate of hunting–gathering populations', in S. Polgar (ed.), *Population, Ecology and Social Evolution*, Paris and The Hague: Mouton, 1975, 27–52; R. W. Casteel, 'Two static maximum population-density models for hunter–gatherers: a first approximation', *World Archaeology*, 4, 1972, 19–39. Some of the socio-spatial processes are explored by R. H. Layton, 'Political and territorial structures among hunter–gatherers', *Man (N.S.)* 21, 1986, 18–33. The relationships of density, exogamy and territory are the subject of the material on theoretical approaches to locational relationships in Palaeolithic society', in R. H. and K. M. Weiss (eds), *The Demographic Evolution of Human Populations*, London: Academic Press, 1976, 49–58; *idem*, 'Boundary conditions for Palaeolithic social systems: a simulation approach', *Amer Antiq* 39, 1974, 147–78; E. M. Wilmsen, 'Interaction, spacing behaviour and the organization of hunting bands', *J Anthrop Res* 29, 1973, 1–31. The question of zoonoses in Amerindian populations is raised by C. Martin, 'Wildlife diseases as a factor in the depopulation of the North American Indian', *Western Historical Quarterly* 7, 1976, 47–62. The supposed idea of a pre-industrial golden age of environmental relations (never held by anybody who had looked at the evidence) is demolished by J. M. Diamond, 'The environmentalist myth', *Nature, Lond.* 324, 1986, 19–20.

Chapter 4 Agriculture and its Impact

The palaeolithic legacy, the emergence of domesticates and of agricultural systems

A number of works cover much of these topics, though few direct themselves to the environmental context. A notable exception is the work of D. R. Harris, especially 'Alternative pathways towards agriculture', in C. A. Reed (ed.), *Origins of Agriculture*, The Hague: Mouton, 1977, 179–243; 'Settling down: an evolutionary model for the transformation of mobile bands into sedentary communities', in J. Freidman and M. J. Rowlands (eds), *The Evolution of Social Systems*, London: Duckworth, 1977, 401–17; 'The environmental impact of traditional and modern agricultural systems', in J. G. Hawkes (ed.), *Conservation and Agriculture*, London: Duckworth, 1978, 61–9; 'The prehistory of human subsistence: a speculative outline', in D. N. Walcher and N. Kretchmer (eds), *Food, Nutrition and Evolution*, New York: Masson, 1981, 15–35; 'Breaking ground: agricultural origins and archaeological explanations', *Bull Inst Archaeol Lond* 18, 1981, 1–20; 'Ethnoecological evidence for the exploitation of wild grasses and forbs: its scope and archaeological implications', in W. Van Zeist and W. A. Casparie (eds), *Plants and Ancient Man*, Rotterdam: Balkema, 1983, 63–9. Some of the many books on agricultural origins are C. A. Reed (ed.), *Agricultural Origins*, The Hague: Mouton, 1977; B. Bender, *Farming in Prehistory*, London: John Barker, 1975; M. N. Cohen, *The Food Crisis in Prehistory*, New Haven and London: Yale UP, 1977; D. Rindos, *The Origins of Agriculture*, Orlando, Fla. and London: Academic Press, 1984; C. Barigozzi (ed.), *The Origin and Domestication of Cultivated Plants*, Amsterdam: Elsevier, 1986, especially D. Zohari, 'The origin and early spread of agriculture in the Old World', 3–20, and J. R. Harlan, 'Plant domestication: diffuse origins and diffusions', 21–34. A new study is D. Zohary and M. Hope, *Domestication of Plants in the Old World*, Oxford: Clarendon Press, 1988. The overviews of K. Butzer are also of great value, *viz.*, *Environment and Archaeology: An Ecological Approach to Prehistory*, Chicago: Aldine, 2nd edn, 1971; and *Archaeology as Human Ecology*, Cambridge: CUP, 1982. A model-based review of causes is in T. F. H. Allen and T. B. Starr, *Hierarchy*, London and Chicago: Chicago UP, 1982. The biological evidence for domesticates is especially clear in the work of J. Harlan, *Crops and Man*, Madison, Wis.: Crop Science Society of America; *idem*, 'Plant and animal distribution in relation to domestication', *Phil Trans R Soc Lond* B 275, 1976, 13–25; and more popularly, 'The plants and animals that nourish men', *Sci Amer* 235, 1976, 89–97; see also H. E. Wright, 'The environmental setting for plant domestication in the Near East', *Science*, 194, 385–9. Modern biologically based accounts of individual species' domestications can be found in N. W. Simmonds (ed.), *Evolution of Crop Plants*, London and New York: Longman, 1976; I. L. Mason (ed.), *Evolution of Domesticated Animals*, London and New York: Longman, 1984.

More specialized material drawn upon includes T. T. Chang, 'The rice cultures', *Phil Trans Roy Soc Lond* B 275, 1976, 143–57; H.-L. Li, 'The origin of cultivated plants in South-east Asia', *Econ Bot* 24, 1976, 3–19; D. Zohary and P. Spiegel-Roy, 'Beginnings of fruit growing in the Old World', *Science* 187, 1975, 319–27; R. Orr Whyte, 'The botanical Neolithic revolution', *Human Ecology* 5, 1977, 209–22; J. L. Angel, 'Ecology and population in the Eastern Mediterranean', *World Archaeol.* 4, 1972, 88–105. For fire see H. T. Lewis, 'The role of fire in the domestication of plants and animals in south-west Asia: a hypothesis', *Man* 7, 1972, 195–222. Mesopotamia and Egypt are the subjects of M. A. J. Williams and H. Faure (eds), *The Sahara and the Nile: Quaternary Environments and Prehistoric Occupation in northern Africa*, Rotterdam: Balkema, 1980; R. McC. Adams and H. J. Nissen, *The Uruk Countryside: The Natural Setting of Urban Societies*, Chicago and London: Chicago UP, 1972; K. W. Butzer, *Early Hydraulic Civilization in Egypt: A Study of Cultural Ecology*, Chicago and London: Chicago UP, 1976; J. Rzoska, *Euphrates and Tigris: Mesopotamian Ecology and Destiny*, The Hague, Boston and London: Junk, Monog. Biol. vol. 38, 1980; R. McC. Adams, *Heartland of Cities: Surveys of Ancient Settlements and Land Use on the Central Flood Plain of the Euphrates*, London and Chicago: Chicago UP, 1981. Sedimentation after forest clearance is documented in M. Bell, 'Valley sediments and environmental change', in M. Jones and G. W. Dimbleby (eds), *The Environment of Man: the Iron Age to the Anglo-Saxon period*, Oxford: B. A. R. British Series no 87, 1981, 75–91; M. A. Robinson and G. H. Lambrick, 'Holocene alluviation and hydrology in the Upper Thames basin', *Nature, Lond.* 308, 809–14; M. B. Davis, 'Erosion rates and land-use history in Southern Michigan', *Environ Cons* 3, 1976, 139–48. The classic descriptions of shifting cultivation in SE Asia are C. Geertz, *Agricultural Involution: The Processes of Ecologic Change in SE Asia*, Berkeley and Los Angeles: Univ. of California Press, 1963; and J. E. Spencer, *Shifting Cultivation in South-eastern Asia*, Univ. of California Pubs in Geography 19, 1966. For Africa the standard work is P. H. Nye and D. J. Greenland, *The Soil under Shifting Cultivation*, Harpenden: Commonwealth Bureau of Soils Technical Communication no 51, 1960. More recent publications used as sources were: D. R. Harris, 'The ecology of swidden cultivation in the Upper Orinoco rain forest, Venezuela', *Geogr Rev* 61, 1971, 475–95; W. M. Denevan, 'Campo subsistence in the Gran Pajonal, eastern Peru', *Geogr Rev* 61, 1971, 496–518; F. O. Adedeji, 'Nutrient cycles and successional changes following shifting cultivation practice in moist semi-deciduous forests in Nigeria', *Forest Ecology and Management* 9, 1984, 87–99; W. M. Denevan et al., 'Indigenous agroforestry in the Amazon: Bora Indian management of swidden fallows', *Interciencia* 9, 1984, 346–57; B. K. Mishra and P. S. Ramakrishnan, 'Slash and burn agriculture at higher elevations in north-eastern India. I. Sediment, water and nutrient losses', *Agric Ecosysts and Env* 9, 1983, 69–82; *idem*, II. Soil fertility changes, ibid., 83–96. A pioneering synthesis of archaeology and geomorphology is A. Gilman and J. B. Thornes, *Land-Use and Prehistory in South-East Spain*, London: Allen and Unwin, The London Research Series in Geography 8, 1985. See also A. Sherratt, 'Plough and pastoralism: aspects of the secondary products revolution', in I. Hodder, G. Ll. Isaac and N. Hammond (eds), *Patterns of the Past: Studies in Honour of David Clarke*, London: CUP, 1981, 261–305; S. Lee and D. G. Bates, 'The origins of specialized nomadic pastoralism: a systemic model', *Amer Antiq* 39, 1974, 187–93. Many data for the Pacific are given in J. Allen, J. Golson and R. Jones (eds), *Sunda and Sahul: Prehistoric studies in Southeast Asia, Melanesia and Australia*, London: Academic Press, 1977; and an interpretation of Pacific colonization patterns by agriculturalists in J. M. Diamond and W. F. Keegan, 'Supertramps at sea', *Nature* 311, 1984, 704–5. For the Middle East under Islam, see A. M. Watson, *Agricultural Innovation in the Early Islamic World: the Diffusion of Crops and Farming Techniques 700–1100*, Cambridge: CUP, 1983. The role of animals in the protein budgets of tropical forest peoples is discussed in the previous chapter; for pigs in PNG see R. A. Rappaport, 'Ritual regulation of environmental relations among a New Guinea people', *Ethnology*, 6, 1967, 17–30; *idem*, *Pigs for the Ancestors*, New Haven: Yale UP, 1968.

Spread and development of agriculture

I have relied largely upon the chronology and empirical data of D. B. Grigg, *The Agricultural Systems of the World: an evolutionary approach*, Cambridge: CUP, 1974; and for some processes on his *Dynamics of Agricultural Change*, London: Hutchinson, 1982. Some details of particular crops are from N. W. Simmonds (ed.), *Evolution of Crop Plants*, London: Longmans, 1976; and more specific material about Europe is found in B. H. Slicher van Bath, *The Agrarian History of Western Europe* AD *500–1850*, London: Edward Arnold, 1963. See also W. S. Cooter, 'Ecological dimensions of medieval agrarian systems', *Agricultural History* 52, 1978, 458–77, with dissent by R. S. Loomis, ibid., 478–83. Also of interest are J. L. Angel, 'Ecology and population in the Eastern Mediterranean', *World Archaeology* 4, 1972, 88–105; T. T. Chang, 'The rice cultures', *Phil Trans Roy Soc Lond* B 275, 1976, 143–57; A. W. Crosby, *The Columbian Exchange: Biological and Cultural Consequences of 1492*, Westport, Iowa: Greenwood Press, 1972; *idem*, *Ecological Imperialism: The Biological Expansion of Europe 900–1900*, Cambridge: CUP, 1986. A remarkably comprehensive account of imported fauna and flora (as well as men and other materials) into T'ang (AD 618–907) China is in E. H. Schafer, *The Golden Peaches of Samarkand: A study of T'ang Exotics*, Berkeley and Los Angeles: Univ. of California Press, 1963. For early Japan, see M. Tsukada et al., 'Oldest primitive agriculture and vegetational environments in Japan', *Nature, Lond.* 322, 1986, 632–4. Also, J. D. Jennings (ed.), *The Prehistory of Polynesia*, Cambridge, Mass. and London: Harvard UP, 1979. The detail about Tamil India is from S. Singaravelu, *Social Life of the Tamils – the Classical Period*, Kuala Lumpur: Univ. Malaya Press, 1966, and that about maize in Venezuela from N. J. van der Meruwe, et al., 'Isotopic evidence for prehistoric subsistence change at Parmana, Venezuela', *Nature, Lond.* 292, 1981, 536–8. The main data on terracing come from J. E. Spencer and G. A. Hale, 'The origin, nature and distribution of agricultural terracing', *Pacific Viewpoint* 2, 1961, 1–40; this paper elicits some controversy including comments by A. C. S. Wright in the same journal 3, 1962, 97–101, and a paper by P. Wheatley, 'Discursive scholia on recent papers on agricultural terracing and on related matters pertaining to northern Indochina and to neighbouring areas', *Pacific Viewpoint* 6, 1965, 123–44. For a more detailed study, see R. A. Donkin, *Agricultural Terracing in the Aboriginal New World*, Tucson: University of Arizona Press, Viking Fund pubs in Anthropology vol. 56, 1979. There are numerous books and papers on Maya subsistence. The present account is derived mainly from D. R. Harris, 'The agricultural foundations of lowland Maya civilization: a critique', in P. D. Harrison and B. L. Turner (eds), *Prehistoric Maya agriculture*, Albuquerque: University of New Mexico Press, 1978, 301–23; and F. M. Wiseman, 'Agricultural and historical ecology of the Maya lowlands', ibid., 63–115. These sources have been supplemented by N. Hammond, *Ancient Maya Civilization*, London: CUP, 1982; R. T. Matheney, 'Ancient lowland and highland Maya water and soil conservation strategies', in K. V. Flannery (ed.), *Maya Subsistence: Studies in Memory of Dennis E. Puleston*, London, etc.: Academic Press Studies in Archaeology, 1982, 157–78; B. L. Turner, 'Prehistoric intensive agriculture in the Mayan lowlands', *Science* 185, 1974, 118–24; *idem*, 'Prehispanic terracing in the central Maya lowlands: problems of agricultural intensification', in N. Hammond and G. R. Willey (eds), *Maya Archaeology and Ethnohistory*, Austin and London: University of Texas Press, 1979, 103–15; B. L. Turner and P. D. Harrison, 'Prehistoric raised-field agriculture in the Maya lowlands', *Science* 213, 1981, 399–405; R. E. W. Adams, W. E. Brown and T. P. Culbert, 'Radar mapping, archaeology and ancient Maya land use', *Science* 213, 1981, 1457–63. Other useful references are: R. T. Matheny, 'Maya lowland hydraulic systems', *Science* 193, 1976, 639–46; B. L. Turner, *Once Beneath the Forest: Prehistoric Terracing in the Rio Bec Region of the Maya Lowlands*, Boulder, Colo.: Westview Press Dellplain Latin American Studies No. 13, 1983, esp. ch. 5; *idem*, 'Constructional inputs for major agroecosystems of the ancient Maya', in J. P. Darch (ed.), *Drained Field Agriculture in Central and South America*, Oxford: BAR International Series 189, 1983, 11–26; M. Pohl (ed.), *Prehistoric Lowland Maya Environment and Subsistence Economy*, Cambridge, Mass.: Harvard UP, 1985.

Four thousand years of agriculture: impact

The material on Ancient Greece and the Roman Empire is from J. D. Hughes, *Ecology in Ancient Civilisations*, Albuquerque: University of New Mexico Press, 1975. Further material on the Mediterranean includes J. V. Thirgood, *Man and the Mediterranean Forest: A history of resource depletion*, London etc.: Academic Press, 1981; and C. Delano Smith, *Western Mediterranean Europe: A Historical Geography of Italy, Spain and Southern France since the Neolithic*, London: Academic Press, 1979. Some of the material on the Levant in J. M. Wagstaff, *The Evolution of Middle East Landscapes: An Outline to 1840 AD*, London and Sydney: Croom-Helm, 1985, is relevant. The more positive view of traditional landscape practices in the Mediterranean is associated with the numerous works of Z. Naveh, and is put into an historical context in Z. Naveh, 'Mediterranean landscape evolution and degradation as multivariate biofunctions: theoretical and practical implications', *Landscape Planning* 9, 1982, 125–46. Continuity of historical and modern environmental impact can be found in two reports of the UNESCO Man and the Biosphere (MAB) programme: *Regional Meeting on integral ecological research and conservation activities in the northern Mediterranean countries*, Paris: MAB report series no. 36, 1977; *Mediterranean forests and maquis: ecology, conservation and management*, Paris: MAB Technical Notes 2, 1977. Much of the impact details of traditional agricultural systems are inferred from management practices described in G. A. Klee (ed.), *World Systems of Traditional Resource Management*, London: Edward Arnold, 1980; especially chapters by J. L. Beyer, 'Africa', 5–37; B. J. Murton, 'South Asia', 67–99, J. Whitney, 'East Asia', 101–129; W. M. Denevan, 'Latin America', 217–43; and 'Oceania' by G. A. Klee, 245–81. The classic qualitative account of padi rice as an ecosystem is C. Geertz, *Agricultural Involution: The process of ecological change in Indonesia*, Berkeley and Los Angeles: University of California Press, 1963, 28–37. An interesting study of impact in China is P. C. Perdue, *Exhausting the Earth, State and Peasant in Hunan 1500–1850*, Cambridge, Mass. and London: Harvard UP East Asian Monographs 130, 1987. Historical data can be found in chapter 1 of D. H. Grist, *Rice*, London: Longmans Green and Co., 1953; and chapter 6 of D. B. Grigg, *The Agricultural Systems of the World: An Evolutionary Approach*, Cambridge: CUP, 1974. Basic inadequacies in current knowledge of nitrogen flows are discussed in I. Watanabe, E. T. Craswell and A. A. App, 'Nitrogen cycling in westland rice fields in south-east and east Asia', in R. Wetselaar, J. R. Simpson and T. Rosswell (eds), *Nitrogen Cycling in South-east Asian Wet Monsoonal Ecosystems*, Canberra: Australian Academy of Sciences/SCOPE, 1981, 4–17. A general paper on Africa is J. O. Adejuwon, 'Human impact on African environmental systems', in C. G. Knight and J. L. Newman (eds), *Contemporary Africa: Geography and Change*, New York: Prentice Hall, 1976, 140–58; and an African example of traditional soil conservation systems is R. McC. Netting, *Hill Farmers of Nigeria*, Seattle and London: University of Washington Press, 1968, reprinted 1973. The introduction of European thought is described in M. Stocking, 'Soil conservation in colonial Africa', *Agricultural History* 59, 1985, 148–61. Further detail on Papua New Guinea comes from J. M. Powell, 'The history of plant use and man's impact on the vegetation', in J. L. Gressit (ed.), *Biogeography and Ecology of New Guinea*, The Hague, Boston and London: Junk,Monographicae Biologicae 42, vol. 1, 207–27; and S. Bulmer, 'Human ecology and cultural variation in prehistoric New Guinea', in J. L. Gressit, ibid., 169–206. See also P. V. Kirch, 'Subsistence and ecology', in J. D. Jennings (ed.), *The Prehistory of Polynesia*, Cambridge, Mass.: Harvard UP, 286–307; A. G. Anderson, 'Man and landscape in the insular Pacific', *Landscape Planning* 4, 1977, 1–28; from Hawaii see S. L. Olson and H. F. James, 'Fossil birds from the Hawaiian islands: evidence for wholesale extinction before Western contact', *Science* 217, 1982, 633–5; *idem*, *Prodromus of the Fossil Avifauna of the Hawaiian Islands*, Washington, DC: Smithsonian Contributions to Zoology no. 365, 1982; R. Lewin, 'Polynesians' litter gives clues to islands' history', *Science* 231, 1986, 453–4. Australia is dealt with in D. A. Adamson and M. D. Fox, 'Change in Australasian vegetation since European settlement', in J. M. B. Smith (ed.), *A History of Australasian Vegetation*, Sydney: McGraw-Hill, 1982, 109–49, and J. T. Salmon, 'The influence of man on the biota', in G. Kuschel (ed.), *Biogeography and Ecology in New Zealand*, The Hague: Junk, 1975, Monographiae

Biologicae vol. 27, 643–61; see also J. J. Parsons, 'Human influences on the pine and laurel forests of the Canary Islands', *Geogr Rev* 71, 1981, 253–71; S. W. Tam, 'Causes of environmental deterioration in eastern Barbados since colonization', *Agriculture and Environment* 5 (1980/81), 285–308; and D. Watts, *Man's Influence on the vegetation of Barbados, 1627–1800*, Hull Univ. Occ. Papers in Geography no 4, 1966. For mountains see S. B. Brush, 'Man's use of an Andean ecosystem', *Human Ecology*, 4, 1976, 147–66; R. McC. Netting, *Balancing on an Alp: Ecological Change and Continuity in a Swiss Mountain Community*, Cambridge: CUP, 1981, and L. W. Price, *Mountains and Man*, Berkeley, California and London: University of California Press, 1981, especially 421–5. A review paper by R. McC. Netting, 'Agrarian ecology', *Ann Rev Anthropol* 3, 1974, 21–56 contains some useful data, including an item on Chinese soot (p. 37). A review of mountain agroecology occurs in G. D. Berreman, 'Ecology demography and domestic strategies in the Western Himalayas', *J Anthrop Res* 34, 1978, 326–68. The information on soil erosion in Russia comes from I. Stebelsky, 'Agriculture and soil erosion in the European forest-steppe', in J. A. Bater and R. A. French (eds), *Studies in Russian Historical Geography*, London: Academic Press, 1983, vol. 1, 45–63. A good deal of material about agricultural history (as well as pastoralism, hunting, gardens, forestry and land reclamation) is in M. S. Randhawa (ed.), *A History of Agriculture in India*, New Delhi: Indian Council on Agricultural Research, 1982, 3 vols. Detail about land management in a Sabbatical year is from I. Klein (trans and ed.), *The Code of Maimonides, Book Seven: The Book of Agriculture*, New Haven and London: Yale UP, Judaica Series vol XXI, 1979, ch. 1.

Agroecosystems as energy traps

Some of the ideas in the second section are from D. C. Coleman, R. Andrews, J. E. Ellis and J. S. Singh, 'Energy flow and partitioning in selected man-management and natural ecosystems', *Agro-Ecosystems* 3, 1976, 45–54; and general material from D. Pimentel, 'Energy flow in agroecosystems', in R. Lowrance, B. R. Stinner and G. J. House (eds), *Agricultural Ecosystems: Unifying Concepts*, New York: Wiley, 1984, 121–44; and D. Pimentel and C. W. Hall (eds), *Food and Energy Resources*, Orlando, Fla.: Academic Press, 1984. A comparative outline account of several of the examples used here can be found in chapter 5 (pp. 46–84) of M. A. Little and G. E. B. Morren, *Ecology, Energetics and Human Variability*, Dubuque, Iowa: W. C. Brown, 1976. Figures for NPP of major world biomes are from H. Lieth and R. H. Whittaker (eds), *Primary Productivity of the Biosphere*, Berlin, Heidelberg and New York: Springer-Verlag Ecological Studies no. 14, 1975. The shifting cultivators and simple semi-intensive systems comes from C. Uhl and P. Murphy, 'A comparison of productivity and energy values between slash and burn agriculture and secondary succession in the upper Rio Negro region of the Amazon basin', *Agro-Ecosystems* 7, 1981, 63–83, in Little and Morren, op. cit.; additional material from R. A. Rappaport, 'The flow of energy in an agricultural society', *Sci Amer* 224(3), 1971, 104–15; and M. J. Meggitt, 'System and subsystem: the *Te* exchange cycle among the Mae Enga', *Human Ecology* 1, 1972, 111–24. See also G. E. B. Morren, 'From hunting to herding: pigs and the control of energy in Montane New Guinea', in T. Bayliss-Smith and R. Feachem (eds) *Subsistence and Survival. Rural Ecology in the Pacific*. London: Academic Press, 1977, 273–315. There is a more extended discussion of the energetic value of cleared forest in C. F. Jordan (ed.), *Amazonian Rain Forests: Ecosystem Disturbance and Recovery*, NY: Springer-Verlag, 1987, esp. ch. 9. See also D. G. McGrath, 'The role of biomass in shifting cultivation', *Human Ecology* 15, 1987, 221–42. A general overview, including some comparisons with other economies, is in C. Clark and M. Haswell, *The Economics of Subsistence Agriculture*, London: Macmillan, 1970, 4th edn, esp. ch. III. Traditional Chinese vegetable growing is part of K. Newcombe's paper 'The energetics of vegetable production in Asia old and new', *Search* 7, 1976, 423–30; and other Chinese data from Wen Dazhong and D. Pimentel, 'Energy use in crop systems in northeastern China', in D. Pimentel and C. W. Hall, *Food and Energy Resources*, op. cit., 91–120. The tropical comparisons of irrigated and rainfed agriculture are in M. J. T. Norman, 'Energy inputs and outputs of subsistence cropping systems in the tropics', *Agro-Ecosystems* 4, 1978, 355–66. Philippine rice data are in N. V. Nguu and R. K. Palis, 'Energy input and output of a modern and a traditional cultivation system in lowland rice culture', *Philippines J Biol* 6, 1977, 1–8. The Nile valley data come from G. Stanhill, 'The Egyptian agro-ecosystem at the end of the 18th century – an analysis based on the "Description de l'Egypte"', *Agro-Ecosystems* 6, 1981, 305–14. Contemporary China is evaluated by V. Smil, 'China's agro-ecosystem', *Agro-Ecosystems* 7, 1981, 27–46. The role of animals in energy flow is discussed in J. E. Ellis, C. H. Jennings and D. M. Swift, 'A comparison of energy flow among the grazing animals of different societies', *Human Ecology* 7, 1979, 135–49, with extra material from R. and N. Dyson-Hudson, 'Subsistence herding in Uganda', *Sci Amer* 220(2), 1969, 76–89, and S. Odend'hal, 'Energetics of Indian cattle in their environment', *Human Ecology* 1, 1973, 3–22. The Wiltshire (England) farm referred to twice is in T. Bayliss-Smith, *The Ecology of Agricultural Systems*, Cambridge: CUP, 1982. The Amish data are from W. A. Johnson et al., 'Energy conservation in Amish agriculture', *Science* 198, 1977, 373–8. The Tunisian example is from W. H. Bedoian, 'Human use of the pre-Saharan ecosystem and its impact upon desertization', in N. L. Gonzalez (ed.), *Social and Technological Management in Dry Lands*, Boulder, Colo.: Westview Press, AAAS Selected Symposia 10, 1978, 61–109.The Nuñoa example is found in Little and Morren, *Ecology, Energetics and Human Variability*, op. cit., and in R. B. Thomas, *Human Adaptation to a High Andean Energy Flow System*, Occ. Papers in Anthropol. 7, Penn. State Univ., University Park, Pa, 1973. The Indian examples are from U. Pandey and J. S. Singh, 'Energetics of hill agro-ecosystems: a case study from central Himalaya', *Agricultural Systems* 13, 1984, 83–95; B. K. Mishra and P. S. Ramakrishnan, 'The economic yield and energy efficiency of hill agro-ecosystems at higher elevations of Meghalaya in North-Eastern India', *Acta Œcologica/Œcologica Applicata* 2, 1981, 369–89. An introduction to the ecological energetics of Indian agriculture is R. Mitchell, *The Analysis of Indian Agro-Ecosystems*, New Delhi: Interprint Environmental Science Series, 1979. The horses of Paris are swept up into the paper by G. Stanhill, 'An urban agro-ecosystem: the example of nineteenth century Paris', *Agro-Ecosystems* 3, 1977, 269–84. The closing considerations of food security are from M. J. T. Norman, 'Energy inputs', op. cit.

Hunting

General works referred to are: M. Brandon, *Hunting and Shooting, From Earliest Times to the Present Day*, London: Weidenfeld & Nicolson, 1971;

K. Clark, *Animals and Men*, NY: Morrow, 1977; Edward, Duke of York, *The Master of Game* (trans. of *Livre de Chasse* or *Gaston Phoebus* by Gaston III, Count of Foix et Bearn), 1410. For one regional account, see J. M. Gilbert, *Hunting and Hunting Reserves in Medieval Scotland*, Edinburgh: John Donald, 1979. The Chinese example is from Niu Zhongxun and Pu Hanxin, 'The history of the rise and decline of Weichang as an enclosure for the Emperor's hunting, and protection and destruction of natural resources' (in Chinese). Data on hunting and falcony in the Middle East are from M. J. S. Allen and R. Smith, 'Some notes on hunting techniques and practices in the Arabian peninsula', *Arabian Studies* 2, 1975, 108–47. For English practices a summary can be found in J. Clutton-Brock, 'The master of game: the animals and rituals of Medieval venery', *Biologist* 31(3), 1984, 167–71. Details of the types of hunting parks in India are to be found in R. K. Winters, *The Forest and Man*, NY: Vantage Press, 1974.

Pastoralism

A brief introduction to subsistence pastoralism is given by G. W. Cox and M. D. Atkins, *Agricultural Ecology*, San Francisco: Freeman, 1979, pp. 113–15 and material on social and ecological organization by N. and A. Dyson-Hudson, 'Nomadic pastoralism', *Ann Rev Anthropol* 9, 1980, 15–16. An early attempt at comparative studies of pastoralists is J. E. Ellis, C. H. Jennings and D. M. Swift, 'A comparison of energy flow among the grazing animals of different societies', *Human Ecology* 7, 1979, 135–49. The Turkana data are from M. B. Coughenour et al., 'Energy extraction and use in a nomadic pastoral ecosystem', *Science* 230, 1985, 619–25. Useful general material is found in L'Equipe Ecologie et Anthropologie des Sociétés Pastorales, *Pastoral Production and Society*, Cambridge University Press/CNRS (Paris), 1981, and J. Swift, 'Sahelian pastoralists: underdevelopment, desertification and famine', *Ann Rev Anthropol* 6, 1977, 457–78; B. Spooner, *The Cultural Ecology of Pastoral Nomads*, Reading, Mass.: Addison-Wesley, 1973; UNESCO/UNEP/FAO, *Tropical Grazing Land Ecosystems*, Paris: UNESCO, Natural Resources Research XVI, 1979. The special case of Islamic sacrifice is documented by P. J. Stewart, 'Islamic law as a factor in grazing management: the pilgrimage sacrifice', *Proc 1st Int Rangeland Congr* 1978, 119–20. Some traditional societies are discussed in L. S. Leshnik and G.-D. Sontheimer (eds), *Pastoralists and Nomads in South Asia*, Wiesbaden: Otto Harrassowitz, 1975 (see especially the papers by B. Spooner on Baluchistan, 171–82, and R. O. Whyte for recent ecological impact in northern India, 220–34). The ancient world is discussed by J. D. Hughes, *Ecology in Ancient Civilisations*, Albuquerque, NM: UNM Press, 1975, and Z. Naveh, 'Mediterranean landscape evolution and degradation as multivariate biofunctions: theoretical and practical implications', *Landscape Planning*, 9, 1982, 125–46. A considerable store of information is to be found in D. R. Harris (ed.), *Human Ecology in Savanna Environments*, London: Academic Press, 1979, especially the chapter by P. A. Jewell on African savannas, 353–81.

Other regional material occurs in E. E. Bacon, 'Types of pastoral nomadism in Central and Southwest Asia', *South Western J Anthropol* 10, 1954; 44–68; S. Webster, 'native pastoralism in the south Andes', *Ethnology* 12, 1973, 115–33. F. Fraser Darling's paper is 'Pastoralism in relation to population of men and animals', in J. B. Cragg and N. W. Pirie (eds), *The Numbers of Men and Animals*, Edinburgh and London: Oliver and Boyd,

1955, 121–8. The quotation about forest fire is from the Vth Pillar Edict of As'oka, on pp. 103–9 of G. Sarinavasta Murti and A. N. Khrishna Aiyangar (trans. and eds), *Edicts of As'oka*, Madras: The Adyar Library, 1951. Two basic references linking agricultural intensification with the origin of nomadic pastoralism are S. Lee and D. Bates, 'The origins of specialized nomadic pastoralism: a systemic model', *Amer Antiq* 39, 1974, 187–93; and A. Sherratt, 'Plough and pastoralism: aspects of the secondary products revolution', in I. Hodder, G. Ll. Isaac and N. Hammond (eds), *Pattern of the Past: Studies in Honour of David Clarke*, London: CUP, 1981, 261–305. See also D. L. Browman, 'Pastoral nomadism in the Andes', *Current Anthropology* 15, 1974, 188–96; B. Allchin, 'Hunters, pastoralists and early agriculturalists in South Asia', in J. V. S. Megaw (ed.), *Hunters, Gatherers and First Farmers beyond Europe*, Leicester University Press, 1977, 127–44, and many of the contributions in part one of D. R. Harris (ed.), *Human Ecology in Savanna Environments*, London: Academic Press, 1981. The experimental exclosures on Hawaii are described in D. Mueller-Dombois and G. Spatz, 'The influence of feral goats on the lowland vegetation in Hawaii Volcanoes National Park', *Phytocoenologia* 3, 1975, 1–29. The domestication, role and spread of the horse is dealt with in detail in H. B. Barclay, *The Role of the Horse in Man's Culture*, London and New York: J. A. Allen, 1980.

Gardens

I have relied very largely upon Edward Hyams, *A History of Gardens and Gardening*, London: Dent, 1971, supplemented by J. S. Berrall, *The Garden: An Illustrated History*, Harmondsworth: Penguin Books, 1978, and E. Hyams, *Capability Brown and Humphrey Repton*, New York: Scribner, 1971. Hyams's *History* is especially well illustrated and contains a lot of botanical information. For a highly cultural (if not indeed eschatological) interpretation of one type of garden see E. B. Moynihan, *Paradise as a Garden in Persia and Mughal India*, London: Scolar Press, 1980. Related to this approach are D. N. Wilber, *Persian Gardens and Garden Pavilions*, Washington DC: Dumbarton Oaks, 1979; and M. Holborn, *The Ocean in the Sand. Japan: from Landscape to Garden*, London: Gordon Fraser, 1978.

Woodland management

Perhaps the most thorough examination of woodland management in medieval Europe is O. Rackham's *Ancient Woodland: Its History, Vegetation and Uses in England*, London: Edward Arnold, 1980; a more popular version of much of the same material is given in his *Trees and Woodlands in the British Landscape*, London: Dent, 1976. Two standard works on the importance of ship timber (though with a variable amount of ecological detail) are R. G. Albion, *Forests and Sea Power: The Timber Problems of the Royal Navy 1652–1862*, Cambridge, Mass.: Harvard UP, 1926; and P. W. Bamford, *Forests and French Sea Power 1660–1789*, Toronto: University of Toronto Press, 1956.

The progress of deforestation is discussed in many places. This account relies largely upon, for the prehistory of Britain, the appropriate chapters in I. G. Simmons and M. J. Tooley (eds), *British Prehistory: the Environment*, London: Duckworth, 1981. George Perkins Marsh, *Man and Nature, or Physical Geography as Modified by Human Action* (New York: Scribner;

London: Sampson Low, Son and Marston, 1864), is available in an edition of the Belknap Press of Harvard University Press, Cambridge, Mass., 1956, ed. D. Lowenthal. Other studies mentioned are M. Mikesell, 'The deforestation of Mount Lebanon', *Geogr Rev* 59, 1969, 1–28; and H. C. Darby, 'The clearing of the woodland in Europe', in W. L. Thomas (ed.), *Man's Role in Changing the Face of the Earth*, Chicago: Chicago UP, 1956, 183–216. The Cistercian practices are quoted by C. Glacken, *Traces on the Rhodian Shore*, Berkeley and Los Angeles: University of California Press, 1967, 309. Details of the Alpine sequence are in H. Ellenberg, *Vegetation Mitteleuropas mit den Alpen in Okologischer Sicht*, Stuttgart: Verlag Eugen Ulmer, 2nd edn, 1978, esp. 34–72. The forest decline of European Russia comes from R. A. French, 'Russians and the forest', in J. H. Bater and R. A. French (eds), *Studies in Russian Historical Geography*, London: Academic Press, 1983, vol. 1, 23–44. The data for forest loss in N. America, the Punjab and China are from F. Ramade, *Ecologie des Ressources Naturelles*, Paris: Masson, 1981, 248–9; farmed parkland is graphically described by R. A. Pullan, 'Farmed parkland in West Africa', *Savanna* 3, 1974, 119–51. Detail about India's forests is scattered through M. S. Randhawa (ed.), *A History of Agriculture in India*, New Delhi: Indian Council of Agricultural Research, 1980, 3 vols. The detailed example of the importance of the forests in the economy of Himalayan agriculture is in A. Macfarlane, *Resources and Population: a study of the Gurungs of Nepal*, Cambridge: CUP, 1976, esp. ch. 4. The less specific data on forest conservation come mostly from the appropriate regional chapters of G. A. Klee (ed.), *World Systems of Traditional Resource Management*, London: Edward Arnold; NY: Halstead Press, 1980. The material on classical times comes from J. D. Hughes, *Ecology in Ancient Civilizations*, Albuquerque: UNM Press, 1975, chs 8 and 10; J. D. Hughes and J. V. Thirgood, 'Deforestation in Ancient Greece and Rome: a cause of collapse?', *The Ecologist* 12, 1982, 196–208; and on tropical mountains in the Andes from H. Ellenberg, 'Man's influence on tropical mountain ecosystems in South America', *J Ecol* 67, 1979, 401–16. Guatemala is discussed by T. T. Veblen, 'Forest preservation in the Western highlands of Guatemala', *Geogr Rev* 68 (4), 917–34. The remarks by P. Sears are in his 'The importance of forests to man', in S. Haden-Guest, J. K. Wright and E. M. Teclaff (eds), *A World Geography of Forest Resources*, NY: Garden Press, American Geographical Society Special Publication no 33, 1956, 3–12. The effects of deforestation upon landforms and hydrology are from Marsh, *Man and Nature*, op. cit.; A. Goudie, *The Human Impact*, Oxford: Basil Blackwell, 2nd edn, 1986,and upon climate from H. H. Lamb, *Climate, History and the Modern World*, London and New York: Methuen, 1982, 320–1.

Nature conservation

Glacken quotes Emil Male on p. 173 of *Traces on the Rhodian Shore.* The account of the Roman circuses is from J. D. Hughes, *Ecology in Ancient Civilizations*, op. cit., and of attitudes to, and practice of, conservation in early China, in E. H. Schafer, 'The conservation of nature under the T'ang dynasty', *J Econ Soc Hist Orient* 5, 1962, 279–308. Pillar Edict V of As'oka is translated and commented upon in G. Srinavasa Murti and A. N. Krishna Aiyangar, *Edicts of As'oka*, op. cit. The sacred fig and lotus are discussed in I. H. Burkill, 'On the dispersal of the plants most intimate to Buddhism', *J. Arnold Arboretum* 27, 1946, 327–39. The African example comes from D. M. Chavunduka, 'African attitudes to conservation', *Rhodesian Agric. J*

75, 1978, 61–3. Deer parks in England are discussed in L. M. Cantor and J. Hatherly, 'The medieval parks of England', *Geography* 64, 1979, 71–85; and L. M. Cantor, 'Forests, chases, parks and warrens', in L. M. Cantor (ed.), *The English Medieval Landscape*, London and Canberra: Croom-Helm, 1982, 56–85.

Mining and quarrying

Prehistoric mining is dealt with in a sketchy and rather patchy fashion by R. Shepherd, *Prehistoric Mining and Allied Industries*, London: Academic Press, 1980; the classical period by J. D. Hughes, *Ecology in Ancient Civilization*, op. cit.; medieval England in appropriate places in H. C. Darby (ed.), *A New Historical Geography of England before 1600*, Cambridge: CUP, 1976. An overview of mining in general in the West is in T. K. Derry and T. I. Williams, *A Short History of Technology*, Oxford: The Clarendon Press, 1960, ch. 4. The data on Nevada mining come from R. O. Clemner, 'The piñon-pine – old ally or new pest? Western Shoshone Indians vs the Bureau of Land Management in Nevada', *Env. Review* 9, 1985, 131–49.

Industry

An overview can be found in E. L. Jones, 'The environment and the economy' in B. Purke (ed.), *The New Cambridge Modern History*, vol. XIII, ch. II, CUP, 1979. Light on the ancient world is shed by J. G. Landels, *Engineering in the Ancient World*, Berkeley and Los Angeles: University of California Press, 1978; and detail of the Netherlands in J. A. Faber, H. A. Diederiks and S. Hart, 'Urbanisering, industrialisering en milieuaantasting in Nederland in de periode van 1500 tot 1800', *AAG Bijdragen* 18, 1973, 251–71. More diffuse information can be found in H. C. Darby (ed.), *A New Historical Geography of England*, Cambridge: CUP, 1976, and T. K. Derry and T. I. Williams (eds), *A Short History of Technology*, Oxford: Clarendon Press, 1960.

Cities

Information about ancient Greece and Rome comes from J. D. Hughes, op. cit.; in Europe generally from E. L. Jones, op. cit., and on London's air pollution problems from W. H. Te Brake, 'Air pollution and fuel crises in preindustrial London 1250–1650', *Technology and Culture* 16, 1965, 337–59. The notion of the emergence of the compact city after 3000 BC is taken from P. Wheatley, *The Pivot of the Four Quarters*, Edinburgh: Edinburgh UP, 1971, 326–7.

War

Examples from ancient times are mentioned in J. D. Hughes, op. cit., and the whole issue of *Ambio* vol. 4, nos 5–6, 1975 is devoted to the environmental consequences of warfare, mostly modern though with some historical material. See also T. Roseberry, 'Biological warfare: some historical considerations', *Bull Atom Sci* 16, 1960, 227–36. A slightly tangential (and depressing) book by A. Ferrill, *The Origins of War: From the Stone Age to Alexander the Great*, London: Thames and Hudson, 1985, argues that a burst of organized warfare accompanied the Neolithic Revolution in the Near East.

Reclamation

General material on Europe is in C. T. Smith, *An Historical Geography of Western Europe before 1800*, London: Longmans, 1967, ch. 9; for England generally, M. Williams, 'Marshland and waste', in L. Cantor (ed.), *The English Medieval Landscape*, London and Canberra: Croom Helm, 1982, 86–125; and as a specialist study of one region: H. E. Hallam, *Settlement and Society: a study of the early agrarian history of South Lincolnshire*, Cambridge: CUP, 1965. For one minor technique, see R. A. Dodgshon and C. A. Jewell, 'Paring and burning and related practices with particular reference to the Southwestern countries of England', in A. Gailey and A. Fenton (eds), *The Spade in Northern and Atlantic Europe*, Belfast: Ulster Folk Museum and Institute of Irish Studies of Queen's University, 1969, 74–87. See also D. J. Maguire, 'The ecological impact of medieval agriculture: a study of Holne Moor, Dartmoor, U.K.', in H. I. Shural (ed.), *Developments in Ecology and Environmental Quality*, Rehorot and Philadelphia: Balaban International Science Services, 1983, vol. II, 83–90; P. H. Armstrong, 'Changes in the land use of the Suffolk sandlings: a study in the disintegration of an ecosystem', *Geography*, 58, 1973, 1–8.

Impact on the oceans

E. L. Jones's views on whaling are found in his 'The environment and the economy' (see above under 'Industry'); a concise ecological and social history of fishing is C. L. Smith and W. Q. Wick, 'Fishing peoples', in R. A. Ragotzkie (ed.), *Man and the Marine Environment*, Boca Raton, Fla.: CRC Press, 1983, 21–44.

Population, resources and environment of agriculturalists

Some very helpful general themes are explored by D. B. Grigg, *The Dynamics of Agricultural Change*, Cambridge: CUP, 1982. The role of climate in human economic affairs in the past is explored in M. J. Ingram, G. Farmer and T. M. L. Wigley, 'Past climates and their impact on Man: a review', in T. M. L. Wigley, M. J. Ingram and G. Farmer (eds), *Climatic and History. Studies in Past Climates and their Impact on Man*, Cambridge: CUP, 1981, 3–50; and M. J. Bowden (and 7 others), 'The effect of climatic fluctuation on human population: two hypotheses', ibid., 479–513. R. Klee's book on traditional resource uses has already been extensively cited. The data which compare energy use in rural India with the USA are from R. Revelle, 'Energy use in rural India', *Science* 192, 1976, 969–75, and famines are chronicled spatially and temporally by W. A. Dando, *The Geography of Famine*, London: Edward Arnold; New York: V. H. Winston, 1980. See also G. M. Ward et al., 'Animals as an energy source in Third World agriculture', *Science* 208, 1980, 570–4. An interesting account of how political change affected 'harvested-yield power density' in a Caribbean country is G. A. Antonini, K. C. Ewel and H. M. Tupper, *Population and Energy: A Systems Analysis of Resource Utilization in the Dominican Republic*, Gainsville, Fla.: Univ of Florida Press, 1975. Especially useful on intensification is H. C. Brookfield, 'Intensification and disintensification in Pacific agriculture: a theoretical approach', *Pacific Viewpoint* 13, 1972, 30–48; see also M. Margolis, 'Historical perspectives on frontier agriculture as an adaptive strategy', *Amer Ethnol* 4, 1977, 42–64. On islands, see T. Bayliss-Smith,

'Constraints on population growth: the case of the Polynesian outlier atolls in the precontact period', *Human Ecology* 2, 1974, 259–93; D. Watts, 'Cycles of famine in islands of plenty: the case of the colonial West Indies in the pre-emancipation period', *Geojournal Supplement* 6, 1984, 49–70. The ideas about Easter Island came from a public lecture by Dr. J. Flenley of the University of Hull. The general remarks about late medieval population–impact relations are in D. Herlihy, 'Ecological conditions and demographic change, in R. L. DeMolen (ed.), *One Thousand Years: Western Europe in the Middle Ages*, Boston: Houghton Mifflin, 1974, 3–43.

Chapter 5 Industrialists

Introduction

A basic outline of the very early use of fossil fuels as well as of the early phases of the industrial revolution itself is given in T. K. Derry and T. I. Williams, *A Short History of Technology*, Oxford: The Clarendon Press, 1960, and by E. Ayres, 'The age of fossil fuels', in W. L. Thomas (ed.), *Man's Role in Changing the Face of the Earth*, Chicago: Chicago UP, 1956, 367–81. Accounts of the English scene are found in R. Lawton, 'Historical geography: the industrial revolution', in J. Wreford Watson and J. B. Sissons (eds), *The British Isles: A Systematic Geography*, London: Nelson, 1964, 221–44; H. C. Prince, 'England circa 1800', and J. B. Harley, 'England circa 1850', in H. C. Darby (ed.), *A New Historical Geography of England after 1600*, Cambridge: CUP, 1976, 89–164 and 227–94 respectively. Detail on historical energy consumption is in W. S. Humphrey and J. Stanislaw, 'Economic growth and energy consumption in the UK, 1700–1975', *Energy Policy* 7, 1979, 29–42. The course of energy consumption is used in an historically explanatory fashion by R. N. Adams, *Paradoxical Harvest: Energy and Explanation in British History 1870–1914*, Cambridge: CUP, 1982, The Arnold and Caroline Rose Monograph Series of the American Sociological Association. Specialized argument by economic historians about technical change and resource scarcities can be found in A. Maczak and W. N. Parker (eds), *Natural Resources in European History*, Washington, DC: RFF Research Paper R-13, 1978. The argument for the depletion of wood in England as the main factor in the development of coal is summarized in J. U. Neff, 'An early energy crisis and its consequences', *Sci Amer* 237, 1977, 140–51. The energy return on investment measure is discussed in C. J. Cleveland et al., 'Energy and the U.S. economy: a biophysical perspective', *Science* 225, 1984, 890–7. For a possibly more inclusive calculation which yields similar rankings of ratios but lower numbers (e.g. the fossil fuels at 6:1 rather than 20–30:1) see M. J. Burnett, 'A methodology for assessing net energy and abundance of energy resources', in W. J. Mitsch, R. W. Bosserman and J. Klopatek (eds), *Energy and Ecological Modelling*, Amsterdam, Oxford and New York: Elsevier Developments in Environmental Modelling, vol. 1, 1981, 703–9. An interesting anthology of prose, poetry and songs from the industrial revolution in England can be found in A. Clayre (ed.), *Nature and Industrialization*, Oxford: The University Press/ Open University Press, 1977; and an eyewitness account of nineteenth century Coalbrookdale in *Coalbrookdale 1801: A contemporary description*, Telford: Ironbridge Gorge Museum Trust, booklet no. 20.03, 1979. An accessible account of industrializing England in the nineteenth century with the emphasis suggested by its title is B. Trinder, *The Making of the Industrial Landscape*, London: Dent, 1982.

The spread of industrialism

The basic spatial data on industrial growth and spread can be found in G. Barraclough (ed.), *The Times Atlas of World History*, London: Times Books Ltd, 1978, especially 212–13 and 218–19. The progress of industrialism's spread in the world is recorded by W. A. Lewis, *Growth and Fluctuations 1870–1913*, London: Allen and Unwin, 1978; and W. A. Lewis (ed.), *Tropical Development 1880–1913*, London: Allen and Unwin, 1970. An interesting discussion of the idea, and the spatial spread, of the modern world-system as a structuring of world trade, economy and society on a core-periphery basis is R. A. Dodgshon, 'A spatial perspective', *Peasant Studies* 6, 1977, 8–19. A case for wood and water as the important power sources for the early development of industry in the USA is in M. V. Melosi, 'Energy transitions in the Nineteenth-century economy', in G. H. Daniels and M. H. Rose (eds), *Energy and Transport: Historical Perspective on Resource Issues*, Beverley Hills, London and New Delhi: Sage, 1982, 55–69. An overview of Russia in this era can be found in W. L. Blackwell, 'The historical geography of industry in Russia during the nineteenth century', in J. H. Bater and R. A. French (eds), *Studies in Russian Historical Geography*, London: Academic Press, 1983, vol. 2, 387–422. The material on southern Nigeria is from A. L. Mabogunje and M. B. Gleave, 'Changing agricultural landscape in Southern Nigeria. The example of Egba division, 1850–1950', *Nigerian Geographical Journal* 7, 1964, 1–15. See also S. Rajaratnam, 'The growth of plantation agriculture in Ceylon, 1886–1931', *Ceylon J Hist Soc Stud* 7, 1964, 1–20. The study of the Amazonian group referred to is P. Henley, *The Panare: Tradition and Change on the Amazonian Frontier*, New Haven and London: Yale UP, 1982.

Impact of early industrialization

The environmental effects of coal in Europe are listed (but only partially quantified) in Environmental Resources Ltd, *Environmental Impact of Future Coal Production and Use in the EEC*, London: Graham and Trotman, for the Commission of the European Communities, 1983. Readers outside Britain will want to know that Shortlands is the country house of the colliery owner in D. H. Lawrence's *Women in Love*; *Coronation Street* is a TV soap-opera of present-day industrial working-class life in a district of houses clearly of nineteenth-century origin, 'the Captains and the Kings depart' is part of a line from Rudyard Kipling's poem *Recessional*, whose refrain is 'Lest we forget'. All a bit self-consciously literary, but in general quite harmless.

Magnitudes of energy use

Most of the numerical data came from one of four sources: H. T. Odum, *Environment, Power and Society*, New York: Wiley, 1971; H. T. and E. C. Odum, *Energy Basis for Man and Nature*, New York: McGraw-Hill, 2nd edn, 1981; E. Cook, *Man, Energy, Society*, San Francisco: Freeman, 1976; G. Foley, *The Energy Question*, Harmondsworth: Pelican Books, 2nd edn, 1981. Additional information is from standard statistical sources and from the reports of the triennial World Energy Conferences. Pre-1980s China is from V. Smil, 'China's agro-ecosystem', *Agro-Ecosystems* 7, 1981, 27–46. T. Bayliss-Smith's data are in his book, *The Ecology of Agricultural Systems*, Cambridge: CUP, 1982.

Energy and spatial structures are dealt with in several contributions to D. R. Cope, P. Hills and P. James (eds), *Energy Policy and Land Use Planning: An International Perspective*, Oxford: Pergamon Press, 1984.

The framework for consideration of mineral industries is by T. B. Denisova, 'The environmental impact of mineral industries', *Soviet Geography* 18, 1977, 646–59 (translated from *Izvestiya Akademii Nauk SSR*, seryia geograficheskya, 1976, no 6, 55–66), with other data from I. Douglas, *The Urban Environment*, London: Edward Arnold, 1983. The lead-mine example is from M. G. Macklin, 'Flood-plain sedimentation in the upper Axe Valley, Mendip, England', *Trans Inst Brit Geog* NS 10, 1985, 235–44. See also L. A. Martinelli et al. 'Mercury contamination in the Amazon: a gold rush consequence', *Ambio* 17, 1988, 252–54. Use trends in the USA are from R. C. Marlay, 'Trends in industrial use of energy', *Science* 226, 1984, 1277–83. The calculations of energy flux in natural and man-made processes are from J. F. Alexander, 'A global systems ecology model', in R. A. Fazzolare and C. B. Smith (eds), *Changing Energy Use Futures*, New York and Oxford: Pergamon Press, 1979, vol. II, 443–56.

Energy production and transformation: the city

A basic source, though more often qualitative than quantitative, is E. E. El-Hinnawi, *The Environmental Impacts of Production and Use of Energy*, Dublin: Tycooly Press for UNEP, 1981. Here this has been supplemented by data from P. N. Cheremisinoff and A. C. Morresi, *Environmental Assessment and Impact Statement Handbook*, Ann Arbor, Michigan: Ann Arbor Science, 1977; J. Golden et al., *Environmental Impact Data Book*, Ann Arbor, Mich.: Ann Arbor Science, 1979, and by Environmental Resources Ltd, *Environmental Impact of Future Coal Production and Use in the EEC*, London: Graham and Trotman, for the EC Commission, 1982. Specific detail is from G. W. Land, 'Energy use and production: agriculture and coal mining', *Landscape Planning* 10, 1983, 3–13; M. Hill and R. Alterman, 'Power plant site evaluation: the case of the Sharon plant in Israel', *J Env Management* 2, 1974, 179–96, and C. G. Lind and W. J. Mitsch, 'A net energy analysis including environmental cost of oil shale development in Kentucky', in W. J. Mitsch, R. W. Bosserman and J. M. Klopatek (eds), *Energy and Ecological Modelling*, Amsterdam, Oxford and New York: Elsevier Scientific Developments in Environmental Modelling, vol. I, 1981, 689–96. For the USA, water consumption for fuel production is given in D. Pimentel et al., 'Water resources in food and energy production', *BioScience* 32, 1982, 861–7. The ranking of fluxes, including fossil fuels, is from J. F. Alexander, op. cit. In the case of the city, see I. Douglas, op. cit., for a comprehensive survey. Other material of use can be found in R. W. Burchell and D. Listokin (eds), *Energy and Land Use*, Piscataway, NJ: Rutgers Univ. Center for Urban Policy and Research, 1982. Data for residential energy consumption are from J. G. Rau and D. C. Wooten, *Environmental Impact Analysis Handbook*, New York: McGraw-Hill, 1980. The nutrient runoff from urban areas is compared with other land uses in W. M. Lewis et al., *Eutrophication and Land Use: Lake Dillon, Colorado*, NY: Springer-Verlag Ecological Studies 46, 1984; and the effect of city growth on small streams in D. Arthington et al., 'Effects of urban development and habitat alterations on the distribution and abundance of native and exotic freshwater fish in the Brisbane region, Queensland', *Aust J Ecol* 8, 1983, 87–101. The Hong Kong study is conveniently summarized in a paper by K. Newcombe et al., 'The metabolism of a city: the case of Hong Kong', *Ambio* 7, 1978, 3–15, and in a book, S. Boyden, S. Millar, K. Newcombe and B. O'Neill, *The Ecology of a City and its People: the Case of Hong Kong*, Canberra: ANU Press, 1981. (The detailed papers on energy flows are in the journal *Urban Ecology* during the 1970s, but see also K. Newcombe, 'Nutrient flow in a major urban settlement: Hong Kong', *Human Ecology* 5, 1977, 179–208; K. Newcombe and E. H. Nichols, 'An integrated ecological approach to agricultural policy-making with reference to the urban fringe: the case of Hong Kong', *Agricultural Systems* 4, 1979, 1–27.) See also R. U. Cooke, *Geomorphological Hazards in Los Angeles*, London: Allen and Unwin, The London Research Series in Geography 7, 1984. The energy equity and efficiency of land use patterns, with examples from Australia, is discussed by R. Sharpe, 'Energy efficiency and equity of various urban land use patterns', *Urban Ecology* 7, 1982, 1–18. The idea of an upper limit comes from R. T. Jaske, 'An evaluation of energy growth and use trends as a potential upper limit in metropolitan development', *The Science of the Total Environment* 2, 1973, 45–60; see also W. Sassin, 'Urbanization and the global energy problem', in P. Laconte et al. (eds), *Human and Energy Factors in Urban Planning: a Systems Approach*, The Hague: Martinus Nijhoff, 1982, 207–33. City size and energy use (in the USA only) are discussed in J. Zuchetto, 'Energy and the future of human settlement patterns: theory, models and empirical considerations', *Ecological Modelling* 20, 1983, 85–111. The scenario for Java is in R. L. Meier, 'A stable urban ecosystem', *Science* 192, 1976, 962–8. He expands some of these ideas in 'Energy and habitat: Developing a sustainable urban ecosystem', *Futures* 16, 1984, 351–71. Overviews of the ecological approach to cities are A. Whyte, 'Ecological approaches to urban systems: a retrospective and prospective look', *Nature and Resources* 21(1), 1985, 13–20; L. R. Brown and J. L. Jacobson, *The Future of Urbanization: Facing the Ecological and Economic Constraints*, Washington DC: Worldwatch Institute Paper 77, 1987.

Agriculture

A global view of agriculture's recent position is given by J. P. O'Hagan, 'World food and agriculture: Elements of the system to the year 2000', *Interdisciplinary Science Reviews* 7, 1982, 230–41. The effect of industrialization on world agriculture is discussed in the relevant parts of two books by D. B. Grigg, *The Agricultural Systems of the World: An Evolutionary Approach*, Cambridge: CUP, 1974; *idem*, *The Dynamics of Agricultural Change*, London: Hutchinson, 1982. The energy structure of modern agriculture is laid out by I. G. Simmons in 'Ecological functional approaches to agriculture in geographical contexts', *Geography* 65, 1980, 305–16, and at greater historical depth and detail (as in the Wiltshire examples) by T. P. Bayliss-Smith, *The Ecology of Agricultural Systems*, Cambridge: CUP, 1982. Minor sources of data were W. A. Dekkers, 'Energy production and use in Dutch agriculture', *Neth J Agric Sci* 22, 1974, 107–18; K. S. Finison and R. B. Thomas, 'Agricultural Change in 19th century New England: an energetic analysis', in R. A. Fazzolare and C. B. Smith (eds), *Changing Energy Use Futures*, NY: Pergamon Press, 1979, vol. II, 494–501.

The question of recent extensions of the world's arable area was explored by C. J. Robertson, 'The expansion of the arable area', *Scot Geogr Mag* 72, 1956, 1–19 and more reliably by D. B. Grigg, 'The growth and distribution of the world's arable land 1870–1970', *Geography* 59, 1974, 104–10. D. B. Grigg also discusses in the two books quoted above the industrial technology which made possible the transformation of the mid-latitude grasslands. The subsequent alterations to the flora of remaining grassland in these regions are the subject of two essays in W. L. Thomas (ed.), *Man's Role in Changing the Face of the Earth*, op. cit., 1956. They are J. T. Curtis, 'The modification of mid-latitude grasslands and forests by man', 721–36, and A. H. Clark, 'The impact of exotic invasion on the remaining New World mid latitude grass lands', 737–62. Nobody interested in the High Plains should fail to read the classic by Walter Prescott Webb, *The Great Plains: A Study in Institutions and Environment*, Boston: Ginn & Co, 1931; nor the interpretations of the historian J. C. Malin, as in his contribution to *MRCFE*, 'The grassland of North America: its occupance and the challenge of continuous reappraisals', 350–65, and *The Grassland of North America: Prolegomena to its History*, Lawrence, Kansas: The Author, 1947.

The best source of discussion of irrigation in LDCs is A. Makhijani and A. Poole, *Energy and Agriculture in the Third World*, Cambridge, Mass.: Ballinger, 1975. Some LDC information and a good deal of US data are in the various contributions to W. Lockeretz (eds), *Agriculture and Energy*, Boulder, Colo.: Westview Press, 1977. Minor sources were K. Parikh and F. Rabir (eds), *Food for All in a Sustainable World: The I.I.A.S.A. Food and Agriculture Program*, Laxenburg, Austria: IIASA, 1981; and A. K. Biswas (ed.), *United Nations Water Conference: Summary and Main Documents*, Oxford: Pergamon Press Water Development, Supply and Management, vol. 2, 1978.

The data and ideas discussed in the section on energy and agricultural intensification start from I. G. Simmons, 'Ecological–functional approaches', op. cit., and for US data draw on W. Lockeretz (ed.), *Agriculture and Energy*, op. cit., 1977, and V. Smil, *Energy Analysis and Agriculture: An application to U.S. Agriculture*, Boulder, Colo.: Westview Press, 1983. Comparative data from the Antipodes are in M. Watt, 'An energy analysis of the Australian food system', *Energy in Agriculture* 3, 1984, 279–88, and M. G. Patterson, 'Energy use in the New Zealand food system', *Energy in Agriculture* 3, 1984, 289–304. The UK data for intensive animal production are from C. R. W. Spedding et al., 'Energy and economies of intensive animal production', *Agro-Ecosystems* 8, 1983, 169–81. Developing nations form the material of V. Smil, 'Energy flows in the developing world', *Amer Sci* 67, 1979, 522–31; the sisal industry of Brazil comes in D. R. Gross and B. A. Underwood, 'Technological change and caloric costs: sisal agriculture in Northeastern Brazil', *Amer Anthropol* 73, 1971, 725–40. Comparisons of rice-producing methods are found in S. M. Freedman, 'Modifications of traditional rice production practices in the developing world: an energy efficiency analysis', *Agro-Ecosystems* 6, 1980, 129–46; and the comparisons with Egypt in G. Stanhill, 'A comparative study of the Egyptian agro-ecosystem', *Agro-Ecosystems* 5, 1979, 213–30. See also B. S. Pathak and D. Singh, 'Energy returns in agriculture, with specific reference to developing countries', *Energy* 3, 1979, 119–26; B. S. Pathak, 'Energy demand growth in Punjab agriculture and the changes in agricultural production', *Energy in Agriculture* 4, 1985, 67–78. S. M. Freedman, 'The use of rice crop residues as a non-commercial energy source in the developing world: the energy and environmental implications', *Agr Ecosysts and Env* 10, 1983, 63–74. Animals in LDC agricultural development are discussed by G. M. Ward et al., 'Animals as an energy source in Third World agriculture', *Science* 208, 1980, 570–4. The French agro-ecosystem is discussed by J. P. Deleage et al., 'Eco-energetics analysis of an agricultural system: the French case in 1970', *Agro-Ecosystems* 5, 1979, 345–65, and for Hong Kong by K. Newcombe, 'Energy use in the Hong Kong food system', *Agro-Ecosystems* 2, 1975, 253–76. Associated material of interest is found in D. Pimentel (ed.), *CRC Handbook of Energy Utilization in Agriculture*, Boca Raton, Fla.: CRC Reviews, 1980; G. Singh and W. Chancellor, 'Energy inputs and agricultural production under various regimes of mechanisation in northern India', *Trans Amer Soc Agric Eng* 18, 1975, 252–9; A. J. Singh and S. S. Miglani, 'An economic analysis of energy requirements in Punjab agriculture', *Indian J Agric Econ* 31, 1976, 165–73; K. Newcombe, 'The energetics of vegetable production in Asia old and new', *Search* 7, 1976, 423–30; V. Smil, 'China's agroecosystem', *Agro-ecosystems* 7, 1981, 27–46; Wen Dahzong and D. Pimentel, 'Energy inputs in agricultural systems of China', *Agr Ecosysts and Env* 11, 1984, 29–35; Wu Wen and Chen En-Jian, 'Our views on the resolution of China's rural energy requirements', *Biomass* 3, 1983, 287–312; D. L. Plucknett and N. J. H. Smith, 'Agricultural research and third world food production', *Science* 217, 1982, 215–20 and D. H. Janzen, 'Tropical agroecosystems', *Science* 182, 1973, 1212–19. Do not forget S. B. Hill and J. A. Ramsay, 'Limitations of the energy approach in defining priorities in agriculture', in W. Lockeretz (ed.), *Agriculture and Energy*, NY: Academic Press, 1977, 713–31. A collection of papers on the technical aspects of energy accounting is G. Stanhill (ed.), *Energy and Agriculture*, Berlin: Springer-Verlag Advanced Series in Agricultural Sciences 14, 1984.

General works which include discussion of the environmental relations of industrialized agriculture include E. P. Eckholm, *Losing Ground: Environmental Stress and World Food Prospects*, New York: W. W. Norton, 1977; M. W. Holdgate et al. (eds), *The World Environment 1972–82*, Dublin: Tycooly International, Natural Resources and the Environment series, vol. 8, 1982, ch. 7; G. O. Barney (Director), *The Global 2000 Report to the President*, vol. 2, ch. 13, Harmondsworth: Penguin Books, 1982; J. G. Hawkes (ed.), *Conservation and Agriculture*, London: Duckworth, 1978; J. Lenihan and W. W. Fletcher (eds), *Food, Agriculture and the Environment*, Glasgow: Blackie, 1975, Environment and Man, vol. 2; on soil erosion see J. Tivy in Lenihan and Fletcher, ibid., ch. 2.

For irrigation, E. B. Worthington, *Arid Land Irrigation in Developing Countries: Environmental Problems and Effects*, Oxford: Pergamon Press, 1977; W. M. Adams and R. C. Carter, 'Small-scale irrigation in sub-Saharan Africa', *Prog Phys Geog* 11, 1987, 1–27. Also useful are V. A. Kovda, 'Soil loss: an overview', *Agro-Ecosystems* 3, 1977, 205–24, and R. A. Brink et al., 'Soil deterioration and the demand for food', *Science* 197, 1977, 625–30. An interesting example of the secondary environmental changes caused by cash crop farming is R. R. Thaman and P. M. Thomas, 'Cassava and change in Pacific island food systems', in D. J. Cattle and K. H. Schwerin (eds), *Food Energy in Tropical Ecosystems*, NY: Gordon and Breach, 1985, Food and Nutrition in History and Anthropology, vol. 4, 189–223. Data on pesticide resistance are from M. Sun, 'Pests prevail despite pesticides', *Science* 226, 1985, 1293; a wider view of the relationships is F. Chaboussou, 'How pesticides increase pests', *The Ecologist* 16(1), 1984, 29–35.

Genetic loss is discussed in UNESCO, *Conservation of Natural Areas and of the Genetic Material They Contain*, Paris: UNESCO, 1973, MAB Report series 12. An optimistic view of this problem is D. N. Duvick, 'Genetic diversity in major farm crops on the farm and in reserve', *Econ Bot* 38, 1984, 161–78. A comprehensive non-technical paper on biocides is D. Pimentel and C. A. Edwards, 'Pesticides and ecosystems', *BioScience* 32(7), 1982, 595–60. The data on landscape change in England come from R. Westmacott and T. Worthington, *New Agricultural Landscapes*, Cheltenham: Countryside Commission, 1974, and on fauna from J. G. de Molenaar, 'The influence of agriculture on the fauna', *Semaine d'Etude Agriculture et Environment*, Centre de Recherches Agronomiques Gembloux, 1974, 397–410. The last paragraph is informed by FAO, *Agriculture: Toward 2000*, Rome: FAO, 1981, Economic and Social Development Series No 23; J. P. O'Hagan, 'World food and agriculture: Elements of the system to the year 2000', *Interdisciplinary Science Reviews* 7(3), 1982, 230–41; and M. A. Altieri, D. K. Letourneau and J. R. Davis, 'Developing sustainable agroecosystems', *BioScience* 33, 1983, 45–9. See also the papers by D. J. Janzen and by Plucknett et al., op. cit., above. The quotation by E. L. Jones is from his 'The environment and the economy', in B. Purke (ed.), *The New Cambridge Modern History*, Cambridge: CUP, 1979, vol. XIII, *Companion Volume*, 15–42.

Pastoralism

A good introduction for the tropics is W. Sanford and E. Wangari, 'Tropical grasslands: dynamics and utilization', *Nature and Resources* 21(3), 1985, 12–34. Literature on contemporary pastoralism is dominated by discussion of sedentarization and by the Sahel zone of Africa. See P. C. Salzman (ed.), *When Nomads Settle*, NY: Praeger, 1980; *idem*, *Processes of Sedentarization as Adaptation and Response*, NY: Praeger, 1980, especially the chapter by D. R. Aaronson, 'Must nomads settle? Some notes towards policy on the future of pastoralism', 173–84. See also B. S. Spooner and H. S. Mann (eds), *Desertification and Development: Dryland Ecology in Social Perspective*, London: Academic Press; and a useful if dated annotated bibliography, C. Oxby, *Pastoral Nomads and Development*, London: Internation African Institute, 1975. A historical account of Rajasthan (India) is R. P. Dhir, 'The human factor in ecological history', in Spooner and Mann, op. cit., 311–31. On desertification overall, see the WWF/IUCN/UNEP World Conservation Strategy, 1980; and a short paper by G. Novikoff, 'Desertification by overgrazing', *Ambio* 12, 1983, 102–5, whose scientific approach is exemplified in H. Bremen and C. T. de Wit, 'Rangeland productivity and exploitation in the Sahel', *Science* 221, 1983, 1341–7; a more human-oriented perspective is by J. Swift, 'Sahelian pastoralists: underdevelopment, desertification and famine', *Ann Rev Anthropol* 6, 1977, 457–78. See also the review by M. Kassas, 'Drought and desertification', *Land Use Policy* 4, 1987, 389–400; and UNEP, *Arid Land Development and the Combat against Desertification: An Integrated Approach*. Moscow: Centre for International Projects GKNT, 1986. A wider African perspective comes in P. A. Jewell, 'Ecology and management of game animals and domestic livestock in African savannas', in D. R. Harris (ed.), *Human Ecology in Savanna Environments*, London: Academic Press, 1980, 353–81. Studies of 'modernization' are E. Cruz de Carvallio, '"Traditional" and "modern" patterns of cattle raising in South Western Angola: a critical evaluation of change from pastoralism to ranching', *J Dev Areas* 8, 1974, 199–226; and

C. Humphrey, 'Pastoral nomadism in Mongolia: the role of herdsmen's co-operatives in the national economy', *Development and Change* 9, 1978, 133–60. See also D. G. Bates, 'The role of the state in peasant-nomad mutualism', *Anthrop Quart* 44, 1971, 109–31. Commercial ranching is given its historical context by D. B. Grigg, *The Agricultural Systems of the World*, CUP, 1974, ch. 2. Some amplification is given in C. F. Bennett, 'Savannas and man in middle America', *Geogr Rev* 54, 1964, 580–2. For Australia see G. Bolton, *Spoils and Spoilers: Australians Make their Environment 1788–1980*, Sydney: Allen and Unwin, 1980, The Australian Experience, no. 2. Other examples are found in M. Taghi Farvar and J. P. Miller (eds), *The Careless Technology*, Garden City, NY: Natural History Press, 1972, especially F. Fraser Darling and M. Farvar, 'Ecological consequences of the sedentarization of nomads', 671–82; H. F. Heady, 'Ecological consequences of Bedouin settlements in Saudi Arabia', 683–93; and F. L. Lamprecht, 'Tsetse fly: a blessing or a curse?', 726–41. A selection of work on reindeer herding in Fennoscandia today is G. W. Scotter, 'Reindeer ranching in Fennoscandia', *J Rge Mgmt* 18, 1966, 301–5; L. Müller-Wille and P. J. Pelto, 'Technological change and its impact in Arctic regions: Lapps introduce snowmobiles into reindeer herding', *Polarforschung* 41, 1971, 142–8; P. J. Pelto, *The Snowmobile Revolution: Technology and Social Change in the Arctic*, Menlo Park, Calif.: Cummings, 1973; L. Müller-Wille, 'The snowmobile, Lapps and reindeer herding in Finnish Lapland', in J. D. Ives and R. G. Barry (eds), *Arctic and Alpine Environments*, London: Methuen, 1974, 915–20; H. Beach, *Reindeer-Herd Management in Transition: The case of Tuorpon Saameby in Sweden*, Uppsala: Uppsala Studies in Cultural Anthropology 3, 1981; T. Ingold, 'The rationalization of reindeer management among Finnish Lapps', *Development and Change* 9, 1978, 103–32. Ingold gives a more theoretical account of changes in reindeer economies in his *Hunters, Pastoralists and Ranchers*, Cambridge: CUP, 1980. Progress with Red Deer in the UK is summarized in Rowett Research Institute and the Hill Farming Research Organisation, *Farming the Red Deer*, Edinburgh: HMSO, 1974.

Gardens

The books referred to in the relevant section of the last chapter are still relevant, especially Edward Hyams's, *A History of Gardens and Gardening*, London: Dent, 1971. Plant transfers are described in T. Whittle, *The Plant Hunters*, London: Picador Books, 1975; and Gertrude Jekyll's work in Northern England by M. J. Tooley (ed.), *Gertrude Jekyll: Artist, Gardener, Craftswoman*, Witton-le-Wear, England: Michaelmas Books, 1984. Suburban lawns are subject to analytical treatment in T. R. Detwyler, 'Vegetation of the city', in T. R. Detwyler and M. G. Marcus (eds), *Urbanization and Environment*, Belmont, Calif.: Duxbury Press, 1972, 3–25; J. H. Falk, 'Energetics of a suburban lawn ecosystem', *Ecology* 57, 1976, 141–50. The output data for garden food are in R. H. Best and D. Ward, *The Garden Controversy*, Wye, Kent: Wye College (Univ. of London) Papers in Agricultural Economics, 1956. A lovely book is E. Hyams, *Great Botanical Gardens of the World*, London: Bloomsbury Books, 1985.

Hunting

A discussion of the role of animal protein in the cultural ecology of Amazonia is by L. E. Sponsel, 'Yanomama warfare, protein capture and cultural ecology: a critical analysis of the arguments', *Interciencia* 8, 1983, 204–10. The impact of the shotgun is described in R. B. Haines, 'A comparison of the efficiencies of the shotgun and the bow in Neotropical forest hunting', *Human Ecology* 7, 1979, 219–52. The Inuit example is from W. B. Kemp, 'Energy flows in a hunting society', *Sci Amer* 224, 1971, 104–15. Other relevant material is in J. Sonnenfeld, 'Changes in an Eskimo hunting technology: an introduction to implement geography', *AAAG* 50, 1960, 172–86; W. Townsend, 'Stone to steel use in New Guinea society', *Ethnology* 8, 1969, 199–206. Data on the percentage of sport hunters in national populations for Europe and the USA are in R. C. Bigalke, 'Utilizing wild species', in D. J. A. Cole and G. C. Brander (eds), *Bioindustrial Ecosystems*, Amsterdam: Elsevier, 1986, Ecosystems of the World, vol. 21, 255–74. The management of North American habitats for deer hunters is discussed in, for example, L. W. Wing, *Practice of Wildlife Conservation*, New York: Wiley, 1951; R. F. Dasmann, *Wildlife Biology*, New York: Wiley, 1964. See also B. Dalrymple, *The Complete Book of Deer Hunting*, New York: Winchester Press, 1973. A general book, well-illustrated, on the history of hunting is M. Brander, *Hunting and Shooting: From Earliest Times to the Present Day*, London: Weidenfeld & Nicolson, 1971. The rest of the material can perhaps be most kindly described as scavengings by the author.

Species conservation

Another basic document is the IUCN/UNEP/WWF *World Conservation Strategy*, Gland, Switzerland: IUCN, 1980; the parallel UNESCO programme is described in anon., 'The environment and natural resources: a six-year plan', *Nature and Resources* 18(4), 1982, 18–24, and a useful summary of the recent past in chapter 6, 'Terrestrial biota', of M. W. Holdgate, M. Kassas and G. F. White (eds), *The World Environment 1972–82*, Dublin: Tycooly International Publishing, Natural Resources and the Environment Series, vol. 8, 210–47. Biosphere reserves are discussed in M. Batisse, 'The biosphere reserve: a tool for environmental conservation and management', *Environmental Conservation* 9, 1982, 101–10; anon., 'Looking at biosphere reserves – a 1983 perspective', *Nature and Resources*, 1983, 22–5. Detail of natural World Heritage Sites is in *The World's Greatest Natural Areas: An Indicative Inventory of Natural Sites of World Heritage Quality*, Gland, Switzerland: IUCN, 1982, with a broader context in J. A. McNeeley and K. R. Miller, *National Parks, Conservation, and Development: The Role of Protected Areas in Sustaining Society*, Washington DC: Smithsonian Institution Press, 1984. An overview of species and habitat conservation is N. Myers, *The Sinking Ark*, Oxford: Pergamon Press, 1979; *idem*, *A Wealth of Wild Species*, Boulder, Colo.: Westview Press, 1983. See also J. M. Diamond, 'How many unknown species are yet to be discovered', *Nature, Lond.* 315, 1985, 538–9. A temporal and ecological perspective is J. M. Diamond, 'Historic extinctions: a Rosetta stone for understanding prehistoric extinctions', in P. S. Martin and R. G. Klein (eds), *Quarternary Extinctions: A Prehistoric Revolution*, Tucson: University of Arizona Press, 1984, 824–62; and E. C. Wolf, *On the Brink of Extinction: Conserving the Diversity of Life*, Washington DC: Worldwatch Institute Paper 78, 1987. A specialized paper on crop diversity is R. and C. Prescott-Allen, 'The case for in situ conservation of crop genetic resources', *Nature and Resources* 18(1), 1982, 15–20. A particular example is discussed in T. T. Chang, 'Conservation of rice genetic resources: luxury or necessity?',

Science 224, 1984, 251–6; *idem*, 'Crop history and genetic conservation: rice – a case study', *Iowa State J. of Research* 59, 1985, 425–55. A great deal of useful information (including that on captive propagation) is in M. A. Soulé and B. A. Wilcox (eds), *Conservation Biology: An Evolutionary–Ecological Perspective*, Sunderland, Mass.: Sinauer 1980. See also R. F. Noss, 'A regional landscape approach to maintain diversity', *BioScience* 33, 1983, 700–6; early research on the practicalities of reserve size is descsribed in R. Lewin, 'Parks: how big is big enough?', *Science* 225, 1984, 611–12; and management problems in J. B. Hall, 'Positive management for strict natural reserves: reviewing effectiveness', *Forest Ecology and Management* 7, 1983/84, 57–66; P. R. Ehrlich and H. A. Mooney, 'Extinction, substitution and ecosystem services', *BioScience* 33, 1983, 248–53; R. T. Corlett, 'Bukit Timah: the history and significance of a small rain-forest reserve', *Env. Conservation* 15; 1988, 37–44; L. D. Harris, *The Fragmented Forest*, Chicago: Chicago UP, 1984. A world overview of the problems and effectiveness of national parks is G. E. Machlis and D. L. Tichnell, *The State of the World's Parks: An International Assessment for Resource Management, Policy, and Research*, Boulder, Colo., and London: Westview Press, 1985. A useful paper on the rationale for protection is D. W. Ehrenfeld, 'The conservation of non-resources', *American Scientist* 64, 1976, 648–56. Other interesting work on e.g. on genetic resources is in O. H. Frankel and E. Bennett (eds), *Genetic Resources in Plants – their Exploration and Conservation*, Oxford: Blackwell Scientific, 1980; N. R. Farnsworth and R. W. Morris, 'Higher plants: the sleeping giant of drug development', *Amer J Pharm* 146, 1976, 46–52; E. E. Storrs, 'The nine-banded armadillo: a model for leprosy and other biomedical research', *Int J of Leprosy* 39, 1971, 703–14. The effect of introduced species on the biota of one continent is described in G. Seddon, 'Biological pollution in Australia', *Resource Management and Optimization* 2, 1983, 243–58. See also the journals *Environmental Conservation* and *Biological Conservation*.

Forests and forestry

The ecological and physiological underpinning of forest products is given in E. D. Ford, 'The potential production of forest crops', in P. F. Wareing and J. P. Cooper (eds), *Potential Crop Production*, London: Heinemann, 172–85 and a general contextual overview of forestry is in E. P. Eckholm, *Planting for the Future: Forestry for Human Needs*, Washington DC: Worldwatch Papers no. 26, 1979. The LDCs are the subject of D. E. Earl, *Forest Energy and Economic Development*, Oxford: Clarendon Press, 1975. The LDC fuelwood problem is treated in E. P. Eckholm, *The Other Energy Crisis: Firewood*, Washington DC: Worldwatch Paper no. 1, 1975; G. Foley, P. Moss and L. Timberlake, *Stoves and Trees*, Washington DC and London: IIED, 1984; W. H. Matthews and T. A. Siddiqi, 'Energy and environmental issues in the developing countries', in M. Chatterji (ed.), *Energy and Environment in the Developing Countries*, Chichester: Wiley, 1981, 53–67; L. Horne, 'The demand for fuel: ecological implications of socio-economic change', in B. Spooner and H. S. Mann (eds), *Desertification and Development: Dryland Ecology in Social Perspective*, London: Academic Press, 201–15; V. Shiva and J. Bandyopadhyay, 'Eucalyptus: a disastrous tree for India', *The Ecologist* 13, 1983, 184–7; A. K. Singh, M. K. Singh and O. A. J. Mascarenhas, 'Community forestry for revitalising rural ecosystems: a case study', *Forest Ecology and Management* 10, 1985, 209–32. Afforestation in the tropics is from J. Evans, *Plantation Forestry in the Tropics*, Oxford: Clarendon Press, 1982; amplifying material can be seen in R. D. Hart, 'A natural ecosystem analog approach to the design of a successional crop system for tropical forest environments', *Biotropica* 12, 1980, 73–82; C. P. Diaz, 'Socio-economic thrusts in an integrated management system: the Philippine case', in E. G. Hallsworth (ed.), *Socio-economic Effects and Constraints in Tropical Forest Management*, Chichester: Wiley, 107–22; A. Conacher, 'Environmental management implications in an indigenous forest ecosystem: a case study from South-western Australia' in T. O'Riordan and R. K. Turner (eds), *Progress in Resources Management and Environmental Planning*, 4, 1983, 117–40; L. H. MacDonald (ed.), *Agro Forestry in the African Humid Tropics*, Tokyo: UNU, 1982; L. R. Brown and J. Jacobson, 'Assessing the future of urbanization', in L. R. Brown (ed.), *State of the World 1987*, New York and London: Norton, 1987. A chapter on forestry in Africa is to be found in L. R. Brown and E. C. Wolf, *Reversing Africa's Decline*. The recent deforestation picture for one African country is given in C. A. Hamilton, *Deforestation in Uganda*, Nairobi: OUP, 1984; and in a wider perspective in L. Brown and E. Wolf, *Reversing Africa's Decline*, Washington DC: Worldwatch Institute Paper no 65, 1985, 46–54. A concise and comprehensive overview of the TMF management problem is given by N. Myers, 'Tropical moist forests: over-exploited and under-utilized?', *Forest Ecology and Management* 6, 1983, 59–79; here supplemented by material from E. R. C. Reynolds and P. J. Wood, 'Natural vs man-made forests as buffers against environmental degradation', *Forest Ecology and Management* 1, 1977, 83–96; a special issue of *The Ecologist* (vol. 10, 1–2, 1980, 1–56) devoted to TMFs; and V. Plumwood and R. Routley, 'World rainforest destruction: the social factors', *The Ecologist* 12(1), 1982, 4–22; G. J. Osemedoo, 'The human causes of forest depletion in Nigeria' *Env Conservation* 15, 1988, 17–28; J. Davidson et al. (eds), *The Future of Tropical Rain Forests in South East Asia*, IUCN Commission on Ecology Papers no. 10, Gland, Switzerland: IUCN, 1985. See also P. Bunyard, 'World climate and tropical forest destruction', *The Ecologist* 15, 1985, 125–36; W. B. J. Jonkers and P. Schmidt, 'Ecology and timber production in tropical rain forest in Suriname', *Interciencia* 9, 1984, 290–7; H. O'R. Sternberg, 'Aggravation of floods in the Amazon river as a consequence of deforestation?', *Geografiska Annaler* ser. A, 69, 1987, 201–19; T. J. Goreau and W. Z. de Mello, 'Tropical deforestation: some effects on atmospheric chemistry', *Ambio* 17, 1988, 275–81. A suggestion for forest-preserving Amazonian cultivation is J. J. Nicholaides et al., 'Agricultural alternatives for the Amazon basin', *BioScience* 35, 1985, 279–85. The intensification of use in DCs is discussed by E. Jaatinen, 'Energy accounting in forestry-summary studies at the Finnish Forest Research Institute', in C. O. Tamm (ed.), *Man and the Boreal Forest,* Ecol Bull (Stockholm) 21, 1976, 87–94; J. P. Kimmins, 'Evaluation of the consequences for future tree productivity of the loss of nutrients in whole-tree harvesting', *Forest Ecology and Management* 1, 1977, 169–83; R. I. Van Hook, D. W. Johnson, D. C. West and L. K. Mann, 'Environmental effects of harvesting forests for energy', *Forest Ecology and Management* 4, 1982, 79–94; F. B. Goldsmith, J. Harding, A. Newbould and N. Smart, 'An ecological basis for the management of broadleaved forest', *Quart J. Forestry* 76, 1982, 237–47. A detailed history of deciduous woodland history, management and conservation in the UK is G. F. Peterken, *Woodland Conservation and Management*, London and New York: Chapman and Hall, 1981. The role of forestry in socio-economic development around the world is critically reviewed by J. Westoby in his collection *The Purpose of Forests: Follies of Development*, Oxford: Basil Blackwell, 1987.

Land reclamation

Some material on the reclamation of mining waste is found in C. G. Down and J. Stocks, *Environmental Impact of Mining*, London: Applied Science Publishers, 1977, but the major text, with international examples, is A. D. Bradshaw and M. J. Chadwick, *The Restoration of Land: The Ecology and Reclamation of Derelict and Degraded Land*, Oxford: Blackwell Scientific Publications, Studies in Ecology, vol. 6, 1980. The Example from China is Liu Xingtu et al., 'Regional development, environmental change, and improved resource management in the Sanjiang Plain', in K. Ruddle and Wu Chanjin (eds), *Land Resources of the Peoples' Republic of China*, Tokyo: UNU Resource Systems Theory and Methodology Series no. 5, 1983, 47–62.

The use of inland waters

The literature on hydrology and water resources is so large that it is easy to feel every molecule of water has been studied at some time. Useful overviews are the appropriate chapter in M. W. Holdgate et al. (eds), *The World Environment 1972–82*, Dublin: Tycooly International; R. J. Chorley (ed.), *Water, Earth and Man*, London: Methuen, 1969; T. Dunne and L. B. Leopold, *Water in Environmental Planning*, San Francisco and Reading: Freeman, 1978. See also A. K. Biswas (ed.), *United Nations Water Conference: Summary and Main Documents*, Oxford: Pergamon Press for the UN, 1978, Water Development, Supply and Management vol. 2. There have been special issues of journals devoted to water: *Ambio* 6(1), 1977 and *Ekistics* 43(254), 1977, and there is an Open University Unit (S26 block 5, 1974) devoted to *Water Resources*, largely in the UK. A short overview is S. Postel, *Water: Rethinking Management in an Age of Scarcity*, Washington DC: Worldwatch Institute Paper 62, 1984. An account (qualitative only) of the impact of a dam is in ch. 6 of J. E. Heer and D. J. Hagerty, *Environmental Assessments and Statements*, NY: Van Nostrand Reinhold, 1977, and two useful collections on human activity in watersheds are A. M. Gower (ed.), *Water Quality in Catchment Ecosystems*, Chichester: Wiley, 1980, and G. E. Hollis (ed.), *Man's Impact on the Hydrological Cycle in the United Kingdom*, Norwich: Geo Abstracts Ltd, 1979. The Chinese example is from A. K. Biswas, Zuo Dakang, J. E. Nickum and Liu Changming (eds), *Long-Distance Water Transfer: A Chinese Case Study and International Experiences*, Dublin: Tycooly International for the UN University, 1983. The depletion of the Ogalala aquifer is described in P. Beaumont, 'Irrigated agriculture and ground-water mining on the High Plains of Texas, USA', *Environmental Conservation* 12, 1985, 119–30. Sunday afternoon in Australia comes from E. Woolmington and J. S. Burgess, 'Hedonistic water use and low-flow runoff in Australia's national capital', *Urban Ecology* 7 (1982–83), 215–27, and channelization is straightened out by A. Brookes, K. J. Gregory and F. H. Dawson, 'An assessment of river channelization in England and Wales', *The Science of the Total Environment* 27, 1983, 97–111; and A. Brooks, 'River channelization: traditional engineering methods, physical consequences and alternative practices', *Prog Phys Geog* 9, 1985, 1–15. An unusually complete assessment of the human-induced changes in a river is in K. E. Limburg, M. A. Moran and W. H. McDowell, *The Hudson River Ecosystem*, New York: Springer-Verlag, 1986, Springer Series in Environmental Management: An LDC view can be seen in R. Bowonder and R. Chettri, 'Urban water supply in India: environmental issues', *Urban Ecology* 8, 1984, 295–311. For a conspectus of the most dammed nation in the world, see P. Beaumont, 'Water resource management in the USA: a case study of large dams', *Applied Geography* 3, 1983, 259–75. A window on the Venetian problem is in A. Ghetti and M. Batisse, 'The overall protection of Venice and its lagoon', *Nature and Resources* 19(4), 1983, 7–19. Domestic WC recycling is treated in e.g. C. E. Parker, 'Treatment of water closet flush water for recycle and reuse', *Int J Environ Studs* 25, 1985, 87–108. A good but concise account of sinking lands is D. R. Coates, 'Large-scale subsidence', in R. Gardner and H. Scoging (eds), *Mega-Geomorphology*, Oxford: Clarendon Press, 1983, 212–33. A short general review is L. Carbognin, 'Land subsidence: a worldwide environmental hazard', *Nature and Resources* 21(1), 1985, 2–12. See also D. Nir, *Man, a Geomorphological Agent*, Jerusalem: Keter Publishing, 1983; E. H. Brown, 'Man shapes the earth', *Geog J* 136, 1970, 74–85.

Warfare and the environment

The UN Environmental Programme has addressed itself to this issue and two fruits of this are chapter 16, 'Peace and security' of M. W. Holdgate et al. (eds), *The World Environment 1972–82*, Dublin: Tycooly Publishing for UNEP 1982, and E. El-Hinnawi and M. U.-H. Hashmi, *Global Environmental Issues*, Dublin: Tycooly International for UNEP, 1982, ch. 2. Some historical data are found in A. E. Cowdray, 'Environments of war', *Environmental Review* 7, 1983, 155–64. The issue of *Ambio* 4, 1975, was devoted to warfare in its environmental context, and comprehensive data are given in A. H. Westing, *Weapons of Mass Destruction and the Environment*, London: Taylor and Francis 1977, and in the most relevant volume of all, A. H. Westing, *Warfare in a Fragile World: Military Impact on the Human Environment*, London: Taylor and Francis, 1980. The second Indo-China war is treated in G. H. Orians and E. W. Pfeiffer, 'Ecological effects of the war in Vietnam', *Science* 168, 1970, 544–54; A. H. Westing, *Ecological Consequences of the second Indo-China War*, Stockholm: Almqvist and Wiksell, 1976; E. A. Carlson, 'International symposium on herbicides in the Vietnam war: an appraisal', *BioScience* 33, 1983, 507–12. The White Sands example is in M. W. Gilliland and P. G. Risser, 'The use of systems diagrams for environmental impact assessment: procedures and an application', *Ecol. Modelling* 3, 1977, 183–209, and the World War II cases are cited in A. D. Merriam and W. Clark, 'Energy and war – a survival strategy for both', in W. J. Mitsch, R. W. Bosserman and J. M. Klopatek (eds), *Energy and Ecological Modelling*, Amsterdam: Elsevier, 1981, Developments in Ecological Modelling 1, 797–825. The Honduras–El Salvador conflict is analysed in W. H. Durham, *Scarcity and Survival in Central America: Ecological Origins of the Soccer War*, Stanford: Stanford UP, 1979. Some general ideas on the industrialization of war are given in W. H. McNeill, *The Pursuit of Power: Technology, Armed Force and Society since* AD *1000*, Chicago: Chicago UP/Oxford: Basil Blackwell, 1982.

Recreation and tourism

There is a large literature on tourism, recreation and environment, mostly, I suspect, because it provides a respectable and possibly funded excuse to go to nice places. The two UNEP volumes already cited several times (M. W. Holgate et al., 1982; E. El-Hinnawi and M. U.-H. Hashmi, 1982) both have chapters in tourism, and a basic book of facts, though inevitably a bit dated, is H. Robinson, *A Geography of Tourism*, London: Macdonald and Evans, 1976. The sociology and anthropology of leisure and its tourism outgrowth are treated in respectively: K. Roberts, *Contemporary Society and the Growth of Leisure*, London and NY: Longman, 1978; and V. L. Smith (eds), *Hosts and Guests: The anthropology of tourism*, Oxford: Basil Blackwell, 1977. Two useful short books on tourism with an adequate treatment of the environment among other topics are, A. Mathieson and G. Wall, *Tourism: Economic, Physical and Social Impacts*, London and New York: Longman, 1982; D. Pearce, *Tourist Development*, London and New York: Longman, Topics in Applied Geography, 1981. See also the monograph by G. Wall and C. Wright, *The Environmental Impact of Outdoor Recreation*, Waterloo, Ont.: University Dept of Geography Publication Series no 11, 1977. There is some treatment of the environment in J. Pigram, *Outdoor Recreation and Resource Management*, London: Croom-Helm, 1983. An interesting if inconclusive discussion of secondary rural effects is F. J. Belisle, 'Tourism and food production in the Caribbean', *Annals of Tourism Research* 10, 1983, 497–513. A series of short regional accounts can be found in the UNEP serial *Industry and Environment*, 9(1), 1986, 3–26. Two papers of a reasonably general nature are J. J. Pigram, 'Environmental implications of tourism development', *Ann Tourism Res* 7, 1980, 554–83 (though most of the examples are Australian), and E. Cohen, 'The impact of tourism on the physical environment', *Ann Tourism Res* 5, 1978, 215–37. See also E. Inskeep, 'Tourism planning: an emerging specialisation', *J Amer Planning Assn* 54, 1988, 360–72. More specialized work includes G. Thode Roos, *The Impact of Tourism on some Breeding Wader species on the Isle of Vlieland in the Netherlands' Wadden See*, Wageningen: PUDOC, 1981; D. Vreugdenhil, 'Management of the Dutch Waddensee', *Nat Sci Tech* 16, 1984, 485–95; R. H. Webb and H. G. Wilshire (eds), *Environmental Effects of Off-Road Vehicles: Impacts and Management in Arid Regions* New York etc.: Springer-Verlag, 1983; G. Wall and I. M. Ali, 'The impact of tourism in Trinidad and Tobago', *Ann Tourism Res* 5, 1977, 43–9; B. Hyma and G. Wall, 'Tourism in a developing area: The case of Tamil Nadu, India', *Ann Tourism Res* 6, 1979, 338–50. Skiing and floods are connected by P. Simons, 'Aprés ski le déluge', *New Scientist* 117, 1988, 49–52.

The great and wide sea

The impact of human activities upon the oceans is gathered into a compact, generally non-quantitative, form in the UNEP 1972–82 volume, edited by M. W. Holdgate et al. Other general studies include D. H. Cushing and J. J. Walsh (eds), *The Ecology of the Seas*, Oxford: Blackwell Scientific, 1976; and D. H. Cushing, *Science and the Fisheries*, London: Edward Arnold Studies in Biology no. 85, 1977. An unusually wide-ranging general text, despite its ungrammatical title, is D. A. Ross, *Opportunities and Uses of the Ocean*, New York etc.: Springer-Verlag, 1980. Cushing writes on the industrialization of fisheries in his *Climate and Fisheries*, London:

Academic Press, 1982, ch. 1. A detailed global review is J. A. Gulland, 'World resources of fisheries', in O. Kinne (ed.), *Marine Ecology*, vol. 5, part 2: *Ecosystems and Organic Resources*, 839–1061. The New Zealand example of fishing decline comes from G. Struik, 'Commercial fishing in New Zealand: an industry bent on extinction', *The Ecologist* 13(6), 1983, 213–21, and notes on Japan from H. Befu, 'Political ecology of fishing in Japan: techno-environmental impact of industrialization in the Inland Sea', *Research in Economic Anthropology* 3, 1980, 323–47. From time to time, data on whales are collected by Friends of the Earth in a *Whale Manual*, London: FOE, 1978. A general book with data on whaling and its 'management' is D. E. Gaskin, *The Ecology of Whales and Dolphins*, London and Exeter, NH: Heinemann, 1982, see esp. table 8.1, chs 9 and 10. See also G. A. Knox, 'The key role of krill in the ecosystem of the southern ocean with special reference to the contention on the conservation of Antarctic marine living resources', *Ocean Management*, 9, 1984, 113–56. A volume with a more cultural approach is R. Anderson, *North Atlantic Maritime Cultures: Anthropological Essays on Changing Adaptations*, The Hague: Mouton, 1979; this contains useful information on *inter alia* the Faeroes, Sweden and Newfoundland. See also M. E. Smith (ed.), *Those Who Live from the Sea: A study in maritime anthropology*, St. Paul, Minn.: West Publishing Co., 1977. For Melanesia, there is a section of 7 chapters on 'Fishing and the environment' in J. Winslow (ed.), *The Melanesian Environment*, Canberra: ANU Press, 1977, 176–235. There is a great deal of material on pollution of the oceans: a long book is R. Johnston (ed.), *Marine Pollution*, London: Academic Press, 1977; two shorter ones are S. A. Patin, *Pollution and the Biological Resources of the Oceans*, London: Butterworth Scientific, 1982 (first published in Russian in 1979); and S. A. Gerlach, *Marine Pollution: Diagnostics and Therapy*, Berlin etc.: Springer-Verlag, 1981 (first published in German in 1976). An interesting study of the long-term effects of a large oil spill can be found in E. R. Gundlach et al., 'The fate of Amoco Cadiz oil', *Science* 221, 1983, 122–9. A concise account of recent moves towards maritime management organizations is H. D. Smith, 'The management and administration of the sea', *Area* 17, 1985, 109–15.

The environmental linkages of wastes

The specialized literature on environmental wastes and pollution is itself a danger to the load-bearing properties of any library shelf, yet good syntheses at a reasonable length are not easy to find: the UNEP studies, for example, fragment the topic under various resource-related headings. Arguably the clearest overall treatment is M. W. Holdgate, *A Perspective of Environmental Pollution*, Cambridge: CUP, 1979, though inevitably some of the data and interpretations have been superseded. Another comprehensive treatment is in H. M. Dix, *Environmental Pollution: Atmosphere, Land, Water and Noise*, Chichester: Wiley, 1981. An attempt to encompass the whole area also occurs in my own *The Ecology of Natural Resources*, London: Edward Arnold, 2nd edn, 1981, ch. 11, but this too (like its author) suffers from age. The carbon dioxide problem has excited a number of volumes as well as many papers in primary scientific journals. Among the former, the accessible and apparently authoritative, though scarcely conclusive, are W. Bach, J. Pankrath and S. Schneider (eds), *Food–Climate Interactions*, Dordrecht etc.: D. Reidel, 1981, including a paper by D. Pimentel, 'Food, energy and climatic change', 303–23; W. Bach, J. Pankrath and W. Kellogg (eds),

Man's Impact on Climate, Amsterdam: Elsevier, 1979, Developments in Atmospheric Science, 10; J. R. Trabalka and D. E. Reichle (eds), *The Changing Carbon Cycle: A Global Analysis*, New York etc.: Springer-Verlag, 1986. The wider biogeochemical context is explained in S. Postel, *Altering The Earth's Chemistry: Assessing the Risks*, Washington DC: Worldwatch Institute Paper no. 71, 1986. See also W. C. Clark (ed.), *Carbon Dioxide Review: 1982*, Oxford: Clarendon Press; and for summaries in easily accessible sources of the science in G. J. Macdonald (ed.), *The Long-term Impacts of Increasing Atmospheric Carbon Dioxide Levels*, Cambridge, Mass.: Ballinger, 1982; and of the social science in R. S. Chen, E. Boulding and S. H. Schneider (eds), *Social Science Research and Climatic Change*, Dordrecht etc.: D. Reidel, 1980. See as well W. Bach, *Our Threatened Climate: Ways of Averting the CO_2 Problem through Rational Energy Use*, Dordrecht: Reidel, 1983. A very concise summary indeed of ozone, carbon dioxide and other gases in the atmosphere is in ch. 2, 'Air quality and atmospheric issues', of E. E. El-Hinnawi and M. H. Hashmi (eds), *The State of the Environment*, London: Butterworth for UNEP, 1987, 5–34. The role of phytoplankton is in R. J. Charlson et al., 'Oceanic Phytoplankton, atmospheric sulphur, cloud albedo and climate', *Nature, Lond.* 326, 1987, 655–61. For a discussion of possible rapid changes, see W. S. Broecker, 'Unpleasant surprises in the greenhouse?', *Nature, Lond.* 328, 1987, 123–6, and W. S. Broecker, D. M. Peteet and D. Rind, 'Does the ocean–atmosphere system have more than one stable mode of operation?', *Nature, Lond.* 315, 1985, 21–6. A succinct overview of the science (as distinct from the politics) is in W. Bach, 'Carbon dioxide and climatic change: an update', *Prog Phys Geog* 8, 1984, 82–93. A very useful review of various scales of interaction is J. Jäger, *Climate and Energy Systems: A review of their interaction*, Chichester: Wiley, 1983. Streamflow predictions are made in S. B. Idso and A. J. Brazel, 'Rising atmospheric carbon dioxide concentrations may increase streamflow', *Nature, Lond.* 312, 1984, 51–3; see also R. A. Houghton et al., 'Net flux of carbon dioxide from tropical forests in 1980', *Nature, Lond.* 316, 1985, 617–20, and S. B. Idso, 'Industrial age leading to the greening of the earth', *Nature, Lond.* 320, 1986, 22. Wastes in the seas get a synthetic treatment in O. Kinne (ed.), *Marine Ecology*, vol. 5: *Ocean Management*, parts 3 and 4, *Pollution and Protection of the Seas*, Chichester: Wiley, 1984; see also I. W. Duedall et al., 'Energy wastes in the marine environment: an overview', in that volume an historic perspective on heavy metals is given in J. O. Nriagu, *Lead and Lead Poisoning in Antiquity*, NY: Wiley, 1983; and one source of information on the entirely modern acid rain problem in Europe is Environmental Resources Ltd., *Acid Rain*, London: Graham and Trotman, 1983. A study of trends in eastern North America which shows the recent history of the problem is J. H. Gibson (ed.), *Acid Deposition: Long-Term Trends*, Washington DC: National Academy Press, 1986; national data and policies in Europe and North America can be found in T. Schneider (ed.), *Acidification and its Policy Implications*, Amsterdam: Elsevier, 1986, Studies in Environmental Science, 30; a longer-term historical context is given by E. B. Cowling, 'Acid precipitation in historical perspective', *Environ Sci Technol* 16, 1982, 110A–123A. See also R. W. Battarbee et al., *Lake Acidification in the United Kingdom 1800–1986*. London: Ensis Publishing 1988; C. O. Tamm and L. Hallbäcken, Changes in soil acidity in two forest areas with different acid deposition: 1920's to 1980's, *Ambio* 17, 1988, 56–61.

Population, resources and environment

Statistics and population, environment and resource are available in a convenient form in a regular UNEP publication, *Environmental Data Report*, Oxford: Basil Blackwell, 1987. Similar data, though with more commentary, come annually (1986–) from IIED/WRI, *World Resources*, New York: Basic Books. The data for human use of NPP are from J. M. Diamond, 'Human use of world resources', *Nature, Lond.* 328, 1987, 479–80.

Chapter 6 The Nuclear Age

The new energy source

A clear introduction to the technology of nuclear power is in W. Patterson, *Nuclear Power*, Harmondsworth: Pelican Books, 1976; the rest of the book is not, however, unbiased. A detailed account of environmental links is in E. E. El-Hinnawi (ed.), *Nuclear Energy and the Environment*, Oxford: Pergamon, 1982, Environmental Sciences and Applications, vol. 11. For the UK only, there is some amplification of parts of El-Hinnawi in the British Nuclear Energy Society's *The Environmental Impact of Nuclear Power*, London: BNES, 1981. Those who want highly technical detail need to consult the appropriate agencies of the various governments, e.g. ERDA, EPA, NRC and USAEC in the USA, UKAEA, CEGB and NRPB in the UK, and AIEA (International-Vienna). A useful summary document is the National Academy of Science (of the USA), *Risks Associated with Nuclear Power*, Washington DC: NAS, 1979; and the possible impacts of fusion power are explained in J. P. Holdren, 'Environmental implications of the use of fusion power', *Environ Cons* 7, 1980, 289–94. An interesting paper on the complexities of calculating the E_r of nuclear power installations is P. F. Chapman, 'Energy analysis of nuclear power stations', *Energy Policy* 3, 1975, 285–98. A technical but comprehensible introduction to the fate of radionuclides in the environment is ch. 5 of F. Ward Whicker and V. Schultz, *Radioecology: Nuclear Energy and the Environment*, Roca Baton, Fla.: CRC Press, 1982, vol. I. As yet no country has completely cleared away a full-scale nuclear plant: an accessible account of the approach to the first at Windscale (now renamed Sellafield), England, is R. Milne, 'Breaking up is hard to do', *New Sci* 112, 1986, 34–7. An account of Chernobyl and its consequences is C. Flavin, *Reassessing Nuclear Power: The Fallout from Chernobyl*, Washington DC: Worldwatch Paper 75, 1987.

Military nuclear power and the environment

The assertion about the number of victims of nuclear weapons production is from R. Bertell, *No Immediate Danger: Prognosis for a Radioactive Earth*, London: The Women's Press, 1985; these data were quoted in a review in *The Ecologist* 15, 1985, 302–3. A very useful source-book on arsenals, the Japanese experience, and testing is J. Petersen (ed.), *Nuclear War: the Aftermath*, Oxford: Pergamon, 1983 (a reprint of *Ambio* 11(2), 1982). A popular treatment, emphasizing the USA, is J. Schell, *The Fate of the Earth*, London: Pan Books, 1982. These predate the 'nuclear winter' scenarios which are described in outline, non-technical form by A. Ehrlich, 'Nuclear winter', *Bull Atom Sci* 40(4), 1984, 35–145. Her data are from specialized papers such as

R. P. Turco et al., 'Nuclear winter: global consequences of multiple nuclear explosions', *Science* 222, 1983, 1283–92; P. R. Ehrlich et al., 'Long-term biological consequences of nuclear war', *Science* 222, 1983, 1293–1300; C. Covey et al., 'Global atmospheric effects of massive smoke injections from a nuclear war: results from general circulation model simulations', *Nature, Lond.* 308, 1984, 21–5; J. E. Penner, 'Uncertainties in the smoke source term for "nuclear winter" studies', *Nature, Lond.* 324, 1986, 222–6. General caution about the assumptions made in the prediction is expressed in the National Academy of Sciences', *The Effects on the Atmosphere of a Major Nuclear Exchange*, Washington DC: National Academy Press, 1985, though they concede that it is a 'clear possibility'. The most comprehensive documents currently available are A. B. Pittock et al., *Environmental Consequences of Nuclear War*, vol. 1: *Physical and Atmospheric Effects*, Chichester: Wiley, 1986, SCOPE 28; M. A. Harwell and T. C. Hutchinson, ibid., vol. 2: *Ecological and Agricultural Effects*, Chichester: Wiley, 1985, SCOPE 28, with a popularization by L. Dotto, *Planet Earth in Jeopardy. The Environmental Consequences of Nuclear War*, Chichester: Wiley, 1986. The June 1988 issue (vol 30(5)) of *Environment* is given over to this topic.

Hi-sci, hi-tech

The speculative material about geophysical warfare and about micro-electronics is general knowledge not derived from any particular source; for a short overview of biotechnology see J. E. Smith, *Biotechnology*, London: Edward Arnold, Studies in Biology no. 136, 1981; more detailed and often technical accounts of recent advances can be found in the issue of *Science* for 11 Feb. 1983 (vol. 219, no. 4585), which includes a summary by P. H. Abelson, 611–13. This type of genetic manipulation is put into context alongside the genetic potential of 'the wild' by N. Myers, *A Wealth of Wild Species: Storehouse for Human Welfare*, Boulder, Colo.' Westview Press, 1983, esp. ch. 13. A speculative review is S. A. Levin and M. A. Harwell, 'Potential ecological consequences of genetically engineered organisms', *Env Mgmt* 10, 1986, 495–513.

Softer paths

The quotation from R. S. Thomas is from 'Cynddylan on a tractor', *Song at the Year's Turning*, London: Rupert Hart-Davis, 1955, 54, and also to be found on a picture postcard published by Oriel, Cardiff CF1 4ED. For a taste of recent thinking on the revaluation of women, see J. H. Momsen and J. G. Townsend (eds), *Women's Role in Changing the Face of the Developing World*, Durham: IBG Women and Geography study group, 1984; available from the Department of Geography of the University of Durham. The whole subject of 'alternatives' arouses strong passions and the literature is strewn with optimistic boosterism on the one hand and vested-interest cold water on the other. Standing back are E. El-Hinnawi and A. K. Biswas (eds), *Renewable Sources of Energy and the Environment*, Dublin: Tycooly International, 1981, Natural Resources and the Environment series vol. 6; and the UK-oriented Energy Technology Support Unit, *The Environmental Impact of the Renewable Energy Sources*, London: Department of Energy, 1979. An advocative but very carefully documented book is A. B. Lovins, *Soft Energy Paths*, London: Penguin Books, 1977. A case for small-scale generation even of electricity is in C. Flavin, *Electricity's Future: The Shift to Efficiency and*

Small-Scale Power, Washington DC: Worldwatch Institute Worldwatch Paper, no. 61, 1984; *idem*, *Electricity for a Developing World: New Directions*, Washington DC: Worldwatch Institute Worldwatch Paper, no. 70, 1986. For potential sources but little environmental comment see J. E. H. Mustoe, *An Atlas of Renewable Energy Sources in the United Kingdom and North America*, Chichester: Wiley, 1984. Large-scale solar trapping seems most fruitful via photovoltaic systems and a progress report is by E. A. DeMeo and R. W. Taylor, 'Solar voltaic power systems: an electric utility R and D perspective', *Science* 224, 1985, 245–51. See also C. Flavin, 'The worldwide potential of photovoltaics', *Interciencia* 10, 1985, 87–93. A general overview, with numerous examples from LDCs, is C. Stamboulis (ed.), *Renewable Energy Sources for Developing Countries*, London etc.: Heliotechnic Press, 1981. An examination of the potential energy sources of a cool temperate region like Ulster is made by P. J. Newbould, 'Energy resources', in J. G. Cruickshank and D. N. Wilcock (eds), *Northern Ireland: Environment and Natural Resources*, Belfast: Queen's University/Coleraine: New University of Ulster, 1982, 241–63. Biomass energy is the subject of thorough and highly cautionary examination by V. Smil, *Biomass Energies: Resources, Links, Constraints*, NY and London: Plenum Books, 1983; and further detailed discussion by D. Pimentel et al., 'Biological solar energy conversion and U.S. energy policy', *BioScience* 28, 1978, 376–82; *idem*, 'Environmental and social costs of biomass energy', *BioScience* 34, 1984, 89–94; R. P. Moss and W. B. Morgan, *Fuelwood and Rural Energy*, Dublin: Tycooly International, 1981. A thorough but largely American state-of-the-art review is D. L. Klass, 'Energy from biomass and wastes: a review and 1983 update', *Resources and Conservation* 11, 1985, 157–239. A useful set of reprints in the field of energy and resource recovery ('recycling') is M. Grayson (ed.), *Recycling, Fuel and Resource Recovery: Economic and Environmental Factors*, New York: Wiley-Interscience, 1984, Encyclopedia Reprint Series. See also C. Pollock, *Mining Urban Wastes: The Potential for Recycling*, Washington DC: Worldwatch Papers 76, 1987. An investigation of one facet of biofuels is given in S. Matsuda and H. Kubota, 'The feasibility of national fuel-alcohol programs in South-east Asia', *Biomass* 4, 1984, 161–82. A brief introduction to the whole topic is by C. Lewis, *Biological Fuels*, London: Edward Arnold Studies in Biology no. 153, 1983. The study of Indian irrigation mentioned is K. S. Malhotra, 'Potential for pumping irrigation water with renewable sources of energy in Indian arid zone', *Energy in Agriculture* 3, 1984, 245–51. The Rajasthan example details are in K. S. Malhotra and P. B. Lal Chaurasia, 'A plan for energy management in 110 villages of the Silora block in Rajasthan, India', *Energy* 6, 1981, 591–601. Energy problems in LDCs and the role of 'alternatives', particularly in Latin America, are discussed in H. Krugmann and J. Goldemburg, 'The energy cost of satisfying basic human needs', *Technological Forecasting and Social Change* 24, 1983, 45–60; V. R. Vanin et al., 'A note on the energy consumption of urban slums and rural areas in Brazil', *Technological Forecasting and Social Change* 23, 1983, 299–303; J. Goldemburg, 'Energy problems in Latin America', *Science* 223, 1984, 1537–62. Some useful amplifications of the 'alternative agriculture' debate can be found in G. H. Heichel, 'Agricultural production and energy resources', *American Scientist* 64, 1976, 64–72; A. P. A. Vink, *Landscape Ecology and Land Use* (ed. D. A. Davidson), London: Longman, 1983; J. A. Langley, E. O. Heady and K. D. Olsen, 'The macro implications of a complete transformation of U.S. agricultural production to organic farming practices', *Agr Ecosys and Env* 10, 1983,

323–33; S. R. Cliesmann et al., 'The ecological basis for the application of traditional agricultural technology in the management of tropical ecosystems', *Agro-Ecosystems* 7, 1981, 173–85; J. J. Schahczenski, 'Energetics and traditional agricultural systems: an overview', *Agric. Systems* 14, 1984, 31–44; S. W. T. Batra, 'Biological control in agroecosystems', *Science* 215, 1982, 134–9; F. H. Buttel and I. G. Youngberg, 'Implications of biotechnology for the development of sustainable agricultural systems', in W. Lockeretz (ed.), *Environmentally Sound Agriculture*, NY: Praeger Scientific, 1983, 377–400; E. C. Wolf, *Beyond the Green Revolution: New Approaches for Third World Agriculture*, Washington DC: Worldwatch Institute Worldwatch Paper 73, 1986; R. Dudal, 'Land resources and production potential for a growing world population', *Proc 12th Congr Int Potash Inst, Bern*, 1982, 277–88. A short guide to environmental problems of LDCs is B. Bowonder, 'Environmental problems in developing countries', *Prog Phys Geog* 11, 1987, 246–59. Some diverse material on energy conservation is to be found in the *Technical Papers* of the 11th World Energy Conference, 1980, vol. 1B, London: WEC 1980. Energy is one of the selected foci in J. Naisbitt, *Megatrends: Ten New Directions Transforming Our Lives*, London and Sydney: Macdonald, 1984.

Soft energy detail

The data on Hawaii come from J. W. Shupe, 'Energy self-sufficiency for Hawaii', *Science* 216, 1982, 1193–9. For the UK's low energy scenario see G. Leach et al., *A Low Energy Strategy for the United Kingdom*, London: IIED Science Reviews, 1979. It is discussed in a critical framework along with other strategies by S. C. Littlechild and K. G. Vaidya, *Energy Strategies for the U.K.*, London: Allen and Unwin, 1982. A similar strategy for the USA, disavowed by the Reagan administration, is in the Ford Foundation's Energy Policy Project's final report, *A Time to Choose: America's Energy Future*, Cambridge, Mass.: Ballinger, 1974. The potential role of renewables, and the scenarios including them, are in J. E. H. Mustoe, *An Atlas of Renewable Energy Resources in the United Kingdom and North America*, Chichester: Wiley, 1984. A recent overview is in C. S. Shea, *Renewable Energy: Today's Contribution, Tomorrow's Promise*, Washington DC: Worldwatch Paper 81, 1988. The PNG example is from K. Newcombe, 'Energy conservation and diversification of energy sources in and around the city of Lae, Papua New Guinea', in F. DiCastri, F. W. G. Baker and M. Hadley (eds), *Ecology in Practice*, Dublin: Tycooly Press/Paris: UNESCO, 1984; vol. 2 *The Social Response*, 88–111. For examples of greater energy efficiency in industry and transport see W. U. Chandler, *Energy Productivity: Key to Environmental Protection and Economic Progress*, Worldwatch Paper 63, Washington DC: Worldwatch Institute, 1985.

Population, resources and environment in the 'post-modern' years

Population data are from the annual *World Population Data Sheet* of the Population Reference Bureau, New York, and a good general account is G. McNicholl, 'Consequences of rapid population growth: an overview and assessment', *Population and Development Review*, 10, 1984, 177–240. The full reference for the World Conservation Strategy is given on p. 431. The

Brundtland Report is *Our Common Future*, Oxford: OUP, 1987. A short argument in favour of CO_2 and toxins being global threats is G. M. Woodwell, 'The blue planet: of wholes and parts and man', in H. A. Mooney and M. Godron (eds), *Disturbance and Ecosystems: Components and Response*, Berlin: Springer-Verlag Ecological Studies, vol. 44, 1983, 2–10.

Chapter 7 Conclusions

Overview

Magnitudes discussed at the outset are also reviewed in a general essay by K. Newcombe, *A Brief History of Concepts of Energy and the Use of Energy by Mankind*, Canberra: ANU Centre for Resource and Environmental Studies publication, HEG 1–76, 1976. A review of current environmental pathologies in the DCs is J. W. Moore, *The Changing Environment*, New York: Springer-Verlag, 1986. The trajectories are common to a great deal of writing in this field and can be found in this form in my *The Ecology of Natural Resources*, London: Edward Arnold, 1981, 2nd edn, Part III. See also S. Cotgrove, 'Environmentalism and utopia', *Sociological Review* 24, 1976, 23–42, reprinted in T. O'Riordan and R. K. Turner (eds), *An Annotated Reader in Environmental Planning and Management*, Oxford: Pergamon Press Urban and Regional Planning Series, vol. 30, 1983, 18–29. The decoupled domes fantasy is in N. Calder, *The Environment Game*, London: Secker & Warburg, 1967, and the 2000-storey building in J. H. Fremlin, 'How many people can the world support?', *New Sci* 24, 1964, 285–7. The collected papers of a major environmentalist thinker are in T. Vale (ed.), *Progress Against Growth: Daniel B. Luten on the American Landscape*, New York and London: Guildford Press, 1986. Both concepts and examples of inner and outer limits are explored in W. H. Matthews (ed.), *Outer Limits and Human Needs*, Stockholm: Almqvist and Wiksell for the Dag Hammerskjöld Foundation, 1976. The cited historical example is J. D. Hughes and J. V. Thirgood, 'Deforestation in ancient Greece and Rome: a cause of collapse?', *The Ecologist* 12, 1982, 196–208. Simple accounts of the human intervention in global biogeochemical cycles can be seen in P. A. Furley and W. W. Newey, *Geography of the Biosphere*, London: Butterworths, 1983, and a more comprehensive view is in SCOPE publication no. 21, *The Major Biogeochemical Cycles and their Interactions* (eds B. Bolin and R. B. Cook), Chichester: Wiley, 1983. Dangers resulting from human impacts on these cycles are discussed in S. Postel, *Altering the Earth's Chemistry: Assessing the Risks*, Washington DC: Worldwatch Institute Worldwatch Paper 71, 1986. An updated overview of the global situation is in E. El-Hinnawi and M. H. Hashmi, *The State of the Environment*, London: Butterworths for UNEP, 1987. The quantity of human intervention in NPP flows is explored in P. M. Vitousek et al., 'Human appropriation of the products of photosynthesis', *BioScience* 36, 1986, 368–73; and J. M. Diamond, 'Human use of world resources', *Nature, Lond.* 328, 1987, 479–80.

The Gaia hypothesis is set out in J. E. Lovelock, *Gaia: A New Look At Life on Earth*, Oxford: OUP, 1979, with an update by Lovelock, 'Gaia: the world as living organism', *New Sci* 112, 1986, 25–8. The idea has also inspired a North American musician, Paul Winter, to write a *Missa Gaia* . . . 'Paul Winter and Paul Winter Consort with the chorus, choristers, and pipe organ of the largest Gothic cathedral in the world, along with voices of wolf, whale, and loon, create a rhythmic, joyous and contemporary celebration of the Earth in the form of the Mass' (Litchfield, Conn.: Living Music Records, cassettes LMRC1–2). Apart from the electrifying use of a wolf howl as the start of the *Kyrie*, it is mostly a musically rather unimaginative vehicle for Mr. Winter on soprano saxophone. Curiously, it lacks a *Credo*.

Regional impact

A continent-by-continent account of the alteration of the main non-urban areas (with maps), albeit at a large scale of treatment, is in C. F. Bennett, *Man and Earth's Ecosystems*, Chichester: Wiley, 1975. A biome-by-biome treatment, only of vegetation and largely of present-day processes (though with historical data in a few instances), is W. Holzner, M. J. A. Werger and I. Ikusima (eds), *Man's Impact on Vegetation*, The Hague: Junk, 1983, Geobotany 5. An interesting, though localized for the UK, checklist of rural impacts is in the Proceedings of the Study Conference, *The Countryside in 1970*, vol. 1, London: HMSO, 1963: pp. 113–51 comprise a 'Chart of human impcts on the countryside'. The kind of diagram of which we need more is given for Panama and Bolivia in J. W. Moore, *The Changing Environment*, New York: Springer-Verlag, 1987, ch. 2. See also the specialist P. Kunstadter, E. C. F. Bird and S. Sabhasri (eds), *Man in the Mangroves*, Tokyo: UNU, 1987; D. H. H. Kühlmann, 'The sensitivity of coral reefs to environmental pollution', *Ambio* 17, 1988, 13–21.

The medium term: energy, entropy, culture

Most of the works referred to have already been cited, including E. Jantsch, *The Self-Organizing Universe*, Oxford: Pergamon, 1980; and N. Georgescu-Roegen, *Energy and Economic Myths*, Oxford: Pergamon, 1976, 3–36; *idem*, *The Entropy Law and the Economic Process*, Cambridge, Mass.: Harvard UP, 1971; *idem*, 'Energetic dogma, energetic economics, and viable technologies', *Adv Econ Energy and Res* 4, 1982, 1–39; *idem*, 'Feasible recipes vs viable technologies', *Atlantic Economic J.* 12, 1984, 21–31; H.-H. Rogner, 'Long-term energy projections and novel energy systems', in J. R. Trabelka and D. E. Reichle (eds), *The Changing Carbon Cycle: A Global Analysis*, New York: Springer-Verlag, 1986, 508–22. Similar views are expressed by T. C. Edens, 'Cassandra and the horn of plenty: ecological and thermodynamic constraints and economic goals', *Urban Ecology* 2, 1976, 15–31. The more socio-political facets are explored by M. Mesarovic and E. Pestel, 'Organic and sustainable growth', *World Futures* 19, 1984, 233–48. A consistently political perspective is F. Sandbach, *Environment, Ideology and Policy*, Oxford: Basil Blackwell, 1980. A popular but sensible overview is J. Naisbitt, *Megatrends: Ten New Directions Transforming our Lives*, London and Sydney: Macdonald, 1984.

Transformations

The first reference is to L. Whyte, 'The historical roots of our ecologic crisis,' *Science* 155, 1967, 1203–7. Rene Dubos's Benedictine leanings are best summed up in his book *Wooing of Earth*, London: Athlone Press, 1980. See also Dubos's 'St. Francis versus St. Benedict', *Psychology Today* 6, 1973, 54–60. It is perhaps worth reminding academic readers with similar sympathies that the basis of Benedictine practice is the observance of a Rule: a flexible set of dispensations but a Rule nevertheless. T. Roszak, *Person/ Planet*, New York: Doubleday, 1978 (London: Gollancz, 1979 and Abacus Books, 1980), is mentioned; and a rather more formal framework is applied by A. R. Drengson ('an avid hiker, climber, and skier'), 'Shifting paradigms: from the technocratic to the person-planetary', *Env. Ethics* 2, 1980, 221–40; *idem*, 'The sacred and the limits of the technological fix', *Zygon* 19, 1984, 259–75. 'Why do we bother?' is a question developed by J. A. Keller, 'Types of motives for ecological concern', *Zygon* 6, 1971, 197–209, and a Christian perspective on the question 'Can we afford tomorrow?' in H. Schwarz, 'The eschatological dimension of ecology', *Zygon* 9, 1974, 323–38. The tying of enduring myths to the human condition is attempted by A. G. Schillinger, 'Man's enduring technological dilemma: Prometheus, Faust, and other macro-engineers', *Technology in Society* 6, 1984, 59–71. See also M. Nicholson, *The New Environmental Age*, Cambridge: CUP, 1987. In the final sentences I acknowledge the influence of the *Tao Te Ching* and of Stephen Toulmin, *The Return to Cosmology*, Berkeley and Los Angeles: Univ. of California Press, 1982, especially 'The fire and the rose', 255–74: 'we have reached the threshold of some painfully difficult and confusing questions, but answering them is a task for the future.' In the same spirit, musical readers might also try Sir Michael Tippett's (1984) oratorio, *The Mask of Time*. I end with four lines (226–9) from T. S. Eliot's 'The Dry Salvages', the third of his *Four Quartets*, which are not dissimilar in spirit from the opening epigraphs of this chapter.

Index